ULTRA-WIDEBAND
ANTENNAS AND PROPAGATION FOR COMMUNICATIONS, RADAR AND IMAGING

ULTRA-WIDEBAND
ANTENNAS AND PROPAGATION FOR COMMUNICATIONS, RADAR AND IMAGING

Edited by

Ben Allen
University of Oxford, UK

Mischa Dohler
France Télécom R&D, France

Ernest E. Okon
BAE Systems Advanced Technology Centre, UK

Wasim Q. Malik
University of Oxford, UK

Anthony K. Brown
University of Manchester, UK

David J. Edwards
University of Oxford, UK

John Wiley & Sons, Ltd

Copyright © 2007   John Wiley & Sons Ltd, The Atrium, Southern Gate, Chichester,
West Sussex PO19 8SQ, England

Telephone   (+44) 1243 779777

Email (for orders and customer service enquiries): cs-books@wiley.co.uk
Visit our Home Page on www.wiley.com

All Rights Reserved. No part of this publication may be reproduced, stored in a retrieval system or transmitted in any form or by any means, electronic, mechanical, photocopying, recording, scanning or otherwise, except under the terms of the Copyright, Designs and Patents Act 1988 or under the terms of a licence issued by the Copyright Licensing Agency Ltd, 90 Tottenham Court Road, London W1T 4LP, UK, without the permission in writing of the Publisher. Requests to the Publisher should be addressed to the Permissions Department, John Wiley & Sons Ltd, The Atrium, Southern Gate, Chichester, West Sussex PO19 8SQ, England, or emailed to permreq@wiley.co.uk, or faxed to (+44) 1243 770620.

Designations used by companies to distinguish their products are often claimed as trademarks. All brand names and product names used in this book are trade names, service marks, trademarks or registered trademarks of their respective owners. The Publisher is not associated with any product or vendor mentioned in this book.

This publication is designed to provide accurate and authoritative information in regard to the subject matter covered. It is sold on the understanding that the Publisher is not engaged in rendering professional services. If professional advice or other expert assistance is required, the services of a competent professional should be sought.

Other Wiley Editorial Offices

John Wiley & Sons Inc., 111 River Street, Hoboken, NJ 07030, USA

Jossey-Bass, 989 Market Street, San Francisco, CA 94103-1741, USA

Wiley-VCH Verlag GmbH, Boschstr. 12, D-69469 Weinheim, Germany

John Wiley & Sons Australia Ltd, 42 McDougall Street, Milton, Queensland 4064, Australia

John Wiley & Sons (Asia) Pte Ltd, 2 Clementi Loop #02-01, Jin Xing Distripark, Singapore 129809

John Wiley & Sons Canada Ltd, 6045 Freemont Blvd, Mississauga, ONT, L5R 4J3, Canada

Wiley also publishes its books in a variety of electronic formats. Some content that appears in print may not be available in electronic books.

British Library Cataloguing in Publication Data

A catalogue record for this book is available from the British Library

ISBN-13 978-0-470-03255-8 (HB)
ISBN-10 0-470-03255-3 (HB)

Typeset in 9/11pt Times by TechBooks, New Delhi, India.
Printed and bound in Great Britain by Antony Rowe Ltd, Chippenham, England.
This book is printed on acid-free paper responsibly manufactured from sustainable forestry in which at least two trees are planted for each one used for paper production.

Contents

Editors	xv
Prime Contributors	xvii
Preface	xxi
Acknowledgments	xxvii
Abbreviations & Acronyms	xxix

1 Introduction to UWB Signals and Systems — 1
Andreas F. Molisch
1.1 History of UWB — 1
1.2 Motivation — 3
   1.2.1 Large Absolute Bandwidth — 3
   1.2.2 Large Relative Bandwidth — 5
1.3 UWB Signals and Systems — 6
   1.3.1 Impulse Radio — 6
   1.3.2 DS-CDMA — 8
   1.3.3 OFDM — 9
   1.3.4 Frequency Hopping — 10
   1.3.5 RADAR — 11
   1.3.6 Geolocation — 11
1.4 Frequency Regulation — 12
1.5 Applications, Operating Scenarios and Standardisation — 13
1.6 System Outlook — 15
   References — 16

Part I Fundamentals — 19

Introduction to Part I — 21
Wasim Q. Malik and David J. Edwards

2 Fundamental Electromagnetic Theory — 25
Mischa Dohler
2.1 Introduction — 25
2.2 Maxwell's Equations — 25
   2.2.1 Differential Formulation — 25
   2.2.2 Interpretation — 26
   2.2.3 Key to Antennas and Propagation — 27

	2.2.4 Solving Maxwell's Equations	28
	2.2.5 Harmonic Representation	29
2.3	Resulting Principles	30
	References	30

3 Basic Antenna Elements — 31
Mischa Dohler

3.1 Introduction — 31
3.2 Hertzian Dipole — 31
    3.2.1 Far-Field – Fraunhofer Region — 33
    3.2.2 Near-Field – Fresnel Region — 33
3.3 Antenna Parameters and Terminology — 34
    3.3.1 Polarisation — 34
    3.3.2 Power Density — 35
    3.3.3 Radiated Power — 36
    3.3.4 Radiation Resistance — 37
    3.3.5 Antenna Impedance — 37
    3.3.6 Equivalent Circuit — 37
    3.3.7 Antenna Matching — 38
    3.3.8 Effective Length and Area — 38
    3.3.9 Friis' Transmission Formula — 39
    3.3.10 Radiation Intensity — 39
    3.3.11 Radiation Pattern — 39
    3.3.12 (Antenna) Bandwidth — 41
    3.3.13 Directive Gain, Directivity, Power Gain — 41
    3.3.14 Radiation Efficiency — 42
3.4 Basic Antenna Elements — 42
    3.4.1 Finite-Length Dipole — 42
    3.4.2 Monopole — 44
    3.4.3 Printed Antennas — 45
    3.4.4 Wideband and Frequency-Independent Elements — 45
References — 47

4 Antenna Arrays — 49
Ernest E. Okon

4.1 Introduction — 49
4.2 Point Sources — 49
    4.2.1 Point Sources with Equal Amplitude and Phase — 50
    4.2.2 Point Sources with Equal Amplitude and 180 Degrees Phase Difference — 53
    4.2.3 Point Sources of Unequal Amplitude and Arbitrary Phase Difference — 53
4.3 The Principle of Pattern Multiplication — 55
4.4 Linear Arrays of *n* Elements — 56
4.5 Linear Broadside Arrays with Nonuniform Amplitude Distributions — 58
    4.5.1 The Binomial Distribution — 59
    4.5.2 The Dolph–Tschebyscheff Distribution — 59
4.6 Planar Arrays — 62
    4.6.1 Rectangular Arrays — 62
    4.6.2 Circular Arrays — 63

4.7	Design Considerations	65
	4.7.1 Mutual Coupling	65
	4.7.2 Array Gain	65
4.8	Summary	66
	References	66

5 Beamforming 67
Ben Allen

5.1	Introduction	67
	5.1.1 Historical Aspects	67
	5.1.2 Concept of Spatial Signal Processing	68
5.2	Antenna Arrays	69
	5.2.1 Linear Array	70
	5.2.2 Circular Array	71
	5.2.3 Planar Array	72
	5.2.4 Conformal Arrays	72
5.3	Adaptive Array Systems	73
	5.3.1 Spatial Filtering	73
	5.3.2 Adaptive Antenna Arrays	74
	5.3.3 Mutual Coupling and Correlation	74
5.4	Beamforming	75
	5.4.1 Adaptive Antenna Technology	75
	5.4.2 Beam Steering	78
	5.4.3 Grating Lobes	83
	5.4.4 Amplitude Weights	84
	5.4.5 Window Functions	85
5.5	Summary	86
	References	87

6 Antenna Diversity Techniques 89
Junsheng Liu, Wasim Q. Malik, David J. Edwards and Mohammad Ghavami

6.1	Introduction	89
6.2	A Review of Fading	89
	6.2.1 Signal Fading	90
	6.2.2 Channel Distribution	91
6.3	Receive Diversity	93
	6.3.1 Single Branch without Diversity	93
	6.3.2 General Combining Schemes for Receive Diversity	95
	6.3.3 Maximum Ratio Combining	96
	6.3.4 Equal Gain Combining	98
	6.3.5 Selection Combining and Switched Diversity	99
	6.3.6 Fading Correlation	99
6.4	Transmit Diversity	100
	6.4.1 Channel Unknown to the Transmitter	101
	6.4.2 Channel Known to the Transmitter	102
6.5	MIMO Diversity Systems	102
	References	103

Part II Antennas for UWB Communications — 105

Introduction to Part II — 107
Ernest E. Okon

7 Theory of UWB Antenna Elements — 111
Xiaodong Chen
- 7.1 Introduction — 111
- 7.2 Mechanism of UWB Monopole Antennas — 112
  - 7.2.1 Basic Features of a CPW-Fed Disc Monopole — 112
  - 7.2.2 Design Analysis — 118
  - 7.2.3 Operating Principle of UWB Monopole Antennas — 120
- 7.3 Planar UWB Monopole Antennas — 121
  - 7.3.1 CPW-Fed Circular Disc Monopole — 121
  - 7.3.2 Microstrip Line Fed Circular Disc Monopole — 125
  - 7.3.3 Other Shaped Disc Monopoles — 129
- 7.4 Planar UWB Slot Antennas — 132
  - 7.4.1 Microstrip/CPW Feed Slot Antenna Designs — 132
  - 7.4.2 Performance of Elliptical/Circular Slot Antennas — 134
  - 7.4.3 Design Analysis — 138
- 7.5 Time-Domain Characteristics of Monopoles — 140
  - 7.5.1 Time-Domain Performance of Disc Monopoles — 142
  - 7.5.2 Time-Domain Performance of Slot Antenna — 143
- 7.6 Summary — 144
- Acknowledgements — 144
- References — 144

8 Antenna Elements for Impulse Radio — 147
Zhi Ning Chen
- 8.1 Introduction — 147
- 8.2 UWB Antenna Classification and Design Considerations — 148
  - 8.2.1 Classification of UWB Antennas — 148
  - 8.2.2 Design Considerations — 150
- 8.3 Omnidirectional and Directional Designs — 153
  - 8.3.1 Omnidirectional Roll Antenna — 153
  - 8.3.2 Directional Antipodal Vivaldi Antenna — 155
- 8.4 Summary — 160
- References — 161

9 Planar Dipole-like Antennas for Consumer Products — 163
Peter Massey
- 9.1 Introduction — 163
- 9.2 Computer Modelling and Measurement Techniques — 164
- 9.3 Bicone Antennas and the Lossy Transmission Line Model — 164
- 9.4 Planar Dipoles — 167
  - 9.4.1 Bowtie Dipoles — 167
  - 9.4.2 Elliptical Element Dipoles — 171
  - 9.4.3 Fan Element Dipoles — 173
  - 9.4.4 Diamond Dipoles — 176

9.5	Practical Antennas	178
	9.5.1 Printed Elliptical Dipoles	178
	9.5.2 Line-Matched Monopoles	185
	9.5.3 Vivaldi Antenna	189
9.6	Summary	194
	Acknowledgements	195
	References	195

10 UWB Antenna Elements for Consumer Electronic Applications — 197
Dirk Manteuffel

10.1	Introduction	197
10.2	Numerical Modelling and Extraction of the UWB Characterisation	199
	10.2.1 FDTD Modelling	199
	10.2.2 UWB Antenna Characterisation by Spatio-Temporal Transfer Functions	201
	10.2.3 Calculation of Typical UWB Antenna Measures from the Transfer Function of the Antenna	202
	10.2.4 Example	204
10.3	Antenna Design and Integration	205
	10.3.1 Antenna Element Design and Optimisation	206
	10.3.2 Antenna Integration into a DVD Player	208
	10.3.3 Antenna Integration into a Mobile Device	211
	10.3.4 Conclusion	213
10.4	Propagation Modelling	214
10.5	System Analysis	215
10.6	Conclusions	218
	References	220

11 Ultra-wideband Arrays — 221
Ernest E. Okon

11.1	Introduction	221
11.2	Linear Arrays	221
	11.2.1 Broadside Array	222
	11.2.2 End-fire Array	222
	11.2.3 End-fire Array with Increased Directivity	224
	11.2.4 Scanning Arrays	224
11.3	Null and Maximum Directions for Uniform Arrays	225
	11.3.1 Null Directions	225
	11.3.2 Maximum Directions	226
	11.3.3 Circle Representations	228
11.4	Phased Arrays	230
	11.4.1 Element Spacing Required to Avoid Grating Lobes	231
11.5	Elements for UWB Array Design	232
11.6	Modelling Considerations	234
11.7	Feed Configurations	234
	11.7.1 Active Array	235
	11.7.2 Passive Array	235
11.8	Design Considerations	238

11.9	Summary	239
	References	240

12 UWB Beamforming — 241
Mohammad Ghavami and Kaveh Heidary

12.1	Introduction	241
12.2	Basic Concept	242
12.3	A Simple Delay-line Transmitter Wideband Array	243
	12.3.1 Angles of Grating Lobes	246
	12.3.2 Inter-null Beamwidth	248
12.4	UWB Mono-pulse Arrays	249
	12.4.1 Problem Formulation	249
	12.4.2 Computed Results	251
12.5	Summary	257
	References	258

Part III Propagation Measurements and Modelling for UWB Communications — 259

Introduction to Part III — 261
Mischa Dohler and Ben Allen

13 Analysis of UWB Signal Attenuation Through Typical Building Materials — 265
Domenico Porcino

13.1	Introduction	265
13.2	A Brief Overview of Channel Characteristics	267
13.3	The Materials Under Test	270
13.4	Experimental Campaign	272
	13.4.1 Equipment Configuration	275
	13.4.2 Results	278
13.5	Conclusions	281
	References	281

14 Large- and Medium-scale Propagation Modelling — 283
Mischa Dohler, Junsheng Liu, R. Michael Buehrer, Swaroop Venkatesh and Ben Allen

14.1	Introduction	283
14.2	Deterministic Models	284
	14.2.1 Free-space Pathloss – Excluding the Effect of Antennas	284
	14.2.2 Free-space Pathloss – Considering the Effect of Antennas	287
	14.2.3 Breakpoint Model	291
	14.2.4 Ray-tracing and FDTD Approaches	296
14.3	Statistical-Empirical Models	297
	14.3.1 Pathloss Coefficient	297
	14.3.2 Shadowing	301
14.4	Standardised Reference Models	303
	14.4.1 IEEE 802.15.3a	304
	14.4.2 IEEE 802.15.4a	304

14.5	Conclusions	306
	References	306

15 Small-scale Ultra-wideband Propagation Modelling — 309
Swaroop Venkatesh, R. Michael Buehrer, Junsheng Liu and Mischa Dohler

15.1	Introduction	309
15.2	Small-scale Channel Modelling	310
	15.2.1 Statistical Characterisation of the Channel Impulse Response	310
	15.2.2 Deconvolution Methods and the Clean Algorithm	312
	15.2.3 The Saleh-Valenzuela Model	312
	15.2.4 Other Temporal Models	316
15.3	Spatial Modelling	321
15.4	IEEE 802.15.3a Standard Model	324
15.5	IEEE 802.15.4a Standard Model	325
15.6	Summary	327
	References	327

16 Antenna Design and Propagation Measurements and Modelling for UWB Wireless BAN — 331
Yang Hao, Akram Alomainy and Yan Zhao

16.1	Introduction	331
16.2	Propagation Channel Measurements and Characteristics	332
	16.2.1 Antenna Element Design Requirements for WBAN	332
	16.2.2 Antennas for UWB Wireless BAN Applications	333
	16.2.3 On-Body Radio Channel Measurements	335
	16.2.4 Propagation Channel Characteristics	338
16.3	WBAN Channel Modelling	345
	16.3.1 Radio Channel Modelling Considerations	346
	16.3.2 Two-Dimensional On-Body Propagation Channels	349
	16.3.3 Three-Dimensional On-Body Propagation Channels	350
	16.3.4 Pathloss Modelling	351
16.4	UWB System-Level Modelling of Potential Body-Centric Networks	353
	16.4.1 System-Level Modelling	353
	16.4.2 Performance Analysis	354
16.5	Summary	355
	References	358

17 Ultra-wideband Spatial Channel Characteristics — 361
Wasim Q. Malik, Junsheng Liu, Ben Allen and David J. Edwards

17.1	Introduction	361
17.2	Preliminaries	361
17.3	UWB Spatial Channel Representation	362
17.4	Characterisation Techniques	363
17.5	Increase in the Communication Rate	364
	17.5.1 UWB Channel Capacity	364
	17.5.2 Capacity with CSIR Only	365

	17.5.3 Capacity with CSIT	366
	17.5.4 Statistical Characterisation	366
	17.5.5 Experimental Evaluation of Capacity	367
17.6	Signal Quality Improvement	370
	17.6.1 UWB SNR Gain	371
	17.6.2 SNR Gain with CSIR Only	371
	17.6.3 SNR Gain with CSIT	371
	17.6.4 Statistical Characterisation	372
	17.6.5 Experimental Evaluation of Diversity	372
	17.6.6 Coverage Range Extension	375
17.7	Performance Parameters	375
	17.7.1 Spatial Fading Correlation	375
	17.7.2 Eigen Spectrum	377
	17.7.3 Angular Spread	379
	17.7.4 Array Orientation	379
	17.7.5 Channel Memory	380
	17.7.6 Channel Information Quality	380
17.8	Summary	381
	References	381

Part IV UWB Radar, Imaging and Ranging — 385

Introduction to Part IV — 387
Anthony K. Brown

18 Localisation in NLOS Scenarios with UWB Antenna Arrays — 389
Thomas Kaiser, Christiane Senger, Amr Eltaher and Bamrung Tau Sieskul

18.1	Introduction	389
18.2	Underlying Mathematical Framework	394
18.3	Properties of UWB Beamforming	398
18.4	Beamloc Approach	401
18.5	Algorithmic Framework	403
18.6	Time-delay Estimation	404
18.7	Simulation Results	406
18.8	Conclusions	410
	References	410

19 Antennas for Ground-penetrating Radar — 413
Ian Craddock

19.1	Introduction	413
19.2	GPR Example Applications	413
	19.2.1 GPR for Demining	413
	19.2.2 Utility Location and Road Inspection	414
	19.2.3 Archaeology and Forensics	416
	19.2.4 Built-structure Imaging	418
19.3	Analysis and GPR Design	419
	19.3.1 Typical GPR Configuration	419
	19.3.2 RF Propagation in Lossy Media	420

	19.3.3 Radar Waveform Choice	423
	19.3.4 Other Antenna Design Criteria	424
19.4	Antenna Elements	425
	19.4.1 Dipole, Resistively Loaded Dipole and Monopoles	425
	19.4.2 Bicone and Bowtie	426
	19.4.3 Horn Antennas	428
	19.4.4 Vivaldi Antenna	428
	19.4.5 CPW-fed Slot Antenna	429
	19.4.6 Spiral Antennas	429
19.5	Antenna Measurements, Analysis and Simulation	430
	19.5.1 Antenna Measurement	430
	19.5.2 Antenna Analysis and Simulation	432
19.6	Conclusions	433
	Acknowledgements	434
	References	434

20 Wideband Antennas for Biomedical Imaging   437
Ian Craddock

20.1	Introduction	437
20.2	Detection and Imaging	437
	20.2.1 Breast Cancer Detection Using Radio Waves	437
	20.2.2 Radio-wave Imaging of the Breast	438
20.3	Waveform Choice and Antenna Design Criteria	440
20.4	Antenna Elements	441
	20.4.1 Dipoles, Resistively Loaded Dipoles and Monopoles	441
	20.4.2 Bowtie	442
	20.4.3 Horn Antennas	443
	20.4.4 Spiral Antennas	443
	20.4.5 Stacked-patch Antennas	444
20.5	Measurements, Analysis and Simulation	445
	20.5.1 Antenna Measurement	445
	20.5.2 Antenna Analysis and Simulation	446
20.6	Conclusions	447
	Acknowledgements	448
	References	448

21 UWB Antennas for Radar and Related Applications   451
Anthony K. Brown

21.1	Introduction	451
21.2	Medium- and Long-Range Radar	452
21.3	UWB Reflector Antennas	453
	21.3.1 Definitions	453
	21.3.2 Equivalent Aperture Model for Impulse Radiation	454
	21.3.3 Parabolic Antenna	456
21.4	UWB Feed Designs	459
	21.4.1 Feed Pattern Effects	460
	21.4.2 Phase Centre Location	460
	21.4.3 Input Impedance	460

	21.4.4 Polarisation	460
	21.4.5 Blockage Effects	461
21.5	Feeds with Low Dispersion	461
	21.5.1 Planar Spiral Antennas	461
	21.5.2 TEM Feeds	462
	21.5.3 Impulse Radiating Antenna (IRA)	466
21.6	Summary	468
	References	468

Index   **471**

Editors

Ben Allen completed his MSc and PhD degrees at the University of Bristol, U.K., in 1997 and 2001 respectively. Having undertaken post-doctorial research in the areas of smart antennas and MIMO wireless systems, he then became a lecturer at the Centre for Telecommunications Research, King's College London where he co-founded the UWB research group. He is now with the Department of Engineering Science, University of Oxford. He has published numerous journal and conference papers in the above areas as well as a book on smart antennas. He has been in receipt of the IEE J Langham Thomson Premium and the ARMMS Best Paper Award, both for publications relating to UWB. He is a senior member of the IEEE, chartered engineer, member of the IEE, and a member of the IEE's Professional Network Executive Committee on Antennas and Propagation.

Mischa Dohler obtained his MSc degree in Telecommunications from King's College London, UK, in 1999, his Diploma in Electrical Engineering from Dresden University of Technology, Germany, in 2000, and his PhD from King's College London in 2003. He was a lecturer at the Centre for Telecommunications Research, King's College London, until June 2005. He is now a Senior Research Expert in the R&D department of France Telecom working on cognitive and sensor networks. Prior to Telecommunications, he studied Physics in Moscow. He has won various competitions in Mathematics and Physics, and participated in the 3rd round of the International Physics Olympics for Germany. He is a member of the IEEE and has been the Student Representative of the IEEE UKRI Section, member of the Student Activity Committee of IEEE Region 8 and the London Technology Network Business Fellow for King's College London. He has published over 50 technical journal and conference papers, holds several patents, co-edited and contributed to several books, and has given numerous international short courses. He has been a TPC member and co-chair of various conferences and is an editor of the EURASIP journal, the IEEE Communication Letters, and the IEEE Transactions on Vehicular Technology.

Ernest E. Okon received the PhD degree in Electronic Engineering from King's College London in 2001 and the MSc (with distinction) and BSc (honours) degrees in Electrical Engineering from the University of Lagos in 1996 and 1992 respectively. His research interest is in electromagnetic modelling techniques, wide band antennas and arrays, sensor networks and RF circuits and devices. He taught undergraduate and postgraduate courses on antennas and propagation whilst at King's College London. He joined BAE Systems Advanced Technology Centre UK in 2001 and is currently a research scientist working on electromagnetic problems, MEMS, antennas and arrays. He has written numerous reports, and published journal and conference papers. He is a member of the IEE, IEEE and Optical Society of America. He is also listed in Who's Who in the World, Marquis USA.

Wasim Q. Malik received his DPhil degree in Communications Engineering from the University of Oxford, UK, in 2005. Since then, he has been a Research Fellow at the University of Oxford, where his research focuses on ultrawideband propagation, antenna array systems, cognitive radio, and nanoscale sensors. He also holds a Junior Research Fellowship in Science at Wolfson College, Oxford, where he researches microwave tomographic imaging. Dr. Malik has published over 50 research papers in refereed journals and conferences, and has delivered keynote and invited talks at a number of conferences. He is a Guest Editor for the *IEE Proceedings on Microwaves Antennas and Propagations* forthcoming special issue on "Antenna systems and propagation for future wireless communications". He has also been the General Co-Chair and Technical Program Committee Member at several international conferences. Dr. Malik received the Best Paper Award in the ARMMS RF and Microwave Conf., UK, Apr. 2006, the Recognition of Service Award from the Association for Computing Machinery (ACM) in 1997, and won the National Inter-University Computer Science Contest, Pakistan, in 1998. He is a member of the IEEE and the IET, and serves on the UK Task Group on Mobile and Terrestrial Propagation.

Anthony K. Brown is a Professor in Communications Engineering and leads the Microwave and Communication Systems research group at the University of Manchester (UK). He joined academia in 2003 having spent 28 years in industry, most recently for Easat Antennas Ltd where he is retained as company Chairman. He is a recognised expert in antennas and propagation as applied to radar and communications systems. Professor Brown is a member of the Technical Advisory Commission to the Federal Communication Commission (USA)- and is a UK representative to the EU's COST Action 284 Management Committee. He has advised various international bodies including in Canada, Malaysia and USA. He has been a Steering Board member of the Applied Computational Electromagnetics Society (ACES USA), and is past recipient of the Founders Award from that organisation. He has served on many national and international committees (including for IEEE and IEE, EUROCAE and ARINC). He was a founder member of the EPSRC Communications College. Professor Brown is a frequent invited lecturer on antennas and related topics, most recently including application of such techniques to Ultra Wide Band communications. He is a listed expert on UWB systems by the Paris Ultra Wide Band Organisation (http://timederivative.com/pubs.html). Prof Brown is a Fellow of the IEE and the IMA and is a Charted Engineer and Mathematician.

David J. Edwards has been an academic for 17 years after 12 years spent in the industry (British Telecom). He has a strong record of innovation in communications systems, electromagnetic measurements, ground probing radar and subsurface imaging radar. He has authored or co-authored in excess of 200 publications in his time as an academic. He has been in receipt of a number of awards and prizes (IEE Prize for Innovation, NPL Metrology award, IEE Mountbatten Premium (2 papers) and IEEE Neil Sheppy prize) for his work and has been extremely well supported by funding from research councils, industry and government agencies. He has a track record of wide collaboration within the UK and internationally. Prof. Edwards is serving and has served on a range of international committees in communications and related fields. He is a Fellow of the Institution of Electrical Engineers and a Fellow of the Royal Astronomical Society.

Prime Contributors

R. Michael Buehrer received the BSEE and MSEE degrees from the University of Toledo in 1991 and 1993 respectively. He received a Ph.D. from Virginia Tech in 1996 where he studied under the Bradley Fellowship. From 1996–2001 Dr. Buehrer was with Bell Laboratories in Murray Hill, NJ and Whippany, NJ. While at Bell Labs his research focused on CDMA systems, intelligent antenna systems, and multi-user detection. He was named a Distinguished Member of Technical Staff in 2000 and was a co-winner of the Bell Labs President's Silver Award for research into intelligent antenna systems. In 2001 Dr. Buehrer joined Virginia Tech as an Assistant Professor with the Bradley Department of Electrical Engineering where he works with the Mobile and Portable Radio Research Group. His current research interests include position location networks, Ultra Wideband, spread spectrum, multiple antenna techniques, interference avoidance, and propagation modeling. In 2003 he was named Outstanding New Assistant Professor by the Virginia Tech College of Engineering. Dr. Buehrer has co-authored approximately 22 journal and 65 conference papers and holds 10 patents in the area of wireless communications. He is currently a Senior Member of the IEEE, and an Associate Editor for IEEE Transactions on Wireless Communications, IEEE Transactions on Vehicular Technologies and IEEE Transactions on Signal Processing.

Xiaodong Chen obtained his BEng in electronic engineering from the University of Zhejiang, Hangzhou, China in 1983, and his PhD in Microwave Electronics from the University of Electronic Science and Technology of China, Chengdu in 1988. He joined the Department of Electronic Engineering at King's College, University of London in September 1988, initially as a Postdoctoral Visiting Fellow and then a Research Associate. He was appointed to an EEV Lectureship at King's College London in March 1996. In September 1999 he joined the Department of Electronic Engineering at Queen Mary and Westfield College, University of London. He was promoted to a Readership at the same institution in October 2003. His research interests are in microwave devices, antennas, wireless communications and bio-electromagnetics. He has authored and co-authored over 170 publications (book chapters, journal papers and refereed conference presentations). He has involved in the organisation of many international conferences. He is currently a member of UK EPSRC Review College and Technical Panel of IEE Antennas and Propagation Professional Network.

Zhi Ning Chen received his BEng., MEng., Ph.D, and DoE degrees all in electrical engineering. During 1988–1997, he conducted his research and teaching in Institute of Communications Engineering, Southeast University, and City University of Hong Kong, China. During 1997–2004, he worked in and visited University of Tsukuba, Japan as a research fellow and visitor under JSPS foundation and Invitation Fellowship Program. In 2004, he worked in Thomas J. Watson Research Center, IBM, USA as Academic Visitor. Since 1999, he has been with the Institute for Infocomm Research, Singapore as Lead Scientist, Lab Head, and Department Manager. He is concurrently holding the adjunct appointments in universities. He founded the IEEE International Workshop on Antenna Technology (iWAT) and organised the first iWAT2005 as General Chair. He is also very active in international events by giving many keynote addresses and talks, organing and chairing several workshopd and sessions. Since 1990, he has authored and co-authored over 160 technical papers published in international journals and presented at international conferences as well as one book entitled "Broadband Planar Antennas" published by Wiley & Sons in 2005. Of his 15 filed patents, three were granted.

Ian Craddock obtained his PhD from the University of Bristol, UK in 1995. His PhD studies on novel FDTD techniques for Antenna Modelling led to post-doctoral work on the analysis of phased arrays. He joined the academic staff at Bristol in 1998 and is currently a Reader in the Centre for Communications Research (CCR). He has led a number of projects at Bristol, including an investigation of wideband radar for landmine detection and, most recently UWB radar for breast cancer detection – work for which he received in 2005 the IEE's J. A. Lodge Award. He has published a large number of papers in refereed journals and proceedings and contributed invited papers to the leading conferences in the field. He is leading a workpackage on Ultra Wideband (UWB) antennas for Surface-Penetrating Radar in the EU Framework 6 Antennas Network of Excellence.

Mohammad Ghavami is a reader in UWB communications with the CTR, King's College London, University of London. He is the leading author of the book "Ultra Wideband Signals and Systems in Communication Engineering", the co-author of the book "Adaptive Array Systems: Fundamentals and Applications", both published by John Wiley & Sons, Ltd and has published over 80 technical papers in major international journals and conferences on areas related to digital communications, adaptive filters and biomedical signal processing. He completed his Ph.D. degree with Great Distinction in electrical engineering at the University of Tehran, in 1993. During his Ph.D. studies he was awarded the DAAD scholarship at the University of Kaiserslautern, Germany from 1990 to 1993. From 1998 to 2000 he was a JSPS Postdoctoral fellow in Yokohama National University, Japan. From 2000 to 2002 he was a researcher at the Sony Computer Science Laboratories, Inc. in Tokyo, Japan. He has been actively researching various aspects of mobile and personal wireless communications and has two invention awards from Sony Ltd.

Yang Hao received his Ph.D. degree from the Centre for Communications Research (CCR) at the University of Bristol, U.K. in 1998. From 1998 to 2000, he was a postdoc research fellow at the School of Electrical and Electronic Engineering, University of Birmingham, UK. In May 2000, he joined the Antenna Engineering Group, Queen Mary College, University of London, London, U.K. first as a lecturer and now a reader in antenna and electromagnetics. Dr. Hao has co-edited a book, contributed two book chapters and published over 60 technical papers. He was a session organiser and chair for various international conferences and also a keynote speaker at ANTEM 2005, France. Dr. Hao is a Member of IEE, UK. He is also a member of Technical Advisory Panel of IEE Antennas and Propagation Professional Network and a member of Wireless Onboard Spacecraft Working Group, ESTEC, ESA. His research interests are computational electromagnetics, on-body radio propagations, active integrated antennas, electromagnetic bandgap structures and microwave metamaterials.

Thomas Kaiser received his Ph.D. degree in 1995 with distinction and the German habilitation degree in 2000, both from Gerhard-Mercator-University Duisburg, Germany, in electrical engineering. From April 2000 to March 2001 he was the head of the Department of Communication Systems at Gerhard-Mercator-University and from April 2001 to March 2002 he was the head of the Department of Wireless Chips & Systems (WCS) at Fraunhofer Institute of Microelectronic Circuits and Systems, Germany. In summer 2005 he joined Stanford's Smart Antenna Research Group (SARG) as a visiting professor. Now he is co-leader of the Smart Antenna Research Team (SmART) at University of Duisburg-Essen. Dr. Kaiser has published more than 80 papers and co-edited four forthcoming books on ultra-wideband and smart antenna systems. He is the founding Editor-in-Chief of the IEEE Signal Processing Society e-letter. His current research interest focuses on applied signal processing with emphasis on multi-antenna systems, especially its applicability to ultra-wideband systems.

Junsheng Liu received his BSc and MSc degrees in electrical and electronics engineering from Shanghai Jiao Tong University in 2002 and from King's College London in 2003, respectively. After his graduation from King's College, he joined the Centre for Telecommunications Research as a PhD research student. His current research interests lie in MIMO UWB and correlation analysis for the UWB propagation channels.

Dirk Manteuffel received his Dipl.-Ing. degree and the Dr.-Ing. degree in electrical engineering from the University of Duisburg-Essen in 1998 and 2002 respectively. Since 1998 he is with the IMST in Kamp-Lintfort, Germany. As a project manager he is responsible for industrial antenna development and advanced projects in the field of antennas and EM modelling. In 2004, he received the innovation award of the Vodafone foundation for science (Vodafone Stiftung für Forschung) for his research on integrated mobile phone antennas with emphasis on the interaction with the user. Dr. Manteuffel is lecturer at the Technical Academy Esslingen (TAE) and member of the Network of Excellence ACE (Antenna Centre of Excellence) of the European Union. He is author and co-author of more than 40 scientific journal and conference papers and co-authored a book on EM modeling. He holds six national and international patents.

Andreas F. Molisch received a Ph.D. from the Technical University Vienna (Austria) in 1994. From 1991 to 2000, he was with the TU Vienna, becoming an associate professor there in 1999. From 2000–2002, he was with the Wireless Systems Research Department at AT&T (Bell) Laboratories Research in Middletown, NJ. Since then, he has been a Senior Principal Member of the Technical Staff with Mitsubishi Electric Research Labs, Cambridge, MA. He is also professor and chairholder for radio systems at Lund University, Sweden. Dr. Molisch has carried out research in the areas of SAW filters, radiative transfer in atomic vapors, atomic line filters, smart antennas, and wideband systems. His current research interests are UWB, MIMO systems, measurement and modeling of mobile radio channels, and sensor networks. Dr. Molisch has authored, co-authored or edited four books (among them the recent textbook "Wireless Communications", Wiley-IEEE Press), eleven book chapters, some 85 journal papers, and numerous conference contributions. Dr. Molisch has been editor and guest editor for IEEE journals, and has been general chair, TPC vice chair, or TPC member for numerous international conferences. He has also been the chair or vice chair of various standardization groups. He is a Fellow of the IEEE and recipient of several awards.

Swaroop Venkatesh received his B.Tech and M.Tech degrees in electrical engineering from the Indian Institute of Technology (IIT) Madras, India in 2003. He is currently working toward the Ph.D. degree at the Mobile and Portable Radio Research Group (MPRG) at Virginia Tech. Since 2003, he has been a research assistant at MPRG, working mainly on Ultra-Wideband (UWB) position-location networks, UWB communication systems and radar applications. His other interests lie in the areas of wireless communication system design and modelling, multiple-antenna systems and multiple-access communication.

Preface

Yet another book on UWB? Not quite!

Although ultra-wideband (UWB) is not an entirely new discovery, it is still considered an emerging field in the civil areas of data communication, imaging and radar. Research advances on a daily basis, and, hopefully, first commercially viable UWB solutions will be on the market soon. A proper design of such UWB devices, however, requires a profound understanding of the UWB communication channel and the antenna elements. Hence, the aim of this book is to give an in-depth and universal understanding of both the UWB channel and antennas, which differ quite significantly from their narrowband counterparts. We also aim to clarify some myths related to ultra-wideband systems.

The theoretical approach, coupled with practical examples, makes this book a complete and systematic compendium in the area of UWB antennas and propagation channels. It serves a wide spectrum of readership, ranging from college students to professional engineers and managers. The healthy mix of academic and industrial contributors to this book guarantees that the emerging issue of UWB communication systems is treated comprehensively. Ideally, this book should to be read from the beginning to the end; however, each chapter can be read on a stand-alone basis.

We have endeavoured to encompass the following readership. We hope that this book will be of great benefit to students who, we trust, will understand and appreciate the challenges associated with current UWB system design; to academic staff who, we hope, will comprehend the analytic behaviour of UWB channels and antennas and use this knowledge to perpetuate UWB research; to industrial designers and researchers who, we believe, will find this book to be of considerable value when designing robust UWB wireless systems and hence make the UWB dream come true. We also believe that managers and policy-setters will benefit from reading this book since a firm appreciation of the subject of antennas and propagation is required in order to assist key management and policy decisions.

The book synopsis is as follows, and we have organised it into four parts:

I. fundamentals,
II. UWB antennas,
III. propagation, and
IV. radar and geolocation.

Part I describes the fundamentals of antenna theory, covering electromagnetic theory, a description of basic antenna elements, and various issues relating to antenna arrays. This section sets the scene for the remainder of the book that builds on these basics within the context of UWB signals and systems.

Chapters 2 and 3 are dedicated to the very fundamentals of antennas and propagation; for instance, they explain why antennas are capable of transmitting and receiving an electromagnetic signal. Surprisingly, transmission and reception are facilitated by a small asymmetry in Maxwell's equations, which triggers an electromagnetic wave to decouple from (transmission) and couple into (reception) a medium carrying free electric charges (antenna). The application of Maxwell's equations to the most fundamental radiation element, the Hertzian dipole, is then explained. It allows some common antenna terminology to be defined, which is not only confined to Hertzian dipoles, but is also applicable to practical UWB antenna elements. This brings us then to the description of typically occurring narrowband and wideband antenna elements, which themselves might be part of larger antenna arrays.

Chapter 4 deals with the fundamental concepts for antenna arrays. The combination of similar antenna elements to form an array is the most popular way to improve the directional properties of antennas without altering the size of the elements. In addition, the array offers more degrees of freedom than single element antennas. There are a number of controls to shape the overall pattern of the array: the relative displacement between the elements, relative amplitude, relative excitation phase between elements, individual patterns of the elements and the overall geometric configuration. This chapter reviews basic array properties, starting with the case of a two-element isotropic source. The principle of pattern multiplication is also addressed. Linear arrays involving amplitude and phase variations are then discussed, followed by a look at planar array types. Finally, design considerations for ultra-wideband arrays are highlighted.

Chapter 5 describes the concept of beamforming. Its emphasis is on narrowband beamforming, i.e., with a fractional bandwidth of less than 1%. The concepts described in this chapter set the scene for ultra-wideband beamforming, which is explored later in this book. Initially, the history of beamforming (adaptive antennas) for wireless systems is described. It then introduces the principal building blocks of a beamformer and the related operational concepts. Following this, the signal model is described and beam- and null-steering functions are explained. The use of window functions in beamformers concludes this chapter.

The focus of *Chapter 6* is on basic analysis of diversity techniques and how they can improve the BER performance of a communications system. There are many kinds of diversity techniques; since antennas can be located in different positions, antenna diversity is a means of achieving the spatial diversity. Antenna diversity can be implemented at the receiver, at the transmitter, or at both, corresponding to receive diversity, transmit diversity and MIMO diversity. MRC is the optimum combining scheme to make use of all the diversity branches and its performance is also the easiest to be theoretically analysed. Thus, most analysis of antenna diversity is based on MRC. The mathematics of the theoretical analysis for the receive diversity, transmit diversity and MIMO diversity are very similar.

Part II is devoted to specific antenna structures for UWB communications and aims to highlight the difference between conventional and UWB antenna design.

Chapter 7 surveys existing UWB antennas especially for pulsed-based systems, and classifies UWB antennas in terms of their geometry and radiation characteristics. Special design considerations for UWB antennas for impulse radio applications are also discussed. This chapter then gives case studies of two designs, namely omni-directional roll antennas and directional antipodal Vivaldi antennas and its modified version for UWB applications.

Chapter 8 addresses the theory of UWB antenna elements. Existing UWB antenna theories, such as Rumsey's principle and Mushiake's relation, cannot be readily used to explain the operation of a range of planar UWB dipoles and monopoles being developed recently. Further insights into the operation of this type of UWB antennas are obtained through a study of a planar disc monopole. Also, a range of UWB planar monopole antennas are investigated to illustrate different features in their operations. It is established that the overlapping of resonances and supporting travelling wave for good impedance matching are two universal ingredients in the operation of the planar UWB antennas. At the same time, some quantitative guidelines have been derived for the designing of the UWB monopole antennas.

Chapter 9 is dedicated to planar dipole-like antennas. Such antennas can be produced with low cost, and thus are promising solutions for consumer applications. This chapter considers the amplitude variations and phase linearity of the radiated signal and the matching range of both idealised structures and practical antennas. Examples include bowtie, elliptical and fan elements, diamond dipoles, Vivaldi antennas, and some commercial designs where the antenna elements are mounted on PCB-like substrates. It is shown that, for many types, the qualitative behaviour of the reflection coefficient with frequency is very similar, converging to a high frequency limit value for the antenna impedance. A lossy transmission line model for the radiation mechanism explains this behaviour. For the smaller designs, the radiation pattern is shown to be strongly affected by the feed cable. Therefore, it may be expected that the performance of these smaller antennas depend on the structures that they are installed in. The chapter concludes with an attempt to compare the relative merits of the antennas considered.

Chapter 10 focuses on the integration of the antennas into the chassis of different applications. As a first step, a method of extracting the spatio-temporal UWB antenna characterisation from a FDTD simulation is given. Thereafter, the shape of a planar monopole is optimised to provide broadband matching. This is followed by the integration of the antenna into a DVD player and a mobile device such as a PDA or a mobile phone. The impact of this integration on the antenna performance, in terms of traditional antenna measures like *ringing* and *delay spread*, is evaluated. Finally, the transfer function of the complete system is extracted and used for indoor propagation modelling in an example home living environment. The investigations show that the antenna integration into the DVD-chassis results into a directive radiation pattern with increased frequency dependency and ringing. In the final step, the complete system, including the integrated antennas and the indoor channel is characterised by means of the transfer function and the impulse response for different link scenarios.

Chapter 11 examines UWB arrays for communication systems. Antenna arrays are often used in communication and sensor systems to realise increased sensitivity and range extension. Emphasis is put on compact and cost-effective designs in realising UWB arrays for communication systems. The chapter describes simple design parameters for array scanning and pattern synthesis. It then examines antenna elements that are applicable to UWB array design, and presents wideband array design parameters. Modelling techniques for UWB arrays are also examined. Subsequently, various wideband array types for passive and active configurations are considered. The final section highlights design considerations for UWB arrays.

In *Chapter 12* we study two beamforming techniques for UWB signals. The tapped delay-line structure employs adjustable delay lines and profits from its simplicity. On the other hand, a quite different concept of UWB mono-pulse arrays includes ultra-wideband sources radiating coded sequences of ultra-narrow Gaussian mono-pulses with very low duty cycle. The characteristics of both methods are illustrated by several examples.

Part III gives a description of UWB propagation channels, including measurement techniques and channel models derived from them. This section aims to provide an insight into the fundamental differences between conventional and UWB signal propagation.

Chapter 13 describes a study of path attenuation due to UWB radiowaves propagation through common building materials. Very few documents have analysed in detail the characteristics of signal strength reduction due to direct penetration through common building materials. The topic of signal attenuation in common environments is important for automated indoor wireless installation/optimisation systems, but also fundamental for spectrum coexistence analysis. This chapter introduces a simple methodology for attenuation estimation and a set of results from measurements which lead to empirical attenuation factors for twelve different construction materials. The results can be directly applied to link budgets or detailed coexistence analysis.

Chapter 14 is dedicated to the modelling of large and medium scale propagation effects in the UWB channel, i.e., pathloss and shadowing. It will be shown that traditional narrowband approaches fail when modelling the ultra-wideband channel. The discussion of large-scale modelling includes an examination of the dependence of pathloss on frequency and the impact of antenna structure. We develop some deterministic and statistical modelling approaches, leading to free-space and breakpoint pathloss models. They also form the basis for the UWB reference pathloss and shadowing models, standardised by the IEEE 802.15.3a and 802.15.4a standardisation groups.

Chapter 15 discusses fundamental issues in the modelling of small-scale fading effects in UWB propagation channels. Small-scale channel modelling for ultra-wideband signals based on measurements in an indoor office environment is examined in detail. The underlying measurements were performed in an indoor office environment as part of the DARPA NETEX Program. The chapter initially gives an overview of the channel measurement procedures. The chapter then presents a brief description of traditional small-scale modelling methods and various temporal models that can be used to model indoor UWB channels, with particular attention given to the Saleh-Valenzuela model. This is followed by a thorough discussion of the IEEE 802.15.3a and 802.15.4a channel models. Finally, we discuss spatial extensions of available temporal models.

Chapter 16 describes the propagation channels for UWB body area networks (BANs), which consist of a number of nodes and units placed on the human body or in close proximity such as on everyday clothing. Since low power transmission is required for body worn devices, the human body can be used as a communication channel between wireless wearable devices. The wireless body-centric network has special properties and requirements in comparison to other available wireless networks. These additional requirements are due to the rapid changes in communication channel behaviour on the body during the network operation. This raises some important issues regarding the characteristics of the propagation channel, radio systems compatibility with such environments, and the effect on the human wearer.

Chapter 17 addresses UWB systems with multiple-antenna arrays. It commences by presenting a framework for the analysis of UWB spatial channels, followed by a description of UWB MIMO performance measurement techniques. The impact of UWB frequency diversity on small-scale fading and antenna diversity is discussed. Next, the chapter evaluates the data-rate improvement achieved with various multiple-antenna configurations. Results from MIMO channel measurements in the FCC-allocated UWB frequency band are included to quantify the rate and diversity improvement. The impact of various system- and environment-related factors, such as multipath propagation, fading correlation, eigenvalue distribution, and channel estimation is discussed.

Finally, **Part IV** discusses UWB radar and imaging, and therefore highlights antenna and propagation related challenges pertaining to these systems. This section describes several applications of UWB antennas and serves as case studies of the concepts described in previous sections of this book.

Chapter 18 is devoted to medium-range radars. A brief introduction discusses applications and the definition of UWB in this context. A wide frequency band, which is channelised, and the use of a narrow instantaneous band and a true broadband antenna are compared. The difference between military and more standard civil applications are also discussed, especially in terms of gain, power handling, and beamwidth control. A brief introduction to TEM horns, log periodic structures and spirals antennas follows, including the use of feeds in reflectors and the resulting problems for the antenna pattern. A short summary of relevant propagation effects concludes this chapter.

Chapter 19 describes "Ground-Penetrating Radar" (GPR). This acronym has become a catch-all term that encompasses many applications, not all of which actually involve the ground – through-wall imaging being an example. GPR antennas are generally designed to operate over the widest possible bandwidth, and commonly designed to radiate pulse waveforms. The chapter commences with a survey of some of the situations in which GPR has been applied. Although the applications are diverse, most share a number of similarities which result in similar constraints for the antenna designer. A number of wideband antenna designs (including TEM horns, bicones, bow ties, spirals and Vivaldi antennas, and resistively-loaded variants), which attempt to meet these constraints, are presented later in this chapter, along with some notes on antenna measurement and analytical methods.

Chapter 20 considers Biomedical Imaging. UWB Biomedical Imaging tends to be conceptually similar to Ground-Penetrating Radar, although there are notable differences. Due to the huge clinical need for a freely repeatable, low-cost and comfortable breast cancer screening technique, the application, which generates the most interest, is imaging of the human breast. A number of groups worldwide are addressing this application and achieving a compact, wideband, reasonably efficient and non-dispersive antenna design is one of the major difficulties. Consideration is given to the various antenna designs employed, including horns, stacked-patches and resistively-loaded dipoles and monopoles.

Chapter 21 deals with beamforming aspects of UWB ranging. One interesting property of UWB beamforming is a "double dB gain", meaning 6 dB sidelobe level reduction when doubling the number of antennas. This makes UWB beamforming seem to be helpful for indoor geolocalisation in NLOS. The remaining problem is that the transmitter's and the receiver's main lobe should steer towards each other, which can be accomplished by a unidirectional communication between them. In our contribution, the whole approach will be explained, the mathematical framework will be set up, and preliminary simulations with a rather simplified channel model will demonstrate the principal feasibility.

We hope that the above contributions will form an interesting and complete compendium on UWB communication systems.

Enjoy reading.

Dr. Ben Allen	Dr. Mischa Dohler	Dr. Ernest E. Okon
University of Oxford	France Télécom R&D	BAE Systems
Oxford, UK	Grenoble, France	Chelmsford, UK
Dr. Wasim Malik	Prof. Anthony Brown	Prof. David Edwards
University of Oxford	University of Manchester	University of Oxford
Oxford, UK	Manchester, UK	Oxford, UK

Acknowledgments

As the editors of this book, we would first of all like to express our sincere gratitude to our knowledgeable co-authors, without whom this book would not have been accomplished. It is their incredible expertise combined with their timely contributions that have facilitated this high quality book to be completed and published on time. We would like to convey a special thank you to Professor Andreas Molisch for writing a comprehensive and insightful introduction to our book. We would like to acknowledge the time dedicated by all the contributors – thank you.

We would like to thank Olivia Underhill, Mark Hammond and Sarah Hinton at Wiley, for their continuous support throughout the preparation and production of this book. We are also grateful to the reviewers for their suggestions in improving the contents of this book. Similarly, we are grateful for the comments received from the copy-editor and type-setter that played a vital role in preparing a professional and coherent finish to the book.

We owe special thanks to our numerous colleagues, with whom we had lengthy discussions related to the topic of UWB propagation and antennas. The discussions have been fundamental in developing the ideas and techniques contained in this book. There are too many individuals to mention, but we thank them all!

Finally, we are immensely grateful to our families for their understanding and support during the time we devoted to writing and editing this book.

Ben Allen wishes to thank his wife, Louisa, for her understanding and support during the preparation of this book. He would also like to thank his son, Nicholas, for being an understanding new-born that enabled Ben to dedicate just enough time to this book! Ben would also like to thank Mischa Dohler for providing invaluable assistance in ensuring the delivery of the manuscript whilst Ben was on paternity leave – a very big thank you!

Mischa Dohler wishes to particularly thank Gemma, his wife, for her kind understanding and support during the final preparations of this book. Furthermore, he would like to thank his colleagues at Tech/Idea, France Télécom R&D, Grenoble, for creating such a fantastic working environment, and in particular Marylin Arndt, Dominique Barthel, Patrice Senn and Yamina Allaoui, all at France Télécom R&D, Grenoble, in giving sufficient freedom for this work to be completed. He would also like to thank his co-editors for having him taken on board and trusting him with the final arrangements for this book.

Ernest E. Okon wishes to specially thank all the contributors and editors for their respective roles in taking what started off as a small idea and producing a compendium that promises to be a useful and insightful contribution to UWB communications, radar and imaging. He is also grateful for the numerous stimulating discussions with various contributors.

Wasim Q. Malik wishes to thank his father for his ceaseless encouragement, support and guidance in infinite ways. He is grateful to David J. Edwards, Ben Allen, Christopher J. Stevens and Dominic O'Brien (University of Oxford), Moe Z. Win (Massachusetts Institute of Technology), Andreas F. Molisch (Mitsubishi Electric Research Labs), Mischa Dohler (France Télécom R&D), and other collaborators for numerous enlightening discussions on UWB communications over the years.

Anthony K. Brown wishes to thank his family for support during the preparation of this work. He also wishes to thank his colleagues at University of Manchester and in particular Mr Yongwei Zhang for discussions on Ultra Wide Band technology. Finally, Ben and Mischa for their help and support in text preparation.

David J. Edwards would like to thank Dr. Ben Allen for organising this book, Dr. Mischa Dohler (France Télécom R&D) for his efforts, and Dr. Wasim Malik for bearing the brunt of the day-to-day effort that has gone into this book in its final stages. He would also like to acknowledge the other members of the Oxford, Manchester and King's College London teams and numerous colleagues in communications groups around the world who have contributed much in the ultrawideband field over the years. Finally the contributions of the authors, without whose efforts this book would not have come to fruition, is gratefully acknowledged.

Abbreviations & Acronyms

2D	Two Dimensional
AWGN	Additive White Gaussian Noise
A/D	Analog/Digital
AP	Access Point
BAN	Body Area Network
BER	Bit Error Rate
BLAST	Bell Labs Layered Space-Time
BPSK	Binary Phase Shift Keying
BTW	Balanced Transmission-Line Wave
CDF	Cumulative Density Function
CDMA	Code Division Multiple Access
CIR	Channel Impulse Response
CM	Channel Model
CMOS	Complementary Metal Oxide Semiconductor
CPW	Coplanar Waveguide
CSIR	Channel State Information at Receiver
CSIT	Channel State Information at Transmitter
CW	Continuous-Wave
dB	Decibel
DEW	Directed Energy Weapons
DoA	Direction-of-Arrival
DoD	Direction-of-Departure
DP	Direct Path
DS	Direct-Sequence
DVD	Digital Video Disc
ECM	Electronic Countermeasures
EGC	Equal Gain Combining
EIRP	Effective Isotropic Radiated Power
EM	Electromagnetic
ESM	Electronic Surveillance Monitoring
ETSA	Exponential Tapered Slot Antenna
FB	Fractional Bandwidth
FCC	Federal Communications Commission
FDTD	Finite Difference Time Domain
FE	Finite Element
FEM	Finite Element Method

FMCW	Frequency Modulated Continuous Wave
FNBW	First Null Beamwidth
GPR	Ground-Penetrating Radar
GSSI	Geophysical Survey Systems, Inc.
HDR	High Data Rate
HDTV	High Definition Television
HPBW	Half Power Beamwidth
HR	High Rate
HSCA	Horn Shaped Self-Complementary Antenna
ICR	Integrated Cancellation Ratio
IDFT	Inverse Discrete Fourier Transform
IEEE	Institute of Electrical and Electronics Engineers
IFFT	Inverse Fast Fourier Transform
INBW	Inter-Null Beamwidth
IR	Impulse Radio
IRA	Impulse Radiating Antenna
IRA	Impulse Radiating Antenna
LDR	Low Data Rate
LM	Lock Mode
LOS	Line Of Sight
LR	Low Rate
LTI	Linear Time-Invariant
MAI	Multiple-Access Interference
MB	Multi-Band
MDF	Medium Density Fibreboard
MF	Matched Filtering
MGF	Moment Generating Function
MIMO	Multiple-Input Multiple-Output
MISO	Multiple-Input Single-Output
MoM	Method of Moments
MPC	Multipath Components
MRC	Maximal-Ratio Combining
NLOS	Non-Line Of Sight
OFDM	Orthogonal Frequency Division Multiplexing
OLOS	Obstructed Line Of Sight
PAM	Pulse Amplitude Modulation
PAN	Personal Area Network
PC	Personal Computer
PCB	Printed Circuit Board
PDA	Personal Digital Assistant
PDF	Probability Density Function
PDP	Power-Delay Profile
PHY	Physical Layer
PML	Perfectly Matched Layer
PN	Pseudo-Random
PPM	Pulse Position Modulation
PSM	Pulse Shape Modulation
QoS	Quality of Service

RF	Radio Frequency
RMS	Root Mean Square
RT	Ray Tracing
Rx, RX	Receiver
SBR	Shooting and Bouncing Rays
SIMO	Single-Input Multiple-Output
SIR	Signal-to-Interference Ratio
SISO	Single-Input Single-Output
SL	Straight Line
SLL	Sidelobe Level
SMA	Subminiature Version A
SNR	Signal-to-Noise Ratio
TDL	Tapped Delay Line
TDPO	Time Domain Physical Optics
TEM	Transversal Electromagnetic
TF	Transfer Function
TG	Task Group
TH	Time-Hopping
TM	Transversal Magnetic
ToA	Time of Arrival
TR	Transmitted-Reference
TV	Television
Tx, TX	Transmitter
UCA	Uniform Circular Array
ULA	Uniform Linear Array
USB	Universal Serial Bus
UTD	Uniform Theory of Diffraction
UWB	Ultra-Wideband
VNA	Vector Network Analyser
VSWR	Voltage Standing Wave Ratio
WBAN	Wireless Body Area Network
WPAN	Wireless Personal Area Network

1

Introduction to UWB Signals and Systems

Andreas F. Molisch

The word 'ultra-wideband' (UWB) commonly refers to signals or systems that either have a large relative, or a large absolute bandwidth. Such a large bandwidth offers specific advantages with respect to signal robustness, information content and/or implementation simplicity, but lead to fundamental differences from conventional, narrowband systems. The past years have seen a confluence of technological and political/economic circumstances that enabled practical use of UWB systems; consequently, interest in UWB has grown dramatically. This book gives a detailed investigation of an important part of this development, namely UWB antennas and propagation. The current chapter is intended to place this part in the bigger picture by relating it to the issues of system design, applications and regulatory rules.

1.1 History of UWB

UWB communications has drawn great attention since about 2000, and thus has the mantle of an 'emerging' technology. It is described in popular magazines by monikers such as 'one of ten technologies that will change your world'. However, this should not detract from the fact that its origins go back more than a century. Actually, electromagnetic communications started with UWB. In the late 1800s, the easiest way of generating an electromagnetic signal was to generate a short pulse: a spark-gap generator was used, e.g., by Hertz in his famous experiments, and by Marconi for the first electromagnetic data communications [1]. Thus, the first practical UWB systems are really more than 100 years old. Also, theoretical research into the propagation of UWB radiation stems back more than a century. It was the great theoretician, Sommerfeld, who first analysed the diffraction of a short pulse by a half-plane – one of the fundamental problems of UWB propagation [2].

However, after 1910, the general interest turned to narrowband communications. Part of the reason was the fact that the spectral efficiency of the signals generated by the spark-gap transmitters was low – the signals that were generated had a low bit rate, but occupied a large bandwidth. In other words, those signals

Ultra-wideband Antennas and Propagation for Communications, Radar and Imaging   Edited by B. Allen,
M. Dohler, E. E. Okon, W. Q. Malik, A. K. Brown and D. J. Edwards
© 2007 John Wiley & Sons, Ltd

had a large spreading factor. At that time, it was not known how to exploit such signal spreading; it was simply seen as a deficiency. On the other hand, narrowband communications, which allowed frequency division multiplexing, offered an easy way of transmitting multiple signals in a finite bandwidth. Thus, UWB research fell dormant.

It was revived in the 1960s in a different context, namely military radar, where spectral efficiency was not a major consideration. Rather, the point was to improve the spatial resolution; in other words, improve the accuracy with which the runtime from the radar transmitter to a specific object, and back to the receiver, can be determined. It follows from elementary Fourier considerations that this can be done the better, the shorter the transmitted radar pulses are. Increased interest in this work coincided with the invention of the sampling oscilloscope, which allowed the experimental analysis of short-duration signals in the time domain. A key component for UWB radar systems was the design of high-power, short-pulse generators, which was investigated by the military in both the USA and the Soviet Union.

Ultra-wideband communications started to receive renewed interest in the 1970s [3]. At this time, it was called 'baseband' or 'carrier-free' communications. Around 1973, it was recognised that short pulses, which spread the signal over a large spectrum, are not significantly affected by existing narrowband interferers, and do not interfere with them, either. However, the problem of multiple-access interference (MAI) of unsynchronised users remained, so that in the 1970s and 1980s, UWB communications continued to be mostly investigated in the military sector, where spectral efficiency was of minor importance. The MAI problem was solved by the introduction of time-hopping impulse radio (TH-IR) in the early 1990s, where the pioneering work of Win and Scholtz [4], [5], [6] showed that impulse radio could sustain a large number of users by assigning pseudorandom transmission times to the pulses from the different users. This insight, coupled with advances in electronics device design, spawned the interest of commercial wireless companies into UWB.

Another key obstacle to commercial use of UWB was political in nature. Frequency regulators all over the world assign narrow frequency bands to specific services and/or operators. UWB systems violate those frequency assignments, as they emit radiation over a large frequency range, including the bands that have already been assigned to other services. Proponents of UWB tried to convince the frequency regulator in the USA, the FCC (Federal Communications Commission), that the emissions from UWB devices would not interfere with those other services. After a lengthy hearing process, the FCC issued a ruling in 2002 that allowed intentional UWB emissions in the frequency range between 3.1 and 10.6 GHz, subject to certain restrictions for the emission power spectrum [7] (for more details, see Section 1.4).

If the introduction of time-hopping impulse radio had created a storm of commercial UWB activities, the FCC ruling turned it into a hurricane. Within two years, more than 200 companies were working on the topic. Recognising this trend early on, the IEEE (Institute of Electric and Electronics Engineers) established a working group (IEEE 802.15.3a) with the task of standardising a physical layer for high-throughput wireless communications based on UWB. High-data rate applications showed the greatest initial promise, largely due to the immediately visible commercial potential. While the process within the IEEE stalled, it gave rise to two industry alliances (Multiband-OFDM Alliance/WiMedia, and the UWB Forum) that were shipping products by 2005. Ironically, neither of those alliances uses impulse radio, but rather more 'mature' technologies, namely OFDM (orthogonal frequency division multiplexing) and DS-CDMA (direct-sequence code division multiple access), respectively (see Section 1.5). UWB is also beneficial for the transmission of data with low rates, using as little energy as possible – the principles of impulse radio are especially suitable in this context. As the goals are greatly different from the high-data rate applications, also a different standardisation group was charged with developing a common specification for such devices, the IEEE 802.15.4a group (see Section 1.5).

Since 2000, the scientific research in UWB communications has gone in a number of different directions. The theoretical performance of time-hopping impulse radio was the first topic that drew widespread interest. The most fundamental problems were solved in [6], [8]; more detailed aspects of multiple access and narrowband interference were treated in [9], [10], [11], [12], [13], [14]. Issues of equalisation and Rake reception are analysed in [15], [16], channel estimation and synchronisation [17], [18], transmitted-reference and differential schemes [19], [20], [21], [22], [23]. Spectral shaping and the design of hopping sequences are the topic of [24], [25], [26], [27]. Practical implementation issues, both for impulse radio, and for alternative implementations, are discussed, e.g., in [28], [29], [30]. The combination of UWB with MIMO (multiple-input multiple-output), i.e., the use of multiple antenna elements at both link ends, is the topic of [31], [32], [33].[1]

It is also interesting to see how the research into UWB antennas and propagation evolved over the years. UWB antennas are more than 100 years old [34]. Some of the very first antennas were biconical antennas and spherical dipoles, which have very good wideband characteristics. They were rediscovered in the 1930s by Carter, who also added broadband transitions from the feed to the radiating elements. It is noteworthy that UWB antenna research never experienced the slump of UWB communications, but rather stayed a popular and important area throughout the past 70 years. A main factor in this development was TV broadcasting: the assigned TV bands extend over a large frequency range, and since it is desirable that a single antenna can transmit/receive all available stations, it implied that the antenna had to be very broadband as well. As UWB communications emerged as a commercially viable option in the 1990s, the development of smaller antennas became a new requirement. Slot antennas and printed antennas have proven to be especially useful in that context.

UWB propagation research, on the other hand, has a more chequered history. Theoretical investigations of the interaction of short pulses with variously shaped objects have abounded since the classical studies of Sommerfeld mentioned above [35]. Practical propagation studies, however, seem to have been limited to radar measurements; as those were typically classified. The development of statistical UWB channel models, which are essential for the development of communications systems, started only recently: the first measurement-based statistical channel model that took UWB aspects explicitly into account was [36], which is valid in the low frequency range (<1 GHz). The IEEE 802.15.3a group established a channel model for residential and office environments [37] in the 3–10 GHz range, based on two measurement campaigns. This model was widely used both by industrial and academic research from 2003 to 2005. More measurement campaigns were performed after 2003, and another standardised model, the IEEE 802.15.4a model, published in 2005, included a larger variety of environments and took more propagation effects into account [38]. However, while considerable progress has been made in understanding UWB channels, as described in Part III of this book, a lot remains to be done; see also [39].

1.2 Motivation

UWB systems can be characterised either by a large relative bandwidth, or a large absolute bandwidth. Each of these has specific advantages, as well as challenges, which we will discuss in the remainder of this section.

1.2.1 Large Absolute Bandwidth

By 'large absolute bandwidth', we usually refer to systems with more than 500 MHz bandwidth, in accordance with the FCC definition of UWB radiation [7]. Such a large bandwidth offers the possibility of very large spreading factors: in other words, the ratio of the signal bandwidth to the symbol rate is

---

[1] The list of references given here is far from complete, and is just intended to exemplify the trends in UWB communications.

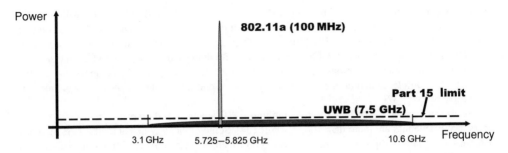

Figure 1.1 Interference between a UWB system and a narrowband (IEEE 802.11a) local area network

very large. For a typical sensor network application with 5 ksymbol/s throughput, a spreading factor of 10^5 to 10^6 is achieved for transmission bandwidths of 500 MHz and 5 GHz, respectively. Spreading over such a large bandwidth means that the power spectral density of the radiation, i.e., the power per unit bandwidth, is very low. A victim legacy (narrowband) receiver will only see the noise power within its own system bandwidth, i.e., a small part of the total transmit power, see Figure 1.1. This implies that the interference to legacy (narrowband) systems is very small. Furthermore, a UWB receiver can suppress narrowband interference by a factor that is approximately equal to the spreading factor. These principles are well understood from the general theory of spread spectrum systems. The distinctive feature of UWB is that it goes to extremes in terms of the spreading factor, and thus brings the power spectral density to such low levels that it does not disturb legacy systems at all under most operating conditions. Ideally, the radiation just increases the noise level seen by the victim receiver by a negligible amount. As an additional advantage, such radiation is almost undetectable for unauthorised listeners. It must be kept in mind that the spreading factor is a function of both the transmission bandwidth and the data rate. Consequently, UWB systems with high data rates (>100 Mbit/s) do not exhibit such a large spreading factor as that in the above example, and are thus more sensitive to interference.

Another important advantage of using a large absolute bandwidth is a high resilience to fading. In conventional narrowband systems, the received signal strength undergoes fluctuations, caused by multipath components (MPCs), i.e., echoes from different scatterers, which interfere with each other constructively or destructively, depending on the exact location of transmitter, receiver and scatterers [40]. The amplitude statistics of the total received signal is typically complex Gaussian, because a large number of (unresolvable) MPCs add up at the receiver. A UWB transceiver receives a signal with a large absolute bandwidth, and can thus resolve many of those MPCs. By separately processing the different MPCs, the receiver can make sure that all those components add up in an optimum way, giving rise to a smaller probability of deep fades. In other words, the many resolvable MPCs provide a high degree of delay diversity. As an additional effect, we will observe in Section 1.3 that the number of actual MPCs that constitute a resolvable multipath component is rather small; for this reason, the fading statistics of each resolvable multipath component does not have a complex Gaussian distribution anymore, but shows a lower probability of deep fades. Due to this, and the delay diversity, UWB systems with large absolute bandwidth need hardly any fading margin for compensating small-scale fading [41], yielding a significant advantages over conventional narrowband systems, see Figure 1.2.

Finally, a large absolute bandwidth also leads to a great improvement of the accuracy of ranging and geolocation. Most ranging systems try to determine the flight time of the radiation between transmitter and receiver. It follows from elementary Fourier considerations that the accuracy of the ranging improves the bandwidth of the ranging signal. Thus, even without sophisticated high-resolution algorithms for the

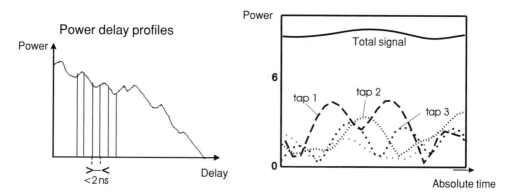

Figure 1.2 Delay diversity in a UWB system

determination of the time-of-arrival of the first path, a UWB system can achieve centimetre accuracy for ranging [42]. Of course, this inherent accuracy can be augmented by additional information, like direction-of-arrival, and/or receive power.

While the large absolute bandwidth gives a number of advantages, it also gives rise to a number of challenges. From a hardware point of view, the accuracy of the local oscillators and/or timing circuits has to be very high. When the absolute bandwidth is 1 GHz, a timing jitter of 1 ns can obviously have catastrophic consequences. Another consequence of the high delay resolution is that a large number of components need to be received and processed. For example, the number of fingers in a Rake receiver that is required to collect 90 % of the available energy can easily reach several tens or even hundreds [43]. Finally, the fine delay resolution can also have drawbacks for ranging: the first, (quasi-) line-of-sight component that needs to be detected by the ranging algorithms can contain little energy, and can thus have a poor signal-to-noise ratio (SNR).

1.2.2 Large Relative Bandwidth

Again, following the FCC definition, we consider systems with a relative bandwidth of larger than 20 % as ultra-wideband. Such a large bandwidth can greatly enhance the signal robustness for data transmission, and can have even more important advantages for radar and ranging. Intuitively, the different frequency components of the signal 'see' different propagation conditions. Thus, there is a high probability that at least some of them can penetrate obstacles or otherwise make their way from transmitter to receiver. This advantage is especially striking in a baseband system, where frequencies from (typically) a few tens of Megahertz, to up to 1 Gigahertz are being used. The low-frequency components can more easily penetrate walls and ground, while the high-frequency components give strongly reflected signals. Consequently, the signal is more robust to shadowing effects (in contrast to the large-absolute-bandwidth systems, which are more robust to interference between MPCs). The good wall and floor penetration is also very useful for radar and geolocation systems.

In many practical cases, large-relative-bandwidth systems are pure baseband systems, i.e., systems where baseband pulses are directly applied to the transmitting antennas (note, though, that no DC components can be transmitted). Such systems have also advantages from an implementation point of view. In particular, they obviate the necessity for RF components such as local oscillators, mixers, etc.

On the other hand, a large relative bandwidth can also lead to considerable complications in the system design. Most devices such as antennas, amplifiers, etc., have inherent narrowband characteristics that are caused by both practical restrictions, and by fundamental principles.

For radar applications, as well as disaster communications systems, a large relative bandwidth is the most important argument for using UWB. For other communications systems, a large absolute bandwidth is typically more important.

1.3 UWB Signals and Systems

The investigation of UWB antennas and propagation is intimately related to the design of UWB transceivers and signal processing. On one hand, it is impossible to design good and efficient systems if we do not know the effective channel (including antennas) that we are designing the system for. On the other hand, we have to know the system design in order to know how antenna design and propagation channel impact the system performance.

In the past years, four methods have emerged for signalling with ultra-wide bandwidths: impulse radio, DS-CDMA, OFDM and frequency hopping.[2] We will also introduce these four signalling schemes in the following sections and also the issues of signalling for geolocation and radar.

1.3.1 Impulse Radio

Impulse radio has many attractive properties, like enabling to build extremely simple transmitters. However, an important problem that plagued impulse radio (baseband transmission) for a long time was the spectral efficiency: it seemed that only a small number of users could be 'on air' simultaneously. Consider the case where one pulse per symbol is transmitted. Since the UWB transceivers are unsynchronised, so-called 'catastrophic collisions' can occur, where pulses from several transmitters arrive at the receiver simultaneously. The signal-to-interference ratio then becomes very bad, leading to a high bit error rate (BER). Win and Scholtz showed that this problem could be avoided by time-hopping impulse radio (TH-IR) [6]. Each data bit is represented by *several* short pulses; the duration of the pulses determines essentially the bandwidth of the system. The transmitted pulse sequence is different for each user, according to a so-called time-hopping (TH) code. Thus, even if one pulse within a symbol collides with a signal component from another user, other pulses in the sequence will not, see Figure 1.3. In other words, collisions can still occur, but they are not catastrophic anymore. TH-IR achieves a multiple-access interference suppression that is equal to the number of pulses in the system. The possible positions of the pulses within a symbol follow certain rules: the symbol duration is subdivided into N_f 'frames' of equal length. Within each frame the pulse can occupy an almost arbitrary position (determined by the time-hopping code). Typically, the frame is subdivided into 'chips', whose length is equal to a pulse duration. The (digital) time-hopping code now determines which of the possible positions the pulse actually occupies.

When all the transmitted pulses have the same polarity, as shown in Figure 1.3, the signal spectrum shows a number of lines. This is highly undesirable, as most spectrum regulators prescribe a maximum power spectral *density* that has to be satisfied, e.g., in each 1-MHz sub-band. Thus, the transmit power of a signal with spectral lines has to be backed off such that the spectral lines satisfy the spectral mask – this

---

[2] A fifth technique, chirping, in which the carrier frequency is linearly changed during transmission, is explicitly forbidden by some frequency regulators, and will not be considered further here.

Introduction to UWB Signals and Systems

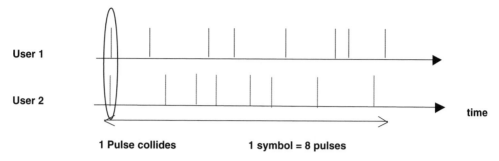

Figure 1.3 Principle of time-hopping impulse radio for the suppression of catastrophic collisions

leads to a considerable loss in SNR. This problem was solved in [24] (see also [25]) by choosing the polarity of the transmit pulses in a pseudorandom way; this process can be undone at the receiver.

The modulation of this sequence of pulses can be pulse-position modulation (PPM), as suggested in [6], or pulse amplitude modulation (PAM) such as BPSK (binary phase shift keying) [40]. PPM has the advantage that the detector can be much simpler – it only needs to determine whether there is more energy at time t_0, or at time $t_0 + \delta$. It allows the use of noncoherent receivers (energy detectors), as well as the use of coherent receivers. For noncoherent receivers, it is required that δ is larger than the delay spread of the channel. BPSK can only be used in conjunction with coherent receivers, however, it gives better performance than PPM since it is an antipodal modulation format. The transmit signal for BPSK modulation reads

$$s_{tr}(t) = \sum_{j=-\infty}^{\infty} d_j b_{\lfloor j/N_f \rfloor} w_{tr}\left(t - jT_f - c_j T_c\right) = \sum_{k=-\infty}^{\infty} b_k w_{seq}(t - kT_s), \qquad (1.1)$$

where $w_{tr}(t)$ is the transmitted unit-energy pulse, T_f is the average pulse repetition time, N_f is the number of frames (and therefore also the number of pulses) representing one information symbol of length T_s, and b is the transmitted information symbol, i.e., ± 1; $w_{seq}(t)$ is the transmitted pulse sequence representing one symbol. The TH sequence provides an additional time shift of $c_j T_c$ seconds to the j-th pulse of the signal, where T_c is the chip interval, and c_j are the elements of a pseudorandom sequence, taking on integer values of between 0 and $N_c - 1$. To prevent pulses from overlapping, the chip interval is selected to satisfy $T_c \leq T_f/N_c$. The polarity randomisation is achieved by having each pulse multiplied by a (pseudo) random variable, d_j, that can take on the values of $+1$ or -1 with equal probability. The sequences d_j and c_j are assumed to be known at transmitter and receiver.

Coherent reception requires the use of Rake receivers in order to collect the energy of the available resolvable MPCs. The Rake can be implemented as a bank of correlators, where correlation is done with the transmit waveform, and the sampling time of each finger is matched to the delay of one resolvable MPC. Note, however, that for systems with a large absolute bandwidth, the number of resolvable MPCs can become very large. Since the number of fingers in practical Rake receivers is limited, only a subset of the available MPCs can be received [16].

Furthermore, Rake receivers operating with UWB signals can cause distortion of the MPCs. For conventional wireless systems, the impulse response of the channel can be written as

$$h(t, \tau) = \sum_{i=1}^{N} a_i(t) \cdot \chi_i(t, \tau) \otimes \delta(\tau - \tau_i), \qquad (1.2)$$

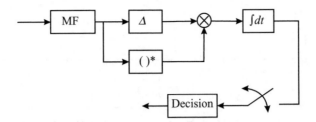

Figure 1.4 Block diagram of a transmitted-reference receiver [49]. Reproduced by permission of © 2004 IEEE.

where N is the number of resolvable MPCs, and the $a_i(t)$ are the complex amplitudes of the resolvable MPCs. For a UWB system, the impulse response must be written as

$$h(t,\tau) = \sum_{i=1}^{N} a_i(t) \cdot \delta(\tau - \tau_i), \qquad (1.3)$$

where $\chi_i(t,\tau)$ denotes the (time-varying) distortion of the i-th echo due to the frequency selectivity of the interactions with the environment, the reasons for which will be described in Part III (see also [39]). The distortions can be significant, especially in systems with large relative bandwidth. For optimum reception, a Rake receiver needs to know the functions $\chi_i(t,\tau)$. Alternatively, the receiver must sample at the Nyquist rate and process all the sample values – the number of which can be significantly higher than the number of MPCs. As a further important conclusion, we find that the matched filter has to take the distortions of the waveform, w_{tx}, by the antennas into account. If that is not possible, it is desirable that the antennas distort w_{tx} as little as possible.

Since coherent reception of impulse radio can be challenging, alternative demodulation schemes have been investigated. Noncoherent reception is the simplest approach, and works very well when the delay spread of the channel is small and the signal-to-noise ratio is high. For large delay spreads, the receiver has to integrate the received energy over a long period, which also means that it picks up a lot of noise in the process. Furthermore, the contribution of the noise–noise cross-terms in the squared signal leads to an additional deterioration of the performance. Finally, noncoherent detection is more sensitive to interference.

As a compromise between coherent and noncoherent schemes, transmitted-reference (TR) schemes are often used. In TR, we first transmit a reference pulse of known polarity (or position), followed by a data pulse whose polarity (position) is determined by the information bit. At the receiver, we then have to multiply the received signal with a delayed version of itself, see Figure 1.4. This scheme has an SNR that is worse than that of a coherent receiver (due to the occurrence of noise–noise crossterms) and comparable to a noncoherent receiver. It is not sensitive to distortions by antennas and channels, because both the data pulse and the reference pulse undergo the same distortions. Furthermore, it is less sensitive to interference than noncoherent detection.

1.3.2 DS-CDMA

Although UWB has long been associated with impulse radio, it is not the only possibility of spreading a signal over a large bandwidth. More 'classical' spreading methods, as discussed, e.g., in [44], can be used as well. In particular, DS-CDMA can be used in a straightforward way to generate UWB signals.

Introduction to UWB Signals and Systems

DS-CDMA spreads the signal by multiplying the transmit signal with a second signal that has a very large bandwidth. The bandwidth of this total signal is approximately the same as the bandwidth of the wideband spreading signal. Conventionally, the spreading sequence consists of a sequence of ±1s. *m*-sequences (*M*aximum-length sequences (*m*-sequences) generated by shift-registers with feedback are the most popular of these sequences, although there are many others. The transmit signal is thus

$$s_{tr}(t) = \sum_{j=-\infty}^{\infty} d_j b_{\lfloor j/N_f \rfloor} w_{tr}(t - jT_c) = \sum_{k=-\infty}^{\infty} b_k w_{seq}(t - kT_s), \tag{1.4}$$

where the symbols have the same meaning as in Equation (1.1).

The difference between a conventional (e.g., cellular) DS-CDMA system and a UWB signal is the chip rate, i.e., $1/T_c$. Consequently, both the theoretical underpinnings and the implementation aspects of DS-CDMA are well understood; this facilitates their use for UWB systems. For example, the high-data-rate UWB system proposed by the UWB Forum industrial group is such a DS-CDMA system.

When comparing Equations (1.4) and (1.1), we find some important similarities: both TH-IR and DS-CDMA transmit a bit by multiplying it with a spreading sequence $w_{seq}(t)$, and the bandwidth is essentially determined by the duration and shape of a basis pulse, $w_{tx}(t)$. The major difference lies in the nature of the spreading sequence. For the DS-CDMA case, it consists only of binary values, ±1, while in the impulse radio (IR) case, it consists of many zeroes, with several ±1's located at pseudorandom positions. As a consequence, DS-CDMA signals can be more difficult to generate: it is not just a matter of generating short pulses at large intervals, but rather it requires the continuous generation of those pulses. Furthermore, DS-CDMA as described above does not allow noncoherent (energy detection) reception since a correlation process is required to recover the original data.

1.3.3 OFDM

OFDM transmits information in parallel on a large number of subcarriers, each of which requires only a relatively small bandwidth. This approach, first suggested for wireless applications by Cimini [45], has become popular for high-data-rate transmission in conventional systems, e.g., the IEEE 802.11a/g wireless standards, and its theory and implementation are now well understood. The block diagram of a typical system is shown in Figure 1.5. The data stream is first serial-to-parallel converted, and then modulated onto subcarriers that are separated by a frequency spacing W/N, where W is the total transmission bandwidth, and N is the number of subcarriers. The modulation process can be done either in the analogue domain (Figure 1.5(a)), or digitally, by performing an inverse fast Fourier transform (IFFT) on the data (Figure 1.5(b)). The latter approach does not need multiple local oscillators, and is thus the one in use today. However, it requires an IFFT and analogue-to-digital converters operating at high speed (clock speed of approximately W).

OFDM transmits each information symbol on one carrier, and thus does not exploit the frequency diversity inherent in a UWB system. This problem can be circumvented by the use of appropriate coding and/or by the use of multicarrier-CDMA, which spreads each modulation symbol over a number of subcarriers [46].

The impact of channels and antennas on UWB-OFDM systems is similar to that on conventional OFDM systems. In either case, the receiver determines the distortion (attenuation and phase shift) on each subcarrier, and compensates for it. The choice of subcarrier spacing depends mostly on the channel characteristics, especially the *maximum excess delay*, and not on the total system bandwidth. Rather, the total number of tones increases approximately linearly with the total bandwidth. Furthermore, the FFT

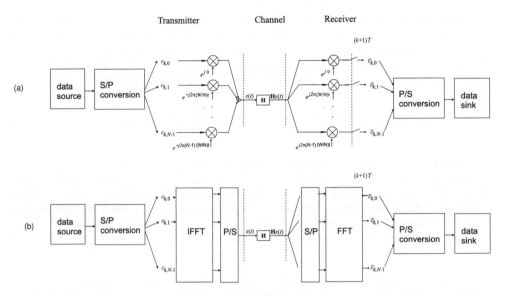

Figure 1.5 Principle of OFDM: analogue implementation (a) and digital implementation (b). Source [40], reproduced by permission of © 2005 John Wiley & Sons, Ltd.

has to operate with a clock speed that is approximately equal to the bandwidth. For these reason, OFDM becomes impractical for bandwidths above 500 MHz (at least with the technology available at the time of writing, 2006).

1.3.4 Frequency Hopping

Fast frequency hopping changes the carrier frequency several times during the transmission of one symbol; in other words, the transmission of each separate symbol is spread over a large bandwidth. *Slow frequency hopping* transmits one or several symbols on each frequency. Frequency hopping has a multiple-access capability. Different users are distinguished by different hopping sequences, so that they transmit on different frequencies at any given time.

For more details, we have to distinguish between the case of synchronised and unsynchronised users. In the synchronised case, all users can use the same hopping pattern, but with different offsets, see Figure 1.6(a). In the unsynchronised case (Figure 1.6(b)), we have no control over the relative timing between the different users. Thus, the hopping sequences must make sure that there is little multiple-access interference for all possible timeshifts between the users; otherwise, catastrophic collisions between users could occur. The situation is analogous to TH-IR, where we need to find time-hopping sequences that avoid catastrophic collisions.

Frequency hopping can be used either as a multiple-access scheme of its own, or it can be combined with other schemes. In the latter case, we divide the available frequency band into sub-bands, and transmit (e.g., with OFDM) in different sub-bands at different times. This approach simplifies implementation, as the sampling and A/D conversion now has to be done only with a rate corresponding to the width of the sub-band instead of the full bandwidth. The UWB channel is thus converted into a number of narrowband channels, because most propagation effects in a 500-MHz channel are the same as those in conventional

Introduction to UWB Signals and Systems

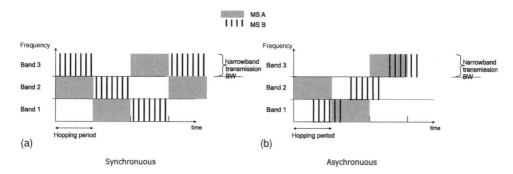

Figure 1.6 Frequency hopping multiple access with synchronous (a) and asynchronous users (b). Source [40], reproduced by permission of © 2005 John Wiley & Sons, Ltd.

(wireless) channels. However, the different sub-bands undergo different attenuations. In a similar manner to conventional OFDM systems, it is also essential that coding/interleaving across different frequency bands is performed.

1.3.5 RADAR

While conventional radar systems work with modulated carriers with a bandwidth of no more than 10 %, UWB radars transmit short, high-powered pulses. The product of the speed of light with the pulse duration should be less than the physical dimensions of the observed objects; it is also often smaller than the dimensions of the used antennas. As a consequence, the signal shape of the signal is distorted by transmission from the antennas, by reflection from the observed objects and by reception at the receive antenna. This situation is similar to the distortion of each separate MPC as discussed in Section 1.3.1. Thus, again, the received signal has an unknown shape, and matched filtering, the mainstay of conventional radar detection theory, cannot be applied [47].

It is also important to recognise that the pulse shape distortions at the antenna depend on the direction of the radiation. As a consequence, the compensation for antenna signal distortion depends on the direction of arrival. When radars with synthetic aperture arrays are used, many of the well-known high-resolution direction-finding algorithms do not work anymore, since they depend on the narrowband signal assumption.

Quite generally, new signal-processing algorithms need to be developed to extract all the available information about target shape, distance and movement from the received signals. The correct modelling of the distortion of the pulses caused by the antenna and object is the *conditio sine qua non* for those algorithms.

1.3.6 Geolocation

For sensor networks and similar applications, ranging and geolocation has become an important function.[3] While (active) ranging shows some similarities to radar, it also has some important differences. A ranging

---

[3] By 'ranging' we mean the determination of the distance between two devices. By 'geolocation', we mean the determination of the absolute position of a device in space. Geolocation of a device can be achieved if the range of

system tries to determine the time-of-arrival of the *first* MPC in the transmission from another active device. Together with knowledge of the absolute time when the transmitter sent out the signal, this allows us to determine the runtime of the signal between the two devices.[4] A major challenge in geolocation is the determination of the first arriving MPC in the presence of other MPCs as well as noise. This is made more difficult by the fact that – due to the very high delay resolution – the first resolvable MPC carries less energy than in conventional systems, especially in non-line-of-sight situations. The actual propagation conditions, especially the attenuation of the (quasi-) line-of-sight, have thus an important influence on the accuracy of UWB ranging.

1.4 Frequency Regulation

When designing a UWB system, the first step is to decide the frequency range over which it should operate. The transmit signals have to satisfy the frequency regulations in the country in which the device operates. Until the turn of the century, frequency regulators the world over prohibited the intentional emission of broadband radiation (and put strict limits on unintentional radiation), because it can interfere with existing, narrowband communications systems. It was pointed out by UWB advocates that UWB systems minimise this interference by spreading the power over a very large bandwidth. After lengthy deliberations, the FCC issued its 'report and order' in 2002, which allowed the emission of intentional UWB emissions [7], subject to restrictions on the emitted power spectral density.[5]

The 'frequency masks' depend on the application and the environment in which the devices are operated. For indoor communications, a power spectral density of −41.3 dBm/MHz is allowed in the frequency band between 3.1 and 10.6 GHz. Outside of that band, no intentional emissions are allowed, and the admissible power spectral density for spurious emissions provides special protection for GPS and cellular services (see Figure 1.7). Similarly, outdoor communications between mobile devices is allowed in the 3.1–10.6 GHz range, though the mask for spurious emissions is different. For wall-imaging systems and ground-penetrating radar, the operation is admissible either in the 3.1–10.6 GHz range, or below 960 MHz; for through-wall and surveillance systems, the frequency ranges from 1.99–10.6 GHz, and below 960 MHz are allowed. Furthermore, a number of military UWB systems seem to operate in that range, though exact figures are not publicly available. The frequency range from 24–29 GHz is allowed for vehicular radar systems.

In the autumn of 2005, the Japanese and European frequency regulators issued first drafts of rulings. These would indicate that operation is allowed in the frequency range between 3.1 and 4.8 GHz, as well as between 7–10 GHz, i.e., omitting the band around 5 GHz. For the 3.1–4.8 GHz range, a 'detect-and-avoid' mechanism is required, i.e., a UWB device must determine whether there are narrowband (victim) receivers in the surroundings, and avoid emissions in the frequency range of those victim devices. Further details are unknown at the time of this writing.

---

this device to a number of other devices with known positions can be determined. Direction-of-arrival information can be used to make this process more accurate.

[4] A variety of techniques can be used to exchange knowledge about the absolute transmission time of the signal, e.g., timestamps on the transmitted signal, or 'ping-pong' schemes, where device A sends a signal, device B receives it, and after a certain time replies with a signal of its own. After determining the arrival time of this signal, device A knows the total roundtrip time of a signal between the two devices.

[5] The ruling also restricts: (i) the admissible peak power; (ii) the location of deployment (fixed installations of transmitters are prohibited outside of buildings); and (iii) the applications for which the products can be used (e.g., UWB transmitters in toys are prohibited).

Introduction to UWB Signals and Systems

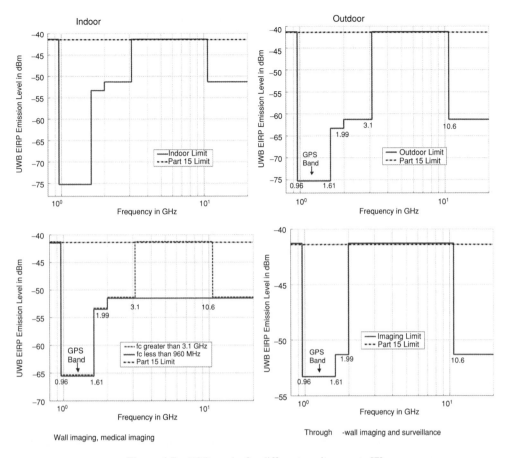

Figure 1.7 FCC masks for different environments [7]

Further restrictions in the useful frequency range arise from the current technological possibilities. Semiconductor devices are available that cover the whole spectrum assigned to UWB. However, complementary metal oxide semiconductor (CMOS) technology, which is by far the most appealing process for high-volume commercial applications, is currently only available for frequencies up to about 5 GHz.

1.5 Applications, Operating Scenarios and Standardisation

For the antennas and propagation researcher, it is important to understand the application and the deployment and operating scenario for which a UWB device is used. The usage of the systems determines their location (which has a big impact on propagation conditions), and on their size (which determines, e.g., the admissible size of the antennas).

One of the most popular applications of UWB is data transmission with a very high rate (more than 100 Mbit/s). Given the large bandwidth of UWB, such high rates can be easily achieved, but the spreading factor is small. The combination of small spreading factor and low admissible power spectral density limits the range of such systems to some 10 m. Networks that cover such a short range are often called *personal area networks* (PANs). High-data-rate PANs are used especially for consumer electronics and personal computing applications. Examples include the transmission of HDTV (high definition television) streams from a set-top box or a DVD player to the TV requires high data rates and wireless USB (universal serial bus), which aims to transmit data at 480 Mbit/s between different components of a computer. For these applications, UWB is in a competition with wireless local area networks (WLANs) based on multiple-antenna technology, such as the emerging 802.11n standard, which also aims to achieve high data rates. UWB has the advantage of possibly lower costs and higher data rates, while the WLANs can achieve longer ranges. In order to further increase the data rate, the combination of UWB with multiple antennas is currently being deliberated. This has important consequences for the antenna research (the design of suitable antenna arrays becomes an issue), signal processing (the appropriate processing of the data from the multiple antennas is different from the narrowband case) and propagation (the directional information of the MPCs at the two link ends is relevant).

In 2002, the IEEE established a standardisation body, the working group 802.15.3a, to write the specifications for high-data-rate PANs. Soon, two major proposals emerged, one based on DS-CDMA (Section 1.3.2), the other on a combination of frequency hopping (Section 1.3.4) and OFDM (Section 1.3.3). Both of those proposals use only the frequency range between 3.1 and 5 GHz. Although the IEEE group has been deadlocked since 2003, both of the proposals are the basis of industry consortia that started to ship products in 2005: the WiMedia consortium, which merged with the MBOA (Multiband OFDM Alliance) consortium, and uses the OFDM-based physical layer specifications; and the UWB Forum, which adopted the DS-CDMA system. The ultimate winner will emerge from a battle in the marketplace; the outcome will also have an impact on the requirements for antennas for a considerable percentage of the UWB market, as discussed in Sections 1.3.2 and 1.3.3.

Another important application area is sensor networks. Data from various sensors are to be sent to a central server, or to be exchanged between different sensors. The volume of data is typically small, so that average data rates of a few kbit/s or less are common. Size restrictions can be stringent, because the transceiver has to be collocated with the sensors. Also the requirements for energy consumption can be very stringent, since the devices are often battery operated [48]. The location of the devices can vary greatly, and include positions where propagation conditions are very unfavourable. Thus, the propagation conditions for such applications can differ significantly from both high-rate UWB devices, and from classical cellular and WLAN applications. Low-data-rate systems are also envisaged for emergency communications, e.g., between people within a collapsed building and rescue workers. In this case, the signal robustness that stems from a large relative bandwidth and the possibility of floor and wall penetration is especially important. Consequently, these systems tend to operate at lower bandwidths. Low-rate systems are currently being standardised by the IEEE 802.15.4a group. In contrast to the deadlocked 802.15.3a group, the 802.15.4a group is expected to produce a standard in 2006, based on an impulse radio approach.

Even at low data rates, the range that can be covered by a single UWB link is rather limited – 30 to 100 m seem to be the maximum. Longer ranges can be achieved by relaying the messages between different nodes, until they arrive at their destination. The appropriate design of the routing and optimisation of the energy spent at each node are some recent research topics that have drawn attention in the academic community. However, from an antenna and propagation point of view, it is sufficient to consider a single link.

Body area networks (BANs) consist of a number of nodes and units placed on the human body or in close proximity such as on everyday clothing. A major drawback of current wired BANs is the inconvenience for the user. While smart (prewired) textiles have been proposed, they imply the need for a special garment to be worn, which may conflict with the user's personal preferences. Wireless body-centric network presents the apparent solution.

In sensor networks, geolocation of the nodes can be of great importance. This is a major argument for UWB, which allows a much more precise location of the devices than narrowband schemes. This has an important impact on propagation research, as discussed in Section 1.3.6. When direction-of-arrival information is used to increase the accuracy of the location estimation, antenna arrays are also important system components.

UWB radars have developed into an important market niche, used mainly for two purposes: (i) high-performance radars that have smaller 'dead zones', and (ii) radars for close ranges that can penetrate walls and ground. The second application is useful for surveillance, urban warfare and landmine detection. Most of the applications in this area are classified, as they serve military or law-enforcement purposes. A commercial application is the vehicular collision avoidance radar. Such a radar typically operates in the microwave range (24–29, or around 60 GHz). Propagation conditions are usually straightforward (line-of-sight); antenna research concentrates on antennas that can be easily integrated into the chassis of cars. Another promising application is biological imaging, e.g., for cancer detection.

Naturally, the above enumeration of applications is not complete, and it can be anticipated that in the future, even more ways to use UWB will be discovered.

1.6 System Outlook

In an area as active as UWB, one question arises naturally: where is it going? From a commercial point of view, the initial hype has somewhat abated: statements like 'UWB is the ultimate solution to all problems in wireless' (a quote from a trade journal in 2002) nowadays sound absurd not only to researchers, but also to all people working in the area. At the same time, many more concrete visions for the application of UWB techniques have been developed, as discussed in Section 1.5. It has also been recognised that there is no single solution for all the different applications. The techniques required by a high-data-rate, short-range system attached to a DVD player are completely different from the ones required by a battery-powered sensor node. There is always a trade-off between cost, power consumption, data rate and range, and different applications require different solutions. The more the market develops, the more diverse products will be established.

In order to create all-CMOS devices (where both the digital signal processing and the radiofrequency electronics can be manufactured in this most cost-effective technology), restrictions on the admissible frequency range have to be accepted. At the moment, 5 GHz represents the upper frequency range that can be achieved with that technology. This brings the chip manufacturers on a collision course with frequency regulators in Asia and Europe, who want to move UWB devices to the 7–10 GHz range.

In terms of antenna research, the main goal is a reduction of the antenna size, while still keeping the antenna efficiency at reasonable levels and keeping the manufacturing costs low. Due to the rules of frequency regulators, it is desirable that antennas have the same radiation pattern at all frequencies. Furthermore, the antenna aperture (and not the antenna gain) should be as independent of frequency as possible. Due to the increasing role of antenna arrays both for direction finding and for increasing the capacity of the systems, the design of suitable arrays is becoming increasingly important.

From a propagation point of view, there are still many theoretical as well as practical open issues. Our understanding of the frequency dependence of different propagation processes such as diffuse scattering, and how to include it in deterministic channel prediction tools, is still incomplete. The set up and evaluation of directionally resolved measurement campaigns is also an area of active research. But most importantly, almost all existing propagation channel models are based on a single measurement campaign, and many of the parameters that are used in those models have no statistical reliability. Thus, extensive measurement campaigns will be a key part of future propagation research.

This completes the introductory chapter on UWB, where we discussed its history, various forms of realising a UWB transceiver, as well as regulatory and standardisation aspects. We move now to the four technical parts of this book, dealing with various issues related to UWB antennas and UWB signal propagation.

References

[1] G. Weightman, *Signor Marconi's Magic Box: The Most Remarkable Invention of the 19th Century and the Amateur Inventor Whose Genius Sparked a Revolution*, da Capo Press, 2003.

[2] R. Qiu, Propagation effects, in M.G. Di Benedetto *et al.*, (ed.), *UWB Communications Systems: A Comprehensive Overview*, EURASIP publishing, 2005.

[3] H.F. Harmuth, *Nonsinusoidal Waves for Radar and Radio Communication*, Academic Press, 1981.

[4] R.A. Scholtz, Multiple access with time-hopping impulse modulation, *Proc. IEEE MILCOM*, 447–50, 1993.

[5] M.Z. Win and R.A. Scholtz, Impulse radio: how it works, *IEEE Comm. Lett.*, **2**, 36–8, 1998.

[6] M.Z. Win and R.A. Scholtz, Ultra-wide bandwidth time-hopping spread-spectrum impulse radio for wireless multiple-access communications, *IEEE Trans. Comm.*, **48**, 679–91, 2000.

[7] Federal Communications Commission, First report and order 02–48, 2002.

[8] M.Z. Win and R.A. Scholtz, Characterization of ultra-wide bandwidth wireless indoor channels: a communication-theoretic view, *IEEE J. Selected Areas Comm.*, **20**, 1613–27, 2002.

[9] F. Ramirez-Mireles, Performance of ultra-wideband SSMA using time hopping and M-ary PPM, *IEEE J. Selected Areas Comm.*, **19**(6), 1186–96, 2001.

[10] L. Zhao and A.M. Haimovich, Performance of ultra-wideband communications in the presence of interference, *IEEE J. Selected Areas Comm.*, **20**, 1684–92, 2002.

[11] X. Chu and R.D. Murch, The effect of nbi on UWB time-hopping systems, *IEEE Trans. Wireless Comm.*, **3**, 1431–6, 2004.

[12] I. Bergel, E. Fishler and H. Messer, Narrowband interference mitigation in impulse radio, *IEEE Trans. Comm.*, **53**, 1278–82, 2005.

[13] L. Piazzo and F. Ameli, Performance analysis for impulse radio and direct-sequence impulse radio in narrowband interference, *IEEE Trans. Comm.*, **53**, 1571–80, 2005.

[14] S. Gezici, H. Kobayashi, H.V. Poor and A.F. Molisch, Performance evaluation of impulse radio UWB systems with pulse-based polarity randomization, *IEEE Trans. Signal Processing*, **53**, 2537–49, 2005.

[15] L. Yang and G.B. Giannakis, A general model and SINR analysis of low duty-cycle UWB access through multipath with narrowband interference and Rake reception, *IEEE Trans. Wireless Comm.*, **4**, 1818–33, 2005.

[16] D. Cassioli, M.Z. Win, A.F. Molisch and F. Vatelaro, Performance of selective Rake reception in a realistic UWB channel, *Proc. ICC 2002*, 763–7, 2002.

[17] V. Lottici, A. D'Andrea and U. Mengali, Channel estimation for ultra-wideband communications, *IEEE J. Selected Areas Comm.*, **20**, 1638–45, 2002.

[18] Y.G. Li, A.F. Molisch and J. Zhang, Practical approaches to channel estimation and interference suppression for OFDM-based UWB communications, *Proc. IEEE 6th Circuits and Systems Symposium on Emerging Technologies*, 21–4, 2004.

[19] J.D. Choi and W.E. Stark, Performance of ultra-wideband communications with suboptimal receivers in multipath channels, *IEEE J. Selected Areas Comm.*, **20**(9), 1754–66, 2002.

[20] T. Q.S. Quek and M.Z. Win, Analysis of UWB transmitted-reference communication systems in dense multipath channels, *IEEE J. Selected Areas Comm.*, **23**, 1863–74, 2005.

[21] R. Hoctor and H. Tomlinson, Delay-hopped transmitted reference RF communications, *IEEE Conf. on Ultra-Wideband Systems and Technologies 2002*, 265–70, 2002.

[22] K. Witrisal, G. Leus, M. Pausini and C. Krall, Equivalent system model and equalization of differential impulse radio UWB systems, *IEEE J. Selected Areas Comm.*, **23**, 1851–62, 2005.

[23] F. Tufvesson, S. Gezici and A.F. Molisch, Ultra-wideband communications using hybrid matched filter correlation receivers, *IEEE Trans. Wireless Comm*, submitted 2006.

[24] Y.P. Nakache and A.F. Molisch, Spectral shape of UWB signals: influence of modulation format, multiple access scheme and pulse shape, *Proc. VTC 2003 spring*, 2510–14, 2003.
[25] Y.P. Nakache and A.F. Molisch, Spectral shaping of UWB signals for time-hopping impulse radio, *IEEE J. Selected Areas Comm.*, in press, 2006.
[26] R. Fuji-Hara, Y. Miao and M. Mishima, Optimal frequency hopping sequences: a combinatorial approach, *IEEE Trans. Information Theory*, **50**, 2408–20, 2004.
[27] W. Chu and C.J. Colbourn, Sequence designs for ultra-wideband impulse radio with optimal correlation properties, *IEEE Trans. Information Theory*, **50**, 2402–7, 2004.
[28] C.J. Le Martret and G.B. Giannakis, All-digital impulse radio with multiuser detection for wireless cellular systems, *IEEE Trans. Comm.*, **50**, 1440–50, 2002.
[29] A. Batra, J. Balakrishnan, G.R. Aiello, J.R. Foerster and A. Dabak, Design of a multiband OFDM system for realistic UWB channel environments, *IEEE Trans. Microwave Theory Techn.*, **52**, 2123–38, 2004.
[30] R.J. Fontana, Recent system applications of short-pulse ultra-wideband (UWB) technology, *IEEE Trans. Microwave Theory Techn.*, **52**, 2087–104, 2004.
[31] W.P. Siriwongpairat, M. Olfat and K.J.R. Liu, Performance analysis of time hopping and direct sequence UWB space-time systems, *IEEE Globecom 2004*, 3526–30, 2004.
[32] N. Kumar and R.M. Buehrer, Application of layered space-time processing to ultra-wideband communication, *Proc. 45th Midwest Symp. Circuits Systems*, 597–600, 2002.
[33] M.Chamchoy, S. Promwong, P. Tangtisanon and J. Takada, Spatial correlation properties of multiantenna UWB systems for in-home scenarios, *Proc. IEEE Int. Symp. Comm. Information Techn. 2004*, 1029–32, 2004.
[34] H. Schantz, A brief history of UWB antennas, *Proc. Ultra Wideband Systems and Technologies*, 209–13, 2003.
[35] H.L. Bertoni, L. Carin and L.B. Felsen, *Ultra-Wideband Short-Pulse Electromagnetics*, Plenum, 1993.
[36] D. Cassioli, M.Z. Win and A.F. Molisch, A statistical model for the UWB indoor channel, *Proc. 53rd IEEE Vehicular Technology Conference*, **2**, 1159–63, 2001.
[37] A.F. Molisch, J.R. Foerster and M. Pendergrass, Channel models for ultra-wideband personal area networks, *IEEE Personal Communications Magazine*, **10**, 14–21, 2003.
[38] A.F. Molisch, K. Balakrishnan, C.C. Chong, D. Cassioli, S. Emami, A. Fort, J. Karedal, J. Kunisch, H. Schantz and K. Siwiak, A comprehensive model for ultra-wideband propagation channels, *IEEE Trans. Antennas Prop.*, submitted, 2006.
[39] A.F. Molisch, Ultra-wideband propagation channels – theory, measurement, and models, *IEEE Trans. Vehicular Techn., special issue on UWB*, invited paper, 2005.
[40] A.F. Molisch, *Wireless Communications*, John Wiley & Sons, Ltd, 2005.
[41] M.A. Win and R.A. Scholtz, On the energy capture of ultra-wide bandwidth signals in dense multipath environments, *IEEE Comm. Lett.*, **2**, 245–7, 1998.
[42] S. Gezici, Z. Tian, G.B. Giannakis, H. Kobayashi, A.F. Molisch, H.V. Poor and Z. Sahinoglu, Localization via ultra-wideband radios: a look at positioning aspects for future sensor networks, *IEEE Signal Processing Magazine*, **22**, 70–84.
[43] A.F. Molisch *et al.*, IEEE 802.15.4a channel model – final report, Tech. Rep. Document IEEE 802.15-04-0662-02-004a, 2005.
[44] M.K. Simon, J.K. Omura, R.A. Scholtz and B.K. Levitt, *Spread Spectrum Communications Handbook*, McGraw-Hill, 1994.
[45] L.J. Cimini, Analysis and simulation of a digital mobile channel using orthogonal frequency division multiplexing, *IEEE Trans. Comm.*, **33**, 665–75.
[46] L. Hanzo, M. Muenster, B.J. Choi and T. Keller, *OFDM and MC-CDMA for Broadband Multi-User Communications, WLANs and Broadcasting*, John Wiley & Sons, Ltd, 2003.
[47] James D. Taylor, ed., *Introduction to Ultra-Wideband Radar Systems*, CRC Press, 1995.
[48] B. Allen, Ultra-wideband wireless sensor networks, *IEE UWB Symposium*, 2004.
[49] F. Tufvesson and A.F. Molisch, Ultra-wideband communication using hybrid-matched filter correlation receivers, *Proc. 59th IEEE Vehicular Techn. Conf.*, spring, 1290–4, 2004.

Part I

Fundamentals

Introduction to Part I

Wasim Q. Malik and David J. Edwards

The robust design of an ultra-wideband wireless communications system is a multidisciplinary process, with elements of signal processing, transceiver design, antenna element design and radio propagation characterisation. The glue that binds these elements together is provided by the system engineer, who specifies the requirements for each, so that the complete UWB system will operate according to the required performance criteria. To successfully complete such a task, it is of paramount importance to comprehend the role of the individual components and the underlying physical processes.

The chapters in this section hence deal with the various factors that enable the design of the UWB component technologies. They are an important prerequisite for the remainder of this book, where the fundamentals are expended to include an understanding of the theoretical and practical issues arising in the context of UWB system design.

The applications of UWB technology are many and varied. In order to obtain a complete understanding of the behaviour of these systems in the context of their environment it is necessary to address each component of the system and understand the physical interaction between the technology of each part of the system, its ability to launch and propagate an electromagnetic wave and the mitigation techniques that can be employed to improve or enhance the basic performance.

The propagation environment. The first area of concern is the propagation environment. In contrast to narrowband systems, ultra-wideband electromagnetic behaviour is not constant across the band for any given environment. Essentially, the interaction between objects and electromagnetic waves is determined by the relative scale of the object and the wavelength of the radiation. Consequently it is that at the lower frequencies (long wavelengths) a given object may be electrically small and behave as such causing diffractive processes to occur in a near-field behavioural model. However, at the higher frequencies (shorter wavelengths) those same objects become electrically large and asymptotically approach a geometric optics model. There is therefore a need to start the modelling of the UWB environment at a full electromagnetic level and include all inductive and near-field effects and move towards the short wavelength approximation.

As an illustration, we can consider a metallic object such as a plate, 1 m wide. At 3 GHz this object is equivalent to 10 wavelengths in width and thus electrically small in that dimension. As a rule of thumb,

---

Ultra-wideband Antennas and Propagation for Communications, Radar and Imaging   Edited by B. Allen,
M. Dohler, E. E. Okon, W. Q. Malik, A. K. Brown and D. J. Edwards
© 2007 John Wiley & Sons, Ltd

objects with dimensions of up to 10 wavelengths do not behave as conventional reflectors, and a highly diffractive behaviour is observed. That same object at 10 GHz is, however, over 33 wavelengths wide and acts as a very good reflector. Similar variations are observed in the context of near- and far-field characteristics. Consider a UWB antenna with an aperture of $D = 10$ cm. The Fraunhofer boundary, taken as $2D^2/\lambda$ extends to 2 m at 3 GHz, while at 10 GHz the extent is nearly 7 m.

Another consideration as far as propagation is concerned is that in many wireless applications the environment causes multipath behaviour. In narrowband systems this can cause severe signal fading due to the destructive interference caused by the various path lengths that the multipath signals follow between transmitter and receiver. In UWB systems for any transmitter/receiver disposition very few of the frequency components suffer this fading and most of the power in the signal can be recovered. In pulsed UWB systems this is not a serious problem; however, in multiband OFDM implementations, this may be a problem for certain sub-bands. It has been observed that the channel models in UWB systems cannot be considered to be constant over the whole of the UWB band, and due to frequency-dependent propagation phenomena [1] there is markedly different behaviour between the lowest and the highest sub-bands. The understanding of this variation is therefore fundamental to generating an accurate model of the UWB link and Maxwell's equations cannot be ignored.

Launchers and receptors – the antenna. Following on from the electrical size problem, the antenna suffers similar problems to objects in the environment. Conventional narrowband launchers tend to be highly tuned, optimised for a particular frequency with a radiation performance which is usually considered to be constant or at best slowly varying across its band of operation. For UWB, careful consideration of candidate antennas needs to be undertaken and due regard taken of radiation performance across the band, including matching and radiation pattern variation. This latter point extends not only to the angular diffractive response but also to the efficiency and factors such as the differential phase delay of the different frequency components as they are launched and received [2, 3]. Although much of this is well understood, conventional design techniques are orientated to narrowband philosophies and design approaches do not in general allow for this variable behaviour across the band. In UWB systems, therefore, there are peculiar requirements and candidate antenna elements need to be chosen with great care, otherwise the transmission and reception of UWB waveforms can be severely affected, and impairment of the information content will ensue.

Antenna arrays and beamforming. The problems associated with single-element antennas can to a large extent be mitigated by the use of antenna arrays [4]. This approach, in effect, gives a further degree of freedom in the optimisation of performance and furthermore opens the door to advanced signal-processing techniques and beamforming algorithms. This, in turn, leads to the possibility of the antenna now becoming part of the radio link, and its subsequent optimisation. The growth of digital signal processing now means that by treating the antenna array as a spatial sampling system, the signals from the elements can be combined, processed and optimised to improve the overall behaviour of the system. This approach does have its problems and if we refer to the wavelength range model mentioned above a temporal response of one nanosecond corresponds to 10 wavelengths at 10 GHz, while at 3 GHz this corresponds to 3 wavelengths. Aperture beamforming is therefore extremely difficult and due regard to phase and amplitude excitation at each frequency is essential. The signal-processing overhead to avoid severe signal distortion is therefore imperative for satisfactory array performance.

Link optimisation and diversity. Returning to the problem of multipath signals, we have established that for UWB signals frequency-selective fading is not a major concern. However, it has also been recognised that in a scattering environment, multipath signals arrive at the receiver at varying times. As in narrowband systems, the recovery of as much signal power as possible greatly enhances the link performance in terms of signal-to-noise ratios at the receiver and can also lead to power savings at the transmitter. Therefore many of the techniques applied to narrowband systems can be considered for the

UWB situation, but with due regard to the peculiar requirements of the extremely wide range of temporal, spatial and frequency variation.

Thus we need to review the signal-processing regimes and approaches available to us and consider how best to extend them to the UWB case. In diversity schemes we consider how the existence of multiple copies of the information signal can be utilised both to improve the link budget (recovery of as much power as possible) and also to enhance the bit error rate performance (recovery of information in a situation where a single point reception is insufficient) or quality of service (QoS).

Two cases are considered in this section. In the first, receive diversity is explored and the viability of this approach in UWB systems is analysed. The approach is essentially analogous to the narrowband case but again signal combining is nontrivial due to the extremely wide range of frequencies. Since the multipaths suffer time delays, the core idea in diversity-combining approaches is the wideband co-phasing of signals before a signal combining algorithm is applied.

Transmit diversity also has a role to play and there is a range of schemes available to the system designer. Here multiple copies of the signal are transmitted to spread the power, the direction, the point of launch (spatial diversity) and the time of launch (e.g. cyclic delay diversity). We therefore arrive at the concept of shared communications links and multiple antenna systems. Spatial multiplexing techniques have been extensively researched in the literature [5]. Unlike diversity schemes, which endeavour to utilise multiple copies of the same information stream, spatial multiplexing seeks to utilise independent communications channels which can exist between pairs of antenna elements in a transmit and receive array. The viability of this approach depends on the degree to which the individual elements of the antenna transfer matrix are independent. This independence can be achieved spatially or by means of polarisation, and the exploitation of the scattering environment in which the multiple-antenna system is deployed. Coding schemes also play a part in MIMO operation and space-time coding has shown much promise [6].

In the following chapters we therefore present a review of each of the above areas in the context of wireless communications systems, and seek to address the key parameters that need to be considered in UWB system design. An extensive bibliography is provided for the reader to explore the topics further for specialised applications.

References

[1] A.F. Molisch, Ultrawideband propagation channels - theory, measurement, and modeling, *IEEE Trans. Veh. Technol.*, **54**, September, 2005.
[2] H.G. Schantz, *The Art and Science of Ultra-Wideband Antennas*, Artech House, 2005.
[3] W.Q. Malik, D.J. Edwards and C.J. Stevens, Angular-spectral antenna effects in ultra-wideband communications links, *IEE Proc.-Commun.*, **153**, February, 2006.
[4] B. Allen and M. Ghavami, *Adaptive Array Systems*, John Wiley & Sons, Ltd, 2005.
[5] A.J. Paulraj, D.A. Gore, R.U. Nabar and H. Bolcskei, An overview of MIMO communications – a key to gigabit wireless, *Proc. IEEE*, **92**, February, 2004.
[6] V. Tarokh, H. Jafarkhani and A.R. Calderbank, Space-time block codes from orthogonal designs, *IEEE Trans. Inform. Theory*, **45**, July, 1999.

2

Fundamental Electromagnetic Theory

Mischa Dohler

2.1 Introduction

This first chapter is dedicated to the very fundamentals of antennas and propagation. We endeavour to explain why antennas are capable of transmitting and receiving an electromagnetic signal. As we will demonstrate below, radio signal transmission and reception are facilitated by a small asymmetry in Maxwell's equations which itself triggers an electromagnetic wave to decouple from (henceforth referred to as *transmission*) and couple into (henceforth referred to as *reception*) a medium carrying free electric charges (henceforth referred to as *antenna*).

We will commence with the differential formulation of Maxwell's equations and related constitutive relations, which are then interpreted from a physical point of view. This allows us to explain radiation processes in a simple and intuitive manner. We then give some general mathematical approaches facilitating the solution of Maxwell's equations, which includes the introduction of potentials, wave equations and retarded potentials, as well as a harmonic formulation. We then summarise some major principles resulting from the solution of Maxwell's equations, and briefly put them in context with the UWB system design.

Despite being a chapter dedicated to fundamentals, we assume that the reader is familiar with the concepts of Cartesian, cylindrical and spherical coordinate systems, gradient, divergence, curl, as well as phasors.

2.2 Maxwell's Equations

2.2.1 Differential Formulation*

Starting from some assumptions and observations, James Clerk Maxwell derived a set of mutually coupled equations which paved the way to the field of electrodynamics, part of which allows a proper understanding and design of narrowband and wideband antenna elements and arrays. In differential form,

---

Ultra-wideband Antennas and Propagation for Communications, Radar and Imaging Edited by B. Allen, M. Dohler, E. E. Okon, W. Q. Malik, A. K. Brown and D. J. Edwards
© 2007 John Wiley & Sons, Ltd

the four equations are [1, 2]:

$$\text{div } \mathbf{D}(\mathbf{r}, t) = \rho(\mathbf{r}, t) \tag{2.1}$$

$$\text{div } \mathbf{B}(\mathbf{r}, t) = 0 \tag{2.2}$$

$$\text{curl } \mathbf{E}(\mathbf{r}, t) = -\frac{\partial \mathbf{B}(\mathbf{r}, t)}{\partial t} \tag{2.3}$$

$$\text{curl } \mathbf{H}(\mathbf{r}, t) = -\frac{\partial \mathbf{D}(\mathbf{r}, t)}{\partial t} + \mathbf{J}(\mathbf{r}, t) \tag{2.4}$$

where $\mathbf{E}(\mathbf{r}, t)$ in V/m is the vector representing the *electric field intensity*, $\mathbf{D}(\mathbf{r}, t)$ in C/m² is the *electric flux density*, $\mathbf{H}(\mathbf{r}, t)$ in A/m is the *magnetic field intensity*, $\mathbf{B}(\mathbf{r}, t)$ in T is the *magnetic flux density*, $\rho(\mathbf{r}, t)$ in C/m³ is the *charge density* and $\mathbf{J}(\mathbf{r}, t)$ in A/m² is the *current density*. All of the above electromagnetic field variables depend on the spatial position with respect to some coordinate system, $\mathbf{r}$ in m, and the elapsed time, t in s.

The electric and magnetic field vectors can be related through the *constitutive relations*, i.e.

$$\mathbf{D}(\mathbf{r}, t) = \epsilon_0 \epsilon_r \mathbf{E}(\mathbf{r}, t) \tag{2.5}$$

$$\mathbf{B}(\mathbf{r}, t) = \mu_0 \mu_r \mathbf{H}(\mathbf{r}, t) \tag{2.6}$$

$$\mathbf{J}(\mathbf{r}, t) = \sigma \mathbf{E}(\mathbf{r}, t) \tag{2.7}$$

where $\epsilon_0 \approx 8.85 \times 10^{-12}$ F/m is the *free space permittivity*, ϵ_r is the material-dependent *relative permittivity* (also called the *dielectric constant*), and $\epsilon = \epsilon_0 \epsilon_r$ simply *permittivity*; $\mu_0 \approx 1.257 \cdot 10^{-6}$ H/m is the *free space permeability*, μ_r is the material-dependent *relative permeability*, and $\mu = \mu_0 \mu_r$ simply *permeability*; and σ is the material-dependent *conductivity* expressed in S/m. In the above constitutive relations, because generally not applicable to UWB system design, we have omitted the terms expressing electric and magnetic polarisation, as well as any possible external current sources. Note that any nonlinear phenomena can also be described with this set of equations through any nonlinearity that possibly exists in the material-dependent variables.

Finally, the *div* operation characterises how much a vector field linearly diverges and the *curl* operation characterises the strength of the curl (rotation) in the field. Both relate to spatial operations, i.e. they do not involve any operations with respect to time.

2.2.2 Interpretation*

To understand the meaning of the mathematical formulation of the above equations, let us scrutinise them one by one.

div $\mathbf{D}(\mathbf{r}, t) = \rho(\mathbf{r}, t)$ means that static or dynamic charges in a given volume are responsible for a diverging electric field. That implies that there must be a distinct source and sink for the electric field since a field cannot possibly (linearly) diverge and start and end in the same location.

div $\mathbf{B}(\mathbf{r}, t) = 0$ means that there is no physical medium which makes a magnetic field diverge. This equation comes from the observation that there are no magnetic charges known to physics. Note that magnetic charges are sometimes introduced in theoretical electrodynamics so as to simplify and beautify the derivation of certain theories.

curl $\mathbf{E}(\mathbf{r}, t) = -\partial \mathbf{B}(\mathbf{r}, t)/\partial t$ means that a spatially varying (curling) electric field will cause a time-varying magnetic field. Alternatively, it can be rewritten as $-\partial \mathbf{B}(\mathbf{r}, t)/\partial t = $ curl $\mathbf{E}(\mathbf{r}, t)$, i.e. a time-varying electric field will cause a curl in the magnetic field.

Finally, and most importantly as we will see shortly, curl $\mathbf{H}(\mathbf{r}, t) = \partial \mathbf{D}(\mathbf{r}, t)/\partial t + \mathbf{J}(\mathbf{r}, t)$ can be read as follows. A spatially varying (curling) magnetic field will cause a time-varying electric field and, if existent, also a current through a medium capable of carrying a flow of electric charges. The equation can also be read as either a current flow through a medium or as a time-varying electric field producing a spatially curling magnetic field.

2.2.3 Key to Antennas and Propagation*

The first two equations yield separately an insight into the properties of the electric and magnetic field, respectively. The remaining two equations, however, show that both fields are closely coupled through spatial (curl) and temporal ($\partial/\partial t$) operations. We can also observe that the equations are entirely symmetric – apart from the current density $\mathbf{J}(\mathbf{r}, t)$. It turns out that this asymmetry is responsible for any radiation process occurring in nature, including the transmission and reception of electromagnetic waves. For the ease of explanation, we will rewrite the last two Maxwell equations as

$$\text{curl } \mathbf{E}(\mathbf{r}, t) = -\mu_0 \mu_r \frac{\partial \mathbf{H}(\mathbf{r}, t)}{\partial t} \qquad (2.8)$$

$$\text{curl } \mathbf{H}(\mathbf{r}, t) = -\epsilon_0 \epsilon_r \frac{\partial \mathbf{E}(\mathbf{r}, t)}{\partial t} + \mathbf{J}(\mathbf{r}, t) \qquad (2.9)$$

Let us assume first that there is a static current density $\mathbf{J}(\mathbf{r})$ available which, according to Equation (2.9), causes a spatially curling magnetic field $\mathbf{H}(\mathbf{r})$; however, it fails to generate a temporally varying magnetic field which means that $\partial \mathbf{H}(\mathbf{r})/\partial t = 0$. According to Equation (2.8), this in turn fails to generate a spatially and temporally varying electric field $\mathbf{E}(\mathbf{r})$. Therefore, a magnetic field is only generated in the location where we have a current density $\mathbf{J}(\mathbf{r})$ present. Since we are interested in making a wave propagating in a wire*less* environment where no charges (and hence current densities) can be supported, a static current density $\mathbf{J}(\mathbf{r})$ is of little use.

Our observations, however, change when we start generating a time-varying current density $\mathbf{J}(\mathbf{r}, t)$ which, according to Equation (2.9), generates a spatially and temporally varying magnetic field $\mathbf{H}(\mathbf{r}, t)$. Clearly, $\partial \mathbf{H}(\mathbf{r}, t)/\partial t \neq 0$ which, according to (2.9), generates a spatially and temporally varying electric field $\mathbf{E}(\mathbf{r}, t)$, i.e. $\partial \mathbf{E}(\mathbf{r}, t)/\partial t \neq 0$. With reference to (2.9), this generates a spatially and temporally varying magnetic field $\mathbf{H}(\mathbf{r}, t)$, even in the *absence* of a current density $\mathbf{J}(\mathbf{r}, t)$. And so on.

A wave is hence generated where the electric field stimulates the magnetic field and vice versa. From now on, we will refer to this wave as an electromagnetic (EM) wave, since it contains both magnetic and electric fields. From the above it is clear that such a wave can now propagate in space without the need of a charge-bearing medium; however, such a medium can certainly enhance or weaken the strength of the electromagnetic wave by means of an actively or passively created current density $\mathbf{J}(\mathbf{r}, t)$.

In summary, to make an electromagnetic wave decouple from a transmitting antenna, we need a medium capable of carrying a time-varying current density $\mathbf{J}(\mathbf{r}, t)$. A medium which achieves this with a high efficiency is called an *antenna*. As simple as that! An antenna can hence be anything: rod, wire, metallic volumes and surfaces, etc.

Remember that $\mathbf{J}(\mathbf{r}, t) = \partial Q/\partial t$, where Q in C is the electric charge; therefore, if we want a time-varying current density for which $\partial \mathbf{J}(\mathbf{r}, t)/\partial t \neq 0$, we need to ensure that $\partial^2 Q/\partial t^2 \neq 0$, i.e. that we need to accelerate or decelerate charges. Such acceleration is achieved by means of e.g. a harmonic carrier frequency as deployed in traditional wireless communication systems or a carrierless Gaussian pulse as deployed in UWB systems. In both cases, the second-order differential exists and is different from zero, thereby facilitating a decoupling of the signal from the antenna.

Sometimes, however, such acceleration happens unintentionally, e.g. in bent wires where electrons in the outer radius move faster than the ones in the inner radius of the bend. Therefore, any nonstraight piece of wire will emit electromagnetic waves which, in the case of antennas, is desirable, but, in the case of wiring between electric components in a computer, not at all. The later category of radiation opens up the interesting subject of electromagnetic compatibility [3], which is, albeit not unrelated, clearly beyond the scope of this book on UWB.

Once the wave is airborne, i.e. decoupled from the antenna, it propagates through space, according to the laws dictated by Equations (2.8) and (2.9), until it impinges upon and interacts with an object or reaches the intended receiver antenna.

We thus observe that the creation of UWB pulses, their transmission from an antenna, their propagation through a channel, their reception by an antenna, and their processing are all governed by the same law – Maxwell's equations. A ubiquitous design approach, however, is too complicated because each UWB system part will yield different intricate solutions to Maxwell's equations.[1] It is hence a common engineering practice to split the design approach into the baseband signalling part, the RF stages, the feeding lines, the transmit and receive antennas, and the channel modelling, etc. While this book aims at exposing analysis related exactly to these parts, it shall not be forgotten that they are the result of simplifications applied, at one stage or another, to Maxwell's equations.

2.2.4 Solving Maxwell's Equations

After having formulated the equations, interpreted them and understood their physical meaning, it is time to attack the problem of solving Maxwell's equations. This has occupied some of the leading mathematicians and physicians for centuries – it is hence a theory well understood and documented. We will only give some sketchy outline of the procedure to solve Maxwell's equations, so the unequipped reader gets the gist of the proceedings and the equipped reader a gentle reminder.

The problem is clearly to obtain the electric and magnetic field components in explicit form from the coupled equations (2.1)–(2.4). This, if brute-force mathematics was to be applied, would lead to intricate differentio-integral equations, which are cumbersome to deal with.

It turns out that proceedings are greatly facilitated by introducing two auxiliary concepts [1, 2], the *magnetic vector potential* $\mathbf{A}(\mathbf{r}, t)$ defined such that

$$\text{curl } \mathbf{A}(\mathbf{r}, t) = \mathbf{B}(\mathbf{r}, t) \tag{2.10}$$

and the *electric scalar potential* $\Phi(\mathbf{r}, t)$ defined such that

$$-\text{grad } \Phi(\mathbf{r}, t) = \mathbf{E}(\mathbf{r}, t) + \frac{\partial \mathbf{A}(\mathbf{r}, t)}{\partial t} \tag{2.11}$$

When plugged into Maxwell's equation (up to the reader to verify), the two potentials manage to decouple Equations (2.3) and (2.4), given that the following normalisation is maintained between both auxiliary potentials:

$$\text{div } \mathbf{A}(\mathbf{r}, t) + \mu\epsilon \frac{\partial \Phi(\mathbf{r}, t)}{\partial t} \equiv 0. \tag{2.12}$$

This normalisation condition is also often referred to as the *Lorentz gauge*. When applied, it leads to a set of decoupled differential equations,

$$\nabla^2 \mathbf{A}(\mathbf{r}, t) - \frac{1}{c^2} \frac{\partial^2 \mathbf{A}(\mathbf{r}, t)}{\partial t^2} = -\mu \cdot \mathbf{J}(\mathbf{r}, t), \tag{2.13}$$

$$\nabla^2 \Phi(\mathbf{r}, t) - \frac{1}{c^2} \frac{\partial^2 \Phi(\mathbf{r}, t)}{\partial t^2} = -\epsilon^{-1} \cdot \rho(\mathbf{r}, t), \tag{2.14}$$

where $c = 1/\sqrt{\epsilon\mu}$ is the speed of light in the material under consideration, and ∇^2 is the Laplace operator, the exact mathematical description of which depends on the coordinate system of choice.

---

[1] While the formulation of Maxwell's equations is a pleasure for the eye, the explicit solution under all material boundary conditions w.r.t. electric and magnetic field components can easily occupy several pages!

Fundamental Electromagnetic Theory

The two equations (2.13) and (2.14) are often referred to as the *wave equations*, and solved in a fairly standard manner in dependency of prevailing sources and boundary conditions for $\mathbf{J}(\mathbf{r}, t)$ and $\rho(\mathbf{r}, t)$, to arrive at [1, 2]

$$\mathbf{A}(\mathbf{r}, t) = \frac{\mu}{4\pi} \oint_{V'} \frac{\mathbf{J}(\mathbf{r}', t - |\mathbf{r} - \mathbf{r}'|/c)}{|\mathbf{r} - \mathbf{r}'|} dV', \tag{2.15}$$

$$\Phi(\mathbf{r}, t) = \frac{1}{4\pi\epsilon} \oint_{V'} \frac{\rho(\mathbf{r}', t - |\mathbf{r} - \mathbf{r}'|/c)}{|\mathbf{r} - \mathbf{r}'|} dV', \tag{2.16}$$

which are often referred to as *retarded potentials*. The reason for this nomenclature is that the effects of current density and charge time t and position $\mathbf{r}'$ are felt at position $\mathbf{r}$ after a time delay of $|\mathbf{r} - \mathbf{r}'|/c$, which is exactly the time the electromagnetic wave needs to propagate.

Once the two retarded potentials $\mathbf{A}(\mathbf{r}, t)$ and $\Phi(\mathbf{r}, t)$ are obtained, either in closed or numerical form, the electric and magnetic field components are obtained from Equations (2.10) and (2.11) by means of simple differential operations. This in turn will determine radiated power, efficiency, etc., and is thus of paramount interest to the UWB system designer.

2.2.5 Harmonic Representation

The above-outlined mathematical proceedings constitute a general approach in solving any electromagnetic radiation problems; however, since we can resolve an arbitrary signal of any bandwidth into its spectral harmonics by means of the Fourier transform, we will show here that the analysis indeed simplifies.

To this end, it is straightforward to show that a source current density of the form $\mathbf{J}(\mathbf{r}, t) = \mathbf{J}(r) \cdot \exp(-j\omega t)$ and a source charge density of the form $\rho(\mathbf{r}, t) = \rho(\mathbf{r}) \cdot \exp(-j\omega t)$, leads to electric and magnetic field components of the form $\mathbf{E}(\mathbf{r}, t) = \mathbf{E}(\mathbf{r}) \cdot \exp(-j\omega t)$ and $\mathbf{H}(\mathbf{r}, t) = \mathbf{H}(\mathbf{r}) \cdot \exp(-j\omega t)$. Here, the spatial dependencies are decoupled from the temporal phasor $\exp(-j\omega t)$ of angular frequency ω, expressed in rad/s. For UWB problems, one would consider all ω between the lower and upper angular frequency of the UWB signal.

Inserting the harmonic representations into Equations (2.13) and (2.14), leads to a simplified and time-independent set of wave equations [1, 2],

$$\nabla^2 \mathbf{A}(\mathbf{r}) - k^2 \mathbf{A}(\mathbf{r}) = -\mu \cdot \mathbf{J}(\mathbf{r}), \tag{2.17}$$
$$\nabla^2 \Phi(\mathbf{r}) - k^2 \Phi(\mathbf{r}) = -\epsilon^{-1} \cdot \rho(\mathbf{r}), \tag{2.18}$$

which are also known as the *Helmholtz equations*; here, $k = \omega/c = 2\pi/\lambda$ is the *wave number*. The solution to the Helmholtz equations yields a simplified set of retarded potentials, i.e.

$$\mathbf{A}(\mathbf{r}) = \frac{\mu}{4\pi} \oint_{V'} \frac{\mathbf{J}(\mathbf{r}')e^{-jk|\mathbf{r}-\mathbf{r}'|}}{|\mathbf{r} - \mathbf{r}'|} dV', \tag{2.19}$$

$$\Phi(\mathbf{r}) = \frac{1}{4\pi\epsilon} \oint_{V'} \frac{\rho(\mathbf{r}')e^{-jk|\mathbf{r}-\mathbf{r}'|}}{|\mathbf{r} - \mathbf{r}'|} dV'. \tag{2.20}$$

The general proceedings to solving any radiation problem can hence be summarised as:

1. Using the Fourier transform, resolve the current density $\mathbf{J}(\mathbf{r}, t)$ into its spectral harmonics at each spatial location $\mathbf{r}$.
2. For each harmonic with angular frequency ω, determine magnetic vector potential $\mathbf{A}(\mathbf{r})$.

3. From Equation (2.10), determine the magnetic field as $\mathbf{H}(\mathbf{r}) = \text{curl}\,\mathbf{A}(\mathbf{r})/\mu$.
4. From Equation (2.4), determine the electric field as $\mathbf{E}(\mathbf{r}) = \text{curl}\,\mathbf{H}(\mathbf{r})/(j\omega\epsilon)$.
5. With the individual spectral contributions, the total response to an arbitrary temporal excitation is finally obtained by linearly superimposing the individual spectral contributions, i.e. applying the inverse Fourier transform.

This is a mandatory approach to radiation problems occurring in UWB systems, because the extremely wide bandwidth of the signals prohibits a narrowband approach, where traditionally only one spectral component is considered. Note that for the analysis of UWB systems, also other mathematical tools are available, most notably time-domain approaches; this, however, is beyond the scope of this introductory chapter.

2.3 Resulting Principles

For a given current density distribution and its temporal behaviour, any radiation problem can hence be solved numerically if not in closed form, so at least numerically. In the quest to simplify the analysis and synthesis further, however, engineers have identified domains which lead to various approximate solutions. Most notably, these are the areas of:

- **Transmission lines:** these are in fact realised by waves guided along wires; the concept of impedance, circuitry and various fundamental laws, such as Ohm's law or Kirchhoff's laws, can be directly derived from Maxwell's equations when radiation is neglected.
- **Antennas:** the emphasis here is the efficiency related to the coupling and decoupling of electromagnetic waves; various antenna parameters, such as beamwidth, can be derived from Maxwell's equations, as demonstrated in the subsequent chapter.
- **Propagation:** here, Maxwell's equations describe how an electromagnetic wave propagates through free space and, after suitable approximations relating to far-field radiation, the Friis free space transmission formula can be derived; it will also be exposed in subsequent chapters.
- **Interaction with matter:** if a wave with wavelength λ impinges upon matter, Maxwell's equations yield different approximate solutions in dependency of the dimension d of the object; if $\lambda \ll d$, then this leads to Fresnell's reflection and transmission formulas; if $\lambda \approx d$, then scattering gains in importance; in addition, if there is any discontinuity, then the element of diffraction is added.

All these effects are strongly dependent on frequency, which means that a signal's frequency component at around 2 GHz will experience a different attenuation and phase rotation than a frequency component at 10 GHz. It is hence evident that a UWB pulse will disperse in time when interacting with electromagnetic objects, which is fundamentally different from the time dispersion caused by delayed multipath components!

References

[1] John D. Kraus and Ronald J. Marhefka, *Antennas*, McGraw-Hill Science/Engineering/Math, 3rd edition, 2001.
[2] Constantine A. Balanis, *Antenna Theory: Analysis and Design*, John Wiley & Sons, Inc., 2nd edition, 1996.
[3] Clayton R. Paul, *Introduction to Electromagnetic Compatibility*, Wiley Series in Microwave and Optical Engineering, Wiley-Interscience, 1992.

---

* Source: *Adaptive Array Systems* B. Allen and M. Ghavami, 2005, © John Wiley & Sons, Ltd. Reproduced with permission.

3

Basic Antenna Elements

Mischa Dohler

3.1 Introduction

For subsequent chapters dealing with UWB antenna and propagation issues in greater detail, some preliminaries are required, which stretch from antenna parameters and terminology to some fundamental radiation behaviour of basic antenna elements. Antennas, be they UWB or not, can be approached from

- an **engineering** point of view, where we are primarily interested in the radiation pattern and efficiency;
- an **analytical** point of view, where we broadly distinguish between wire (dipole, loop, etc.) and aperture antennas (horn, slot, etc.), both types of which are related by the *equivalence theorem*;
- an **deployment** point of view, where we broadly distinguish between basic antenna elements (dipole, horn, etc.) and composite antennas (arrays, reflectors, etc.).

The dipole can be mathematically viewed as the dual structure to a loop antenna and the complementary structure to the slot antenna. In this chapter, we will hence dwell on the application of Maxwell's equations to the most fundamental radiation element, the Hertzian dipole. That allows us to define some common antenna terminology and parameters. At the end of the chapter, we will also introduce some basic radiation structures, some of which are of use to the UWB system design.

3.2 Hertzian Dipole*

A wire of infinitesimal length δl is known as a Hertzian dipole. It plays a fundamental role in the understanding of finite-length antenna elements, because any of these consists of an infinite number of Hertzian dipoles. Antenna arrays, on the other hand, can be represented by means of a plurality of finite-length antenna elements. This justifies the importance of properly understanding the radiation behaviour of a Hertzian dipole.

Ultra-wideband Antennas and Propagation for Communications, Radar and Imaging   Edited by B. Allen,
M. Dohler, E. E. Okon, W. Q. Malik, A. K. Brown and D. J. Edwards
© 2007 John Wiley & Sons, Ltd

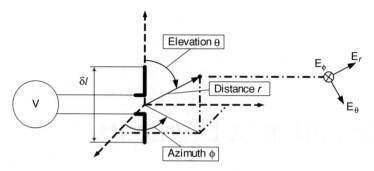

Figure 3.1 Feeding arrangement and coordinate system for Hertzian dipole. Reproduced by permission of © 2005 John Wiley & Sons, Ltd*

The Hertzian dipole can be fed by a current $I(t)$ of any temporal characteristics; however, since we can resolve an arbitrary signal into its spectral harmonics, as already explained in Section 2.2.5, we will concentrate on the analysis of harmonic feeding. We will hence consider currents of the form $I(t) = I_{max} \cdot \exp(j\omega t)$ where I_{max} is the maximum current and ω is the angular frequency. Once the response to a particular spectral component in the UWB signal has been determined, the total response to an arbitrary temporal excitation is obtained by linearly superimposing the individual spectral contributions.

With these assumptions, it is straightforward to show that a feeding current of $I(t) = I_{max} \cdot \exp(j\omega t)$ will cause an EM wave of the form $\mathbf{E}(\mathbf{r}, t) = \mathbf{E}(\mathbf{r}) \cdot \exp(j\omega t)$ and $\mathbf{H}(\mathbf{r}, t) = \mathbf{H}(\mathbf{r}) \cdot \exp(j\omega t)$. We will only concentrate on the spatial dependencies of the field components and henceforth omit the harmonic temporal factor $\exp(j\omega t)$.

A Hertzian dipole with feeding generator is depicted in Figure 3.1. For obvious reasons, the coordinate system of choice is spherical where a point in space is characterised by the azimuth $\phi \in (0, 2\pi)$, elevation $\theta \in (0, \pi)$ and distance from the origin r. The field vectors can therefore be represented as $\mathbf{E}(\mathbf{r}) = (E_\phi, E_\theta, E_r)$ and $\mathbf{H}(\mathbf{r}) = (H_\phi, H_\theta, H_r)$.

To obtain these spectral electromagnetic field contributions radiated by a Hertzian dipole, one needs to solve Maxwell's equations (2.1)–(2.4). If we follow the procedure outlined in Section 2.2.5, we obtain for the non-zero EM field components

$$E_\theta = -\frac{\eta k^2}{4\pi} I\delta l \sin(\theta) e^{-jkr} \left[\frac{1}{jkr} + \left(\frac{1}{jkr}\right)^2 + \left(\frac{1}{jkr}\right)^3 \right] \quad (3.1)$$

$$E_r = -\frac{\eta k^2}{2\pi} I\delta l \cos(\theta) e^{-jkr} \left[\left(\frac{1}{jkr}\right)^2 + \left(\frac{1}{jkr}\right)^3 \right] \quad (3.2)$$

$$H_\phi = -\frac{k^2}{4\pi} I\delta l \sin(\theta) e^{-jkr} \left[\frac{1}{jkr} + \left(\frac{1}{jkr}\right)^2 \right] \quad (3.3)$$

and all remaining components are zero, i.e. $E_\phi = 0$ and $H_\theta = H_r = 0$. Here, $k = \omega/c = 2\pi/\lambda$ is the wave number, c is the speed of light, λ in m is the wavelength, and $\eta = \sqrt{\mu/\epsilon} = k/\omega\epsilon$ is the intrinsic impedance in Ω ($\eta = 120\pi \approx 377\,\Omega$ for free space). These equations can be simplified for distances very near to and far from the Hertzian dipole.

Basic Antenna Elements

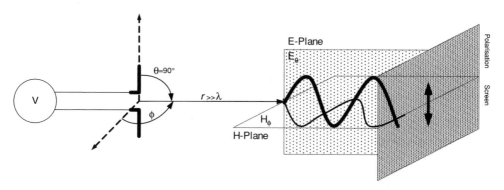

Figure 3.2 Relationship between **E** and **H** in the far-field. Reproduced by permission of © 2005 John Wiley & Sons, Ltd*

3.2.1 Far-Field – Fraunhofer Region

The far-field, also referred to as the *Fraunhofer region*, is characterised by $kr \gg 1$, which allows simplification of Equations (3.1)–(3.3) to yield the following electric and magnetic field components

$$E_\theta = \eta H_\phi \tag{3.4}$$

$$H_\phi = \frac{k^2}{4\pi} I \delta l \frac{\sin(\theta)e^{-jkr}}{kr} \tag{3.5}$$

and $E_\phi = E_r = 0$ and $H_\theta = H_r = 0$. The wave is visualised in Figure 3.2 for $\theta = \pi/2$, i.e. perpendicular to the Hertzian dipole. In the far-field, **E** and **H** are clearly in in-phase, i.e. they have maxima and minima at the same locations and times. Also, the EM wave turns into a *plane wave* which consists of two mutually orthogonal electromagnetic field components.

It is worth investigating whether a UWB receiver is always operated in the far-field for all of its spectral components. Assuming a UWB signal spanning 2–10 MHz, i.e. a wavelength ranging from 15 cm to 3 cm, one can fairly safely assume that even for the lowest spectral components the receiver is most often operated in the far-field. An exception to this rule are UWB-based body area networks, which will be dealt with in Chapter 16.

For the Hertzian dipole, it can be further observed that the field strength is not only inversely dependent on distance, but also on frequency. Therefore, higher frequency components in a UWB signal experience a greater loss in signal strength than the lower ones; the UWB pulse hence undergoes a temporal dispersion. As will be explained in more detail in Chapter 14, the effect of frequency dependency strongly depends on the antenna of choice and is not due to the actual propagation channel.

3.2.2 Near-Field – Fresnel Region

The near-field, also referred to as the *Fresnel region*, is characterised by $kr < 1$, which does not allow clear simplifications to Equations (3.1)–(3.3). At fairly small distances r, one can neglect the first term in (3.1) and (3.3). Furthermore, at very small distances, where $kr \ll 1$, one will obtain the following approximate field components

$$E_\theta \approx -j\frac{\eta k^2}{4\pi} I\delta l \frac{\sin(\theta)e^{-jkr}}{(kr)^3} \qquad (3.6)$$

$$E_r \approx -j\frac{\eta k^2}{2\pi} I\delta l \frac{\cos(\theta)e^{-jkr}}{(kr)^3} \qquad (3.7)$$

and $E_\phi \approx 0$; furthermore, the H-field is negligible compared to the E-field.

Again, the applicability of the near-field region to UWB systems is worth investigating. For instance, if one assumed a UWB signal stretching from 100 MHz to 1 GHz, i.e. a wavelength of 3 m to 30 cm, the far-field conditions are not necessarily obeyed and near-field effects start to play a dominant role.[1] Also, a temporal dispersion of a UWB pulse can be observed in the near-field region for this type of antenna, because the field strength is again inversely dependent on the frequency.

3.3 Antenna Parameters and Terminology*

The preceding brief introduction to the radiation behaviour of an infinitesimal radiating element allows us now to introduce some concepts that are vital for the characterisation of UWB antenna elements and the UWB radio channel.

3.3.1 Polarisation

From Equation (3.4) and Figure 3.2, we observe that in the far-field from the dipole the E-wave oscillates in a plane. We have also seen that the H-wave oscillates perpendicular to the E-wave. Since both waves always occur together, we will, unless otherwise mentioned, from now on refer to the E-wave.

If we cut a plane orthogonal to the direction of propagation, indicated as the polarisation screen in Figure 3.2, we would observe the electric field vector E_θ oscillating on a straight line. This polarisation state is referred to as *linear polarisation* and is further illustrated in Figure 3.3.

If we have two orthogonal dipoles, instead of one, and we feed them with inphase currents, then this will trigger two decoupled EM waves to be orthogonal in the far-field. We would then observe a tilted but straight line on the orthogonally cut screen; this polarisation state is often referred to as *linear tilted polarisation*. If, on the other hand, we feed both dipoles with currents in quadrature phase, i.e. shifted by 90°, then the resulting E-field will be *circularly polarised*. Finally, if the amplitudes of the two decoupled fields is different due to different feeding current amplitudes, then the resulting polarisation will be *elliptical*. In dependency of whether the two feeding currents are $+90°$ or $-90°$ shifted, the polarisation will be left or right circular or elliptical polarised. Note finally, if the two feeding currents deviate by a phase different from 90° and/or more Hertzian dipoles are used, then more complicated polarisation patterns can be obtained.

Because of the fairly short communication distances in UWB systems and predominantly LOS scenarios, the received pulse is likely to arrive in the same polarisation state as the transmitted pulse. NLOS situations, however, may lead to chance in the signal polarization. Polarisation hence plays a vital role in the UWB system design, as both transmitter and receiver need to be electrically aligned.

---

[1] Such a situation may arise if a UWB link was to realise wireless connections in a body area network, where communication distances rarely exceed human dimensions.

Basic Antenna Elements

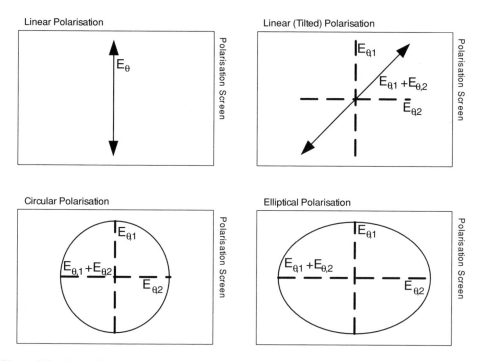

Figure 3.3 Linear, linear tilted, circular and elliptical polarisation states. Reproduced by permission of © 2005 John Wiley & Sons, Ltd*

3.3.2 Power Density

The instantaneous power density, w, in W/m^2 is defined as

$$w = \text{Re}\{\mathbf{E} \times \mathbf{H}^*\} \tag{3.8}$$

where $\text{Re}(x)$ denotes the real part of x, and $\mathbf{H}^*$ is the complex conjugate of $\mathbf{H}$. The average power density can be obtained from Equation (3.8) by assuming that the EM wave is harmonic, which yields

$$\overline{w} = \text{Re}\{\mathbf{S}\} = \text{Re}\left\{\frac{1}{2}\mathbf{E} \times \mathbf{H}^*\right\} \tag{3.9}$$

where $\mathbf{S}$ is referred to as the pointing vector. It is hence clear that the power density is only defined here for harmonic processes. To obtain the power density of a UWB pulse, an averaging over all frequency components is required; this also applies to subsequently introduced physical parameters.

As an example, let us calculate the average power density of an EM wave in the far-field for a Hertzian dipole, where we utilise Equations (3.4) and (3.5) to obtain

$$\overline{w} = \text{Re}\left\{\frac{1}{2}\begin{vmatrix} \mathbf{e}_r & \mathbf{e}_\theta & \mathbf{e}_\phi \\ 0 & E_\theta & 0 \\ 0 & 0 & H_\phi^* \end{vmatrix}\right\} \tag{3.10}$$

$$= \frac{\eta I^2 \sin^2(\theta)}{8r^2}\left(\frac{\delta l}{\lambda}\right)^2 \mathbf{e}_r \tag{3.11}$$

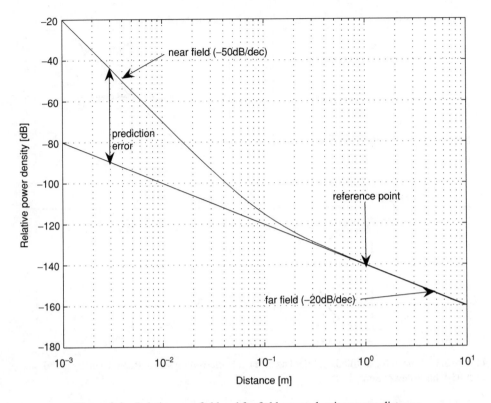

Figure 3.4 Relative near field and far field power density versus distance

where $\mathbf{e}_r$, $\mathbf{e}_\theta$ and $\mathbf{e}_\phi$ are the unit vectors of the perpendicular spherical coordinates. Expressed in decibels,

$$\overline{w}_{dB} = 10 \log_{10} \overline{w} = K - 20 \log_{10} r, \qquad (3.12)$$

a loss of 20 dB per decade distance can be observed; K is some constant.

As a further example, the reader is invited to calculate the average power density of an EM wave in the near-field for a Hertzian dipole. In the near-field, as depicted in Figure 3.4, a loss of up to 50 dB per decade distance can be observed. That has the implication that if a far-field pathloss model is given with respect to a reference point, typically at 1 m, the near-field pathloss is very pessimistic and more power is available at the receiver than predicted. This is an important observation to bear in mind for near-field UWB communication systems.

3.3.3 Radiated Power

Given the power density, we can calculate the total average power passing through a sphere with surface area σ as

$$P = \int_\sigma \mathrm{Re}\,\{\mathbf{S}\}\,d\sigma \qquad (3.13)$$

$$= \int_0^{2\pi} \int_0^\pi \mathrm{Re}\,\{\mathbf{S}\} r^2 \sin(\theta)\,d\theta\,d\phi \qquad (3.14)$$

Basic Antenna Elements

which, for the Hertzian dipole, yields

$$P = \frac{\pi \eta I^2}{3} \left(\frac{\delta l}{\lambda}\right)^2. \quad (3.15)$$

Clearly, the average radiated power is independent of the distance but does depend on the geometrical (δl) and electrical (λ) properties of the dipole; the higher frequency components in a UWB signal are hence radiated at a higher power.

3.3.4 Radiation Resistance

The radiation resistance, R_r, is defined as *the value of a hypothetical resistor which dissipates a power equal to the power radiated by the antenna when fed by the same current I*, i.e.

$$\frac{1}{2} I^2 R_r = P \quad (3.16)$$

With reference to Equation (3.15), this simply yields for the Hertzian dipole:

$$R_r = \frac{2\pi \eta}{3} \left(\frac{\delta l}{\lambda}\right)^2 = 80\pi^2 \left(\frac{\delta l}{\lambda}\right)^2 = 789 \left(\frac{\delta l}{\lambda}\right)^2 \quad (3.17)$$

the units of which are Ω.

3.3.5 Antenna Impedance

The antenna impedance, Z_a, is defined as *the ratio of the voltage at the feeding point V(0) of the antenna to the resulting current flowing in the antenna I for a given frequency*, i.e.

$$Z_a = \frac{V(0)}{I_{\text{antenna}}} \quad (3.18)$$

where

- if $I_{\text{antenna}} = I_{\max}$, then the impedance Z_a is referred to the *loop current*;
- if $I_{\text{antenna}} = I(0)$, then the impedance Z_a is referred to the *base current*;

Referring the impedance to the base current, we can write it as

$$Z_a = \frac{V(0)}{I(0)} = R_a + jX_a \quad (3.19)$$

where X_a is the antenna reactance and $R_a = R_r + R_l$ is the antenna resistance, where R_r is the radiation resistance and R_l is the ohmic loss occurring in the antenna. The antenna impedance, which varies from frequency to frequency, is an important design criterion in UWB systems.

3.3.6 Equivalent Circuit

We have therefore reduced the behaviour of a radiating antenna to that of an equivalent impedance at a given frequency. This is not a coincidence as Maxwell's equations treat both radiated EM waves and

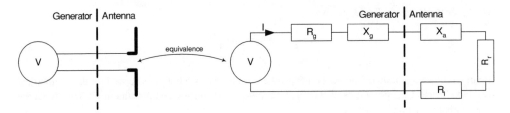

Figure 3.5 Equivalence between physical radiating antenna and linear circuit. Reproduced by permission of © 2005 John Wiley & Sons, Ltd*

guided EM waves equally. The duality allows us to introduce an equivalent circuit as shown in Figure 3.5. It depicts a generator, here a voltage source, with internal impedance consisting of resistance R_g and reactance X_g being connected to the antenna with impedance consisting of resistance R_a and reactance X_a.

3.3.7 Antenna Matching

The equivalent circuit allows us to tune the impedance of the generator so as to maximise the radiated power delivered from the generator at a given frequency. This procedure is also often referred to as *impedance matching*. The average power delivered to the antenna is clearly given as

$$P = \frac{1}{2} I^2 R_a \qquad (3.20)$$

where

$$I = \frac{V_g}{(R_a + R_g) + j \cdot (X_a + X_g)} \qquad (3.21)$$

The power in Equation (3.20) is maximised if conjugate matching is deployed, i.e. $R_g = R_a$ and $X_g = -X_a$, which yields a delivered power of

$$P = \frac{1}{8} \frac{V_g^2}{R_a} \qquad (3.22)$$

Since the matching procedure is different for every spectral component in a UWB signal, such an approach constitutes considerable problems for the system designer. A simple way out is to match the spectral region with the strongest spectral power density and hope that the mismatch in the remaining regions is not dramatic.

3.3.8 Effective Length and Area

The effective length, l_e, characterises the antenna's ability to transform the impinging electric field E into a voltage at the feeding point $V(0)$, and vice versa, and is defined as

$$l_e = \frac{V(0)}{E} \qquad (3.23)$$

which implies that the more the voltage that is induced with less electric field strength, the larger the effective length of the antenna.

Basic Antenna Elements

The effective area, A_e, *characterises the antenna's ability to absorb the incident power density w and to deliver it to the load*, and is defined as

$$A_e = \frac{P_{\text{load}}}{w} \tag{3.24}$$

which implies that the higher the delivered power with respect to the incident power density, the larger the effective area.

3.3.9 Friis' Transmission Formula

With the definition of the effective area, we will state Friis' free space propagation formula, i.e.

$$P_{rx} = \left(\frac{A_{e,tx} \cdot A_{e,rx}}{\lambda^2 \cdot r^2} \right) \cdot P_{tx} \tag{3.25}$$

where P_{tx} and P_{rx} are the transmitted and received power respectively, $A_{e,tx}$ and $A_{e,rx}$ are the (frequency-dependent) effective areas of the transmitting and receiving antennas respectively, λ is the wavelength, and r the distance between transmitter and receiver. Due to the direct dependence on the wavelength and also the indirect dependence through the effective areas, different spectral components in a UWB signal will hence be received with different powers. The UWB pulse will hence arrive dispersed,[2] which needs to be considered when designing optimum UWB receiver architectures. Note that the above formula only holds for perfect transmit and receive antenna matching at given frequency.

3.3.10 Radiation Intensity

The radiation intensity U is defined as *the power P per solid angle Ω*, and it is defined as

$$U = \frac{dP}{d\Omega} \tag{3.26}$$

This can be rewritten as

$$U = \frac{\text{Re}\{S\} d\sigma}{d\Omega} = \frac{\text{Re}\{S\} r^2 d\Omega}{d\Omega} = \text{Re}\{S\} r^2 \tag{3.27}$$

which, for the Hertzian dipole, yields

$$U = \frac{\eta I^2 \sin^2(\theta)}{8} \left(\frac{\delta l}{\lambda} \right)^2 \tag{3.28}$$

and is clearly independent of the distance of observation.

3.3.11 Radiation Pattern

Since the radiation intensity U is independent of distance in the far-field, but only depends upon the antenna's inherent parameters, it can be taken to describe the radiation pattern of an antenna, an example

---

[2] For the case where all frequency dependencies compensate each other so that no pulse dispersion occurs, consult Chapter 14.

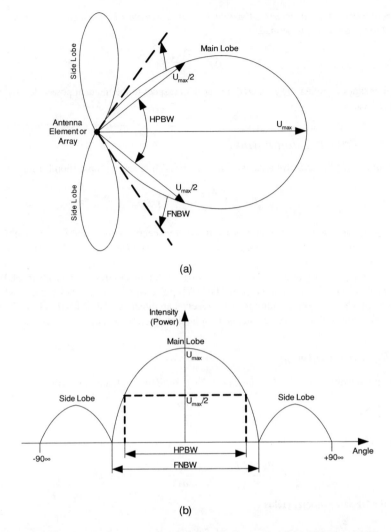

Figure 3.6 Two different representations of a radiation pattern. Reproduced by permission of © 2005 John Wiley & Sons, Ltd*

of which is depicted in Figure 3.6(a). If we roll out this radiation pattern and depict it on a Cartesian coordinate system, we will obtain a graph as depicted in Figure 3.6(b). With reference to the example figure, we define the following:

- *Radiation null*: angle at which the radiated power is zero.
- *Mainlobe*: the angular region between two radiation nulls which contains the angle with the strongest radiation intensity $U_{\max}$.
- *Sidelobes*: the angular regions between two radiation nulls which do not contain the angle with the strongest radiation power.

Basic Antenna Elements

- *Half-power beamwidth (HPBW)*: the angle spanned by the intensity region for which $U_{max}/2 \leq U \leq U_{max}$. The HPBW is associated with the ability of an antenna to direct a beam. The HPBW is often referred to as the 3 dB beamwidth for obvious reasons.
- *First null beamwidth (FNBW)*: the angle spanned by the mainlobe. The FNBW is associated with the ability of an antenna to reject an interferer.

The radiation pattern strongly depends on the frequency, which means that different spectral components in a UWB signal will be radiated into different directions with varying strength [4]. The UWB pulse will hence not only contain different power in different locations, but also exhibit different temporal dispersions. This is a serious challenge to the UWB transceiver designer, because independent of the location, spectral masks have to be obeyed and receivers be made efficient.

3.3.12 (Antenna) Bandwidth

The bandwidth B is defined as the *frequency band ranging from f_{lower} to f_{upper} within which the performance of the antenna, with respect to some characteristics, conforms to a specified standard, e.g. a drop by 3 dB*. Such a definition is fairly broad and includes characteristics such as radiation pattern, beamwidth, antenna gain, input impedance, radiation efficiency. An antenna is said to be *narrowband* if the fractional bandwidth, defined as $FB = (f_{upper} - f_{lower})/f_{center}$ is below 1%; otherwise, it is said to be *wideband*.

For example, we design an antenna such that it radiates a total power of 0 dBm at a centre frequency of 1.8 GHz. We say that the specified standard is a power drop of 3 dB. Therefore, if the radiated power does not drop to -3 dBm or lower between $f_{lower} = 1.782$ GHz and $f_{lower} = 1.8185$ GHz but falls below -3 dBm out of that range, then the antenna is narrowband. If, however, the radiated power does not fall below -3 dBm well beyond 18 GHz, then the antenna is wideband.

Ideally, a UWB antenna ought to have a sufficiently wide bandwidth so as to cover the entire UWB spectral bandwidth. As demonstrated in subsequent chapters, this is not an easy task to achieve.

3.3.13 Directive Gain, Directivity, Power Gain

An isotropic radiator is defined as *a radiator which radiates the same amount of power in all directions*. It is a purely hypothetical radiator used to aid the analysis of realisable antenna elements. For an isotropic radiator, we define the radiation intensity, U_0, to be $U_0 = P/(4\pi)$.

This allows us to define the directive gain, g, as *the ratio of the radiation intensity U of the antenna to that of an isotropic radiator U_0 radiating the same amount of power*, which can be formulated as

$$g = \frac{U}{U_0} = 4\pi \frac{U}{P} \qquad (3.29)$$

For the Hertzian dipole we have

$$g(\theta) = 1.5\sin^2(\theta) \qquad (3.30)$$

which is clearly a function of direction, but not distance from the radiator or frequency.

From the above we define the directivity D as *the ratio of the maximum radiation intensity U_{max} of the antenna to that of an isotropic radiator U_0 radiating the same amount of power*, giving

$$D = \frac{U_{max}}{U_0} = 4\pi \frac{U_{max}}{P} \qquad (3.31)$$

For the Hertzian dipole we now have

$$D = 1.5 \qquad (3.32)$$

which is simply a value and not a function of direction or frequency anymore.

This allows us finally to define the power gain G as *the ratio of the radiation intensity U of the antenna to that of an isotropic radiator U_0 radiating an amount of power equal to the power accepted by the antenna*, i.e.

$$G = \frac{U}{U_{0,input}} = 4\pi \frac{U}{P_{input}} \qquad (3.33)$$

3.3.14 Radiation Efficiency

The radiation efficiency, e, is defined as *the ratio of the radiated power P to the total power P_{input} fed to the antenna*, where $P_{input} = P + P_{loss}$, i.e.

$$e = \frac{P}{P_{input}} = \frac{P}{P + P_{loss}} \qquad (3.34)$$

which can also be related to previous antenna parameters as $e = G/g$.

3.4 Basic Antenna Elements

Any wire antenna can be viewed as the superimposition of an infinite amount of Hertzian dipoles which, theoretically speaking, allows one to calculate the resulting EM field at any point in space by adding the field contributions of each Hertzian dipole. From Equations (3.4) and (3.5) we have seen that to calculate the field contributions we need to know the current (and its direction) through the Hertzian dipole.

For many antenna configurations, we can calculate or estimate the current distribution and hence apply the aforementioned theory. In this case, the antennas are referred to as *wire antennas*. In many other configurations, however, it is difficult to obtain an exact picture of the current distribution. In this case, it is easier to utilise Huygen's principle and deduce the radiated EM field at any point in space from an estimate of the EM field at a well-defined surface; such antennas are also referred to as *aperture antennas*.

Basic wire antennas include dipoles of finite length, loop and helix antennas. Basic aperture antennas include the horn and slot antenna, as well as parabolic dishes. Examples of common antenna elements are shown in Figure 3.7. To obtain a feeling for the properties of these basic antenna elements, we will briefly deal with some of the most important designs in the following sections.

3.4.1 Finite-Length Dipole*

Dipoles of finite length L are of practical interest, an example of which is depicted in Figure 3.7(a). If we assume that we feed the dipole with a sinusoidal voltage generator, then the resulting current in the finite-length dipole will also be (approximately) harmonic with a maximum amplitude of I_{max}. Such a

Basic Antenna Elements

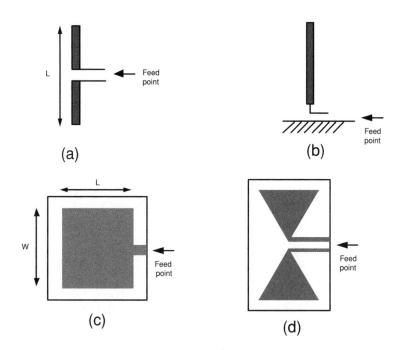

Figure 3.7 Common Types of Antenna Elements, (a) Dipole (b) Mono-pole (c) Square Patch (d) Bow Tie. Reproduced by permission of © 2005 John Wiley & Sons, Ltd*

feeding procedure is referred to as *balanced*, i.e. one feeding wire carries a current that is in anti-phase to the current in the other. That allows us to integrate over the field contributions Equations (3.4) and (3.5) to arrive at [1]

$$E_\theta = \eta H_\phi \qquad (3.35)$$

$$H_\phi = \frac{jI_{max}e^{-jkr}}{2\pi r} \cdot P(\theta) \qquad (3.36)$$

where $P(\theta)$ is the pattern factor given by

$$P(\theta) = \frac{\cos\left(\frac{1}{2}kL\cos(\theta)\right) - \cos\left(\frac{1}{2}kL\right)}{\sin(\theta)} \qquad (3.37)$$

With Equations (3.35) and (3.36), we are in the position to calculate the antenna parameters introduced in Section 3.3.

For example, the radiation resistance of a $\lambda/2$–dipole, i.e. $L = \lambda/2$, can be calculated to be $R_r = 73\,\Omega$ and the directivity is $D = 1.64$. Also, while radiating power uniformly over the azimuth plane, the finite-length dipole develops some interesting radiation patterns over the elevation plane. It can be shown that for $L < 1.1\lambda$ only one mainlobe exists (Figure 3.8(a)), whereas for $L \geq 1.1\lambda$ the radiation pattern develops multilobes (Figure 3.8(b)). From this figure we also observe that the power radiated along the mainlobe decreases once multilobes develop, which is detrimental in most wireless applications. Note that the variation in radiation patterns observed between Figures 3.8(a) and (b) also depicts the change that occurs with frequency, since the radiation characteristics depend on the ratio between the physical length L and the wavelength λ.

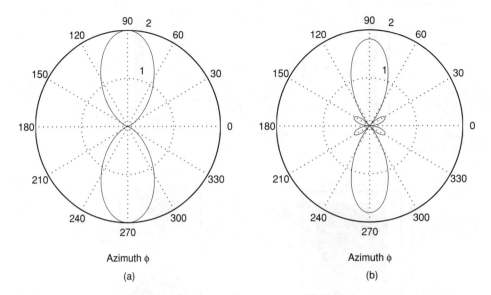

Figure 3.8 Radiation pattern of a dipole of length (a) $L = 1.0\lambda$ and (b) $L = 1.2\lambda$. Reproduced by permission of © 2005 John Wiley & Sons, Ltd*

That has interesting consequences for a UWB system: assume that we have a finite-length dipole antenna of $L = 6$ cm (or its monopole equivalent of 3 cm), then all UWB spectral components below 5 GHz do not exhibit any multilobes, whereas the higher frequency components do. The UWB pulse will hence undergo different temporal dispersions in different radiation directions.

3.4.2 Monopole*

A monopole antenna is often considered as half of a dipole placed above a groundplane, where the groundplane acts as an electric mirror thus creating the other half of the dipole, as illustrated in Figure 3.9.

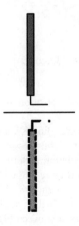

Figure 3.9 Mono-pole above a ground-plane. Reproduced by permission of © 2005 John Wiley & Sons, Ltd*

In contrast to the balanced feed for the dipole, a monopole is fed with a *single-ended* feeder where one wire carries the signal to the antenna and the other is connected to the groundplane. This has the advantage of being directly connected to most receiver and transmitter modules that are often designed around a single-ended grounding system, and is fully compatible with co-axial cables and associated connectors.

Since half of the radiating plane is cut-off by the groundplane, the radiated power (and hence the radiation resistance) of a monopole is only half compared to the dipole with the same current; however, the directivity is doubled. Therefore, a $\lambda/4$ monopole has a radiation resistance of $R_r = 36.5\,\Omega$ and a directivity of $D = 3.28$. The latter, together with the compact spatial realisation and the simple feeding mechanism, is the main incentive to use monopoles.

3.4.3 Printed Antennas*

The antennas discussed so far have required a wire structure as a convenient way of realisation. Alternatively antennas can be produced using printed circuit techniques where one side of a copper-clad dielectric is etched to the desired shape and the other is left as the groundplane. Dipoles can also be produced in this way, but this section focuses on patch antennas.

Patch antennas are practical and popular owing to their ease of manufacturing and the flexibility of design in terms of shape and topology. Many applications require antennas which are capable of conforming to the shape of the surface onto which they are mounted, and are required to have a planar profile for aesthetic or mechanical reasons. Patch antennas are a good solution for such applications. Figure 3.7(c) shows an example of a square-patch antenna. Although the surface conductor can be practically any shape, rectangular and circular tend to be the most common.

The far-field radiation pattern is orientated orthogonal to the surface conductor, so in Figure 3.7(c) it projects towards the user. As a rule of thumb, length L is approximately $\lambda_g/2$ and controls the operating frequency and width W is $0.9\lambda_g$ and controls the radiation resistance [2], where $\lambda_g = 1/\sqrt{\varepsilon_r}$ which is defined as the normalised wavelength in a media with relative permeability ε_r.

Note that contrary to common belief, the surface conductor does not form the radiating element as it does in a dipole. Instead, radiation occurs from along edges L and W, and which edge depends upon the electromagnetic mode of radiation the antenna is operating in. The radiation pattern of a square patch operating in the TE_{10} mode is $E\phi \approx \cos(\theta)$.

It is evident from Figure 3.7(c) that a single-ended feeder is required, as described in Section 3.4.2 for the monopole antenna, and therefore has the characteristics associated with a single-ended feeder. Furthermore, although patch antennas have the advantages mentioned, they have a narrow operating bandwidth, hence making them an unlikely candidate for UWB systems. Advanced structures can improve these parameters, an example of which is the stacked patch, which consists of several layers sandwiched together, where the size of each layer and the distance between the layers is carefully chosen [3].

3.4.4 Wideband and Frequency-Independent Elements*

With reference to the definition given in Section 3.3.12, an antenna is said to be wideband if given characteristics, e.g. input impedance, do not vary over a frequency band from f_{lower} to f_{upper} where $f_{upper} : f_{lower} > 10 : 1$. Simple antenna elements, such as the Hertzian dipole, finite-length dipole and monopole, are not capable of maintaining any characteristics over such wide bandwidths. More sophisticated antenna structures are hence required, an example of which are bowtie antennas and horn antennas.

From previous analysis, we gather that radiation characteristics depend on the ratio between the physical length L and the wavelength λ. From this fact, Rumsey observed that if we design an antenna, which is only described by angles and is itself infinite in length, then it is inherently self-scaling and thus frequency-independent [1].

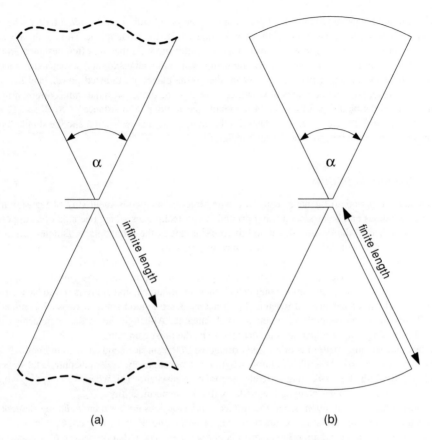

Figure 3.10 A frequency independent (wideband) antenna in a) theoretical realisation and b) practical realisation (bow-tie antenna). Reproduced by permission of © 2005 John Wiley & Sons, Ltd*

An example of a radiating structure obeying Rumsey's principle is depicted in Figure 3.10(a). Since it is impossible to build the depicted radiating structure which is infinite in size, several techniques have been suggested to make them effectively infinite. One of them is simply to truncate the infinite sheet, leading to a radiating structure as depicted in Figure 3.10(b); such an antenna is also referred to as a *bowtie antenna*. Since Rumsey's design properties are violated, the antenna will not be frequency-independent anymore; however, it exhibits parameter stability over a very wide frequency band, where exact numbers depend on the specific antenna realisation.

More sophisticated antenna elements are the *log-period toothed antennas, log-period trapezoidal wire antenna* and *log-period dipole array*. Due to their log-period self-scaling nature, they extend the effective length of the antenna structure. Also, antennas which, by nature, are self-scaling, are fractal antennas [1]. Given the tough requirements imposed by UWB signals, frequency-independent structures are ideal candidates for UWB system deployment.

We have thus completed the introductory text describing some fundamentals issues when analysing and designing radiating antenna elements. We have tried to touch upon issues particular to the UWB antenna design. This has provided us with sufficient material to embark upon more sophisticated topics, such as UWB antenna arrays, beamforming and diversity reception techniques.*

References

[1] Constantine A. Balanis, *Antenna Theory: Analysis and Design*, John Wiley & Sons, Inc., 2nd edition, 1996.
[2] R. Munson, Conformal microstrip antennas and microstrip phased arrays, *IEEE Transactions on Antennas and Propagation*, **22**(1) 74–8, 1975.
[3] J.R. James, P.S. Hall and C. Wood, *Microstrip Antenna: Theory and Design*, IEE Electromagnetic Waves, Series 12, 1986.
[4] W.Q. Malik, D.J. Edawars and C.J. Stevens, Angular-spectral antenna effects in ultra-wideband communications links, *IEEE Proc.-Commun.*, **153**, February, 2006.

---

* Source: *Adaptive Array Systems* B. Allen and M. Ghavami, 2005, © John Wiley & Sons, Ltd. Reproduced with permission.

4

Antenna Arrays

Ernest E. Okon

4.1 Introduction

This chapter deals with some of the fundamental concepts related to antenna arrays. It is often useful to make antennas more directional to improve the gain and range for long-distance communication and sensing. An increase in the electrical size of the antenna element produces a more directional antenna. This approach is commonly used in aperture antennas. The combination of similar antenna elements to form an array is the most popular way to improve the directional properties of antennas without altering the size of the elements. In addition, the array offers more degrees of freedom than single-element antennas. Antenna arrays could be combined in very many ways and they are useful for UWB applications. In an array of geometrically similar elements, there are a number of ways to shape the overall pattern of the array. These are the relative displacement between the elements, relative amplitude, relative excitation phase between elements, individual patterns of the elements and the overall geometrical configuration.

In this chapter, a review of basic array properties is given. The case of a two-element isotropic source is reviewed. The principle of pattern multiplication is then addressed. Linear arrays involving amplitude and phase variations are discussed. Subsequently a treatment of planar array types is given. Finally, design considerations for wideband arrays are highlighted.

4.2 Point Sources

The point source is a convenient and simple model to introduce the concept of arrays. We shall consider three special cases involving a pair of isotropic point sources to introduce the array concept. These are point sources with equal amplitude and phase, sources with equal amplitude but 180 degrees out of phase, and sources with unequal amplitude and phase.

Ultra-wideband Antennas and Propagation for Communications, Radar and Imaging   Edited by B. Allen,
M. Dohler, E. E. Okon, W. Q. Malik, A. K. Brown and D. J. Edwards
© 2007 John Wiley & Sons, Ltd

4.2.1 Point Sources with Equal Amplitude and Phase

Consider two point sources with equal amplitude excitation phases separated by a distance d along a line. The far-field intensity along a direction from the line can be expressed as,

$$\underline{E} = A_o e^{-j\psi/2} + A_o e^{j\psi/2} \tag{4.1}$$

where A_o is the amplitude of the sources, k is the propagation constant and ψ is defined as $\psi = kd \cos\phi$ which is a projection of the displacement along the direction of the field. Also, $k = 2\pi/\lambda$ where λ is the wavelength. We observe from Equation (4.1) that source 1 is retarded in phase by $\psi/2$ while source 2 is advanced in phase by the same amount. We can rewrite Equation (4.1) as

$$\underline{E} = 2A_o(e^{-j\psi/2} + e^{j\psi/2})/2$$

This reduces to

$$\underline{E} = 2A_o \cos(\psi/2) \tag{4.2}$$

Figure 4.1(a) shows the elements and the direction of the field while Figure 4.1(b) depicts the far-field pattern for the normalised field where $2A_o = 1$ and $d = \lambda/2$. The field pattern is known as a pattern on boresight with E plotted for a variation of φ. Equation (4.1) is defined with the phase reference at the centre of the spacing between the point sources. Thus, in Figure 4.1(b), Equation (4.1) reduces to

$$\underline{E} = \cos\left(\frac{\pi}{2}\cos\psi\right)$$

If the phase centre is at the first element, as depicted in Figure 4.2, then the expression for the far-field expression can be expressed as

$$\underline{E} = A_o + A_o e^{j\psi} \tag{4.3}$$

This can be rewritten as

$$\underline{E} = 2A_o e^{j\psi/2}(e^{-j\psi/2} + e^{j\psi/2})/2$$

which reduces to

$$\underline{E} = 2A_o e^{j\psi/2} \cos(\psi/2) \tag{4.4}$$

The field is normalised by using $2A_o = 1$, which yields,

$$\underline{E} = e^{j\psi/2} \cos(\psi/2) \tag{4.5}$$

We observe from Equation (4.4) that the amplitude of the field is exactly the same as that for Equation (4.2). However, there is now a phase term, which indicates that the field has a phase variation when the phase centre is moved from the centre of the array.

If we now consider the phase centre to be at the second element, the expression for the far field becomes

$$\underline{E} = A_o e^{-j\psi} + A_o \tag{4.6}$$

which can be rewritten as

$$\underline{E} = 2A_o e^{-j\psi/2}(e^{-j\psi/2} + e^{j\psi/2})/2$$

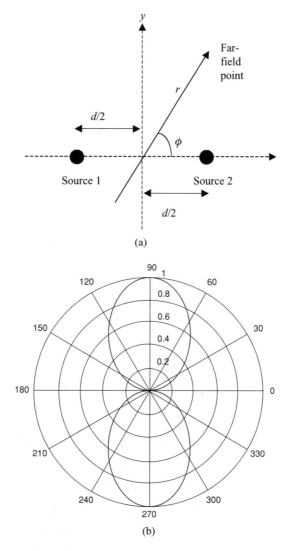

Figure 4.1 (a) Two isotropic point sources of same phase and separation $d = \lambda/2$; (b) field pattern of the sources

and reduces to

$$E = 2A_o e^{-j\psi/2} \cos(\psi/2) \tag{4.7}$$

Normalising the field by setting $2A_o = 1$, Equation (4.7) becomes

$$\underline{E} = e^{-j\psi/2} \cos(\psi/2) \tag{4.8}$$

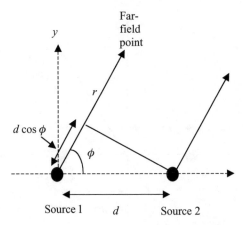

Figure 4.2 Two isotropic point sources with phase reference on source 1

In Equation (4.8) the cosine term gives the amplitude of the field while the phase variation with respect to source 2 is defined by the exponential term. Figure 4.3 shows the phase variation of the sources for the three types of phase centres. The phase variation of the fields with ϕ is plotted for the case $d = \lambda/2$. It is observed that the phase variation shows no change when the phase is referred to the centre of the two sources. However, for the phase referenced to either source 1 or 2, a phase variation of $\psi = (\pi/2)\cos\varphi$

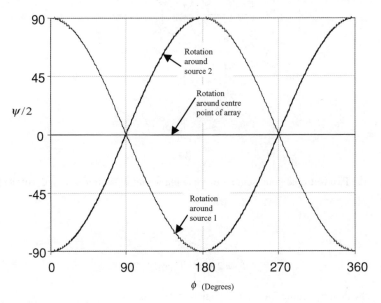

Figure 4.3 Phase of total field as a function of φ for two isotopic point sources of same amplitude and phase with $d = \lambda/2$ spacing

is observed around the array. Thus, an observer at a fixed distance away from the array observes no phase change when the array is rotated with respect to ϕ around its midpoint. A phase variation is, however, observed when the array is rotated with source 1 or 2 at the centre of the array.

In addition, from Figure 4.1(b) when point sources are placed in a linear array, with equal amplitude and phase excitations the resultant field pattern is orientated with the maximum value on boresight and normal to the plane of the array. The array is known as a 'broadside' type of array.

4.2.2 Point Sources with Equal Amplitude and 180 Degrees Phase Difference

Consider a pair of point sources similar to that of Figure 4.1(a) but with a phase difference of 180 degrees. Then, the resultant field in the ϕ direction with phase reference at the centre of the array and at a large distance r is given as

$$\underline{E} = A_o e^{-j\phi/2} - A_o e^{j\phi/2} \tag{4.9}$$

This can be rewritten as

$$\underline{E} = j2A_o \sin\frac{\psi}{2} \tag{4.10}$$

Equation (4.10) shows that the phase of the array is shifted by 90 degrees as a result of the excitation phase reversal of an element in the array. This is indicated by the operator j in Equation (4.10). Putting $j2A_o = 1$ and considering the case for $d = \lambda/2$, Equation (4.10) can be rewritten as

$$\underline{E} = \sin\left(\frac{\pi}{2}\cos\psi\right) \tag{4.11}$$

The field pattern of the array given by Equation (4.11) is depicted in Figure 4.4. The pattern is a relatively broad figure-of-eight with the maximum field along the line joining the sources as shown in Figure 4.1(a). The pattern is in contrast to the broadside array of Figure 4.1(b), which has a maximum field normal to the line, joining the sources. The pattern in Figure 4.4 is known as an 'end-fire' pattern and the two sources in this array format are known as an end-fire array.

4.2.3 Point Sources of Unequal Amplitude and Arbitrary Phase Difference

Consider once more, two isotropic point sources of unequal amplitude and arbitrary phase difference. This is a general case and we assume that the sources are situated as in Figure 4.5(a) with source 1 at the origin. We assume that source 1 has amplitude A_1 and source 2 has amplitude A_2. The total phase difference between the fields is given as

$$\psi = kd\cos\phi + \beta, \tag{4.12}$$

where β is the phase difference between sources 1 and 2. All other quantities are as previously defined. Thus, with source 1 as the reference, Equation (4.12) indicates that source 2 is advanced in phase by β. If the value of β is negative, then the source is retarded in phase by that amount. The total field at a distance r is given as

$$\underline{E} = A_1 + A_2 e^{j\psi} \tag{4.13}$$

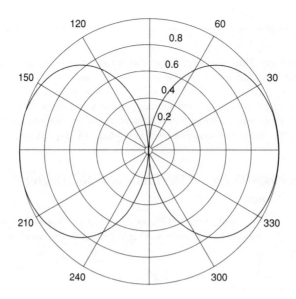

Figure 4.4 Field pattern for two isotropic point sources of same amplitude but opposite phase and spaced $\lambda/2$ apart

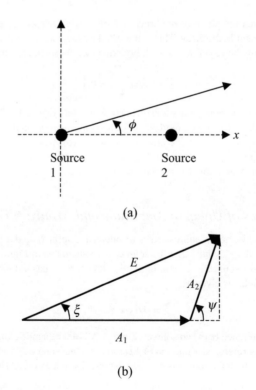

Figure 4.5 Two isotropic point sources of unequal amplitude and arbitrary phase. (a) Layout of the sources in the coordinate system; (b) vector addition of fields

The magnitude and phase of the total electric field is obtained by referring to Figure 4.5(b) and is given explicitly as

$$\underline{E} = \sqrt{(A_1 + A_2 \cos\psi)^2 + (A_2 \sin\psi)^2} \angle \tan^{-1}(A_2 \sin\psi/(A_1 + A_2 \cos\psi)), \quad (4.14)$$

where the resultant field has a phase angle ξ referred to source 1 and ψ is expressed in Equation (4.12).

4.3 The Principle of Pattern Multiplication

In the previous section, isotropic point sources were used to describe arrays. We now extend this concept to a more general situation when the sources are nonisotropic but similar. By similar, we mean that the field variation of the amplitude and phase of the sources (as well as their physical orientation) is the same. In this regard, we consider the case of two dipoles arranged in a linear array with the dipoles orientated vertically as in Figure 4.6. This is similar to the situation in Section 4.2.1 with the modification that the sources have field patterns given as

$$A_o = A |\cos\phi| \quad (4.15)$$

where A is a distance- and frequency-dependent amplitude expression. Patterns of this nature are typical of dipoles parallel to the vertical axis with ϕ varying in an anticlockwise manner as depicted in Figure 4.6. We note that for a dipole orientated in the horizontal axis, a $\sin\phi$ field pattern would be assumed. Substituting Equation (4.15) in Equation (4.2) and normalising by setting $2A = 1$ yields

$$\underline{E} = \cos\phi \cos(\psi/2) \quad (4.16)$$

where $\psi = kd\cos\phi + \beta$. The result above is equivalent to multiplying the pattern of the individual source by the pattern of two isotropic point sources. In a similar manner, for the ease of two unequal sources in Section 4.2.3, the pattern for two dipoles is easily obtained by replacing the amplitude expression by the individual pattern for the dipole. This illustrates the principle of pattern multiplication.

The principle of pattern multiplication states that the field pattern for an array of nonisotropic but similar sources is the product of the pattern of the individual source and the pattern of an array of

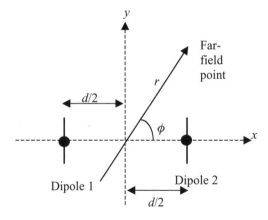

Figure 4.6 Array of two dipole sources with respect to the coordinate system

isotropic point sources (array factor) with the same locations, phase centres, relative amplitudes and phases of the nonisotropic point sources. This is expressed as

$$\underline{E}(total) = \underline{E} \ (individual \ element) \times (array factor) \tag{4.17}$$

The expression above for the field values includes the phase pattern of the total and individual elements.

As an example, we consider the case of the dipole in Figure 4.6 with the spacing between the elements given as $d = \lambda/4$ and $\beta = 0$. Then the total field pattern is obtained from Equation (4.16) as

$$\underline{E} = \cos\phi \cos\left(\frac{\pi}{4}\cos\phi\right) \tag{4.18}$$

This pattern is depicted in Figure 4.7(c) as the product of the individual pattern in Figure 4.7(a) and the array factor in Figure 4.7(b).

The principle of pattern multiplication is widely used in array design for pattern synthesis and multiplication. It has been included here as a method for obtaining desired radiation patterns from arrays, provided the individual pattern and the array factor is known.

4.4 Linear Arrays of n Elements

In the previous section, the concept of arrays and the array factor was introduced. We now generalise further by considering a linear array of n elements. The elements are assumed to be isotropic point sources as generalisations for other elements can be obtained by the use of pattern multiplication.

Consider an array of n isotropic point sources of equal amplitude and spacing as depicted in Figure 4.8. We assume that each succeeding element in the array has a progressive phase lead excitation β relative to the preceding one. An array of identical elements all of identical magnitude and each with a progressive phase shift is known as a uniform array. The total field at a large distance in the direction ϕ and referenced to source 1 is given as

$$AF = 1 + e^{j\psi} + e^{j2\psi} + e^{j3\psi} + \cdots e^{j(n-1)\psi} \tag{4.19}$$

and can be compacted as

$$AF = \sum_{m=1}^{n} e^{j(m-1)\psi}. \tag{4.20}$$

where AF indicates the array factor, ψ is the total phase difference of field from adjacent sources and is expressed as

$$\psi = kd \cos\phi + \beta \tag{4.21}$$

The amplitudes of the sources are assumed to be all equal and of unit value. Equation (4.20) is a geometric series and can be expressed in a simple trigonometric form. To obtain the trigonometric expression, we multiply Equation (4.19) by $e^{jn\psi}$ to give

$$AFe^{j\psi} = e^{j\psi} + e^{j2\psi} + e^{j3\psi} + \cdots e^{jn\psi} \tag{4.22}$$

Antenna Arrays

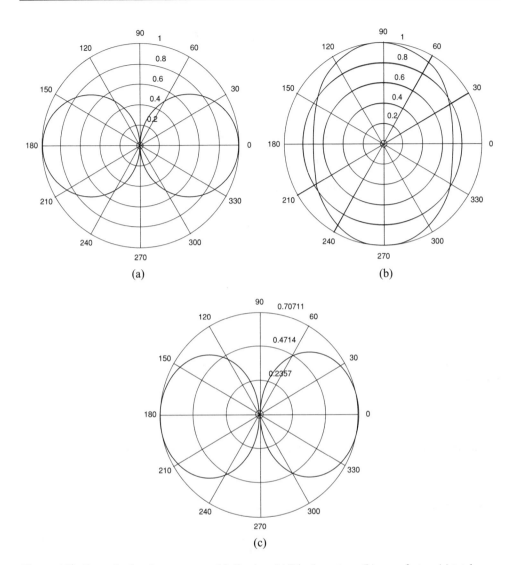

Figure 4.7 Example showing pattern multiplication. (a) Dipole pattern; (b) array factor; (c) total array pattern equivalent to a multiplication of the dipole pattern and the array factor

Subtracting Equation (4.22) from Equation (4.19) and dividing by $1 - e^{j n \psi}$ gives

$$AF = \frac{1 - e^{jn\psi}}{1 - e^{j\psi}} \qquad (4.23)$$

which can be rewritten as

$$AF = \frac{e^{jn\psi/2}}{e^{j\psi/2}} \left(\frac{e^{jn\psi/2} - e^{-jn\psi/2}}{e^{j\psi/2} - e^{-j\psi/2}} \right) \qquad (4.24)$$

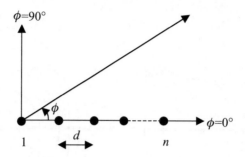

Figure 4.8 Linear array of n elements

Equation (4.24) can be further expressed as

$$AF = e^{j\xi} \frac{\sin(n\psi/2)}{\sin(\psi/2)} \tag{4.25}$$

where

$$\xi = \left(\frac{n-1}{2}\right)\psi \tag{4.26}$$

If the phase reference is chosen as the centrepoint of the array, Equation (4.25) admits a representation given as

$$AF = \frac{\sin(n\psi/2)}{\sin(\psi/2)} \tag{4.27}$$

The maximum value of (4.27) occurs in the limit as $\psi \to 0$. Thus, the maximum array factor is

$$AF_{max} = n \tag{4.28}$$

We can recast the array factor in a normalised form as

$$AF_n = \frac{1}{n}\frac{\sin(n\psi/2)}{\sin(\psi/2)} \tag{4.29}$$

There are a number of cases for the linear array related to various phase values. These are broadside arrays, ordinary end-fire arrays and highly directive end-fire arrays [1], [2].

4.5 Linear Broadside Arrays with Nonuniform Amplitude Distributions

We now consider the more general case of a linear array with nonuniform amplitude distribution. The uniform array is known to provide the maximum possible antenna gain at boresight (with the main beam pointing at the centre and perpendicular to the plane of the array). However, the sidelobe levels of the array are relatively high compared to the main beam. This value translates to power values of about -13.5 dB below the main beam. Thus, a number of techniques have been devised to reduce the sidelobe levels by varying the amplitudes of the array elements. A linear array of varying amplitude is given as

$$AF = A_0 + A_1 e^{j\psi} + A_2 e^{j2\psi} + A_3 e^{j3\psi} + \cdots A_{n-1} e^{j(n-1)\psi} \qquad (4.30)$$

Two popular techniques are the binomial distribution and the Dolph–Tschebyscheff distribution. We shall consider these two techniques.

4.5.1 The Binomial Distribution

The binomial distribution was proposed by John Stone Stone to reduce the sidelobe levels for a linear in-phase array [3]. He proposed that the array elements have amplitudes proportional to a binomial series of the form

$$(a+b)^{n-1} = {}^{n-1}C_0 a^{n-1} + {}^{n-1}C_1 a^{n-2} b + {}^{n-1}C_2 a^{n-3} b^2 + \cdots \qquad (4.31)$$

where C denotes the combination, and n is the number of sources. Thus, for an array of three sources, the amplitudes are given as (1,2,1). For an array of four sources, the amplitudes are (1,3,3,1). For an array of five sources, the amplitudes are (1,4,6,4,1) and so on. The coefficients can easily be obtained from the Pascal's triangle.

4.5.2 The Dolph–Tschebyscheff Distribution

The Dolph–Tschebyscheff distribution is a technique for the reduction of sidelobe levels introduced by Dolph [4]. The technique uses Tschebyscheff polynomials to specify a sidelobe level below the main beam. The optimum amplitude elements are then computed to ensure that all sidelobe levels conform to this specification. The binomial distribution is a special case of the Dolph–Tschebyscheff distribution. We now briefly describe the technique.

Consider a linear array of an even number, n, of isotropic point sources in a linear array of spacing d as shown in Figure 4.9. The individual sources have the amplitudes $A_0, A_1, A_2, \ldots$. The amplitude distribution is symmetrical about the centre of the array. Thus, the total field AF_e is given as

$$AF_e = 2A_0 \cos\frac{\psi}{2} + 2A_1 \cos\frac{3\psi}{2} + \cdots + 2A_k \cos\left(\frac{n_e - 1}{2}\psi\right) \qquad (4.32)$$

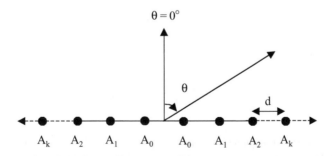

Figure 4.9 Uniform array of even-numbered elements

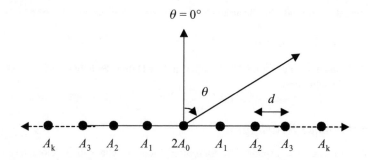

Figure 4.10 Uniform array of odd-numbered elements

where $\psi = kd \sin\theta$. If we let

$$2(k+1) = n_e, k = 0, 1, 2, 3 \ldots$$

we see that

$$\frac{n_e - 1}{2} = \frac{2k+1}{2}$$

Therefore, Equation (4.32) can be written as

$$AF_e = 2 \sum_{k=0}^{k=N-1} A_k \cos\left(\frac{2k+1}{2}\psi\right) \tag{4.33}$$

where $N = n_e/2$.

Next, we consider a linear array of an odd number of isotropic point sources of uniform spacing arranged as in Figure 4.10. The amplitude distribution is symmetrical about the centre source. We assume the amplitude of the centre source is $2A_0$, the next as A_1, the next as A_2 and continuing in that fashion. The total field at a large distance in the θ direction AF_o is then given as

$$AF_o = 2A_0 + 2A_1 \cos\psi + 2A_2 \cos 2\psi + \cdots + 2A_k \cos\left(\frac{n_o - 1}{2}\psi\right). \tag{4.34}$$

Now, if we let

$$2k + 1 = n_o, k = 0, 1, 2, 3 \ldots$$

then we can rewrite Equation (4.34) as

$$AF_o = 2 \sum_{k=0}^{k=N} A_k \cos\left(2k\frac{\psi}{2}\right) \tag{4.35}$$

where $N = (n_o - 1)/2$

The expressions given in Equation (4.34) or (4.35) could be seen to be a finite Fourier series with N terms. Thus, we observe from Equation (4.34) the constant term and the higher harmonics.

We now observe Equations (4.33) and (4.35) and relate this to the Tschebyscheff polynomials. We desire to show that Equation (4.33) and (4.35) can be represented as polynomials of degree $n_e - 1$ and

Antenna Arrays

$n_o - 1$ respectively. This implies polynomials equal to the number of sources less 1. We consider only the broadside array where $\beta = 0$. Hence,

$$\psi = kd \sin\theta \tag{4.36}$$

We now apply de Moivre's theorem

$$e^{jm\psi/2} = \cos m\frac{\psi}{2} + j\sin m\frac{\psi}{2} = \left(\cos\frac{\psi}{2} + \sin\frac{\psi}{2}\right)^m \tag{4.37}$$

On taking the real parts of Equation (4.37) we obtain

$$\cos m\frac{\psi}{2} = \cos^m\frac{\psi}{2} - \frac{m-1}{2!}\cos^{m-2}\frac{\psi}{2}\sin^2\frac{\psi}{2} + \cdots \tag{4.38}$$

Using the substitution $\sin^2(\psi/2) = 1 - \cos^2(\psi/2)$ and for particular values of m, Equation (4.38) becomes

$$\text{For } m = 0, \cos m\frac{\psi}{2} = 1$$
$$\text{For } m = 1, \cos m\frac{\psi}{2} = \cos\frac{\psi}{2}$$
$$\text{For } m = 2, \cos m\frac{\psi}{2} = 2\cos^2\frac{\psi}{2} - 1 \tag{4.39}$$
$$\text{For } m = 3, \cos m\frac{\psi}{2} = 4\cos^3\frac{\psi}{2} - 3\cos\frac{\psi}{2}$$

etc.

If we let

$$x = \cos\frac{\psi}{2} \tag{4.40}$$

we can deduce the Tschebyscheff polynomials by the substitution

$$T_m(x) = \cos m\frac{\psi}{2} \tag{4.41}$$

Thus, for particular values of m, the Tschebyscheff polynomials are obtained from Equation (4.41) as

$$\begin{aligned} T_0(x) &= 1 \\ T_1(x) &= x \\ T_2(x) &= 2x^2 - 1 \\ T_3(x) &= 4x^3 - 3x \end{aligned} \tag{4.42}$$

etc.

Thus, we observe from Equation (4.42) that the degree of the polynomial is identical to the value of m. Also, we note from Equation (4.41) that the roots of the polynomial occur when

$$m\frac{\psi}{2} = (2p-1)\frac{\pi}{2} \tag{4.43}$$

where $p = 1, 2, 3, \ldots$

The roots of the variable x, which we designate x_o is thus given as

$$x_o = \cos\left[(2p-1)\frac{\pi}{2m}\right] \qquad (4.44)$$

We observe that $\cos(m\psi/2)$ has been expressed as a polynomial of degree m. Thus, Equations (4.33) and (4.35) can be expressed as Tschebyscheff polynomials of degree $(2k+1)$ and $2k$ respectively. This is the case as each of these equations is the sum of cosine polynomials of the form $\cos(m\psi/2)$. Thus, for an even number n_e of sources, $n_e - 1 = 2k + 1$, and for an odd number n_o of sources, $n_o - 1 = 2k$. Thus, Equations (4.33) and (4.35), which express the field patterns of symmetric in-phase arrays, are polynomials of degree equal to the number of sources less 1. We can then set the array polynomial of Equations (4.33) and (4.35) to be equal to the Tschebyscheff polynomial of like degree and equate the array coefficients equal to the coefficients of the Tschebyscheff polynomial. Thus, the amplitude distribution given by these coefficients is a Tschebyscheff distribution and the field pattern of the resulting array corresponds to the Tschebyscheff polynomial of degree $n - 1$, where n is the number of elements in the array.

There are other alternatives to the Tschebyscheff distribution, such as the Taylor distribution. The Taylor distribution provides compensation for practical errors encountered when applying the Tschebyscheff approach [2].

4.6 Planar Arrays

Planar arrays are two-dimensional arrays and can be obtained by modifying elements along a line (linear arrays) to obtain elements along a plane. Planar arrays can be of various shapes such as rectangular arrays formed by arranging elements in a rectangular grid or circular arrays formed by the arrangement of elements along a circle. These arrays are more versatile than linear arrays due to the additional variables in a two-dimensional grid. Also, planar arrays can be used to scan the main beam of the antenna towards any point in space. Planar arrays are applicable to UWB systems due to their compactness and inherent low-profile nature. We now consider two representative planar arrays in greater detail. These are the rectangular and circular arrays.

4.6.1 Rectangular Arrays

Depicted in Figure 4.11 is a rectangular array configuration of isotropic elements. In formulating the field pattern (array factor) for the array, we consider the expressions in the x and y dimensions. Thus, if M elements are initially placed along the x-axis, we recall with the help of Equation (4.20), that the field representation for a linear array of n elements can be expressed as

$$AF = \sum_{m=1}^{M} I_m e^{j(m-1)kd_x \sin\theta \cos\phi + \beta_x} \qquad (4.45)$$

where A_m is the excitation coefficient of each element, d_x is the element spacing and β_x is the progressive phase shift along the x-axis. Also, if we consider N linear elements placed along the y-axis with excitation coefficients A_n, element spacing d_y and progressive phase shift β_y the array factor for the entire rectangular array can be written as

$$AF = \sum_{n=1}^{N} A_n \left[\sum_{m=1}^{M} A_m e^{j(m-1)(kd_x \sin\theta \cos\phi + \beta_x)}\right] e^{j(n-1)(kd_y \sin\theta \cos\phi + \beta_y)} \qquad (4.46)$$

Antenna Arrays

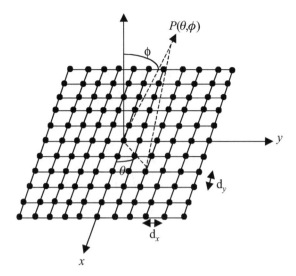

Figure 4.11 Planar array geometry

Thus, Equation (4.46) indicates that the array factor is the product of the array factors of the array in the x and y directions. In addition, if the array has a uniform excitation amplitude for all the elements, where

$$A_o = A_m A_n \tag{4.47}$$

Then, the array factor can be expressed as

$$AF = A_o \sum_{m=1}^{M} e^{j(m-1)(kd_y \sin\theta \cos\phi + \beta_y)} \sum_{n=1}^{N} e^{j(n-1)(kd_x \sin\theta \cos\phi + \beta_x)} \tag{4.48}$$

4.6.2 Circular Arrays

The circular array consists of elements arranged in a circular ring. The array is of great practical interest for applications such as radio direction finding. The array is applicable for UWB systems in areas such as radar and communications.

Figure 4.12 illustrates a circular array consisting of N isotropic elements and spaced a constant distance apart on the circular ring of radius a.

The field at a large distance from the array is known to vary inversely with respect to the distance from the array. Thus, the normalised field can be expressed as

$$\underline{E}_n(r, \theta, \phi) = \sum_{n=1}^{N} a_n \frac{e^{-jkR_n}}{R_n} \tag{4.49}$$

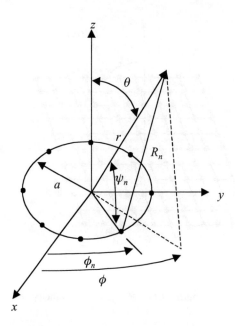

Figure 4.12 Circular array configuration

The distance of the n-th element from the observation point R_n can be expressed using the cosine law as

$$R_n = (r^2 + a^2 - 2ar\cos\psi_n)^{1/2} \qquad (4.50)$$

For $r \gg a$, we obtain an approximation as

$$R_n \cong r - a\cos\psi_n = r - a\sin\theta\cos(\phi - \phi_n) \qquad (4.51)$$

Substituting $R_n \cong r$ in Equation (4.49) and using Equation (4.51) we obtain

$$\underline{E}_n(r, \theta, \phi) = \frac{e^{-jkr}}{r} \sum_{n=1}^{N} a_n e^{+jka\sin\theta\cos(\phi - \phi_n)} \qquad (4.52)$$

where a_n denotes the excitation coefficients of the n-th element and $\phi_n = 2\pi(n/N)$ is the angular position of the n-th element on the $x - y$ plane. If we assume

$$a_n = A_n e^{j\alpha_n} \qquad (4.53)$$

where A_n is the amplitude of the n-th element and α_n is the phase excitation of the n-th element relative to the array centre.

In view of Equation (4.53), we can deduce the array factor for the circular array as

$$AF(\theta, \phi) = \sum_{n=1}^{N} A_n e^{j[ka\sin\theta\cos(\phi - \phi_n) + \alpha_n]} \qquad (4.54)$$

The peak of the main beam can be directed in the direction (θ_o, ϕ_o) by making the exponential term in Equation (4.54) equal to zero. Thus, this leads to

$$\alpha = -ka \sin \theta_o \cos(\phi_o - \phi_n) \tag{4.55}$$

The array factor can thus be written with the aid of Equation (4.55) as

$$AF(\theta, \phi) = \sum_{n=1}^{N} A_n e^{jka[\sin\theta \cos(\phi-\phi_n) - \sin\theta_o \cos(\phi_o-\phi_n)]} \tag{4.56}$$

4.7 Design Considerations

Antenna arrays for UWB applications will need to exhibit broad bandwidths for both radiation pattern and input impedance. The basic concepts we have outlined in this chapter have concentrated on the radiation patterns and on the basic design principles for the array. The input impedance characteristics have not been considered, as this will depend on the characteristics of the elements used in the array. We have assumed simple isotropic elements in the basic description of antenna arrays. In practice, the array characteristics will be affected by mutual coupling and this will affect the array impedance and radiation characteristics.

4.7.1 Mutual Coupling

Mutual coupling arises due to the interaction of two or more elements as a result of the elements being in close proximity to each other. This effect is observed during transmission or reception of elements in the array. The effects of mutual coupling can be modelled by the use of full-wave numeric electromagnetic solvers incorporating methods such as the finite element method, finite difference time domain method and method of moments. The effect of mutual coupling can be reduced by minimising the interaction between elements in the array. One of the ways to achieve this is by adjusting the interelement separation.

4.7.2 Array Gain

The gain of an antenna array can be expressed as the ratio of the maximum radiation intensity of the array to the maximum radiation intensity of a reference antenna, i.e. isotropic antenna, provided the same input power is used for both. The gain at boresight for a planar antenna array of $m \times n$ elements, with planar face similar to Figure 4.12, can be expressed as [5]

$$G_A = \frac{4\pi mn A_e}{\lambda^2} \tag{4.57}$$

where we assume that each element has an elemental area of A_e and λ is the wavelength. The gain for a variation with elevation angle θ is expressed as

$$G_A = \frac{4\pi mn A_e}{\lambda^2} \cos\theta \tag{4.58}$$

In a practical array, the effect of mutual coupling will need to be considered. Thus, to account for mutual coupling, the array gain admits a representation given as

$$G_A = \frac{4\pi n A_e}{\lambda^2} \left(1 - \Gamma_{mn}^2\right) \cos\theta \qquad (4.59)$$

where the reflection coefficient Γ_{mn} accounts for the effects of mutual coupling between the elements in the array.

4.8 Summary

A review of the fundamental concepts for antenna arrays has been presented in this chapter. Isotropic sources and the two-element array have been adopted to describe the basic concepts. Subsequently, the principle of pattern multiplication and the array factor and have been discussed. Various amplitude and phase distributions have been introduced for linear arrays. In addition, a number of array types have been reviewed. Finally, design considerations for arrays have been discussed and factors to be considered in UWB arrays have been briefly highlighted.

References

[1] J.D. Kraus, *Antennas*, McGraw-Hill, 1988.
[2] C.A. Balanis, *Antenna Theory*, John Wiley & Sons, Inc., 1997.
[3] S. A. Schelkunoff, A mathematical theory of linear arrays, *Bell System Technical Journal*, **22**, 80–107, 1943.
[4] C.L. Dolph, A current distribution for broadside arrays which optimises the relationship between beamwidth and side-lobe level, *Proc. IRE Waves and Electrons*, **34**(6), 335–48, 1946.
[5] P.W. Hannan, The element-gain paradox for a phased array antenna, *IEEE Trans. Antennas Propagat.*, **AP-12**, 423–33, 1964.

5

Beamforming

Ben Allen

5.1 Introduction

This chapter describes the concept of beamforming with an emphasis on narrowband systems, i.e., with a fractional bandwidth of less than 1 % as defined by

$$FB = \frac{f_h - f_l}{(f_h + f_l)/2} \times 100 \%, \qquad (5.1)$$

where f_h and f_l are the highest and lowest signal components, respectively. The concepts set the scene for ultra-wideband beamforming, which is explored later in this book.

This chapter initially describes the history of beamforming (adaptive antennas) for wireless systems. It then introduces the principal building blocks of a beamformer and the related operational concepts. Following this, the signal model is described and beam- and null-steering functions are explained. The use of window functions in beamformers is described next, and the chapter is then summarised.

5.1.1 Historical Aspects*

An early simple adaptive antenna was the *adaptive sidelobe canceller*, which was developed in the late 1950s by P. Howells [1] and subsequently by P. Howells and S. Applebaum [2]. Such a system is shown in Figure 5.1, where two omnidirectional antennas are used in a similar manner to microphones in an adaptive noise canceller, i.e., one to provide a reference signal and the other to provide an input to the adaptive filter. Both antennas receive the interfering and wanted signals, but since the antennas are spatially separated the two signals can be distinguished using a suitable filter. This configuration works very well as long as the input *signal-to-interference ratio* (SIR) is low. Indeed, when the SIR is high, the wanted signal can be attenuated! This is because the system directs a null towards a specific signal and under these conditions the null is directed towards the wanted signal. There are several techniques that can reduce this effect such as injecting additional noise, or using a specially modified adaptive algorithm.

Ultra-wideband Antennas and Propagation for Communications, Radar and Imaging Edited by B. Allen, M. Dohler, E. E. Okon, W. Q. Malik, A. K. Brown and D. J. Edwards
© 2007 John Wiley & Sons, Ltd

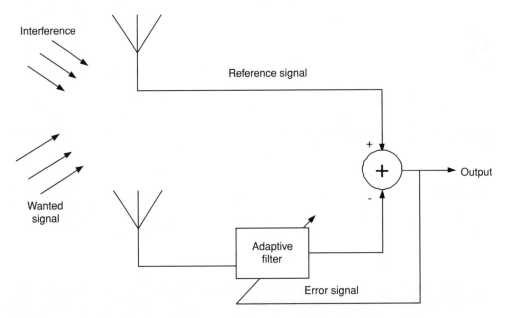

Figure 5.1 Two-element adaptive sidelobe canceller. Reproduced by permission of © 2005 John Wiley & Sons, Ltd*

Another type of adaptive beamformer utilises a pilot signal. This technique was developed by Widrow, Mantey, Griffiths and Goode [3]. It operates by forming a beam towards the wanted signal while simultaneously directing nulls towards interfering signals. This differs from the adaptive sidelobe canceller which can only steer one null. Griffiths [4] and Frost [5] have also developed similar systems which have been shown to be simpler and, in some cases, perform better. In 1976, Compton, Huff, Swantner and Ksienski [6] applied the pilot adaptive beamformer to wireless communication systems, where prior to this radar was the primary application. This new application was key to an overwhelming amount of subsequent research, a cross-section of which is reported in this book.

5.1.2 Concept of Spatial Signal Processing*

Having two ears gives us the ability to sense the direction of audio signals and to adjust our hearing to 'tune' into certain signals and ignore others. Moreover, human hearing covers a frequency range of approximately 5 Hz to 20 kHz. Due to the bandwidth, designing such a system electronically is highly challenging and requires specialised wideband array design techniques that account for the signal characteristics as they change over the operating bandwidth.

The remarkable ability of human hearing to adapt to certain signals forms the basis of a wideband adaptive array. Such an array can be designed for audio signals or electromagnetic signals and therefore has applications in sonar, radar and any other system that can benefit from the spatial processing of wideband signals.

One such area that has blossomed recently is multimedia communications where bandwidth limitations have meant that advanced techniques such as adaptive antenna systems are required to provide the necessary channel capacity [7].

Traditionally, wireless signals have occupied bandwidths of up to 20 MHz. However, ultra-wideband wireless systems, which occupy several Gigahertz, have recently become a candidate for delivery of unprecedented data rates over short ranges [8], and this has provided a demand for beamformers to operate over these bandwidths.

Adaptive arrays can also be applied to radar and sonar systems that enable the direction of targets to be determined. One key advantage adaptive antennas have is that they eliminate the need to mechanically rotate or move the sensors. However, rotating radar antennas are still a common sight on ships and at airports.

The study of adaptive antenna arrays is multidisciplinary, demanding knowledge of signal processing, transceiver design, antenna design and propagation. Without any one of these component disciplines, it becomes very difficult to design and commission an effective and robust system. As a consequence, it has attracted the attention of many of the world's top researchers and enormous investment from a broad range of commercial and government organisations. This research has been supported by the rapid growth of the electronics industry over the last few decades. As a consequence, cost-effective components, such as analogue-to-digital converters and signal-processing technology have become available. Also, the opening up of the mobile communications marketplace has been a key driver for adaptive antenna development.

Smart antennas have been considered as a key technology to provide the services demanded by mobile network users. This is due to their ability to control interference levels, thereby allowing additional services to be provided without causing unacceptable additional interference levels. Although smart antennas have not been widely integrated into cellular systems as many have predicted, there have been many examples of their use, particularly in rural areas where they are used to increase base station coverage. Also, a possible future application of smart antennas is as an enabling technology for executing spectrum liberalisation. Spectrum liberalisation, which is currently being considered by spectrum planning authorities, is the deregulating of the civil radio spectrum to allow users to operate on any frequency (within certain constraints) assuming the interference levels are of an acceptable level. This is in contrast to existing policy where users are assigned a specific operating frequency. In this context, smart antennas will control interference levels received and generated by the transceiver, and would operate in conjunction with software-definable radios and associated protocols.

5.2 Antenna Arrays*

The half-power beamwidth (HPBW) of a Hertzian dipole is 90°. In most wireless terrestrial and space applications, a narrower HPBW is desired because it is desirable to direct power in one direction and no power into other directions. For example, if a Hertzian dipole is used for space communication, only a fraction of the total power would reach the satellite and most power would simply be lost (space applications require HPBWs of well below 5°).

With a finite-length dipole, it would be possible to decrease the HPBW down to 50° by increasing the length to $L \approx 1.1\lambda$; a further increase in length causes the HPBW to decrease further; however, unfortunately multilobes are generated hence further decreasing the useful power radiated into the specified direction.

It turns out that with the aid of antenna arrays, it is possible to construct radiation patterns of arbitrary beamwidth and orientation, both of which can be controlled electronically. An *antenna array* is by definition a radiating configuration consisting of more than one antenna element. The definition does not specify which antenna elements are used to form the array, nor how the spatial arrangement ought to be. This allows us to build antenna arrays consisting of different elements, feeding arrangements and spatial placement, hence resulting in radiating structures of different properties.

Example antenna arrays consisting of several patch antennas are depicted in Figure 5.2, and briefly dealt with below.

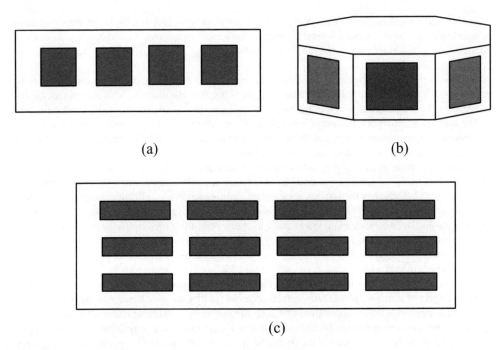

Figure 5.2 Common types of antenna arrays: (a) Linear array; (b) circular array; (c) planar array. Reproduced by permission of © 2005 John Wiley & Sons, Ltd*

5.2.1 Linear Array*

The most common and most analysed structure is the *linear antenna array*, which consists of antenna elements separated on a straight line by a given distance. Although each single antenna element can have a large HPBW, the amplitude and phase of the feeding current to each element can be adjusted in a controlled manner such that power is transmitted (received) to (from) a given spatial direction.

If adjacent elements are equally spaced then the array is referred to as a *uniform linear array* (ULA). If, in addition, the phase α_n of the feeding current to the nth antenna element is increased by $\alpha_n = n\alpha$, where α is a constant, then the array is a *progressive phase shift array*. Finally, if the feeding amplitudes are constant, i.e., $I_n = I$, then this is referred to as a *uniform array*.

The uniform array, depicted in Figure 5.2(a), with an interelement spacing of $\lambda/2$ is the most commonly deployed array as it allows simple feeding, beam-steering and analysis. For example, the power radiated in azimuth at an elevation of $\theta = \pi/2$ for such an array consisting of $\lambda/2$-dipoles can be calculated as

$$P(\theta = \pi/2, \phi) \propto \left| \frac{1}{N} \frac{\sin(\frac{1}{2}N(\pi \cos(\phi) + \alpha))}{\sin(\frac{1}{2}(\pi \cos(\phi) + \alpha))} \right|^2, \quad (5.2)$$

where α is the progressive phase shift, N is the total number of antenna elements, and it has been assumed that $I_n = 1/N$.

Figure 5.3 depicts the dependency of the radiation pattern on the number of antenna elements, where $\alpha = 0$. It can be observed that increasing the number of elements N, also decreases the HPBW where

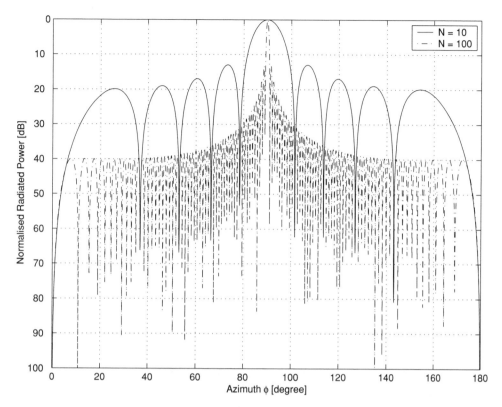

Figure 5.3 Dependency of the radiation pattern on the number of elements N. Reproduced by permission of © 2005 John Wiley & Sons, Ltd*

the width is inversely proportional to the number of elements. It is also observed that, independent of the number of elements, the ratio between the powers of the main lobe and the first sidelobe is approximately 13.5 dB. Therefore, if such a level does not suit the application, antenna arrays different from uniform arrays have to be deployed.

Figure 5.4 illustrates the dependency of the beam direction on the progressive feeding phase, α, where $N = 10$. Increasing α increases the angle between the broadside direction and the main beam, where the dependency is generally not linear.

In summary, by means of a simply employable uniform antenna array, the HPBW can be reduced by increasing the number of antenna elements and steering the main beam into an arbitrary direction by adjusting the progressive phase shift.

5.2.2 Circular Array*

If the elements are arranged in a circular manner as depicted in Figure 5.2(b), then the array is referred to as a *uniform circular array* (UCA). With the same number of elements and the same spacing between them, the circular array produces beams of a wider width than the corresponding linear array. However, it outperforms the linear array in terms of diversity reception.

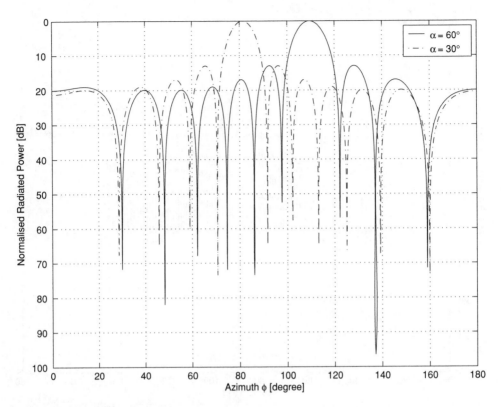

Figure 5.4 Dependency of the beam direction on the progressive feeding phase α. Reproduced by permission of © 2005 John Wiley & Sons, Ltd*

If maximal ratio combining is used, then it can be shown that the UCA outperforms the ULA on average for small and moderate angle spread for similar aperture sizes. However, the ULA outperforms the UCA for near-broadside angles-of-arrival with medium angular spreads. It can also be shown that the central angle-of-arrival has a significant impact on the performance of the ULA, whereas the UCA is less susceptible to it due to its symmetrical configuration.

5.2.3 Planar Array*

Both linear and circular arrangements allow the beam to be steered in any direction in the azimuth plane; however, the elevation radiation pattern is dictated by the radiation pattern of the antenna elements. In contrast, the *planar antenna array* allows one also to steer the beam in elevation, thereby producing so-called *pencil beams*. An example realisation of a planar array is depicted in Figure 5.2(c). For the planar antenna array, the same observations as for the linear array hold.

5.2.4 Conformal Arrays*

The array types discussed so far have all been based upon a regular, symmetrical design. This is suitable for applications where such mounting is possible; however, this is not the case for many scenarios where the surface is irregular or the space is confined. For these situations a *conformal array* is required which,

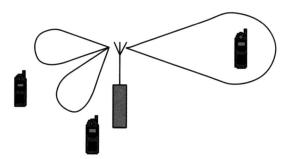

Figure 5.5 Principle of spatial filtering applied to a mobile communication system. Reproduced by permission of © 2005 John Wiley & Sons, Ltd*

as the name suggests, conforms to the surrounding topology. The arrays could be elliptical or follow a more complex topology. The challenge for the design of these antennas is to make sure that the main lobe beamwidth and sidelobe levels fall within the required specification. Specialised antenna elements are also required since they will be mounted on an irregular surface and must therefore follow the surface contours. This particularly applies to patch antennas that usually require a flat surface.

5.3 Adaptive Array Systems

5.3.1 Spatial Filtering*

From the previous section, it has been shown that appropriate feeding allows antenna arrays to steer their beam and nulls towards certain directions, which is often referred to as *spatial filtering*. Spatial filtering is of particular importance in mobile communication systems since their performance tends to be interference limited.

The very first mobile communication systems had base stations with an omnidirectional antenna element, i.e., the transmit power was equally spread over the entire cell and hence serving everybody equally. However, since many such communication cells are placed beside each other, they interfere with each other. This is because a mobile terminal at a cell fringe receives only a weak signal from its own base station, whereas the signals from the interfering base stations grow stronger.

Modern wireless communication systems deploy antenna arrays, as depicted in Figure 5.5. Here, a base station communicates with several active users by directing the beam towards them and it nulls users that cause interference. This has two beneficial effects: first, the target users receive more power compared to the omnidirectional case (or, alternatively, transmit power can be saved); and, second, the interference to adjacent cells is decreased because only very selected directions are targeted.

The above-described spatial filtering can be realised by means of the following mechanisms:

- **Sectorisation**: the simplest way is to deploy antenna elements which inherently have a sectorised radiation pattern. For instance, in current second-generation mobile phone systems, three base station antennas with 120° sectors are deployed to cover the entire cell.
- **Switched beam**: a beam is generated by switching between separate directive antennas or predefined beams of an antenna array. Traditionally, algorithms are in place, which guarantees that the beam with the strongest signal is chosen.
- **Phased antenna array**: by adjusting the feeding phase of the currents, a moveable beam can be generated (see Section 5.4.1 and Figure 5.4). The feeding phase is adjusted such that the signal level is maximised.

- **Adaptive antenna array**: as with the phased array, a main lobe is generated in the direction of the strongest signal component. Additionally, sidelobes are generated in the direction of multipath components and also interferers are nulled. Here, not only the signal power is maximised but also the interference power is minimised, which clearly requires algorithms of higher complexity to be deployed when compared to the switched beam and phased arrays. Since adaptive antenna arrays are currently of great importance and also the main topic of this book, their functioning is briefly introduced below.

5.3.2 Adaptive Antenna Arrays*

Why adaptive? Clearly, adaptability is required to change its characteristics to satisfy new requirements, which may be due to:

- changing propagation environment, such as moving vehicles;
- changing filtering requirements such as new targets or users requiring processing.

This is particularly the case for spatial filtering which can dynamically update the main beam width and direction, sidelobe levels and direction of nulls as the filtering requirement and/or operating environment changes. This is achieved by employing an antenna array where the feeding currents can be updated as required. Since the feeding current is characterised by amplitude and phase, it can be represented by a complex number. For this reason, from now on explicit reference to the feeding current is omitted and instead the term *complex array weights* is used. Such a system with adaptive array weights is shown in Figure 5.6. Clearly, the aim of any analysis related to adaptive antenna arrays is to find the optimum array weights so as to satisfy a given performance criterion.

5.3.3 Mutual Coupling and Correlation*

Any form of beamforming relies on the departure of at least two phase-shifted and coherent waves. If either the phase shift or the coherence is violated, then no beamforming can take place; this may happen as follows.

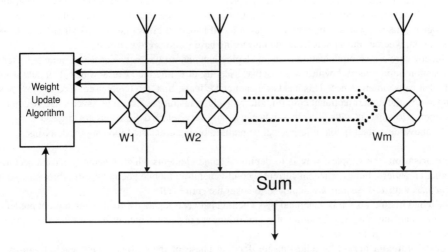

Figure 5.6 Concept of an adaptive antenna array. Reproduced by permission of © 2005 John Wiley & Sons, Ltd*

If two antenna elements are placed very closely together, then the radiated electromagnetic (EM) field of one antenna couples into the other antenna and vice versa and therefore erasing any possible phase shift between two waves departing from these elements; this effect is known as *mutual coupling*. The antenna array designer has to ensure that the mutual coupling between any of the elements of an antenna array is minimised. This is usually accomplished for interelement spacing larger than $\lambda/2$. If, on the other hand, the spacing is too large, then the departing waves do not appear coherent anymore, hence, also preventing an EM wave to form a directed beam.

From a purely antenna array point of view, a suitable interelement spacing for beamforming is thus in the order of $\lambda/2$. Sometimes, however, the antenna array is not utilised as a beamformer, but as a diversity array. This requires the arriving and departing signals to be as decorrelated as possible. It can be shown that if the antenna array is uniformly surrounded by clutter, then the spatial correlation function of the channel requires a minimal interantenna element spacing of $\lambda/2$ for the signals to appear decorrelated. Such ideal clutter arrangement, however, is rarely found in real-world applications where the correlation between antenna elements increases as the angular spread of the impinging waves decreases. For instance, a base station mounted on the rooftop with little clutter around requires spacing of up to 10λ to achieve decorrelation between the elements.

From this short overview, it is clear that the antenna array designer faces many possible realisations where the optimum choice will depend on the application at hand.

5.4 Beamforming*

In the previous sections, antenna arrays of various topologies were introduced and it was shown that they can be used to steer beams to a given direction. Consequently, antenna arrays were introduced where a beam can be steered by multiplying the received signal at each element by a complex weight. The weights are chosen to steer the beam in a certain direction. The weighted signals are finally summed to form a single signal available for the application to process further. The weights are chosen such that the combined signal is enhanced in the presence of noise and interference.

The following section build on this concept by initially introducing essential terminology, then demonstrating the function of the complex weights and how the phase and amplitude weights impact on the control of the array pattern. For example, the main beam can be steered toward a target for signal enhancement, or nulls can be steered toward interferers in order to attenuate them. Grating lobes are also investigated and it is shown how these can produce sporadic effects when the array is used as a spatial filter. It is then shown how the application of window functions can be used to control the sidelobes of the radiation pattern by controlling the trade-off between the main-beam width and the sidelobe level. Thus, very low sidelobes can be achieved in order to give good rejection to signals arriving at these angles at the expense of having a wider main beam.

The focus of this chapter is narrowband beamformers. The term *narrowband* has many definitions. For clarity, it is defined here as an array operating with signals having a fractional bandwidth (FB) of less than 1 % i.e., FB < 1 %, where FB is defined by Equation (5.1). As the bandwidth of the signal is increased the phase weighting computed to steer the beam at a particular angle is no longer correct, therefore a more complex weighting arrangement is required. Ultra-wideband array processing techniques are described later in this book.

5.4.1 Adaptive Antenna Technology*

In order that beamforming techniques can be analysed, fundamental terminology must first be explained. This section introduces basic terms such as *steering vector* and *weight vector*.

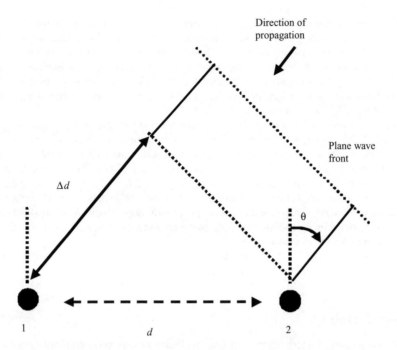

Figure 5.7 Array signal model for a two-element array. d = distance between elements, θ = angle of arrival of wave from boresight, Δd = additional distance of wave. Reproduced by permission of © 2005 John Wiley & Sons, Ltd*

Consider an array of L omnidirectional elements operating in free space where a distant source is radiating sinusoids at frequency f_o, as shown in Figure 5.7 for $L = 2$, i.e., a two-element array. The signals arriving at each of the sensors will be both time and phase delayed with respect to each other due to the increased distance of the sensors from a reference sensor. This is evident from the figure where the wave front will arrive at element 2 before element 1. This means that the phase of the signal induced in element 2 will lead that in element 1. The phase difference between the two signals depends upon the distance between the elements, d, and the angle of the transmitter with respect to the array boresight, shown as θ in the figure. The additional distance the wave has travelled to arrive at element 1, Δd, is given by

$$\Delta d = d \cdot \sin \theta \qquad (5.3)$$

If a planar array is considered with elements in the $x - y$ plane, the signal arriving at each of the sensors will therefore arrive at angles θ and ϕ, and in this more general case, Δd becomes

$$\Delta d = d \cdot \sin \phi \sin \theta \qquad (5.4)$$

and the phase difference becomes

$$\Delta \Phi = 2\pi \frac{d}{\lambda} \cdot \sin \phi \sin \theta \qquad (5.5)$$

where λ is the wavelength. This signal model assumes that the wave arrives at each element with equal amplitude.

After the signals have arrived at the sensors, they are weighted and subsequently summed. Consider a simple array consisting of two elements with unity amplitude and zero phase weighting, the received signal after summation is given by the equation below, where element 1 is the phase reference point.

$$E(\theta, \phi) = 1 + e^{j2\pi \frac{d}{\lambda} \sin\phi \sin\theta} \tag{5.6}$$

Plotting $10\log_{10}|E(\theta,\phi)|$ as a function of θ and ϕ yields the far-field radiation pattern of the array. Note that the two terms in the above equation refer to the received signals from the transmitter at each of the elements with respect to the first element. The terms can be expressed as a vector, referred to as the *received signal vector*, thus the received signal vector for the above example is

$$\mathbf{r} = \begin{bmatrix} 1 & e^{j2\pi \frac{d}{\lambda} \sin\phi \sin\theta} \end{bmatrix} \tag{5.7}$$

For N equally spaced elements, Equation (5.6) is generalised as:

$$E(\theta, \phi) = \sum_{n=1}^{N} e^{j2\pi \frac{d}{\lambda} \sin\phi \sin\theta} \tag{5.8}$$

and can be easily extended to accommodate unequally spaced elements by including the distance of each element with respect to the first.

If several transmitters are detected by the sensors, the received signals can be expressed in the same way for each of the transmitters, thus each of the signals has an associated received signal vector, i.e., if there are K transmitters, K received signal vectors can be determined. This frequently occurs in practice, such as a radar tracking multiple targets, or in a mobile communications system where multiple users are active. The received signal vector of the kth signal is frequently referred to as the *steering vector*, $\mathbf{S}_k$.

The above analysis is generalised for a planar array consisting of $M \times N$ uniformly spaced elements and where the radiation pattern is evaluated in azimuth θ and elevation ϕ as shown in Figure 5.8. The resulting radiation pattern is given by:

$$E(\theta, \phi) = \sum_{m=1}^{M} \sum_{n=1}^{N} e^{j \frac{2\pi}{\lambda} (md_x \cos\theta + nd_y \sin\theta) \sin\phi} \tag{5.9}$$

where d_x and d_y are the element spacing in the x and y dimensions respectively. Note that equating Equation (5.9) to zero and solving for θ and/or ϕ gives the location of the nulls. For a two-element array, nulls occur in the pattern when

$$\cos\left(\pi \frac{d}{\lambda} \sin\theta \sin\phi\right) = 0 \tag{5.10}$$

$$\pi \frac{d}{\lambda} \sin\theta \sin\phi = \frac{2n+1}{2} \pi \tag{5.11}$$

where $n = 0, 1, 2, \ldots$ Assuming $\theta = 90°$,

$$\phi_{nulls} = \sin^{-1}\left(\left[n + \frac{1}{2}\right] \frac{\lambda}{d}\right) \tag{5.12}$$

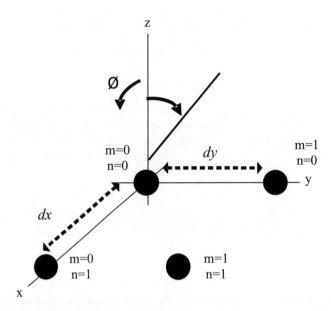

Figure 5.8 Planar array signal model for $M = 2$, $N = 2$ array. d_x = distance between x-axis elements, d_y = distance between y-axis elements, θ = azimuth angle of arrival of wave from boresight, ϕ = elevation angle of arrival of wave from boresight. Reproduced by permission of © 2005 John Wiley & Sons, Ltd*

Furthermore, the impact of applying weights to each of the received signals can also be analysed. The above analysis assumed unity amplitude and zero phase weights applied to each element, thus for a two-element array, the weights can be expressed as the vector $\mathbf{W} = [1.e^{j0}\ 1.e^{j0}]$, which is referred to as the *weight vector*. The values of the weights are computed to give a desired response, i.e., to steer the main beam and nulls in the desired directions and set the sidelobes to the desired level and set the main-beam width. The computation of $\mathbf{W}$ is discussed in the subsequent subsections. $\mathbf{W}$ is a $1 \times N$ vector for a linear array and a $M \times N$ matrix for a planar array. It is incorporated into the analysis by multiplying the array response by $\mathbf{W}$ as shown below.

$$E(\theta, \phi) = \sum_{m=1}^{M} \sum_{n=1}^{N} \mathbf{W}_{mn} e^{j\frac{2\pi}{\lambda}(md_x \cos\theta + nd_y \sin\theta)\sin\phi} \quad (5.13)$$

5.4.2 Beam Steering*

The primary function of an adaptive array is to control the radiation pattern so that the main beam points in a desired direction and to control the sidelobe levels and directions of the nulls. This is achieved by setting the complex weights associated with each element to values that cause the array to respond desirably.

The complex weights consist of real and imaginary components, or alternatively, amplitude and phase components, i.e., $A_n = \alpha_n . e^{j\beta_n}$ where A_n is the complex weight of the nth element, α_n is the amplitude weight of the nth element and β_n is the phase weight of the nth element. The phase components control the angles of the main beam and nulls, and the amplitude components control the sidelobe level and

Beamforming

main-beam width (although when the main beam is steered towards end-fire by phase weighting, the main-beam width will also increase). These are considered in turn in the following sections.

5.4.2.1 Phase Weights

The main beam can be steered by applying a phase taper across the elements of the array. For example, by applying phases of $0°$ and $70°$ to each element of a two-element array with $\lambda/2$ spacing between the elements, the beam is steered to $35°$, as shown in Figure 5.9. The figure also shows the radiation pattern when zero phase weights are applied, i.e., the main beam is steered towards boresight and a gain of 3 dB is shown. The following section describes a method for computing the required weights to steer the main beam to a desired angle.

Phase weights for narrowband arrays, which is the focus of this chapter, are applied by a phase shifter, However, where wideband and ultra-wideband beamforming is considered, it is important to distinguish between phase-shifter arrays and delay arrays. The first generates a phase shift for a certain frequency of operation and can be used for narrowband arrays operating over a very small range of frequencies.

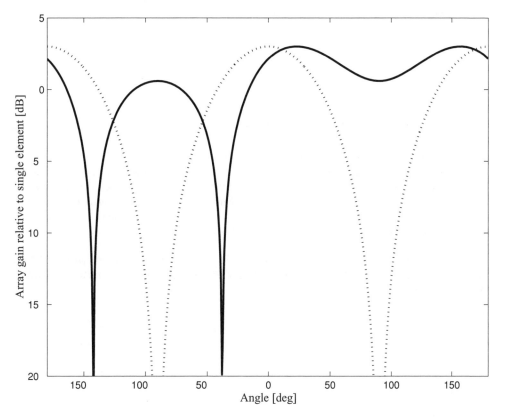

Figure 5.9 Radiation pattern with phase weighting for a two-element array with $\lambda/2$ spacing. Dotted line with no phase tapper, solid line with beam steered to $35°$. Reproduced by permission of © 2005 John Wiley & Sons, Ltd*

On the other hand delay arrays generate a pure time delay for each element depending on the desired angle and produce wideband characteristics. This is because in phase arrays, as the frequency varies, the steering angle of the array also changes, but in delay (wideband) arrays, the steering angle never changes with frequency. Of course, in the delay array sidelobe variations can be observed with frequency. Pure delays can be implemented by digital filters, analogue devices or optical circuits.

5.4.2.2 Main Beam Steering*

A simple beamformer steers the main beam in a particular direction θ, ϕ. The weight vector, $\mathbf{W}$, for steering the main beam is given by

$$\mathbf{W} = \frac{1}{L}\mathbf{S_o} \tag{5.14}$$

where L is the number of elements and $\mathbf{S_o}$ is the steering vector. Due to the normalisation factor $1/L$, this beamformer yields a unity response in the look direction (not considered in Equation (5.13)). In other words, the output power is the same as that of a single-element system. In an environment consisting only of noise, i.e., no interfering signals, this beamformer provides maximum SNR. This can be determined as follows. The autocorrelation function of the noise, $\mathbf{R_N}$, is given by:

$$\mathbf{R_N} = \sigma_N^2 \mathbf{I} \tag{5.15}$$

where σ_N^2 is the noise variance and $\mathbf{I}$ is the identity matrix. Thus the noise power at the output of the beamformer is

$$P_N = \mathbf{W}^H \mathbf{R_N} \mathbf{W} = \frac{\sigma_N^2}{L} \tag{5.16}$$

where $.^H$ is the Hermitian transpose. Thus the noise power is reduced by a factor of L.

Although this beamformer yields a maximum SNR in an environment consisting only of noise, this will not produce a maximum SNR in the presence of directional interference. Such scenarios are very common in radar and sonar deployments where an intentional jammer may be targeting the area, as well as mobile communications where other network users will create unintentional interference. The following section introduces a beamformer for scenarios where the number of directional interfering signals, $N \leq L - 1$.

Note that the above technique for computing the weights to steer the main beam has considered a two-dimensional array, should the beam of a planar array require steering in both azimuth and elevation, the technique can be applied for both planes.

Example 1
A receiver equipped with a four-element linear array is required to enhance the wanted signal arriving from a transmitter located at 20° from the array boresight. Compute the required weight vector for the array to perform this function given that the elements are spaced apart by $d = \lambda/2$.
The weight vector, $\mathbf{W}$, is computed using Equation (5.14). Thus, $\mathbf{S_o}$ is computed to be

$$\mathbf{S_o} = \begin{bmatrix} 1 & e^{j1.0745} & e^{j2.149} & e^{j3.223} \end{bmatrix} \tag{5.17}$$

and from (5.14),

$$\mathbf{W} = \begin{bmatrix} 1 & e^{j1.0745} & e^{j2.149} & e^{j3.223} \end{bmatrix} \tag{5.18}$$

5.4.2.3 Null Steering*

As well as directing the main beam, the nulls in the array pattern can also be directed. This is useful when it is necessary to attenuate unwanted signals arriving at angles other than that of the main beam. This subsequently increases the signal-to-interference ratio at the output of the beamformer. Figure 5.9 illustrates how null steering would improve the performance of this system. With the main beam steered at 35°, the radiation pattern shows that around 3 dB of rejection is achieved for signals arriving in the range of 50° to 100°. Considering the additional cost and complexity of implementing this system, 3 dB rejection is not a significant improvement. However, by steering one of the nulls (currently at −40° and −140° to 90°, 20 dB of rejection can be achieved for signals arriving within this angular range. This is a significant improvement for a small amount of computation.

Assume an array pattern with unity response in the desired direction and nulls in the directions of the interfering signals, the weights of the beamformer are required to satisfy these constraints [9, 10].

Let S_0 be the main beam steering vector and $S_1, \ldots S_k$ are k steering vectors for the k nulls. The desired weight vector is the solution to the following simultaneous equations

$$\mathbf{W}^H \mathbf{S}_0 = 1 \tag{5.19}$$

$$\mathbf{W}^H \mathbf{S}_k = 0 \tag{5.20}$$

Let the columns of matrix $\mathbf{A}$ contain the $\mathbf{k} + 1$ steering vectors and

$$\mathbf{C} = \begin{bmatrix} 1 & 0 & \cdots & 0 \end{bmatrix}^T \tag{5.21}$$

For $\mathbf{k} = L - 1$, where L is the number of elements and $\mathbf{A}$ is a square matrix. The above equations can be written as

$$\mathbf{W}^H \mathbf{A} = \mathbf{C}_1^T \tag{5.22}$$

Thus, the weight vector, $\mathbf{W}^H$, can be found by the equation below.

$$\mathbf{W}^H = \mathbf{C}_1^T \mathbf{A}^{-1} \tag{5.23}$$

For $\mathbf{A}$ to exist, it is required that all steering vectors are linearly independent. Should this not be the case, the pseudo-inverse can be found in its place.

When the number of required nulls is less than $L - 1$, $\mathbf{A}$ is not a square matrix. Under such conditions suitable weights may be given by

$$\mathbf{W}^H = \mathbf{C}_1^T \mathbf{A}^H (\mathbf{A}\mathbf{A}^H)^{-1} \tag{5.24}$$

However, this solution does not minimise the uncorrelated noise at the array output.

Example 2
Compute the weight vector required to steer the main beam of a two-element array towards the required signal at 45° and to minimise the interfering signal that appears at −10°. Assume the element spacing $d = \lambda/2$. Such a scenario is typical of a TV receiver situated on a tall building and is therefore able to receive signals from both a local transmitter and a distant co-channel transmitter. A null is then steered towards the co-channel transmitter to provide attenuation of this unwanted signal.

The weight vector **W** is computed using Equation (5.23). The steering vectors S_0 and S_1 are first required and computed with reference to Equation (5.8) as follows.

$$S_0 = \begin{bmatrix} 1 & e^{j2..221} \end{bmatrix} \tag{5.25}$$

$$S_1 = \begin{bmatrix} 1 & e^{-j0.5455} \end{bmatrix} \tag{5.26}$$

Thus,

$$A = \begin{bmatrix} S_0 & S_1 \end{bmatrix} \tag{5.27}$$

and therefore,

$$A = \begin{bmatrix} 1 & 1 \\ e^{-j2.221} & e^{j0.5455} \end{bmatrix} \tag{5.28}$$

and $C = [\,1\ 0\,]$ and after computing A^{-1}, it is trivial to compute the weight vector to be

$$W = \begin{bmatrix} 0.5 - j0.095 & -0.3782 + j0.3405 \end{bmatrix} \tag{5.29}$$

The resulting radiation pattern with the main beam steered towards 45° and a null at −10° is depicted in Figure 5.10. The null is shown to provide an attenuation in excess of 20 dB. In this example, the angle between the null and the main beam is 55°. This is comparable to the main-beam width of a two-element array. If the null was required to be closer to the main beam it would interfere with the main-beam shape therefore compromising the reception of the wanted signal. Cases such as this can be circumvented by increasing the number of antenna elements, which would narrow the main beam therefore enabling a null to be formed closer to it. Increasing the number of elements would also:

- reduce the sidelobe level (up to a maximum of 13 dB);
- increase the main-beam gain;
- increase the available degrees of freedom, and therefore the number of nulls; and
- also increase the cost.

If the number of elements is increased and only one null is required to be steered, **A** is no longer a square matrix and the array weights would then be computed by means of Equation (5.24).

One of the drawbacks of null steering is the alignment of nulls in the desired directions due to their narrow widths. Exact alignment requires an accuracy of <2°, i.e., from the figure, the null width is 2° for a null depth of 10 dB. Thus, a small misalignment of the null of 0.5° will result in a 5 dB reduction of the attenuation of the co-channel signal. This can be overcome by null broadening which is simply achieved by steering multiple nulls towards the same target, but with a small angular offset between them [11]. This technique is particularly useful in mobile communications where the signal propagation conditions often cause the signals to disperse. Furthermore, a deep null can only be achieved by a well-designed system and a practical system will compromise this due to system imperfections, although system calibration techniques can reduce the impact of imperfection to a certain extent.

The null-steering technique described here jointly steers the main beam and nulls to the desired angles. Modifying the vector **C** enables the existence of nulls and beams (or signal minima and maxima) to be specified according to the prevailing requirements.

Beamforming

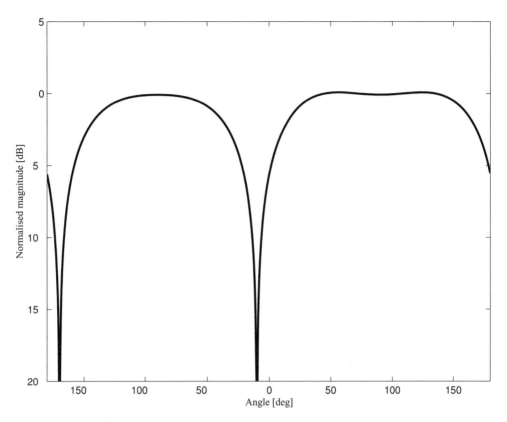

Figure 5.10 Joint beam and null steering: main beam steered to 45°, null steered to −10°. Reproduced by permission of © 2005 John Wiley & Sons, Ltd*

5.4.3 Grating Lobes*

So far the analysis of the behaviour of arrays has considered elements with a spacing of $d = \lambda/2$. What happens when other element spacings are used? Such a scenario occurs when the array is used over a wide frequency range and may also occur when certain physical design constraints are imposed on the array, such as element size. This is explored in the following.

Assuming a two-element array with unity amplitude and zero-phase weights, the radiation pattern is obtained from Equation (5.7). Plotting this expression for $d = 0.5\lambda$, 0.6λ and 2λ gives the radiation patterns shown in Figure 5.11. It can be seen that, as the element spacing becomes wider, the main lobe narrows and, when $d > 0.5\lambda$, additional lobes appear with an increasing amount of energy appearing in them as d is further increased. These additional lobes are referred to as *grating lobes*.

The time-domain analogy of grating lobes is aliasing. Aliasing occurs when the Nyquist sampling theorem is not obeyed. For the $\theta = 90°$ plane, the positions of these lobes are given by:

$$\cos\left(\pi \frac{d}{\lambda} \sin \phi_{max}\right) \tag{5.30}$$

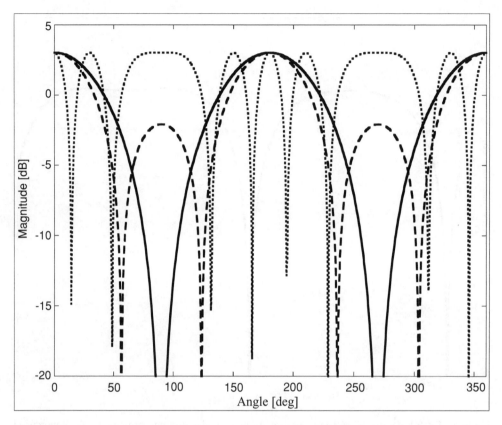

Figure 5.11 Examples of grating lobes for $d = 0.5\lambda$, 0.6λ, 2λ. Solid line $d = 0.5\lambda$, dashed line $d = 0.6\lambda$, dotted line $d = 2\lambda$. Reproduced by permission of © 2005 John Wiley & Sons, Ltd*

where

$$\pi \frac{d}{\lambda} \sin \phi_{max} = n\pi, n = 0, 1, 2 \ldots \quad (5.31)$$

$$\phi_{max} = \sin^{-1}\left(n\frac{\lambda}{d}\right) \quad (5.32)$$

5.4.4 Amplitude Weights

Consider a four-element array with $\lambda/2$ spacing between element and zero phase weights and unity amplitude weights applied. As the amplitude weighting of the two outside elements is reduced towards zero, the main beam width increases. This is illustrated in Figure 5.12 where amplitude weights of [1, 1, 1, 1] (crossed markers), [0.5, 1, 1, 0.5] (dot markers) and [0.05, 1, 1, 0.05] (dashed line) have been used. The radiation pattern of a two-element array with unity amplitude weighting is also plotted (solid line) and it can be seem that the radiation pattern of the four-element array approaches that of the two-element array as the amplitude weights of the outside elements are reduced. This also has the

Beamforming

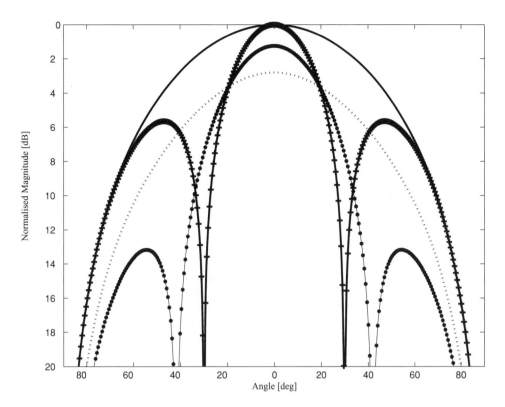

Figure 5.12 Array pattern of a four-element ULA with various amplitude weights. The impact of amplitude weighting on the sidelobe level and main beamwidth can be seen. Array pattern of a two-element array with unity amplitude weights is shown for comparison (solid line). Reproduced by permission of © 2005 John Wiley & Sons, Ltd*

effect of reducing the sidelobe levels where the sidelobe level is <-40 dB when the outer element amplitude weightings are reduced to 0.05 (not visible in the figure). Note that the difference shown in the figure between the amplitude weighted four-element array pattern and the two-element array pattern is in the amplitude of the main beam, where the four-element array will transmit more power, i.e., through the two additional elements. This additional 'gain' can be eliminated by appropriate normalisation to allow direct comparison. Hence, with normalisation little difference is observed between the two main beams.

Window functions enable the amplitude weights to be controlled with certain constraints. The constraints relate to the characteristics of the choice of function, such as desired sidelobe level or main-beam width, or a combination of both. These functions are commonly employed in other areas of signal processing such as temporal filter design and several types of window functions are investigated in the following section.

5.4.5 Window Functions*

Sidelobes can, to some extent, be controlled by employing *window functions*. These are a family of functions that set the amplitude weights of the beamformer. Common window functions are rectangular,

Table 5.1 Half-power beamwidth and sidelobe levels resulting from various window functions applied to an eight-element array. Reproduced by permission of © 2005 John Wiley & Sons, Ltd*

Window function	HPBW	SLL (dB)
Rectangular	40°	−12.5
Bartlett	60°	−25
Blackman	80°	−50
Chebyshev	36°	−20
Hamming	58°	−33
Hanning	50°	−32
Kaiser	36°	−15
Triangular	45°	−30

Hamming, Hanning, Bartlett, Triangular, Kaiser and Dolph-Chebyshev, which have many applications in signal processing as well as beamforming. Each of these window functions are characterised in Table 5.1 in terms of main beamwidth (defined as the *half-power beamwidth* or HPBW) and sidelobe level (SLL). An eight-element uniform linear array with $d = \lambda/2$ is considered for each of the examples. For each example a table of the amplitude weights is given (Table 5.2), which enable the characteristic taper of the weights that result from using a window function to be observed, with the peak occurring at the centre of the array and the minimum at each end.

5.5 Summary

This chapter has introduced the concept of beamforming (beam and null steering) and the functions of these principal operating modes. The function of phase and amplitude weights has been demonstrated along with the concept of grating lobes. It has been shown that as the element spacing increases from $\lambda/2$, these additional lobes appear, which can reduce the spatial filtering performance of the beamformer.

Table 5.2 Amplitude weights for various window functions for an eight-element array. Reproduced by permission of © 2005 John Wiley & Sons, Ltd*

El. No.	1	2	3	4	5	6	7	8
Rectangular	1	1	1	1	1	1	1	1
Bartlett	0	0.2857	0.5714	0.8571	0.8571	0.5714	0.2857	0
Blackman	0	0.0905	0.4592	0.9204	0.9204	0.4592	0.0905	0
Triangular	0.1250	0.3750	0.6250	0.8750	0.8750	0.6250	0.3750	0.1250
Hamming	0.0800	0.2532	0.6424	0.9544	0.9544	0.6424	0.2532	0.0800
Hanning	0.1170	0.4132	0.7500	0.9698	0.9698	0.7500	0.4132	0.1170
Kaiser $\alpha=1$	0.7898	0.8896	0.9595	0.9595	0.9595	0.9595	0.8896	0.7898
Chv −20 dB	0.578	0.659	0.8786	1	1	0.8786	0.659	0.578

This chapter has introduced the fundamentals of narrowband beamformers, and has set the scene for the later chapter that investigates beamforming architectures for UWB signals. For a more complete treatment of adaptive array systems and related topics the interested reader is referred to [12].

Acknowledgements

The Nuffield Foundation is acknowledged for financial support that has facilitated this work. Mischa Dohler and Mohammad Ghavami are also thanked for their contributions to this chapter. Finally, John Wiley and Sons Ltd are acknowledged for allowing reproduction of several sections from Reference [12] that have formed this chapter.

References

[1] P.W. Howells, Intermediate frequency sidelobe canceller, US Patent 3202990, May 1959.
[2] S.P. Applebaum, Adaptive arrays, *IEEE Trans. Antennas Propagat.*, **24**, 585–98, 1976.
[3] B. Widrow, P.E. Mantey, L.J. Griffiths and B.B. Goode, Adaptive antenna systems, *Proc. IEEE*, **55**(12), 2143–59, 1967.
[4] L.J. Griffiths, A simple adaptive algorithm for real-time processing in antenna arrays, *Proc. IEEE*, **57**, 1696–704, 1969.
[5] O.L. Frost, An algorithm for linearly constrained adaptive array processing, *Proc. IEEE*, **60**, 926–35, 1972.
[6] R. Compton, R. Huff, W. Swantner and A. Ksienski, Adaptive arrays for communication systems: an overview of research at the Ohio State University, *IEEE Trans. Antennas Propagat.*, **60**(5), 599–607, 1976.
[7] B. Allen and M. Beach, On the analysis of switched-beam antennas for the W-CDMA downlink, *IEEE Trans. Veh. Technol.*, **53**(3), 569–78, 2004.
[8] M. Ghavami, L.B. Michael and R. Kohno, *Ultra Wideband Signals and Systems in Communication Engineering*, John Wiley and Sons, Ltd, 2004.
[9] H. Assumpcao and G. Moutford, An overview of signal processing for arrays and receivers, *J. Int. Eng. Aust. and IREE Aust*, **4**, 6–19, 1984.
[10] V. Anderson, A realisable adaptive process, *J. Acoust. Soc. Amer.*, **45**, 398–405, 1969.
[11] K. Hugl, L. Laurila and E. Bonek, Downlink performance of adaptive antennas with null broadening, *Proc. IEEE Veh. Technol. Conf.*, 1999.
[12] B. Allen and M. Ghavami, *Adaptive Array Systems*, John Wiley and Sons, Ltd, 2005.

---

* Source: *Adaptive Array Systems* B. Allen and M. Ghavami, 2005, © John Wiley & Sons, Ltd. Reproduced with permission.

6

Antenna Diversity Techniques

Junsheng Liu, Wasim Q. Malik, David J. Edwards
and Mohammad Ghavami

6.1 Introduction

The additive white Gaussian noise (AWGN) channel is the simplest wireless digital communications channel, in which the only distortion to the transmitted signal is caused by Gaussian noise. In realistic wireless communications environments, multipath is inevitable and causes additional distortion. The interference of the various received signal echoes may enhance, weaken or even annul the received signal power due to the differences in the relative phases of the multipath arrivals. In the case of deep fading where little signal power is received, wireless digital communications becomes nearly impossible.

Diversity implementations involve multiple communications branches carrying the same transmitted signal, with the objective of decreasing the outage probability. Outage is defined as the event that the signal level falls lower than the receiver threshold due to fading, so that communication is no longer possible. Diversity techniques thus aim to mitigate fading and improve the link quality [1].

Several diversity techniques are well known in the literature, such as frequency diversity, time diversity, polarisation diversity and spatial diversity, resulting from the diversity branches operating over different carrier frequencies, time slots, polarisation states or spatial positions. Spatial, or antenna, diversity is achieved by multiple-antenna arrays, with the elements placed a certain distance apart.

This chapter aims to demonstrate the effectiveness of antenna diversity techniques in combating fading in wireless channels. A description of fading in random channels requires a statistical treatment in terms of the probability density functions (PDFs) of the channel parameters, which will be undertaken in the next several sections. The analysis is then extended to antenna diversity and its advantages. In order to make the concepts easily understandable, illustrative figures are presented throughout.

6.2 A Review of Fading

Before progressing to the mechanism of antenna diversity, we introduce the problem that it presents a solution to.

Ultra-wideband Antennas and Propagation for Communications, Radar and Imaging Edited by B. Allen,
M. Dohler, E. E. Okon, W. Q. Malik, A. K. Brown and D. J. Edwards
© 2007 John Wiley & Sons, Ltd

6.2.1 Signal Fading

Consider a signal s_i propagated through channel h_i. The received signal y_i can be expressed as

$$y_i = h_i s_i + n_i \qquad (6.1)$$

where n_i is the AWGN. For the convenience of expression and fairness of comparison, the following normalisation is applied to the channel,

$$E\{|h_i|^2\} = 1 \qquad (6.2)$$

The channel includes the effect of multipath, which results in a random variation in the channel coefficient. Fading, caused by the overlapping of the received multipath components arising from various scatterers, is depicted in Figure 6.1. As shown in the figure, the real part and the imaginary part of y_i are formed by the sum of the real and imaginary parts of the multipath components, respectively. If the multipath components interfere destructively, the magnitude of y_i approaches zero and the receiver is thus said to be located in a fade. Time-varying wireless channels experience fast fading, in which the signal levels reaching the receiver suffer fades at some time instants.

For narrowband time-varying fading channels, the magnitude of Y_i varies with time as shown in Figure 6.2. It can be seen that without any mitigation techniques, it is possible that the received signal suffers fading. In order to maintain the wireless link for a certain percentage of the time, the wireless system must overcome fading.

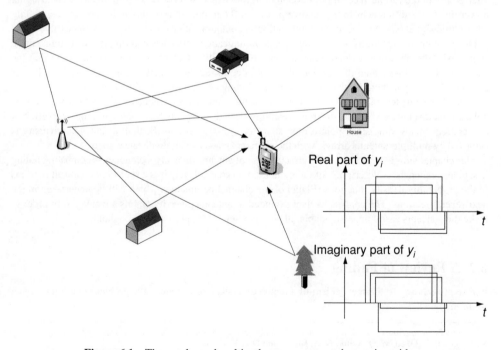

Figure 6.1 The overlapped multipath components at the receiver side

Antenna Diversity Techniques

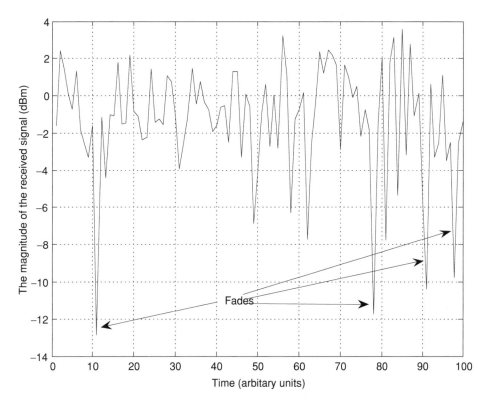

Figure 6.2 Magnitude of the received signal suffering fast fading (the averaged power of the received signal has been normalised to 0 dBm; the time slot can be any time unit suited to the analysis)

There are two approaches to mitigate fast fading. The first, which is related to antenna diversity, makes use of multiple propagation branches, h_i, so that when one of the branches suffers a deep fade, the other may compensate. The other scheme involves narrowing the time duration of the transmitted signal, s_i, which enlarges its bandwidth. If the time duration of s_i is comparable to or less than the differential delay of the multipath components, they are unlikely to overlap with each other. By applying appropiate receiver design, such as rake receivers, fading may be mitigated. This can be achieved by using UWB impulse radio. In this chapter, however, we will focus on the first scheme (spatial diversity), which can be easily applied to UWB transceivers systems, as undertaken in chapter 17 of this book.

6.2.2 Channel Distribution

For narrowband communications, the channel is represented by a complex random variable $h_i = r_i e^{\theta_i}$. The phase of h_i, denoted by θ_i, is often considered to be uniformly distributed. The magnitude of h_i, denoted by r_i, is related to the received signal power. It is important to understand the probability distribution of the received signal before undertaking the analysis of the diversity schemes.

For narrowband non-line-of-sight (NLOS) channels, the signal time spans are large compared to the multipath delays, and r_i is Rayleigh distributed [5] with a PDF given by

$$p_R(r_i) = \frac{r_i}{\sigma_i^2} \exp\left(-r_i^2/2\sigma_i^2\right), \quad r \geq 0 \tag{6.3}$$

With the conditions that the number of the overlapped components is large, the Rayleigh distribution of the narrowband NLOS channel coefficients is supported by the central limit theorem [6].

For the case of narrowband LOS communications, the received signal is the deterministic and relatively stronger direct, direct component overlapped with the Rayleigh distributed NLOS random signal that results from the scattering. In this case, r_i follows the Rician distribution and its PDF is given by

$$p_R(r_i) = \frac{r_i}{\sigma_i^2} \exp\left(-\frac{(r_i^2 + l_i^2)}{2\sigma_i^2}\right) I_0\left(\frac{r_i l_i}{\sigma_i^2}\right), \quad r \geq 0 \tag{6.4}$$

where $l_i^2 = (E\{r_i\})^2$ is the power of the line-of-sight (LOS) component and referred to as the noncentrality parameter.

If the number of the overlapped multipath components is not large enough, the central limit theory cannot be applied, and the received signal is neither Rayleigh nor Rician distributed. A more general distribution, the Nakagami-m distribution, can be applied in this case. The PDF of the Nakagami-m distribution is given as

$$p_R(r_i) = \frac{2}{\Gamma(m_i)} \left(\frac{m_i}{\Omega_i}\right)^{m_i} r_i^{2m_i - 1} \exp\left(-\frac{m_i r_i^2}{\Omega_i}\right) \tag{6.5}$$

where $\Omega_i^2 = E\{r_i^2\}$, $m_i = \frac{\Omega_i^2}{E\{(r_i^2 - \Omega_i)^2\}}$, $m_i \geq 1/2$.

For $m_i = 1/2$, r_i is one-sided Gaussian distributed, while for $m_i = 1$, r_i is Rayleigh distributed.

Several other distributions related to the above three are listed here and will be used later in this chapter. The power amplification coefficient of the Nakagami-m distributed channel, $\gamma_i = r_i^2$, is Gamma distributed with PDF

$$p_\gamma(\gamma_i) = \frac{m_i^{m_i} \gamma_i^{m_i - 1}}{\bar{\gamma}_i^{m_i} \Gamma(m_i)} \exp\left(-\frac{m_i \gamma_i}{\bar{\gamma}_i}\right), \quad \gamma_i \geq 0 \tag{6.6}$$

where $\bar{\gamma}_i = E\{\gamma_i\}$. When $m_i = 1$, Equation (6.6) becomes the exponential distribution for the case of Rayleigh-distributed channel magnitude, i.e.,

$$p_\gamma(\gamma_i) = \frac{1}{\bar{\gamma}_i} \exp\left(-\frac{\gamma_i}{\bar{\gamma}_i}\right), \quad \gamma_i \geq 0 \tag{6.7}$$

The sum of the power of the n identically Rayleigh-distributed received signals, $\gamma = \sum_{i=1}^{n} |r_i|^2$, follows the central chi-square distribution with n degrees of freedom. This is a special case of the Gamma distribution with m_i being an integer and $n = 2m_i$. The PDF of the central chi-square distribution is given as

$$p_\gamma(\gamma) = \frac{1}{\sigma^n 2^{n/2} \Gamma(n/2)} \gamma^{n/2 - 1} \exp\left(-\frac{\gamma}{2\sigma^2}\right), \quad \gamma \geq 0 \tag{6.8}$$

Antenna Diversity Techniques

The sum of the power of the n identically Rician distributed received signals is noncentral chi-square distributed.

6.3 Receive Diversity

It is common to realise a diversity-combining system on the receiver side. This is because the channel information is crucial for diversity combining, and it is easier to achieve it on the receiver side than on the transmitter side. We start with the performance analysis for a single branch without any diversity combining. And only afterwards we will show the benefits of different diversity combining schemes.

6.3.1 Single Branch without Diversity

As the purpose of this chapter is not to compare the diversity performance for different modulation schemes, only the simplest modulation scheme, binary phase shift keying (BPSK), will be analysed.[1] The transmitted signal is $a = b\sqrt{E_b}$, where $b = \pm 1$, and $E_b = |s_i|^2$ is the signal power for one bit. The signal propagates through a fading channel represented by h_i, and the instantaneous signal-to-noise ratio (SNR) of the received signal is given as

$$SNR_i = \frac{r_i^2 E_b}{N_0} = \gamma_i \rho \quad (6.9)$$

where $\rho = E_b/N_0$ is the raw SNR and is the ratio of the power of the transmitted signal to the power of the noise over the whole channel. The bit error rate (BER) for a given SNR level is

$$P_b(SNR_i) = Q(\sqrt{2 \cdot SNR_i}) \quad (6.10)$$

where $Q(x) = \int_x^\infty \frac{1}{\sqrt{2\pi}} \exp\left(-\frac{t^2}{2}\right) dt$ is the Marcum Q-function. The bit error rate averaged over all instantaneous fading conditions as a function of the average SNR is given as

$$\bar{P}_b(\rho) = E\{P_b(SNR_i)\} = \int_0^\infty Q(\sqrt{2\gamma_i \rho}) p_\gamma(\gamma_i) d\gamma_i \quad (6.11)$$

The above integration of the product of a Q-function and a PDF over γ_i can be transformed into the integration of the moment generation function (MGF) associated with γ_i over a variable, say θ, varying from 0 to $\pi/2$, by using an alternative expression for the Q-function [2]. This new integration can often lead to a closed form expression of $\bar{P}_b(\rho)$ for most distributions of γ_i. However, a closed-form expression may not help the analysis of the diversity performance. Instead, the expression for the Q-function $Q(x)$ can be replaced by its tight upper bound. The Chernoff bound is given by

$$Q(x) \leq \exp(-x^2/2) \quad (6.12)$$

---

[1] In fact, for UWB impulse radio, pulse position modulation (PPM) is a very popular modulation scheme. The analysis for PPM is very similar to that for BPSK by using a proper template on the receiver side.

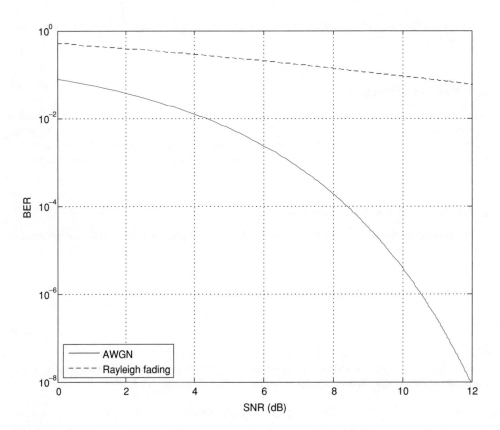

Figure 6.3 BER vs. raw SNR for BPSK signal passing through a Rayleigh fading channel with a single transmitting and a single receiving antenna. The averaged squared channel magnitude is normalised to 1, i.e., $E\{r_i^2\} = \bar{\gamma}_i = 1$

for x being sufficiently large; thus:

$$\bar{P}_b(\rho) \leq \int_0^\infty \exp\left(-\frac{2\gamma_i \rho}{2}\right) p_\gamma(\gamma_i) d\gamma_i = E\{\exp(-\gamma_i \rho)\} = \Psi_{\gamma_i}(\rho) \qquad (6.13)$$

where $\Psi_{\gamma_i}(\rho) = \int_0^\infty e^{-\gamma_i \rho} p_\gamma(\gamma_i) d\gamma_i$ is the MGF associated with γ_i.[2]

For a Rayleigh fading channel, $\Psi_{\gamma_i}(\rho) = (1 + \rho\bar{\gamma}_i)^{-1}$. Figure 6.3 shows the averaged BER vs. the SNR for the Rayleigh fading channel.

It can be shown that Rayleigh fading can severely degrade the BER performance compared to the AWGN channel where there is no fading. In the high SNR region, the BER curve can be approximated as:

$$\bar{P}_b(\rho) = \rho^{-1} \qquad (6.14)$$

---

[2] The MGFs for various distributions of γ_i can be looked up in Chapter 2 of [2]. It should be noted here that in the definition for the MGF in [2], the sign in front of γ_i is + rather than −.

Antenna Diversity Techniques

As shown by Equation (6.14), $\bar{P}_b(\rho)$ decreases at the order of -1 as the SNR, ρ, increases. There are two ways of improving the BER performance. The first is to increase the SNR for one bit, which is equivalent to left-shifting the BER curve when the abscissa is in log-scale. This can be done using precoding for the transmitted bits by, e.g., assigning more power to one bit, or using multiple antennas achieve an array gain for each bit. The second method is to increase the gradient. This can be achieved using diversity combining when multiple diversity branches are available. Mathematically, it is obvious that the second scheme is more effective than the first.

6.3.2 General Combining Schemes for Receive Diversity

In linear fading channels and for the single user case, maximal-ratio combining (MRC) is the most effective scheme for making full use of the signal power from all the branches and the easiest to be analysed mathematically. Equal gain combining (EGC) is simpler than MRC in terms of practical implementation and provides diversity performance comparable to MRC. Selection combining and switched diversity are relatively easier than MRC and EGC to realise and provide suboptimum performance. The analysis for these three combining schemes can be based on the general combining schemes discussed in this subsection.

If the antenna elements are sufficiently separated, as shown in Figure 6.4, they will experience independent fading. The received signal from different antennas, i.e. different diversity branches, can be weighted and summed together to form the decision variable. The more receive antennas we have, the lower probability that all the diversity branches experience fading at the same time, thus the more reliable the decision variable is.

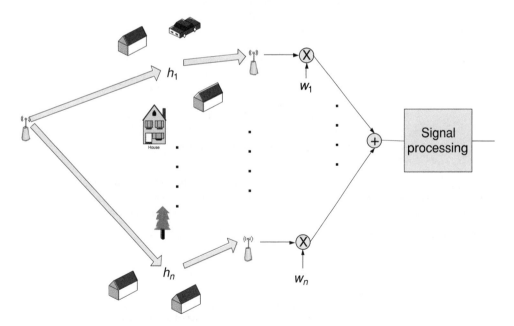

Figure 6.4 A multiple-antenna receiver

The spatial diversity branches for a multiple-antenna receiver are illustrated with the help of the following. The received signal components, experiencing independent fading channels, are weighted and summed before further signal processing for detection. The weighted and summed received signal variable can be written as

$$z = \sum_{i=1}^{N} w_i y_i = \sum_{i=1}^{N} w_i h_i s_i + \sum_{i=1}^{N} w_i n_i \tag{6.15}$$

The choice of the weight $\{w_i\}, i = 1, 2, \ldots n$, depends on the combining scheme chosen and is related to the channel information. It is noticed that for receiver diversity, the channel information is required only at the receiver.

The actual realisations of the weights are different for different combining schemes as discussed next.

6.3.3 Maximum Ratio Combining

For MRC, the weights are chosen as $w_i = h_i^*$, where * stands for complex conjugation. The combined signal is given as:

$$z = \sum_{i=1}^{N} |h_i|^2 s + \sum_{i=1}^{N} h_i^* n_i \tag{6.16}$$

Let us assume that the following conditions are satisfied [4]:

- The channels varies slowly compared to the transmitted signal and noise.[3]
- Different transmitted signals s_i and s_j are independent.
- The noise power is the same for different branches,[4] and is defined as $E\{|n_i|^2\} = N_0$.

By making use of the above assumptions, the instantaneous SNR for the combined signals can be given as

$$\rho_z = \frac{E\left\{\left(\sum_{i=1}^{N}|h_i|^2 s\right)\left(\sum_{i=1}^{N}|h_i|^2 s^*\right)\right\}}{E\left\{\left(\sum_{i=1}^{N}h_i^* n_i\right)\left(\sum_{i=1}^{N}h_i^* n_i\right)^*\right\}} = \frac{\left(\sum_{i=1}^{N}|h_i|^2\right)^2 E\{|s|^2\}}{\left(\sum_{i=1}^{N}|h_i|^2\right) E\{|n_i|^2\}} = \rho \sum_{i=1}^{N} \gamma_i \tag{6.17}$$

where γ_i is the power amplification coefficient for the i-th branch. It can be seen that the instantaneous SNR of the combined signal is the sum of the SNR from different branches. This makes MRC the most efficient way to make use of signal powers from the diversity branches and thus give the best performance. From Equation (6.17), the averaged received combined SNR can be given as $\bar{\rho}_z = \rho \sum_{i=1}^{N} \bar{\gamma}_i$. If we further

---

[3] This is especially true for the indoor environment.
[4] For branches with different noise power, the weight for MRC should be h_i/n_i. The general conclusion for the MRC would be the same for the cases of branches with the same noise power and with different noise power.

assume that the fading statistics for each fading branch are the same, the averaged combined SNR is

$$\bar{\rho}_z = N\bar{\gamma}\rho \tag{6.18}$$

The n diversity branches can be written in vector form as $\vec{H} = \begin{bmatrix} h_1 & \ldots & h_n \end{bmatrix}^T$. The variance matrix for $\vec{H}$ is $R = E\{\vec{H} \cdot \vec{H}^H\}$, and the SNR of the combined signals can be rewritten as $\rho_z = \rho \cdot tr\{R\} = \rho \cdot ||\vec{H}||_F^2$.[5] For Rayleigh fading channels, the averaged bit error rate associated with the combined signal is bounded as [3]:

$$\bar{P}_b(\rho) \leq \exp\left\{-\rho||\vec{H}||_F^2\right\} = \prod_{i=1}^{N_\lambda} \frac{1}{1+\rho\lambda_i} \tag{6.19}$$

where λ_i is the i-th eigenvalue of the variance matrix $\mathbf{R}$, and N_λ is the number of nonzero eigenvalues.

If the fading for the diversity branches is independent and identically distributed (IID), $R = I_n$ and $\lambda_i = 1, I = 1 \ldots N$. Thus Equation (6.19) can be written as:

$$\bar{P}_b(\rho) \leq (1+\rho)^{-N} \tag{6.20}$$

For the high SNR region, $\bar{P}_b(\rho) \leq \rho^{-N}$. It can be seen that when MRC is applied, by using n independent diversity branches, the gradient of $\bar{P}_b(\rho)$ can be increased from 1 in Equation (6.14) to the maximum of n in Equation (6.20). Here n is defined as the diversity order of the system, and is equal to the number of the available independent diversity branches.

The BER performance as a function of the number of diversity branches is shown in Figure 6.5.

It can be seen that the receiver diversity can even outperform the single branch AWGN channel. This can be simply explained as followed. From Equation (6.17), the averaged received SNR for the receiver diversity system is $E\{\rho_z\} = E\{\rho\}E\{\sum_{i=1}^{N}\gamma_i\} = E\{\rho\}N$, which is, on average, n times of the SNR of the received signal over an AWGN channel. The decaying order is ρ for the AWGN channel and n for the receiver diversity system. Before the decaying order dominates as ρ increases, it is possible that the receiver diversity applying MRC can outperform the single-branch AWGN channel with sufficient diversity branches.

The diversity performance for the correlated diversity branches can be analysed by studying the distribution of eigenvalues of the channel covariance matrix. Equation (6.19) gives the framework for the diversity analysis for both correlated and uncorrelated diversity branches.

Although MRC gives the best available performance for the single user diversity system, it is the most complex to realise because it requires the full, complex channel information.[6] When the full channel information is not available at the receiver side, suboptimum combining schemes may be considered.

---

[5] $||M||_F^2$ is the squared Frobenius norm of a $N \times M$ matrix $\mathbf{M}$, and is defined as $||\vec{M}||_F^2 = \sum_{n=1}^{N}\sum_{m=1}^{M}|M_{nm}|^2$.

[6] The delay for each multipath component is also needed for a wideband system.

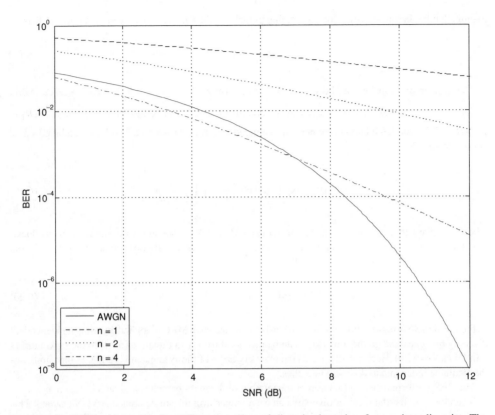

Figure 6.5 BER performance for different number of diversity branches for receiver diversity. The abscissa is the raw SNR, ρ

6.3.4 Equal Gain Combining

For EGC, the weights are set as $w_i = e^{-j\theta_i}$. It can be seen that the information of the channel magnitude is not needed and only phase information is required. According to Equation (6.15), the combined signal for EGC can be given as

$$z = \sum_{i=1}^{N} r_i s + \sum_{i=1}^{N} e^{-j\theta_i} n_i \qquad (6.21)$$

The instantaneous SNR of the combined signal is

$$\rho_z = \frac{E\left\{\left(\sum_{i=1}^{N} r_i\right)^2 s^2\right\}}{E\left\{\left(\sum_{i=1}^{N} e^{-j\theta_i} n_i\right)^2\right\}} = \frac{\left(\sum_{i=1}^{N} r_i\right)^2 E_b}{N \cdot N_0} = r^2 \rho / N \qquad (6.21)$$

where $r = \sum_{i=1}^{N} r_i$. The last equality comes from the reasonable assumption that the noise statistics for each diversity branch are the same.

Let us further assume that the fading statistics for each fading branch are the same. For the case of Rayleigh fading channel, the averaged combined SNR is [2]

$$\bar{\rho}_z = \bar{\gamma}\rho(1 + (N-1)\pi/4) \tag{6.22}$$

Comparing with Equation (6.18), it can be seen that EGC is less effective than MRC in collecting the signal power from the diversity branches.

The general averaged BER performance for BPSK can be given as

$$\bar{P}_b(\rho) = \int_0^\infty Q\left(\sqrt{2\rho r^2/n}\right) p_{r^2}(r^2) dr \leq \int_0^\infty e^{-\rho r^2/N} p_{r^2}(r^2) dr \tag{6.23}$$

This integration involves finding the PDF of the squared combined channel magnitude. Even for the optimum case of independent fading branches, it requires the convolution of the PDFs of the magnitudes of all the diversity branches and is hard to analyse, although closed-form solutions exist. In this chapter, the simulation results are given for the independent Rayleigh fading branches, shown in Figure 6.6.

6.3.5 Selection Combining and Switched Diversity

In selection combining, only the signal from the branch bearing the highest SNR is selected and fed to the decision block on the receiver side. Assuming equal noise power among all the branches, the branch with the largest channel response magnitude can be chosen. The weights in Figure 6.4 for selection combining are defined as

$$w_i = \begin{cases} 1 & \text{for largest } |h_i| \\ 0 & \text{otherwise.} \end{cases} \tag{6.24}$$

For the switched diversity, an SNR threshold is set. The w_i remains 1 for the branch with SNR greater than the threshold while w_i remains 0 for all the other branches. When the SNR of the selected branch falls below the threshold, the receiver is switched to one of the other branches with SNR greater than the threshold and stays there until the next switch occurs. If none of the branches has SNR greater than the threshold, switching does not occur.

Selection combining and switch diversity are less complex to realise than MRC and EGC, but their performance is far more complex to analyse. Readers interested in their exact output statistics and BER performance are referred to [2].

6.3.6 Fading Correlation

In the previous analysis for the diversity combining it is assumed that the diversity branches are independent from each other. This is the optimum case for the diversity system. Increasing the receive antenna separation is an effective means of lowering the correlation among the diversity branches. The available space on a mobile unit, however, may not permit large inter-antenna spacing, resulting in fading correlation between the branches and thus increasing the probability that the fading at different branches occurs at the same time. As a result, the diversity gain for the system is degraded.

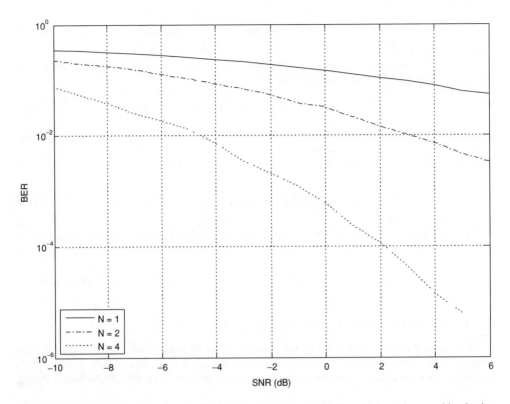

Figure 6.6 BER performance in a Rayleigh fading channel with N diversity branches combined using EGC

With MRC, correlated fading can reduce the rank of the channel covariance matrix from the full rank of N to N_λ to. According to Equation (6.20), the reduced diversity order can slow down the decaying rate of the BER curve as SNR increases in the high SNR region. Even in the low SNR region, partially or perfectly correlated branches can increase the difference between some of the eigenvalues in Equation (6.19). Although the sum of the eigenvalues can be kept the same, the BER performance is deteriorated.

6.4 Transmit Diversity

Receive diversity is not always achievable on the small mobile unit. Transmit diversity is sometimes more practical because there can be enough space for placing the multiple-antenna array at the base station. Signal precoding is required in transmit diversity to assist the signal combination process on the receiver side. The Alamouti scheme [7] is a simple precoding technique to realise the transmit diversity in the case that the channel is unknown to the transmitter; other schemes, such as space–time trellis coding, are also feasible. If, however, the channel is known to the transmitter, the transmit-MRC [8] scheme can be used to achieve diversity.

Antenna Diversity Techniques

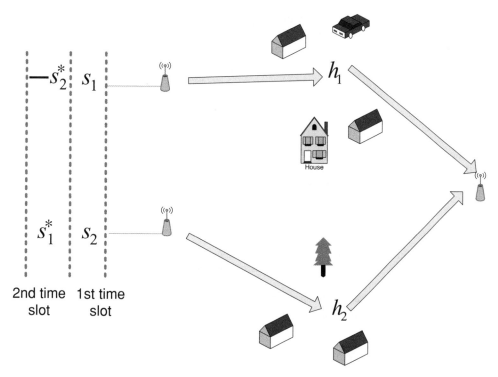

Figure 6.7 Precoding for the transmit diversity using the Alamouti scheme

6.4.1 Channel Unknown to the Transmitter

Let us assume that there are two transmit antennas and one receive antenna. Two consecutive symbols s_1 and s_2 are transmitted through the two antennas. Figure 6.7 shows the transmission strategy for the Alamouti scheme.

The two consecutive received signals are

$$y_1 = h_1 \cdot s_1/\sqrt{2} + h_2 \cdot s_2/\sqrt{2} + n_1 \tag{6.25}$$
$$y_2 = -h_1 \cdot s_2^*/\sqrt{2} + h_2 \cdot s_1^*/\sqrt{2} + n_2 \tag{6.26}$$

The coefficient of $\sqrt{1/2}$ comes from that fact that the total transmit power for all the transmit antennas is constrained as $E_b = E\{|s|^2\}$. When the transmitter does not have channel information, the optimum power allocation strategy is to distribute the power uniformly to all the transmit antennas.

Equations (6.25) and (6.26) can be combined together in a matrix form:

$$\vec{y} = \begin{bmatrix} y_1 \\ y_2^* \end{bmatrix} = \begin{bmatrix} h_1 & h_2 \\ h_2^* & -h_1^* \end{bmatrix} \begin{bmatrix} s_1/\sqrt{2} \\ s_2/\sqrt{2} \end{bmatrix} + \begin{bmatrix} n_1 \\ n_2^* \end{bmatrix} = H_{Alamouti} \cdot \vec{s}/\sqrt{2} + \vec{n} \tag{6.27}$$

At the receiver, the received signal over two time slots can be combined together to form the decision variable by multiplying vector $\vec{y}$ by $H^*_{Alamouti}$:

$$\vec{z} = \begin{bmatrix} z_1 \\ z_2 \end{bmatrix} = \begin{bmatrix} s_1/\sqrt{2} \cdot \sum_{i=1}^{2} |h_i|^2 \\ s_2/\sqrt{2} \cdot \sum_{i=1}^{2} |h_i|^2 \end{bmatrix} + \begin{bmatrix} \tilde{n}_1 \\ \tilde{n}_2 \end{bmatrix} \quad (6.28)$$

where $\tilde{n}_1 = h_1^* n_1 + h_2 n_2^*$, $\tilde{n}_2 = h_2^* n_1 - h_1 n_2^*$.

Comparing Equations (6.16) and (6.28), it can be easily seen that they are almost the same except for a 3 dB power degradation for each branch in the latter one. This is a penalty for the transmit diversity system with no channel information on the transmitter side.

6.4.2 Channel Known to the Transmitter

In order to make full use of channel information, the transmitted signal can be weighed before transmission. The purpose of preweighing the signal is to assign more power to those branches with less fading.

Assume that there are n transmit antennas and one receive antenna. The channel vector can be defined as $\vec{h} = [h_1 \ldots h_N]$. According to the transmit-MRC scheme, the weighting vector is

$$\vec{w} = \sqrt{N} \vec{h}^H / \sqrt{||\vec{h}||_F^2} \quad (6.29)$$

Thus the received combined signal is

$$y = \vec{h} \cdot \vec{w} s / \sqrt{N} + n \quad (6.30)$$

$$y = \sqrt{||\vec{h}||_F^2} \cdot s + n \quad (6.31)$$

with the instantaneous SNR the same as that in Equation (6.17), and no 3 dB penalty compared to Equation (6.27).

6.5 MIMO Diversity Systems

The MIMO system is a communication system with multiple-input and multiple-output antennas at both transmitter and receiver. It can be used to increase the capacity of the system, or the diversity order, or both at the same time. This section briefly introduces the use of MIMO to achieve the maximum diversity order.

Assume that there are two transmit antennas and two receive antennas. Any two of the set of propagation channels between one of the transmit antennas and one of the receive antennas are independent. The channel matrix for the MIMO system can be expressed as

$$H = \begin{bmatrix} h_{11} & h_{12} \\ h_{21} & h_{22} \end{bmatrix} \quad (6.32)$$

Antenna Diversity Techniques

Based on the previous discussion on receive and transmit diversity, it can be stated that this system comprises up to four independent diversity branches. Let us suppose that symbol s_i is sent and z_i is the combiner output at the receiver. Similar to Equations (6.16), (6.28) and (6.31) above, we can have a diversity order of 4 if $z_i = K \cdot \sum_{n=1}^{2} \sum_{m=1}^{2} |h_{nm}|^2 \cdot s_i + n_i$. Now let us explore how to transfer the channel matrix H into $||H||_F^2 = \sum_{n=1}^{2} \sum_{m=1}^{2} |h_{nm}|^2$ via precoding at the transmitter and combining at the receiver.

When the channel information is unknown to the transmitter, the transmission energy is evenly attributed to the two transmit antennas. According to the Alamouti scheme, symbol vector $[s_1/\sqrt{2} \ s_2/\sqrt{2}]^T$ is transmitted in the first timeslot, and $[-s_2^*/\sqrt{2} \ s_1^*/\sqrt{2}]^T$ in the second timeslot. The received vector is $\vec{y} = [y_1 \ y_2]^T$ in the first timeslot, and $\vec{y}' = [y_1' \ y_2']^T$ in the second timeslot. The total received signals over the two timeslots can be written as

$$\bar{y} = \begin{bmatrix} \vec{y}^T \\ \vec{y}'^T \end{bmatrix} = \begin{bmatrix} h_{11} & h_{12} \\ h_{21} & h_{22} \\ h_{12}^* & -h_{11}^* \\ h_{22}^* & -h_{21}^* \end{bmatrix} \cdot \begin{bmatrix} s_1/\sqrt{2} \\ s_2/\sqrt{2} \end{bmatrix} + \begin{bmatrix} n_1 \\ n_2 \\ n_3^* \\ n_4^* \end{bmatrix} = \sqrt{\frac{1}{2}} \cdot H_{Alamouti} \cdot \vec{s} + \vec{n} \qquad (6.33)$$

On the receiver side, the four elements of the vector $\bar{y}$ are weighed and combined together using the channel information to create the two decision variables, i.e.,

$$\vec{z} = \begin{bmatrix} z_1 \\ z_2 \end{bmatrix} = H_{Alamouti}^H \cdot \bar{y} = \begin{bmatrix} ||H||_F^2 \cdot s_1/\sqrt{2} \\ ||H||_F^2 \cdot s_2/\sqrt{2} \end{bmatrix} + \vec{n}' \qquad (6.34)$$

where $\vec{n}' = H_{Alamouti} \cdot \vec{n}$.

It is obvious that the decision variable z_i contains the coefficient $||H||_F^2$. Similar to Equation (6.20), the BER performance of this 2×2 MIMO system is bounded by

$$\bar{P}_b(\rho) \leq (1 + \rho/2)^{-4} \qquad (6.35)$$

The 3 dB penalty of the SNR is due to the unavailability of the channel information on the transmitter side.

For the MIMO system, the instantaneous complete channel information is hard to achieve at the transmitter. But theoretically the channel information can be made full use of at the transmitter side if it is available by employing dominant eigenmode transmission. The interested reader is referred to [3].

In summary, this chapter introduced some basic combining schemes for multi-antenna communication systems to combat multipath fading. It hence paves the way for further treatment in later chapters on how multi-antenna systems are advantageously deployed in UWB systems.

References

[1] S.N. Diggavi, N. Al-Dhahir, A. Stamoulis and A.R. Calderbank, Great expectations: the value of spatial diversity in wireless networks, *Proc. IEEE*, **92**(2), 2004.
[2] M.K. Simon and M.-S. Alouini, *Digital Communication over Fading Channels: A Unified Approach to Performance Analysis*. John Wiley & Sons, Ltd, 2000.
[3] A. Paulraj, R. Nabar and D. Gore, *Introduction to Space–Time Wireless Communications*. Cambridge University Press, 2003.

[4] D.G. Brennan, Linear diversity combining techniques, *Proc. IEEE*, **91**(2), 2003.
[5] J.G. Proakis, *Digital Communications*. 4th ed., McGraw-Hill, 2001.
[6] H. Stark and J.W. Woods, *Probability and Random Processes with Applications to Signal Processing*. 3rd edition, Prentice-Hall, 2002.
[7] S. Alamouti, A simple transmit diversity techniques for wireless communications, *IEEE J. Select. Areas Commun.*, **16**(8), 1998.
[8] T. Lo, Maximal ratio transmission, *IEEE Trans. Commun.*, **47**(10), 1999.

Part II

Antennas for UWB Communications

Introduction to Part II

Ernest E. Okon

Ultra-wideband communications typically refers to the use of very narrow pulses, of duration in the nanosecond or sub-nanosecond range for the transmission of data. This enables the transmission of high data rates in excess of 100 Mb/s. UWB systems will coexist with other traditional communication systems in the same frequency band by using lower power levels. The Federal Communications Commission (FCC) in the USA has designated the 3.1 to 10.6 GHz band with an effective isotropic radiated power (EIRP) of below –40 dBm/kHz for UWB communications. UWB communications may be effected using either of the emerging formats of direct sequence (DS-UWB) or multiband orthogonal frequency division multiplexing (MB-OFDM).

The use of UWB systems necessitates efficient antennas to provide acceptable bandwidth requirements, and radiation pattern characteristics throughout the designated UWB spectrum. It is generally accepted that for antennas to be classified as ultra-wideband, the requirement will be to satisfy minimum fractional bandwidths of at least 20 % or 500 MHz or more. This section examines a range of categories within the context of antennas applicable in UWB communication systems. These comprise: the theory of UWB antenna elements; antenna elements for impulse radio; planar dipole-like antennas for consumer products; UWB antenna elements for consumer electronic applications; UWB arrays and UWB beamforming.

In formulating the theory of UWB antenna elements, previous advances in the design of wideband antennas are highlighted in Chapter 7. This includes previous work on frequency-dependent antennas and self-complimentary antennas with constant impedance over the frequency bandwidth. It is observed that in addition to satisfying the impedance bandwidth, the radiation pattern must be constant over the bandwidth to satisfy the requirements of a UWB antenna. The planar antennas such as the disk and slot monopoles have been chosen as sample antenna elements to explain the underlying theory of UWB antennas. Measured and simulated results for various configurations of coplanar and microstrip-fed disc monopoles are presented. Electric and magnetic field radiation patterns, current distributions and input impedance characteristics are depicted for these antenna types and optimal design parameters are highlighted. Frequency and time domain results are also presented. It is observed that the overlapping

---

Ultra-wideband Antennas and Propagation for Communications, Radar and Imaging Edited by B. Allen, M. Dohler, E. E. Okon, W. Q. Malik, A. K. Brown and D. J. Edwards
© 2007 John Wiley & Sons, Ltd

of multiple resonances and the support of travelling wave effects are important aspects for the design of compact planar UWB antennas.

Impulse radio involves the transmission and reception of very short duration pulses without a carrier and is an important aspect for UWB communications. Antenna elements applicable in impulse radio systems are discussed in Chapter 8. These antennas have been classified into two- and three-dimensional antennas as well as directional and omnidirectional antennas. Design considerations are highlighted from a systems perspective. Transfer functions for UWB systems are derived as well as expressions for the system fidelity. Design examples are subsequently given of directional and omnidirectional antennas for base station and mobile applications, respectively. An omnidirectional roll monopole antenna and a directional antipodal Vivaldi antenna have been designed and measured results are presented in the frequency and time domains. It is observed that in impulse radio systems, the phase response is a unique consideration of the system as it determines the nature of the transmitted and received pulse. The system transfer function provides a good way of evaluating the system performance through the use of Fourier transforms in the time and frequency domains.

Planar dipole-like antennas for consumer products are the subject of Chapter 9. Antenna elements for UWB applications need to be compact and cost-effective. Thus, planar structures satisfy these requirements and have relative ease of manufacture compared to other conventional antenna types such as dipoles and horns. Computer simulation methods and measurement techniques are reviewed. Subsequently, the classical bicone antenna has been adopted for introducing the general properties of broadband dipole antennas. Measured and simulated results are presented for a range of planar dipole-like antennas such as the bowtie, circular, elliptical, diamond dipole and Vivaldi designs. The results comprise phase, impedance and radiation patterns of the antennas. Practical designs that can be made from printed circuit boards are subsequently discussed.

In Chapter 10, UWB antennas elements for consumer electronics applications are discussed. A method to extract the space-time characteristics of UWB antennas using a finite-difference time-domain method is given. Subsequently, the design for a planar monopole is optimised to provide broadband matching. The antenna model is then integrated into the model of a DVD player or a mobile device such as a PDA or mobile phone. The impact of this integration is then evaluated. The system transfer function is subsequently extracted and used for indoor propagation modelling in a home environment. The results indicate that the antenna integration into a DVD chassis results in a directive radiation pattern with a frequency dependency. A ray-tracing propagation model simulation indicates significant variation in the radiated power within the room. When the received power is subsequently averaged over a larger bandwidth, the results show a more smooth variation of the power due to the frequency-dependent effects of the radiation pattern and propagation environment. The complete system is subsequently characterised in terms of a transfer function and impulse response for different link scenarios.

Chapter 11 discusses UWB arrays for communication systems. Antenna arrays are often used in communication systems to increase sensitivity and for range extension. The relatively low power levels in UWB systems will enable coexistence with other traditional communication systems. However, for applications requiring longer ranges, UWB arrays will be applicable without the need to increase power levels. Emphasis is placed on compact and cost-effective designs in realising UWB arrays. Thus, basic design parameters for pattern synthesis and array scanning are presented. Antenna elements applicable to UWB array design are then highlighted. Wideband array design parameters are also given. Modelling techniques applicable for UWB array design are provided. Various examples of passive and active wideband array configurations are noted. Finally, design considerations for UWB arrays are highlighted.

UWB beamforming is the subject of Chapter 12. A major challenge of the UWB system is the equalisation of the channel impulse response with a corresponding delay spread in indoor environments among other factors such as range limitations due to low transmit power. The use of multiple antennas in an array format is a promising approach to combat these problems. Beamforming is required to combine these multiple elements coherently. Analogue beamforming requires true-time delays. The chapter reviews the theory of array signal processing for beamforming and presents design formulations for the realisation of practical beamformers applicable in UWB systems. The tapped delay line and the concept of UWB mono-pulse arrays are reviewed and the characteristics of both methods are illustrated by a number of examples.

7

Theory of UWB Antenna Elements

Xiaodong Chen

7.1 Introduction

As is the case in the other electromagnetic (EM) subjects, there exist several schools of theories to explain the operation of UWB antennas being developed over the last several decades.

The most popular one was introduced by Victor Rumsey in the 1950s to explain a family of so-called frequency-independent antennas, this is often was referred to as 'Rumsey's principle' [1]. Rumsey's principle suggests that the impedance and pattern properties of an antenna will be frequency independent if the antenna shape is specified only in terms of angles. To satisfy the equal-angle requirement, the antenna configuration needs to be infinite in principle, but is usually truncated in size in practice. This requirement makes frequency-independent antennas quite large in terms of wavelength. Rumsey's principle has been verified in spiral antennas, conic spiral antennas and some log periodic antennas.

An alternative theory was introduced by Yasuto Mushiake in the 1940s to account for so-called self-complementary antennas, which were invented by Wilhelm Runge in the 1930s [2]. Mushiake discovered that the product of input impedances of a planar electric current antenna (plate) and its corresponding 'magnetic current' antenna (slot) was a real constant. Therefore, an antenna built in a complementary structure of electric and magnetic currents exhibits a real constant impedance. This theory, referred as 'Mushiake's relation', has led to the development of a large family of self-complementary antennas with constant input impedance [3]. Mushiake's relation has relaxed the condition for achieving ultra-wide impedance bandwidth. However, it doesn't guarantee constant radiation patterns over the operation bandwidth. Incidentally, an infinitely long bicone antenna was also demonstrated to exhibit a constant impedance a long time ago, although it is not a self-complementary structure.

Lately, many researchers have started to look into the physical foundation of UWB antennas from a different angle. Based on a type of UWB antenna – spiral-mode microstrip antenna – Johnson Wang has suggested that either frequency-independent or self-complementary antennas are travelling wave antennas in nature [4]. Apparently, this is a very generalised hypothesis, which needs further justification in other types of UWB antennas.

Ultra-wideband Antennas and Propagation for Communications, Radar and Imaging Edited by B. Allen, M. Dohler, E. E. Okon, W. Q. Malik, A. K. Brown and D. J. Edwards
© 2007 John Wiley & Sons, Ltd

With the development of the latest UWB communication systems, especially DS-UWB systems, there has been a surge of research interest into small UWB antennas. Antenna engineers are now facing new challenges in providing antennas for commercial UWB terminals. First, the antennas have to be small enough to be compatible to the UWB unit. Secondly, the omnidirectional radiation patterns are often required for UWB terminal antennas. Finally, a good time-domain characteristic, i.e. a good impulse response with minimal distortion is also required for transmitting and receiving antennas. However, conventional frequency-independent or constant impedance antennas in the geometry of either log periodic, or bicone and/or spiral fail to satisfy most of these three requirements [5].

Fortunately, two families of compact antennas have emerged to provide solutions. The first family is originated from biconal antennas, but in a compact planar configuration, such as bow tie, diamond, circular and elliptical disc dipoles. They have been demonstrated to provide UWB characteristics and also satisfy other requirements imposed by commercial UWB systems. This family of planar dipoles has already been widely used in commercial UWB systems and are well documented in [6]. The second family is due to the further development of broadband monopole antennas, in which planar elements, such as circular, square, elliptical, pentagonal and hexagonal discs appear. They have also been explored for application in UWB terminals [7]–[12]. Lately, the planar types of monopoles have also been demonstrated to provide UWB impedance bandwidth with satisfactory radiation patterns [13]–[17]. Comparing with a planar dipole, a planar monopole does not require a balun feed and can be made even compact in size. The author's group has further extended this electric type of monopole to a magnetic type and have successfully developed a range of UWB slot antennas [18]–[21].

It is interesting to note that these planar UWB dipole/monopole antennas operating in resonances can not be readily explained using the existing UWB antenna theories. There are some attempts to explain this mystery. Hans Schantz has suggested the dipole mode and high-order modes to account for the operation of the planar dipoles being studied by him [6]. However, how exactly a compact UWB dipole or monopole operates across the entire bandwidth, remains an open question. Why does this resonating type of antennas retain a seemingly omnidirectional pattern (less than 10 dB gain variation)? And how well do they behave in the time domain?

In this chapter, we will try to answer these questions and provide further insight into the operation of this type of UWB antennas through a study of a planar disc monopole. Also, a range of UWB planar monopole antennas will be investigated to illustrate different features in their operations. At the same time, we will derive some quantitative guidelines for designing this type of UWB antennas.

7.2 Mechanism of UWB Monopole Antennas

In order to illustrate the mechanism of planar UWB antennas, a coplanar waveguide (CPW) fed circular disc monopole is chosen as the example for analysis because it shows basic features of this type of antenna operation across the whole bandwidth.

7.2.1 Basic Features of a CPW-Fed Disc Monopole

The CPW-fed disc monopole antenna studied in this section has a single-layer metallic structure, as shown in Figure 7.1. A circular disc monopole with a radius of r and a 50 Ω-CPW are printed on the same side of a dielectric substrate. W_f is the width of the metal strip and g is the gap of distance between the strip and the coplanar ground plane. W and $L = 10$ mm denote the width and the length of the ground plane, respectively, h is the feed gap between the disc and the ground plane. In this study, a dielectric substrate with a thickness of $H = 1.6$ mm and a relative permittivity of $\varepsilon_r = 3$ is chosen, so W_f and g are fixed at 4 mm and 0.33 mm, respectively, in order to achieve 50 Ω impedance.

Theory of UWB Antenna Elements

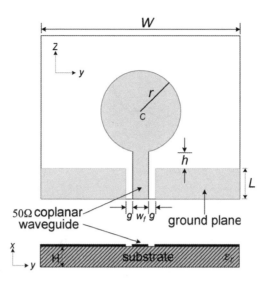

Figure 7.1 The geometry of the CPW-fed circular disc monopole

The simulations were performed using the CST Microwave Studio^{TM} package, which utilises the finite integration technique for electromagnetic computation [22]. Figure 7.2 illustrates the simulated return loss curve of an optimal design of the antenna for covering the Federal Communications Commission (FCC) defined UWB bandwidth. Actually, the simulated −10 dB bandwidth spans an extremely wide frequency range from 2.64 GHz to more than 15 GHz. This UWB characteristic of the proposed CPW-fed circular disc monopole antenna is confirmed in the measurements described in a later section. The corresponding input impedance and Smith chart curves are plotted in Figures 7.3 and 7.4.

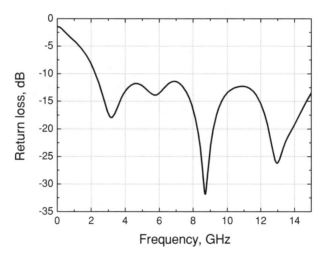

Figure 7.2 Simulated and measured return loss curve of the CPW-fed disc monopole with $r = 12.5$ mm, $W = 47$ mm and $h = 0.3$ mm

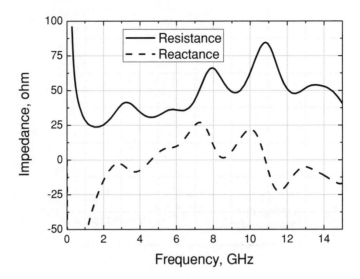

Figure 7.3 Simulated impedance curve of the CPW-fed disc monopole with $r = 12.5$ mm, $W = 47$ mm and $h = 0.3$ mm

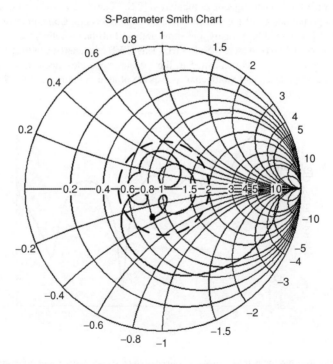

Figure 7.4 Simulated Smith chart of the CPW-fed disc monopole with $r = 12.5$ mm, $W = 47$ mm and $h = 0.3$ mm

It is difficult to identify the resonance modes of the antenna on the impedance or Smith charts in a traditional way, i.e. reactance being equal to zero. However, the return loss curve indicates good impedance matching at some frequencies (dips on the return loss curve), which are regarded as the resonances of the antenna here.

As shown in Figure 7.2, the first resonance occurs at around 3.0 GHz, the second resonance at 5.6 GHz, the third one at 8.6 GHz and the fourth one at 12.8 GHz. It is evident that the overlapping of these resonance modes which are closely distributed across the spectrum results in an ultra-wide −10 dB bandwidth. It is noticed on the Smith chart that the input impedance loops around the impedance matching point within the voltage standing wave ratio (VSWR) = 2 circle, but does not settle down to a real impedance point with the increase of frequency.

The return loss or input impedance can only describe the behaviour of an antenna as a lumped load at the end of feeding line. The detailed EM behaviour of the antenna can only be revealed by examining the field/current distributions or radiation patterns. The typical current distributions on the antenna close to the resonance frequencies are plotted in Figure 7.5.

Figure 7.5(a) shows the current pattern near the first resonance at 3.0 GHz. The current pattern near the second resonance at 5.6 GHz is given in Figure 7.5(b), indicating approximately a second order

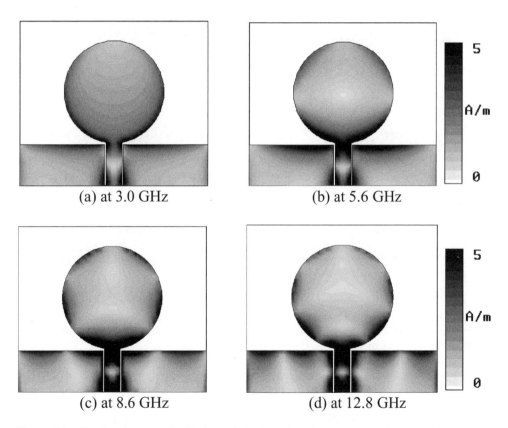

Figure 7.5 Simulated current distributions of the CPW-fed disc monopole with $r = 12.5$ mm, $W = 47$ mm and $h = 0.3$ mm

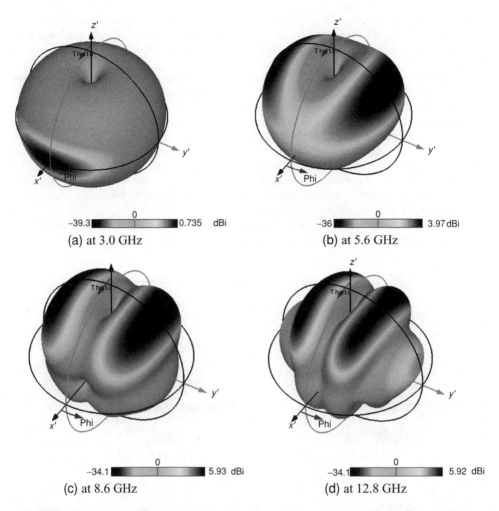

Figure 7.6 Simulated 3D radiation patterns with $r = 12.5$ mm, $W = 47$ mm and $h = 0.3$ mm

harmonic. Figures 7.5(c) and (d) illustrate two more complicated current patterns at 8.6 GHz and 12.8 GHz, corresponding to the third and fourth order harmonics, respectively. These current distributions support that the UWB characteristic of the antenna is attributed to the overlapping of a sequence of closely spaced resonance modes.

The simulated 3D radiation patterns close to these resonances are plotted in Figure 7.6. The radiation pattern looks like a doughnut, similar to a dipole pattern, at the first resonant frequency, as shown in Figure 7.6(a). At the second harmonics, the pattern looks like a slightly pinched donut with the gain increase around $\theta = 45°$ in Figure 7.6(b). When at the third and fourth harmonics, the patterns are squashed in x-direction and humps form in the up-right directions (gain increasing), as shown in Figures 7.6(c)–7.6(d), respectively. It is also noticed that the patterns on the H-plane are almost omnidirectional at lower resonances (first and second harmonics) and become distorted at the higher harmonics.

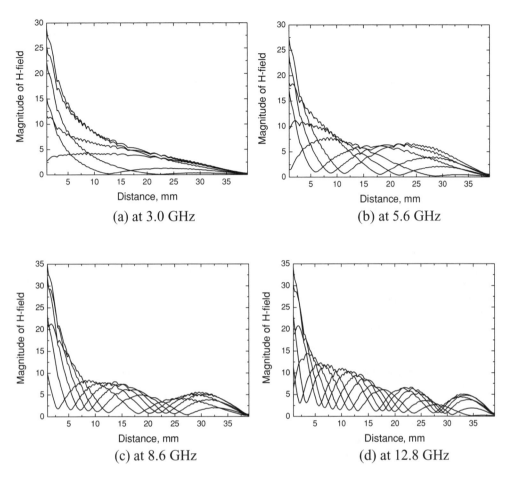

Figure 7.7 Magnetic field distributions along the edge of the half-disc L ($L = 0$–39 mm: bottom to top) at different phases in each resonance

The transition of the radiation patterns from a simple doughnut pattern at the first resonance to the complicated patterns at higher harmonics indicates that this antenna must have gone through some major changes in its behaviour. In order to gain further insight into the antenna operation, the animation of current variation at difference resonances was generated and observed in CST-Microwave Studio. The snapshots of the magnetic field distributions corresponding to the currents along the half-disc edge L ($L = 0$–39 mm: bottom to top) at different phases at each resonance are plotted in Figure 7.7(a)–(d), respectively.

Figure 7.7(a) shows that at the first resonance the current is oscillating and has a pure standing wave pattern along most part of the disc edge. So the disc behaves like an oscillating monopole. But the variation of the currents becomes more complicated at higher resonance harmonics. Figure 7.7(b) indicates a complex current variation patterns at the second harmonic. The current is travelling along the lower disc edge, but oscillating at the top edge. There is a broad current envelope peak formed around

$L = 25$ mm, which corresponds well to the gain increase in the radiation pattern, Figure 7.6(b). Hence, the CPW-fed disc monopole operates in a hybrid mode of standing wave and travelling wave at higher resonance harmonics. In Figure 7.7(c), at the third harmonic, the feature of travelling wave seems more prominent at the lower edge of the disc, while standing wave retains on the top edge with an envelope peak, which again corresponds well to a hump (gain peak) on the radiation pattern in Figure 7.6(c). At the fourth harmonic in Figure 7.7(d), standing wave has two envelope peaks on the top half of the disc edge while travelling wave dominates the lower half of the disc, which is also reflected in the radiation pattern, Figure 7.6(d). So the feature of standing and travelling waves becomes more distinctive at even higher harmonics.

7.2.2 Design Analysis

Having gained some insights into the operation of the antenna, we are now ready to investigate the effects of some design parameters and derive some design rules.

It is noted that the performance of this CPW-fed disc monopole is quite sensitive to the feed gap. Figure 7.8 depicts the simulated return loss curves with different feed gaps ($h = 0.3, 0.7, 1$ and 1.5 mm) when W is fixed at 47 mm and r at 12.5 mm, respectively. It can be seen that the return loss curves have similar shape for the four different feed gaps, but the -10 dB bandwidth of the antenna varies significantly with the change of h. It is noticed that when h becomes bigger the -10 dB bandwidth is getting narrower due to the impedance matching of the antenna getting worse. Looking across the whole spectrum, it seems that a bigger gap does not affect the first resonance very much, but has a much larger impact on the high harmonics. This suggests that the feed gap affects more the travelling wave operation of the antenna. The optimal feed gap is found to be at $h = 0.3$ mm, which is close to the CPW line gap. It makes perfect sense that the optimal feed gap should have a smooth transition to the CPW fed line.

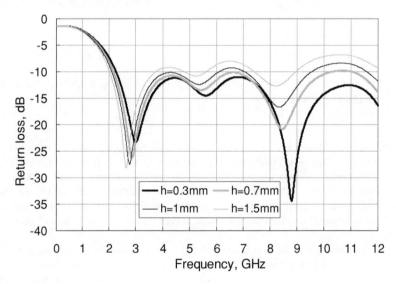

Figure 7.8 Simulated return loss curves for different feed gaps with $W = 47$ mm and $r = 12.5$ mm

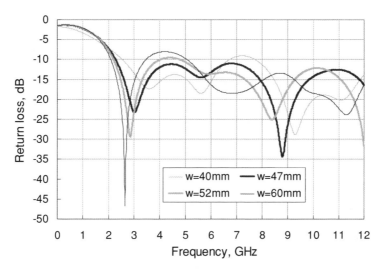

Figure 7.9 Simulated return loss curves for different widths of the ground plane with $h = 0.3$ mm and $r = 12.5$ mm

The second design parameter influencing the UWB characteristic of the antenna is the width of the ground plane. The simulated return loss curves with $r = 12.5$ mm and optimal feed gap h of 0.3 mm for different widths W are presented in Figure 7.9. It can be seen that the variation of the ground plane width shifts all the resonance modes across the spectrum. It is interesting to notice that the -10 dB bandwidth is reduced when the width of the ground is either too wide or too narrow. The optimal width of the ground plane is found to be at $W = 47$ mm. Again, this phenomenon can be explained when the ground plane is treated as a part of the antenna. When the ground plane width is either reduced or increased from its optimal size, so does the current flow on the top edge of the ground plane. This corresponds to a decrease or increase of the inductance of the antenna if it is treated as a resonating circuit, which causes the first resonance mode either to be upshifted or downshifted in the spectrum. Also, this change of inductance causes the frequencies of the higher harmonics to be unevenly shifted. Therefore, the change of the ground plane width makes some resonances not closely spaced across the spectrum and reduces the overlapping between them. Thus, the impedance matching becomes worse (>-10 dB) in these frequency ranges.

It is also noticed that the performance of the antenna is almost independent of the length L of the ground plane. This is understandable by inspecting current distributions in Figure 7.5 that the RF current is mostly distributed on the top edge of ground plane.

It has been established that the relative position and width of the ground plane determine critically the impedance bandwidth of the antenna. Also, there is an interesting phenomenon in Figures 7.7 and 7.9 that the first resonance always occurs at around 3 GHz for different feed gaps and some width of the ground plane when the disc radius is fixed at 12.5 mm. In fact, the quarter-wavelength at the first resonant frequency (25 mm) just equals the diameter of the disc. This suggests that the first resonance occurs when the disc behaves like a quarter-wave monopole.

Here, the CPW-fed circular disc monopoles with different diameters are examined in the simulation to derive some design rules. Figure 7.10 shows the simulated return loss curves for different dimensions

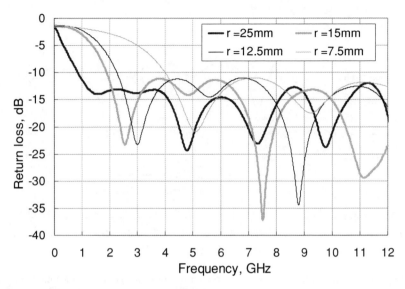

Figure 7.10 Simulated return loss curves for different dimensions of the circular disc in the optimal designs

of the circular disc with their respective optimal designs, which are given in Table 7.1. It can be seen that the ultra-wide impedance bandwidth can be obtained in these designs. The relationship between the disc diameters and the first resonances is also listed in Table 7.1.

Table 7.1 demonstrates that the first resonant frequency is determined by the diameter of the disc, which approximately corresponds to the quarter-wavelength at this frequency. Also, it shows that the optimal width of the ground plane is just less than two times the diameter of the disc, ranging from 1.80 to 1.88. The optimal feed gap h is around 0.3 mm, which is close to the CPW line gap (0.33 mm), with slight variations for the big and small discs. Table 7.1 is a good summary of the design rules for achieving the ultra-wide impedance bandwidth in a CPW-fed disc monopole.

7.2.3 Operating Principle of UWB Monopole Antennas

Now let us look back at what we have learned about the operation of this CPW-fed disc monopole. It is demonstrated that the overlapping of the closely distributed resonance modes in this antenna is

Table 7.1 Optimal design parameters of the CPW-fed disc monopole and relationship between the diameter and the first resonance

Diameter $2r$ (mm)	First resonance f(GHz)	Wavelength λ at f (mm)	$2r/\lambda$	Optimal W (mm)	$W/2r$	Optimal h (mm)
50	1.52	197.4	0.25	90	1.80	0.5
30	2.57	116.7	0.26	56	1.87	0.3
25	3.01	99.7	0.25	47	1.88	0.3
15	5.09	58.9	0.25	28	1.87	0.1

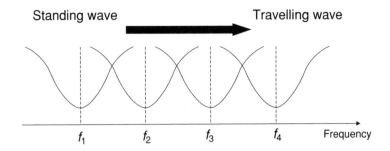

Figure 7.11 Schematic of UWB antenna operation principle

responsible for an ultra-wide −10 dB bandwidth. At the low frequency end (the first resonance) when the wavelength is bigger than the antenna dimension, the EM wave can easily 'couple' into the antenna structure so it operates in an oscillating mode, i.e. a standing wave. With the increase of the frequency, the antenna starts to operate in a hybrid mode of standing and travelling waves. At the high frequency end, the travelling wave becomes more critical to the antenna operation since the EM wave needs to travel down to the antenna structure which is big in terms of the wavelength. For the CPW-fed antenna, the slots formed by the lower edge of the disc and the ground plane with a proper dimension can support the travelling wave very well. So an optimal designed CPW-fed circular disc monopole can exhibit an extremely wide −10 dB bandwidth. The principle of the antenna operation across the whole spectrum is illustrated in Figure 7.11.

Though the operating principle of UWB antenna is derived from a CPW-fed circular disc monopole, it is found in our other studies that it is generally valid to account for other similar types of UWB antennas with some variations. Also, the analyses have been mainly conducted in computer simulation so far. It is necessary to verify this theory with experiments. All these issues will be addressed in the following sections.

7.3 Planar UWB Monopole Antennas

As mentioned in the introductory section of this chapter, there are several types of planar disc monopoles which exhibit ultra-wide impedance bandwidth. Here, three categories of planar disc monopoles will be investigated both numerically and experimentally.

7.3.1 CPW-Fed Circular Disc Monopole

This type of antenna has been extensively studied in simulation in the previous section. A prototype of this antenna in the optimal design, as shown in Figure 7.12, was fabricated and tested in the Antenna Measurement Laboratory at Queen Mary, University of London (QMUL). The return loss was measured in an anechoic chamber by using an HP8720ES network analyser and the measured result was compared with the simulated one. Figure 7.13 illustrates the simulated and the measured return loss curves. Here, the 50 Ω SMA feeding port is taken into account in the simulation and it is noticed that this SMA port mainly affects the third and fourth resonances by shifting their resonant frequencies.

The measured return loss curve agrees very well with the simulated one in most ranges of the frequency band except at around 8 GHz where a resonance occurs in the simulation, but it is not apparent in

Figure 7.12 Photo of the CPW-fed circular disc monopole in the optimal design: $r = 12.5$ mm, $h = 0.3$ mm and $W = 47$ mm

the measurement. For the other three resonances (at around 3.0 GHz, 5.6 GHz and 11.1 GHz), the measured ones are very close to those obtained in the simulation with differences of less than 5 %. Generally speaking, the -10 dB bandwidth spans an extremely wide frequency range in both simulation and measurement. The simulated bandwidth ranges from 2.64 GHz to more than 15 GHz. This UWB characteristic of the CPW-fed circular disc monopole antenna is confirmed in the measurement, with only a slight shift of the lower frequency to 2.73 GHz.

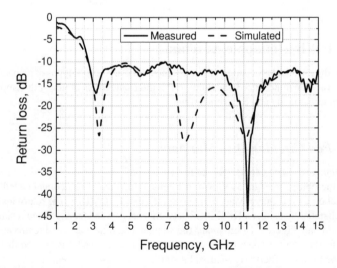

Figure 7.13 Simulated and measured return loss curves of the CPW-fed disc monopole in the optimal design

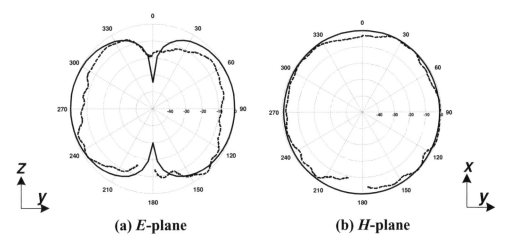

Figure 7.14 Simulated (solid line) and measured (dashed line) radiation patterns of the CPW-fed disc monopole at 3 GHz

The radiation patterns of the antenna at the frequencies close to the resonances have been measured inside an anechoic chamber. The measured and simulated radiation patterns at 3 GHz, 5.6 GHz, 7.8 GHz and 11.0 GHz are plotted in Figures 7.14–7.17, respectively.

As shown in Figures 7.14–7.17, the measured radiation patterns are generally very close to those obtained in the simulation. This has verified the simulated radiation patterns presented in Figure 7.6. The E-plane patterns have large back lobes and look like a doughnut or a slightly pinched doughnut at lower frequencies. With the increase of the frequency, the back lobes become smaller, splitting into many minor ones, while the front lobes start to form humps and notches.

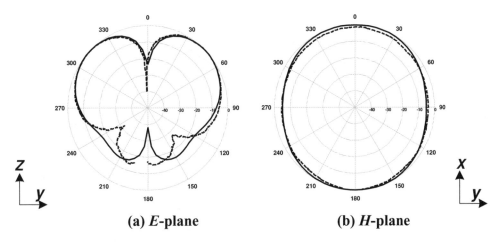

Figure 7.15 Simulated (solid line) and measured (dashed line) radiation patterns of the CPW-fed disc monopole at 5.6 GHz

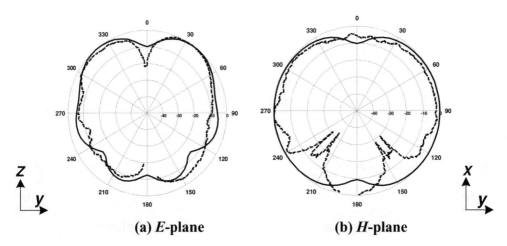

Figure 7.16 Simulated (solid line) and measured (dashed line) radiation patterns of the CPW-fed disc monopole at 7.8 GHz

It is noticed that the H-plane pattern is omnidirectional at lower frequencies (3 GHz and 5.6 GHz) and still close to omnidirectional at the high end of the bandwidth (11 GHz). However, the measured H-plane pattern does not agree well with the simulated one at 7.8 GHz. This discrepancy seems to be due to an enhanced perturbing effect on the antenna performance caused by the feeding structure and cable at this frequency. Although the overall radiation pattern of the antenna has gone through a substantial transformation, the simulated H-plane pattern retains a good omnidirectionality (less than 10 dB gain variation) over the entire bandwidth. This is due to the fact that the dimension of the CPW-fed circular disc monopole is within the wavelengths of the first few resonance harmonics (1:4).

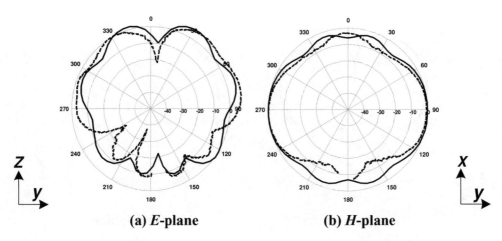

Figure 7.17 Simulated (solid line) and measured (dashed line) radiation patterns of the CPW-fed disc monopole at 11.0 GHz

7.3.2 Microstrip Line Fed Circular Disc Monopole

A circular disc monopole can also be fed by a microstrip line, as illustrated in Figure 7.18. The circular disc monopole with a radius of r and a 50 Ω microstrip feed line are printed on the same side of the FR4 substrate (the substrate has a thickness of 1.5 mm and a relative permittivity of 4.7). L and W denote the length and the width of the printed circuit board (PCB), respectively. The width of the microstrip feed line is fixed at $W_1 = 2.6$ mm to achieve 50 Ω impedance. On the other side of the substrate, the conducting ground plane with a length of $L_1 = 20$ mm only covers the section of the microstrip feed line. h is the feed gap between the feed point and the ground plane.

Figure 7.19 shows the simulated and the measured return loss curves for an optimal design. The measured -10 dB bandwidth ranges from 2.78 GHz to 9.78 GHz, while in simulation from 2.69 GHz to 10.16 GHz. The measurement confirms the UWB characteristic of the proposed printed circular disc monopole, as predicted in the simulation.

Looking at the simulated return loss curve in Figure 7.19, the UWB characteristic of the antenna can be again attributed to the overlapping of the first three resonances which are closely distributed across the spectrum. However, despite considerable efforts being spent in tuning the design parameters, a good overlapping between the third and fourth harmonics cannot be achieved. Thus, the -10 dB bandwidth of this antenna is always limited at the high end around 10 GHz. By viewing the simulated input impedance plotted on the Smith chart, Figure 7.20, it is noted that at the high frequency limit the input impedance loops out of the VSWR = 2 circle, i.e. the impedance matching is getting worse. This can be understood by examining the variation in current distribution with resonance.

Simulated current distributions and magnetic field variations of the microstrip line fed disc monopole at different frequencies are presented in Figure 7.21. Figure 7.21(a) shows the current pattern near the

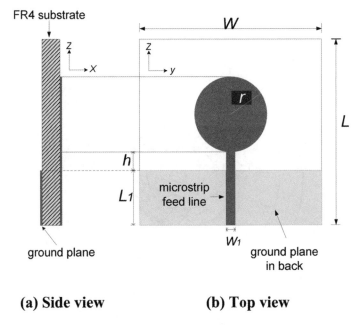

(a) Side view   (b) Top view

Figure 7.18 Geometry of the microstrip line fed circular disc monopole in the optimal design: $r = 10$ mm, $h = 0.3$ mm, $W = 42$ mm and $L = 50$ mm

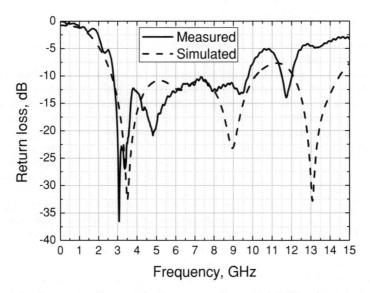

Figure 7.19 Simulated and measured return loss curves of microstrip line fed disc monopole in the optimal design

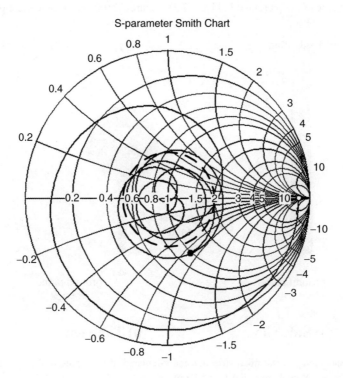

Figure 7.20 Simulated Smith chart of microstrip line fed disc monopole in the optimal design

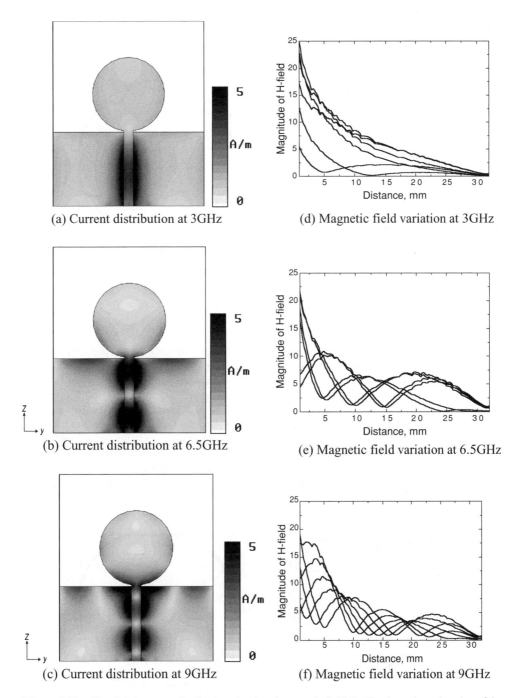

Figure 7.21 Simulated current distributions (a–c) and magnetic field distributions along the edge of the half-disc L ($L = 0$–33 mm: bottom to top) at different phases (d–f) of microstrip line fed disc monopole in the optimal design

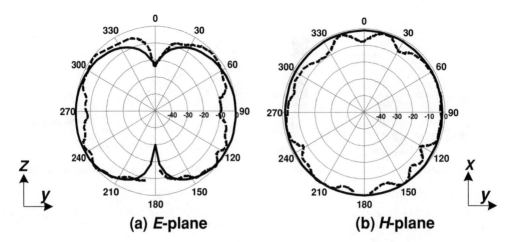

Figure 7.22 Simulated (solid line) and measured (dashed line) radiation patterns of microstrip fed disc monopole in the optimal design at 3 GHz

first resonance at 3 GHz. The current pattern near the second resonance at around 6.5 GHz is given in Figure 7.21(b), indicating approximately a second order harmonic. Figure 7.21(c) illustrates a more complicated current pattern at 9 GHz, corresponding to the third order harmonic. The current variation patterns are similar to those in the CPW-fed disc monopole, as shown in Figures 7.21(d)–(f). The antenna also operates in a hybrid mode of travelling and standing waves at higher frequencies. However, the ground plane on the other side of the substrate cannot form a good slot with the disc to support travelling waves as good as the CPW-fed disc monopole. Therefore, the impedance matching becomes worse for the travelling wave dependent modes at high frequency, as indicated in Figures 7.19 and 7.20.

The measured and the simulated radiation patterns at 3 GHz, 6.5 GHz and 9 GHz are plotted in Figures 7.22–7.24, respectively. The measured H-plane patterns are very close to those obtained in the

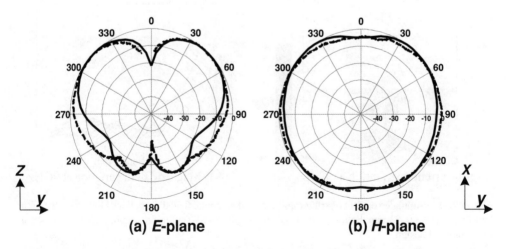

Figure 7.23 Simulated (solid line) and measured (dashed line) radiation patterns of microstrip fed disc monopole in the optimal design at 6.5 GHz

Theory of UWB Antenna Elements

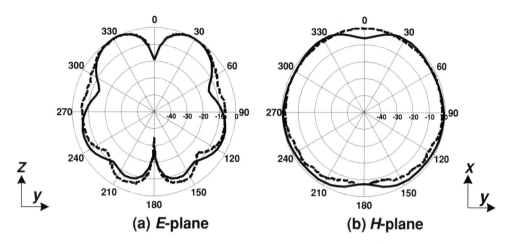

Figure 7.24 Simulated (solid line) and measured (dashed line) radiation patterns of microstrip fed disc monopole in the optimal design at 9 GHz

simulation. It is noticed that the H-plane pattern is omnidirectional at lower frequency (3 GHz) and still retains a good omnidirectionality at higher frequencies (6.5 GHz and 9 GHz). The measured E-plane patterns also follow the shapes of the simulated ones well. The E-plane pattern is like a doughnut at 3 GHz. With the increase of frequency (6.5 GHz and 9 GHz), it starts to form humps and get more directional at around 45 degrees from the z-direction.

The design rules for the microstrip-fed disc monopole are also similar to those of the CPW-fed disc monopole. It is observed in the simulation that, in a manner similar to the CPW-fed disc monopole, the -10 dB bandwidth of the microstrip-fed monopole is critically dependent on the feed gap h, the width of the ground plane W. For the benefit of design, four discs with different diameters have been investigated in simulation. Table 7.2 lists their corresponding first resonance frequencies and associated optimal design parameters, bearing in mind that a different substrate (FR4) is used here.

Again, the first resonance frequency can be estimated by treating the disc as a quarter-wave monopole. The optimal width of the ground plane is found to be around two times of the diameter and the optimal feed gap is still around 0.3 mm.

7.3.3 Other Shaped Disc Monopoles

The UWB operation principle has been derived from and tested in the circular disc monopoles in the previous sections. It has already been demonstrated that the planar disc monopoles in other shapes can

Table 7.2 Optimal design parameters of microstrip-fed disc monopole and relationship between the diameter and the first resonance

Diameter $2r$ (mm)	First resonance f (GHz)	Wavelength λ at f (mm)	$2r/\lambda$	W (mm)	$W/2r$	h (mm)
20	3.51	85.5	0.23	42	2.1	0.3
25	2.96	101.4	0.25	50	2	0.3
30	2.56	117.2	0.26	57	1.9	0.3
40	1.95	153.8	0.26	75	1.9	0.4

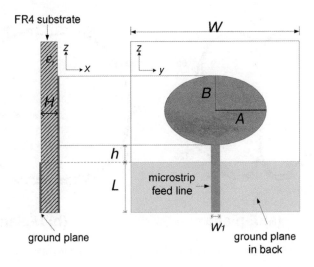

Figure 7.25 Geometry of the printed elliptical disc monopole

also exhibit an UWB characteristic [23, 24]. Hence, it is a natural development to test this principle in other shaped disc monopoles. Here, two examples of planar monopoles, i.e. an elliptical disc and a circular ring disc, will be analysed.

A planar elliptical disc monopole has been designed on the FR4 substrate by using a 50 Ω microstrip line feed to provide an ultra-wide operating bandwidth. A prototype of the proposed elliptical disc monopole in the optimal design, i.e. $h = 0.7$ mm, $W = 44$ mm, $B = 7.8$ mm and $A/B = 1.4$, as shown in Figure 7.25, was fabricated and tested.

The measured and simulated return loss curves, as given in Figure 7.26, show a good agreement. The simulated −10 dB bandwidth ranges from 3.07 GHz to 9.58 GHz. This UWB characteristic is confirmed in the measurement, with only a slight shift of the upper edge frequency to 9.89 GHz.

It can be seen that the UWB characteristic of the antenna is mainly attributed to the overlapping of the first three resonances, which are closely spaced across the spectrum. For the similar reason as that in a microstrip-fed circular disc monopole, the −10 dB bandwidth is limited at the high end of the frequency due to the increase of impedance mismatching.

As shown in Figures 7.5 and 7.21, the current is mainly distributed along the edge of the circular disc. This implies that a ring disc monopole can still achieve ultra-wide bandwidth because the performance of the antenna is independent of the central part of the disc.

A prototype of the CPW-fed circular ring monopole antenna (central cut) in the optimal design was built and tested, as shown in Figure 7.27. The return losses were measured in an anechoic chamber by using an HP8720ES network analyser.

Figure 7.28 illustrates the simulated and the measured return loss curves of the antenna. Here, the complete configuration of the antenna, including the 50 Ω SMA feeding port, was simulated in order to obtain accurate comparison with the measurement. The measured return loss curve is quite close to the simulated one, as shown in Figure 7.28. Furthermore, the measured curve of the CPW-fed ring monopole is almost identical to that of the CPW-fed disc monopole, as shown in Figure 7.13. This confirms that this circular ring monopole can provide a similar UWB performance as its counterpart disc monopole.

Theory of UWB Antenna Elements

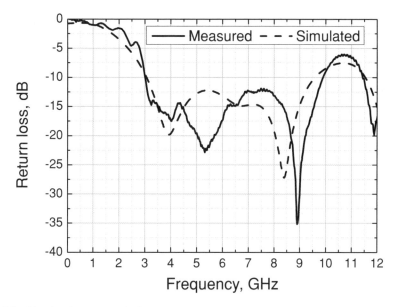

Figure 7.26 Simulated and measured return loss curves with $h = 0.7$ mm, $W = 44$ mm, $B = 7.8$ mm and $A/B = 1.4$

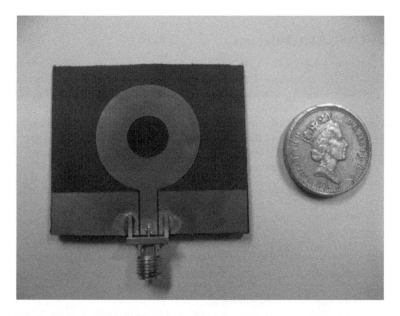

Figure 7.27 Photograph of the CPW-fed circular ring monopole in the optimal design $r = 12.5$ mm, $r_1 = 5$ mm (hole radius), $h = 0.3$ mm and $W = 47$ mm

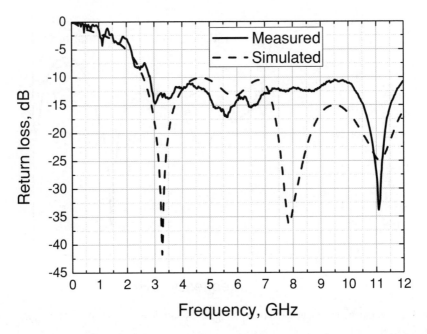

Figure 7.28 Simulated and measured return losses of the CPW-fed circular ring monopole in the optimal design

7.4 Planar UWB Slot Antennas

It has been demonstrated that the slot types of antennas are also capable of operating in an ultra-wide bandwidth [18]–[21]. They can be regarded as the magnetic type of monopoles in view of the EM duality. Therefore, it will be interesting to examine this type of antenna in light of the UWB principle, which has been discussed in the previous sections.

In this section, two types of printed elliptical/circular slot antennas are designed for the UWB operation. One is fed by a microstrip line, and the other by a CPW. In both designs, a U-shaped tuning stub is introduced to enhance the coupling between the slot and the feed line so as to broaden the operating bandwidth of the antenna. Furthermore, an additional bandwidth enhancement can be achieved by tapering the feeding line.

7.4.1 Microstrip/CPW Feed Slot Antenna Designs

The proposed printed elliptical/circular slot antennas with two different feeding structures are illustrated in Figures 7.29(a) and (b), respectively. For the microstrip line-fed elliptical/circular slot antenna, the slot and the feeding line are printed on different sides of the dielectric substrate; for the CPW-fed one, they are printed on the same side of the substrate.

In both designs, the elliptical/circular radiating slot has a long axis radius A and a short axis radius B (for circular slot, $A = B$) and is etched on a rectangular FR4 substrate with a thickness $t = 1.5$ mm and

Theory of UWB Antenna Elements

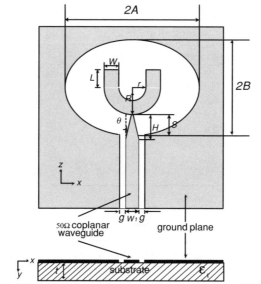

(b) Elliptical/circular slot antenna fed by CPW

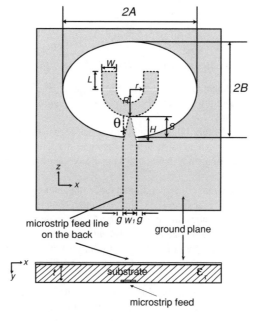

(b) Elliptical/circular slot antenna fed by CPW

Figure 7.29 Geometry of printed elliptical/circular slot antennas

Table 7.3 The optimal dimensions of the printed elliptical/circular slot antennas

	Microstrip line fed		CPW fed	
	Elliptical slot	Circular slot	Elliptical slot	Circular slot
A (mm)	16	13.3	14.5	13.3
B (mm)	11.5	13.3	10	13.3
S (mm)	0.6	0.5	0.4	0.4
R (mm)	5.9	5	5.5	5
r (mm)	2.9	1.8	2.5	1.8
W (mm)	3	3.2	3	3.2
L (mm)	6	6.7	3	4.3
Substrate size (mm^2)	42 × 42	43 × 50	40 × 38	44 × 44

a relative dielectric constant $\varepsilon_r = 4.7$. The feed line is tapered with a slant angle $\theta = 15$ degrees for a length H to connect with the U-shaped tuning stub which are all positioned within the elliptical/circular slot and symmetrical with respect to the short axis of the elliptical/circular slot. The U-shaped tuning stub consists of three sections: the semi-circle ring section with an outer radius R and an inner radius r, and two identical branch sections with equal heights L and equal widths W. S represents the distance between the bottom of the tuning stub and the lower edge of the elliptical/circular slot.

Four printed elliptical/circular slot antennas with the optimal designs were fabricated and tested in the Antennas Laboratory at QMUL. Their respective dimensions are given in Table 7.3.

7.4.2 Performance of Elliptical/Circular Slot Antennas

The return losses of the four antennas were measured by using an HP8720ES vector network analyser. The measured and simulated return loss curves are plotted in Figure 7.30. Their respective −10 dB bandwidths are tabulated in Table 7.4.

As shown in Figure 7.30 and Table 7.4, good agreement has been achieved between the measurement and experiment for each of the antennas. It is demonstrated that all of the four antennas can operate across an extremely wide frequency range and hence meet the bandwidth requirement for UWB antennas.

It is shown in Figure 7.30 that the UWB operation of these slot antennas is all due to the overlapping of the closely spaced resonances over the frequency band. It is interesting to note that the −10 dB bandwidth is always limited at the high frequency end, no matter for a CPW-fed or a microstrip line-fed slot in either elliptical or circular shape. The input impedances for the microstrip line-fed and CPW-fed elliptical slots are plotted on a Smith chart, as shown in Figures 7.31(a) and (b), respectively. It is clearly shown that at the high frequency limit the input impedance spirals out of VSWR = 2 circle, i.e. the impedance matching is getting worse.

The simulated current distributions of the CPW-fed elliptical slot antenna at three frequencies are presented in Figure 7.32, as a typical example. On the ground plane, the current is mainly distributed along the edge of the slot for all of the three different frequencies. The current patterns indicate the existence of different resonance modes, i.e. the first harmonic at 3.3 GHz in Figure 7.32(a), the second harmonic around 5 GHz in Figure 7.32(b) and the fourth harmonic around 10 GHz in Figure 7.32(c). This confirms that the elliptical/circular slot is capable of supporting multiple resonant modes, and the

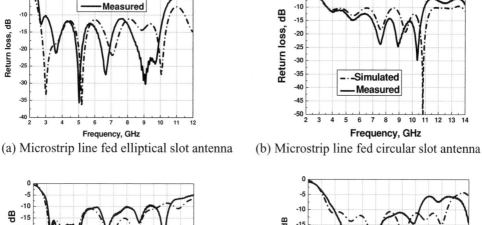

Figure 7.30 Measured and simulated return loss curves of printed elliptical slot antennas

overlapping of these multiple modes leads to the UWB characteristic, as analysed in the disc type of monopoles.

Again, the current variations have been observed in Figures 7.32(d)–(f). The feature is similar to that of the microstrip line-fed disc monopole. The travelling wave at high frequency is not well supported in this enclosed structure. Hence, the impedance matching becomes worse at the high frequency and the -10 dB bandwidth is limited at the high end.

Table 7.4 Measured and simulated bandwidths of printed elliptical/circular slot antennas

Antenna types		Simulated -10 dB bandwidth (GHz)	Measured -10 dB bandwidth (GHz)
Microstrip line fed	Elliptical slot	2.6–10.6	2.6–10.22
	Circular slot	3.45–13.22	3.46–10.9
CPW fed	Elliptical slot	3.0–11.4	3.1–10.6
	Circular slot	3.5–12.3	3.75–10.3

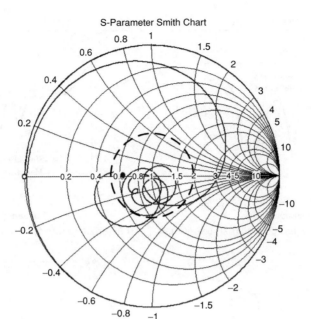

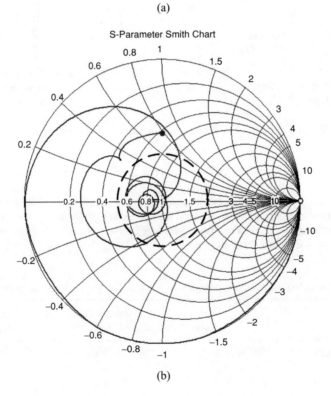

Figure 7.31 (a) Simulated Smith chart of microstrip-fed elliptical slot. (b) Simulated Smith chart of a CPW-fed elliptical slot antenna

Theory of UWB Antenna Elements

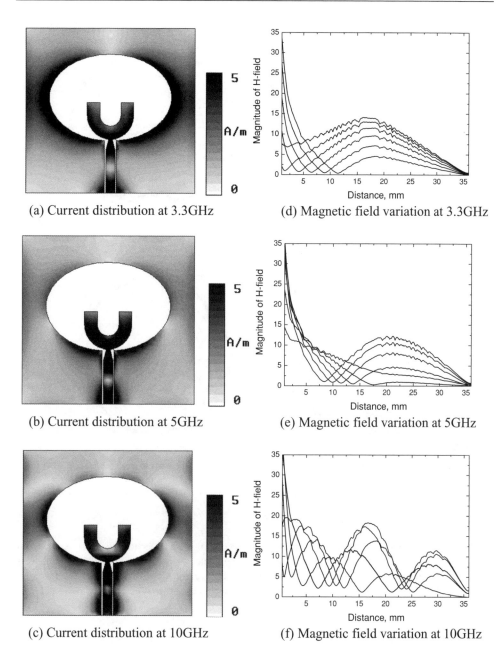

(a) Current distribution at 3.3GHz

(b) Current distribution at 5GHz

(c) Current distribution at 10GHz

(d) Magnetic field variation at 3.3GHz

(e) Magnetic field variation at 5GHz

(f) Magnetic field variation at 10GHz

Figure 7.32 Simulated current distributions (a–c) and magnetic field distributions along the edge of the half-slot L ($L = 0$–36 mm: bottom to top) at different phases (d–f) of a CPW-fed elliptical slot antenna

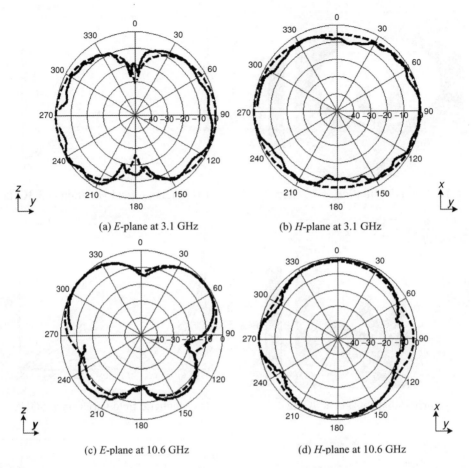

Figure 7.33 Simulated (dotted line) and measured (solid line) radiation patterns of a CPW-fed elliptical slot antenna

The radiation patterns of the antennas were also measured inside an anechoic chamber. As shown in Figure 7.33, the measured patterns agree well with the simulated ones for both elliptical slot antennas. The E-plane pattern transforms from a doughnut shape at lower frequency to a multilobe shape at higher frequency. The H-plane retains a reasonable omnidirectionality over the entire bandwidth.

7.4.3 Design Analysis

Studies in the previous sections have indicated that the ultra-wide bandwidth of the slot antenna results from the overlapping of the multiple resonances introduced by the combination of the elliptical slot and the feeding line with U-shaped tuning stub. Thus, the slot dimension, the distance S and the slant angle θ are the most important design parameters affecting the antenna performance and need to be further investigated.

Theory of UWB Antenna Elements

Table 7.5 The calculated and measured lower edge of −10 dB bandwidth

		A (mm)	B (mm)	Measured f_l (GHz)	Calculated f_l (GHz)
Microstrip line fed	Elliptical slot	16	11.5	2.6	2.74
	Circular slot	13.3	13.3	3.46	3.21
CPW fed	Elliptical slot	14.5	10	3.1	3.17
	Circular slot	13.3	13.3	3.75	3.21

1. **Dimension of elliptical slot.** It is noticed that the dimension of the slot antenna is directly related to the lower edge of the impedance bandwidth. In the case of elliptical disc monopoles, an empirical formula for estimating the lower edge frequency of the −10 dB bandwidth f_l is derived based on the equivalence of a planar configuration to a cylindrical wire [8].

   In this study, the elliptical slot can be regarded as an equivalent magnetic surface. An empirical equation for estimating f_l is expressed as:

$$f_l = \frac{30 \times 0.32}{L + r} \quad (7.1)$$

   where f_l in GHz, L and r in cm. L is the slot width, $2A$, r is equivalent radius of the cylinder given by $2\pi r L = \pi AB$.

   The comparison between the calculated f_l and the measured one for different printed slot antennas are tabulated in Table 7.5. It is shown that the measured f_l matches the calculated one quite well.

2. **Feed gap S.** The simulated return loss curves of CPW-fed elliptical slot antenna for various S ($S =$ 0.15 mm, 0.4 mm, 0.65 mm) with $A = 14.5$ mm, $B = 10$ mm and $\theta = 15$ degrees are illustrated in Figure 7.34. It is seen that the curves for different S have similar shape and variation trend, but the optimal distance is $S = 0.4$ mm, which is close to the CPW line gap of 0.33 mm.

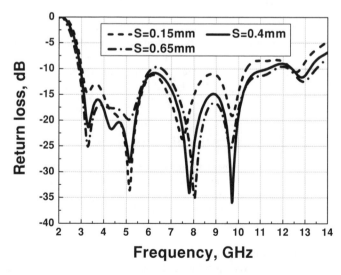

Figure 7.34 Simulated return loss curves of CPW-fed elliptical slot antenna for different S with $A = 14.5$ mm, $B = 10$ mm and $\theta = 15$ degrees

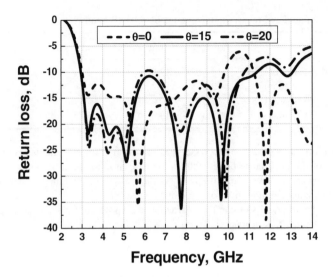

Figure 7.35 Simulated return loss curves of CPW-fed elliptical slot antenna for different θ with $A = 14.5$ mm, $B = 10$ mm and $S = 0.4$ mm

3. **Slant angle θ.** It is also noticed that the bandwidth is quite sensitive to the slant angle θ. In Figure 7.35, the return loss curves of CPW-fed elliptical slot antenna for different slant angles with $A = 14.5$ mm, $B = 10$ mm and $S = 0.4$ mm are plotted. It is observed that the lower edge of the -10 dB bandwidth is independent of θ, but the upper edge is very sensitive to the variation of θ. The optimal slant angle is found to be at $\theta = 15$ degrees, with a bandwidth of 8.4 GHz (from 3.0 GHz to 11.4 GHz).

It is clear that tuning of the design parameters of feed gap S and slant angle θ can affect the wave transmission at the feed point to the slot. A proper adjustment of these parameters can either enhance or shift the antenna resonances to overlap more across the spectrum, hence, a more wide operating bandwidth.

7.5 Time-Domain Characteristics of Monopoles

As mentioned in Section 7.1, a good time-domain characteristic is also required for the UWB terminal antennas, apart from the compactness and omnidirectional radiation pattern. The two types of small UWB monopole antennas, i.e., planar discs and slots being studied in the previous sections, rely on the overlapping of the resonance modes to provide the FCC-defined UWB bandwidth and a reasonable omnidirectionality. It is equally important to assess their performance in the time domain for commercial applications.

In this section, two identical monopoles are used to form an antenna system. The antenna pair is vertically placed; the distance between them is set to be 1.0 m or 1.2 m. Three orientations will be investigated, namely face to face, side by side and face by side, as shown in Figure 7.36.

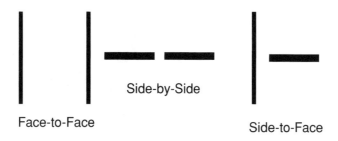

Figure 7.36 Antenna system set up

The first-order Rayleigh pulse, as presented in Equation (7.2), is widely used as the source signal to drive the transmitter.

$$f(t) = \frac{-2(t-1)}{a^2} \exp\left(-\left(\frac{(t-1)}{a}\right)^2\right), \qquad (7.2)$$

where a is the pulse parameter.

Figure 7.37 presents the normalised source pulses with three different pulse parameters. It is generally agreed the pulse of $a = 45$ ps is a good test for the FCC-defined UWB bandwidth.

The antenna system can be modelled as a linear, time-invariant system, so the received signal can be calculated by convolving the input pulse and impulse response of the antenna system. In this study, the transfer function (magnitude together with phase) was measured by using an HP8720ES network analyser, and it will be firstly transformed to the time domain by performing an inverse Fourier transform, then convolving with the input pulse and the measured UWB pulse can be consequently obtained.

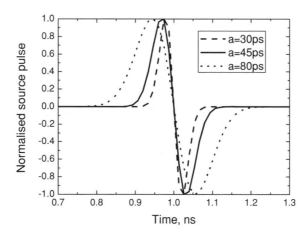

Figure 7.37 Source pulse waveforms with different a

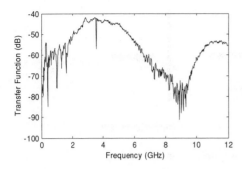 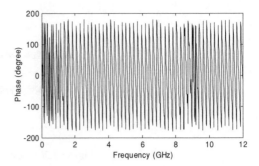

Figure 7.38 Measured magnitude (a) and phase (b) of transfer function for the CPW-fed disc monopoles in face-to-face orientation

7.5.1 Time-Domain Performance of Disc Monopoles

Figure 7.38 displays the measured magnitude and phase of the transfer function (TF) for the CPW-fed disc monopoles (as shown in Figure 7.12) in face-to-face orientation (1.2 m separation).

It is noticed in Figure 7.38 that the magnitude of TF undergoes a null at 9 GHz, and the phase of TF is nonlinear within the 0.5–0.8 GHz and 8.9–9.1 GHz frequency band. These may cause the distortion of the input signal. The measured received pulse is obtained by applying the aforementioned convolution approach.

Also, the complete antenna system as shown in Figure 7.36, including the separation gap, has been modelled using CST Microwave Studio^{TM}. A comparison between simulated and measured results is plotted in Figure 7.39. As can be seen from the figure, good agreement between the simulated and measured received pulses is obtained. Moreover, the ringing effect is found to be very small. The received pulse is just only distorted or differentiated (to be further addressed in Chapter 8), which can be correctly detected by choosing a proper template. Therefore, the CPW-fed disk monopole demonstrates a good performance in time domain.

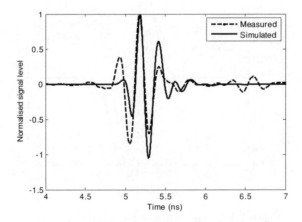

Figure 7.39 Simulated and measured received pulse ($a = 45$ ps)

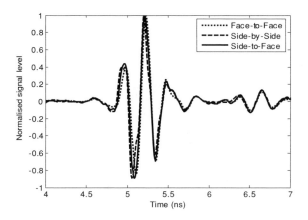

Figure 7.40 Measured received signal ($a = 45$ ps) for different antenna orientations

Apart from face-to-face orientation, side-to-face and side-by-side orientations are also investigated. Figure 7.40 illustrates the measured received signal for different antenna pair orientations. It is evident that the received pulses are almost identical, which indicates the nearly omnidirectional radiation property of the CPW-fed disk monopole.

7.5.2 Time-Domain Performance of Slot Antenna

Figure 7.41 displays the measured magnitude and phase of the transfer function for the CPW-fed elliptical slots (as shown in Figure 7.29(b)) in both face-to-face and side-by-side orientations (1.0 m separation).

It is noticed in Figure 7.41 that the magnitude of TF is almost linear for the face-to-face orientation, but undergoes a null at 8 GHz for the side-by-side orientation. Measured received pulses are obtained again by applying the aforementioned convolution approach. The received pulses in both orientations,

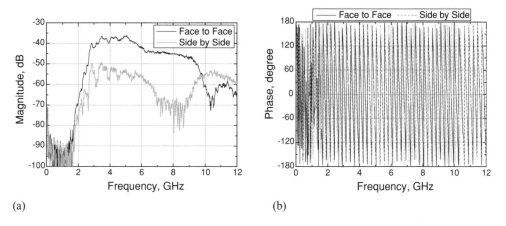

Figure 7.41 Measured magnitude (a) and phase (b) of transfer function for the CPW-fed elliptical slots

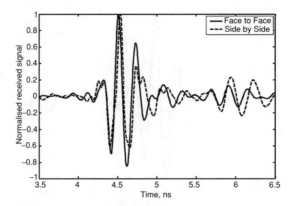

Figure 7.42 Measured received signal ($a = 45$ ps) for different antenna orientations

as shown in Figure 7.42, respectively, are distorted more in comparison to that in the CPW-fed disc monopoles. Also, the ringing is getting slightly worse. This is understandable as the structure of the slot antenna is more complex than the disc monopole. Generally speaking, the performance of the CPW-fed elliptical slot in the time domain is still acceptable.

7.6 Summary

It has been demonstrated in this chapter that the overlapping of closely spaced multiple resonances accounts for the UWB characteristics of a range of planar monopoles. Actually, the antenna behaves like a quarter-wave monopole in the first resonance. It starts to operate in a hybrid mode of standing and travelling waves in the second order harmonics or above. Travelling waves become more critical to the antenna operation with the increase of frequency. So it is essential to design a smooth transition between the feeding line and the antenna for good impedance matching over the operation bandwidth. It happens that the overlapping of the first few harmonics (1–3 or 4) of this range of planar monopoles can cover the FCC-defined UWB bandwidth. Therefore, the radiation pattern retains a good omnidirectionality on H-plane over the UWB bandwidth. The design rules have been derived from this principle for the electric and magnetic types of planar monopoles, as shown in Tables 7.1 and 7.2, and Equation (7.1).

This principle has also been applied to explain the operation of the vertical planar monopoles successfully [11]–[12]. It is believed that overlapping of multiple resonances and supporting of travelling wave for good impedance matching are two universal ingredients for compact planar UWB antennas.

Acknowledgements

The author would like to thank Jianxin Liang, Pengcheng Li and Lu Guo for conducting most of the research work and Peter Massey for a stimulating discussion on the operation of UWB planar antennas.

References

[1] Victor H. Rumsey, Frequency-independent antennas, *IRE National Convention Record*, pt. I, 251–9, 1957.
[2] Wilhelm Runge, *Polarization diversity reception*, U.S. Patent 1,892,221, 27 December, 1932. Y. Mushiake, "Self-Complementary Antennas," *IEEE Antennas and Propagation Magazine*, 34, 6, December 1992, pp. 23–29.

[3] J. Johnson and H. Wang, The physical foundation, developmental history, and ultra-wideband performance of SMM (spiral-mode microstrip) antennas, *2005 IEEE Antennas and Prop. Symp., Washington*, DC, July 2005.
[4] S. Licul, J.A.N. Noronha, W.A. Davis, D.G. Sweeney, C.R. Anderson and T.M. Bielawa, A parametric study of time-domain characteristics of possible UWB antenna architectures, *Proc. Vehicular Technology Conf.*, **5**, 3110–14, 2003.
[5] Hans Schantz, *The Art and Science of Ultrawideband Antennas*, Artech House, Inc, 2005.
[6] S. Honda, M. Ito, H. Seki and Y. Jinbo, A disk monopole antenna with 1:8 impedance bandwidth and omnidirectional radiation pattern, *Proc. Int. Symp. Antennas Propagat.*, 1145–8, 1992.
[7] N.P. Agrawall, G. Kumar and K.P. Ray, Wide-band planar monopole antennas, *IEEE Trans. Antennas Propagat.*, **462**, 294–5, 1998.
[8] Z.N. Chen, M.Y.W. Chia and M.J. Ammann, Optimization and comparison of broadband monopoles, *IEE Proc.-Microw. Antennas Propag.*, **150**(6), 2003.
[9] P.V. Anob, K.P. Ray and G. Kumar, Wideband orthogonal square monopole antennas with semi-circular base, *2001 IEEE Antennas and Propagation Society International Symposium*, Boston, Massachusetts, July 8–13, **3**, 294–7, 2001.
[10] J. Liang, C. Chiau, X. Chen and J. Yu, Study of a circular disc monopole antenna for ultra wideband applications, *2004 International Symposium on Antennas and Propagation*, 17–21 August, 2004.
[11] J. Liang, C. Chiau, X. Chen and C.G. Parini, Analysis and design of UWB disc monopole antennas, *IEE International Workshop on Ultra Wideband Communication Technologies & System Design*, 103–6, July 2004.
[12] J. Liang, C. Chiau, X. Chen and C.G. Parini, Study of a printed circular disc monopole antenna for UWB systems, *IEEE Trans Antennas Propagat.*, 2005 (accepted).
[13] J. Liang, C. Chiau, X. Chen and C.G. Parini, "Printed circular disc monopole antenna for ultra wideband applications", *IEE Electronic Letters*, vol. 40, no. 20, September 30th, 2004, pp.1246–1248.
[14] J. Liang, C. Chiau and X. Chen, Printed circular ring monopole antennas, *Microwave and Optical Technology Letters*, **45**(5), 372–5, 2005.
[15] J. Liang, L.Guo, C.C. Chiau, X. Chen and C.G.Parini, Study of CPW-fed circular disc monopole antenna, *IEE Proceedings Microwaves, Antennas and Propagation*, 2005 (accepted).
[16] J. Liang, C. Chiau, X. Chen and C.G. Parini, CPW-fed circular ring monopole antenna, *2005 IEEE AP-S International Symposium on Antennas and Propagation*, Washington, DC, USA, 3–8 July, 2005.
[17] P. Li, J. Liang and X. Chen, Study of printed elliptical/circular slot antennas for ultra wideband applications, *IEEE Trans Antennas Propagat.*, 2005 (accepted).
[18] P. Li, J. Liang and X. Chen, Ultra-wideband elliptical slot antenna fed by tapered micro-strip line with U-shaped tuning stub, *Microwave and Optical Technology Letters*, **47**(2), 140–3, 2005.
[19] P. Li, J. Liang and X. Chen, Ultra-wideband printed elliptical slot antenna, *2005 IEEE AP-S International Symposium on Antennas and Propagation*, Washington, DC, USA, 3–8 July, 2005.
[20] P. Li, J. Liang and X. Chen, Planar circular slot antenna for ultra-wideband applications, *IEE Seminar on Wideband and Multi-band Antennas and Arrays*, University of Birmingham, UK, 7 September, 2005.
[21] CST-Microwave Studio, *User's Manual*, 4, 2002.
[22] Seong-Youp Suh, Warren L. Stutzman and William Davis, Multi-broadband monopole disc antennas, *2003 IEEE Antennas and Propagation Society International Symposium*, Columbus, Ohio, June 22–27, **3**, 616–19.
[23] Taeyoung Yang and William Davis, Planar half-disk antenna structures for ultra-wideband communications, *2004 IEEE Antennas and Propagation Society International Symposium*, Monterey, California, June 20–25, **3**, 2508–11.

8

Antenna Elements for Impulse Radio

Zhi Ning Chen

8.1 Introduction

In impulse radio systems transmitting and receiving a chain of extremely short pulses (impulses), the ultra-wideband spectrum is occupied by spectra of a single or several impulses. In other words, the spectra of the impulses span an ultra-wide frequency range from, e.g., a few Giga Hertz, up to 7.5 GHz. Moreover, impulse radio systems are carrier-free. Therefore, distortion of the waveforms of the transmitted pulses may significantly degrade quality of wireless communications by impulse radio systems. These features of the impulse transmission differentiate the design considerations of UWB antennas for impulse radio systems from conventional narrowband and even broadband radio systems [1, 2].

In general, UWB antennas are expected to provide impedance matching and gain in desired directions across the operating band. This operating band can be the full band of 3.1 to 10.6 GHz, but can be less as UWB systems often use a subset of the full band. There have been a lot of broadband or frequency-independent antenna designs for a variety of broadband applications. For example, transversal electromagnetic (TEM) horns and self-similar antennas such as spiral antennas have frequency-independent impedance response and stable gain over broad bandwidths [3–9]. However, TEM horns are too bulky for portable devices and their directional radiation is not suitable for mobile applications. Conical spiral antennas suffer from frequency-dependent changes in phase centres, which may severely distort the waveforms of radiated pulses [10]. Broadband biconical and disk-conical antennas feature relatively stable phase centres [11–12]. Impedance bandwidths of cylindrical antennas can be enhanced by resistive loading [13, 14]. In addition, the planar versions of broadband antennas are extensively used in UWB radio systems, which feature broad bandwidth, small size and low cost [15–25].

This chapter gives a brief survey and classification of existing UWB antenna solutions in Section 8.2. Due to the uniqueness of impulse radio systems, special design considerations for UWB antennas are also discussed in Section 8.2. Section 8.3 provides a case study by elaborating two design examples, namely

---

Ultra-wideband Antennas and Propagation for Communications, Radar and Imaging Edited by B. Allen,
M. Dohler, E. E. Okon, W. Q. Malik, A. K. Brown and D. J. Edwards
© 2007 John Wiley & Sons, Ltd

omnidirectional roll antennas and the directional antipodal Vivaldi antenna and its modified version for UWB applications. Section 8.4 summarises UWB antenna design for impulse radio systems.

8.2 UWB Antenna Classification and Design Considerations

8.2.1 Classification of UWB Antennas

Requirements for antennas vary with applications and systems. Methods to categorise antennas can be highlighted in terms of their operating frequency, geometry, function, materials and so on. For UWB radio systems, a lot of antennas have been designed and used for testing and wireless connections. In terms of geometry and radiation characteristics, they may be two- or three-dimensional designs, and omnidirectional and directional designs. Thus they can be roughly classified into four categories as listed in Table 8.1. Some typical designs are exemplified in the table.

Among two-dimensional directional solutions, tapered slot antennas particularly Vivaldi antennas are widely used. These are basically planar notch antennas. They are one type of endfire travelling-wave antennas [26]. A variety of taper profiles of those slots have been presented, such as exponential, tangential, parabolic, linear-constant, exponential-constant, step-constant and linear profiles as shown in Figure 8.1. However, frequency-independent log-periodic and conical spiral antennas with change in their phase centres are not recommended for impulse radio systems. Their nonlinear phase response significantly distorts the waveforms of received pulses by producing undesirable ringing and time delay.

Horn and reflector antennas are typical three-dimensional high-gain directional solutions which have been studied widely, and they are usually employed in test or point-to-point applications [3–9]. Figure 8.2 shows a pyramidal horn antenna with a rectangular cross-section. However, its bulky design is not suitable for applications with size constraint for example, portable devices.

Three-dimensional resistive-loaded cylindrical dipoles, biconical, disc-cone and roll antennas are typical omnidirectional designs with stable impedance (both magnitude and phase responses) and radiation performance across a broad bandwidth [11–14]. They are good candidates for base stations in mobile applications. A biconical antenna is shown in Figure 8.3. It evolves from the biconical transmission structure and features broad bandwidth for impedance matching and radiation performance.

Table 8.1 Category of UWB antennas.

	Directional	Omni-directional
Two-dimensional	Vivaldi Antenna Tapered Slot Antenna Log-periodic Antenna Planar Log-Spiral Antenna Conical Spiral Antenna	Planar Dipole Slot Antenna Printed Antenna on PCB
Three-dimensional	TEM Horn Antenna Ridge Horn Antenna Reflector Antenna	Loaded Cylindrical Dipole Bi-conical Antenna Disc-cone Antenna Roll Antenna

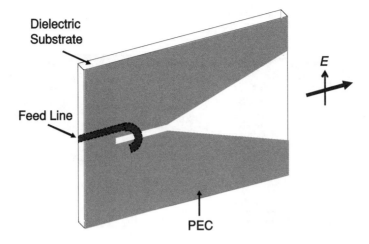

Figure 8.1 Tapered slot antenna in its basic form

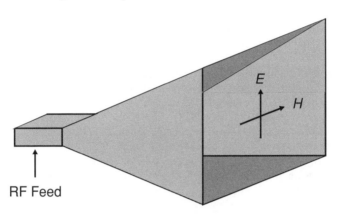

Figure 8.2 TEM horn antenna

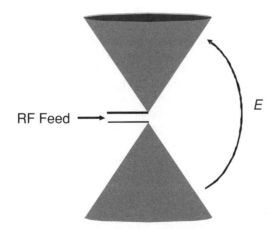

Figure 8.3 Disc-cone antenna

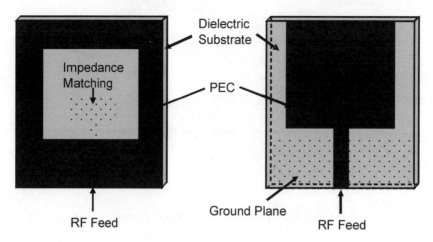

Figure 8.4 Large slot and printed antennas

For portable terminals in mobile applications, antennas should be small in size, lightweight and embeddable [15–25]. So, planar antennas (dipoles and monopoles), slot antennas and antennas printed on printed circuit boards (PCBs) have attracted increasing attention. Figure 8.4 illustrates a large-slot antenna and a rectangular monopole printed on PCBs. A microstrip line excites the slot of the antenna with an impedance matching section at its end. The rectangular radiator is fed by a microstrip line directly. A finite-size ground plane is printed on the opposite surface of the PCB.

In addition, arrays comprising elements mentioned above can be used to enhance gain, configure specific coverage and achieve beam steering. Such arrays have wide applications in impulse reception due to the emission limits.

8.2.2 Design Considerations

Conventionally, performance of antennas can be evaluated in terms of their bandwidths for impedance matching, radiation direction, beamwidth, gain and polarization for narrowband and broadband communication systems. However, UWB systems without any carriers require special considerations for antenna design [25]. From a systems point of view, UWB antenna systems should transmit and receive high-quality signals for high signal-to-noise ratio (SNR) at a receiver. For impulse radio systems, distorted waveforms of received pulses at a receiver will degrade the SNR. Therefore, performance of UWB antennas should be assessed in terms of system transfer function, radiation transfer function and fidelity, which address the characteristics of transmit and receive antennas, as well as antenna system response.

8.2.2.1 System Transfer Function

A transmit/receive antenna system is depicted in Figure 8.5. The frequency-dependent Friis transmission formula relating the output power of the receive antenna to the input power of the transmit antenna is given in Equation (8.1), where it is assumed that each antenna is in the far-field zone of the other. Due to

Antenna Elements for Impulse Radio

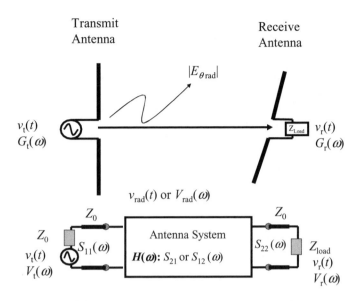

Figure 8.5 Illustration of a transmit/receive antenna system

the ultra-wide bandwidth, the parameters vary with frequency.

$$\frac{P_r(\omega)}{P_t(\omega)} = \left(1 - |\Gamma_t(\omega)|^2\right)\left(1 - |\Gamma_r(\omega)|^2\right) G_r(\omega) G_t(\omega) |\hat{\rho}_t(\omega) \cdot \hat{\rho}_r(\omega)|^2 \left(\frac{\lambda}{4\pi r}\right)^2 \quad (8.1)$$

where:

P_t, P_r: time average input power of the transmit antenna and time average output power of the receive antenna;
Γ_t, Γ_r: return loss at the input of the transmit antenna and the output of receive antenna;
G_t, G_r: gain of the transmit antenna and the receive antenna;
$|\hat{\rho}_t \cdot \hat{\rho}_r|^2$: polarisation matching factor between the transmit antenna and the output of receive antenna;
λ: operating wavelength; and
r: distance between transmit and receive antennas.

The gain of the transmit and receiving antennas, G_t and G_r as well as $|\hat{\rho}_t \cdot \hat{\rho}_r|^2$ are also functions of orientation (θ, ϕ).

Furthermore, a system transfer function $H(\omega)$ is defined to describe the relation between the source signal (voltage) at the input of the transmit antenna and the received signal (voltage) at the output of the receive antenna $\left[\frac{V_t(\omega)}{2}\right]^2 = 2P_t(\omega)Z_0$ and $\frac{V_r(\omega)^2}{2} = P_r(\omega)Z_{\text{load}}$ where P_t is the transmitted power, Z_0 is the transmit antenna impedance, P_r is the received power and Z_{load} is the receiver load, then the formula for $H(\omega)$ can be simplified as Equation (8.2).

$$H(\omega) = \frac{V_r(\omega)}{V_t(\omega)} = \sqrt{\frac{P_r(\omega) Z_{\text{load}}}{P_t(\omega) 4Z_0}} \, e^{-j\phi(\omega)} = |H(\omega)|\, e^{-j\phi(\omega)}; \phi(\omega) = \phi_t(\omega) + \phi_r(\omega) + \omega r/c \quad (8.2)$$

where, c is velocity of light, and $\phi_t(\omega)$ and $\phi_r(\omega)$ are the phase variation due to the transmit and receive antennas, respectively. The system transfer function describes the characteristics of both the transmit and receive antennas which include impedance matching, gain, polarisation matching and distance as well as orientation of the antennas. Therefore, the transfer function $H(\omega)$ can be used to describe antenna systems, which may be dispersive.

The system transfer function can be measured by means of scattering parameters of a two-port network such as S_{11} and S_{21} as shown in Figure 8.5 in the frequency domain. They can also be measured by testing their far-field parameters such as gain against frequency.

It is clear that the waveforms of received pulses are determined by the magnitude and phase responses of the antenna system transfer function, $|H(\omega)|$ and $\phi(\omega)$. UWB antenna design aims to achieve the constant magnitude and linear phase responses of the system across the bandwidth covered by the spectra of source pulses. Otherwise, the waveforms of the received pulses will be distorted by low signal levels and undesired ringings [25].

8.2.2.2 Radiation Transfer Function

The radiation transfer function relating radiated electric fields (pulses) and source pulses employed at the transmit antenna can be defined in Equation (8.3).

$$\vec{E}_{rad}(\omega) = \vec{H}_{rad}(\omega) V_t(\omega) = \hat{a} |H_{rad}(\omega)| e^{-j\phi_{rad}(\omega)} V_t(\omega); \phi_{rad}(\omega) = \phi_t(\omega) + \omega r/c \qquad (8.3)$$

The radiation transfer function $\vec{H}_{rad}(\omega)$ is a vector with its direction determined by the polarisation direction $\hat{a}$ of the transmit antenna, and it also describes the characteristics of the transmit antenna including impedance matching, gain and the orientation of observation point. $V_t(\omega)$ is the spectrum of a source signal (voltage). This transfer function is defined to evaluate the radiated power density spectrum to comply with emission limits by regulators.

From Equation (8.3), it can be found that the spectrum of radiated pulse can be controlled by the antenna and the source pulse. For an antenna covering the operating bandwidth, the spectrum of radiated pulse can be controlled by optimising the spectrum of the source pulse. The transmit antenna can function as a bandpass filter to tailor the spectrum of the radiated pulse if its outband emission is higher than the limits.

The radiation transfer function is also orientation-dependent. Therefore, radiation patterns of a transmit antenna may be defined for a waveform or spectrum of radiated pulses, not only gain or radiated fields. Figure 8.6 shows the radiation pattern for waveforms of radiated pulse. The antenna is a circular dipole excited by a monocycle impulse. The waveforms vary in the different directions because the gain or magnitude phase responses of the transfer function are orientation-dependent.

8.2.2.3 Fidelity

The fidelity of the received pulse at a receiver is proposed to evaluate the quality of a received pulse through the antenna system and RF channel in order to select a proper detection template [25–27]. It is defined in Equation (8.4).

$$F = \max_{\tau} \int_{-\infty}^{\infty} L\left[p_{source}(t)\right] p_{output}(t - \tau) dt, \qquad (8.4)$$

where the source pulse $p_{source}(t)$ and output pulse $p_{output}(t)$ are normalised by their energy, respectively. The fidelity F is the maximum integration by varying time delay τ. The linear operator $L[\bullet]$ operates on

Antenna Elements for Impulse Radio

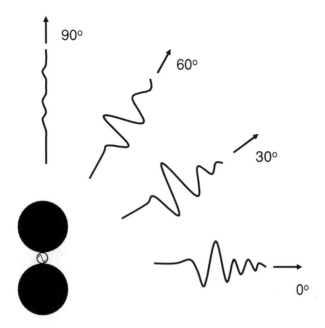

Figure 8.6 Radiation pattern for waveforms

the input pulse $p_{source}(t)$ of a transmit antenna. The optimal template at the output of a receiving antenna will be $L[p_{source}(t)]$, not the original pulse $p_{source}(t)$ for maximum fidelity. The waveforms of the received pulses are not identical with those of source pulses due to dispersive transmission (including antennas and RF channel).

Moreover, with the fidelity, the system performance such as bit error ratio (BER) can be directly evaluated for impulse radio systems if the system uses correlation receivers [28]. The study shows that the broadband planar antenna can lower the BER compared with a narrowband thin-wire antenna.

8.3 Omnidirectional and Directional Designs

Besides the design considerations discussed above, design of UWB antennas is also determined by devices used in applications. For example, antennas for fixed base stations should have high gain and specific coverage, where they usually have less strict space limitation. For portable devices, small, embeddable and omnidirectional designs are desired. This section provides the case study for practical antenna design. Two typical antennas with directional and omnidirectional radiation are exemplified.

8.3.1 Omnidirectional Roll Antenna

A roll monopole antenna has been presented for broadband omnidirectional applications since 2003 [29–30]. Also, they can be used in impulse radio systems in their simple forms. A bi-arm roll monopole with a height of 16 mm and a maximum diameter of 3 mm can be formed by wrapping a 13 × 16 mm rectangular conducting copper sheet as shown in Figure 8.7. The roll monopole has a pair of twisted

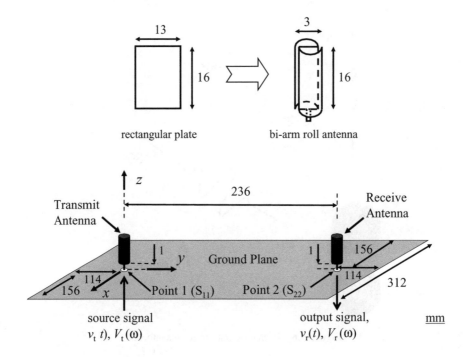

Figure 8.7 Roll antenna and testing setup

arms with semicircular cross-sections and a planar inner portion. Figure 8.7 also shows the testing setup for a transmit/receive antenna system that comprises two identical monopoles, where the monopoles are vertically installed 1 mm above a 312 × 464 mm ground plane with a distance r of 236 mm between their centres, and each monopole is fed at the centre of its bottom by a 1.2 mm diameter 50 Ω probe. The two identical monopole antennas are positioned face to face and side by side to examine their omnidirectional radiation, where the orientations of antennas have 90° difference. A source pulse drives the transmit monopole antenna at point ($x = 0$ mm, $y = 0$ mm, $z = 0$ mm), and $p_r(t)$ indicates the output signal of the receive monopole antenna at point ($x = 0$ mm, $y = 236$ mm, $z = 0$ mm).

The impedance and transfer responses of the roll monopole antenna are experimentally examined for UWB applications as shown in Figure 8.8. The roll monopole antennas with their small size achieve good impedance matching across the UWB band for the return loss less than −7.5 dB. The transfer characteristics are evaluated by the measured $|S_{21}|$. It is seen that the measured $|S_{21}|$ varies around −30 dB. At frequencies higher than about 6.2 GHz, the measured $|S_{21}|$ decreases by around 4 dB, which is mainly due to the change in the direction of maximum radiation, whereas the change in orientations of the antenna does not cause any variation of the return losses ($|S_{11}|$) and transfer characteristics ($|S_{21}|$) [31]. This implies that the roll monopole antennas feature omnidirectional radiation, and are suitable for mobile application scenarios.

For impulse radio systems, it is important to examine the performance of UWB antennas in the time domain in terms of system gain as well as the waveforms of radiated and received pulses [31]. The frequency range of interest is the UWB band of 3.1 GHz to 10.6 GHz, which is subjected to a 10 dB bandwidth. To comply with the emission mask regulated by the FCC, the source signals are properly

Antenna Elements for Impulse Radio

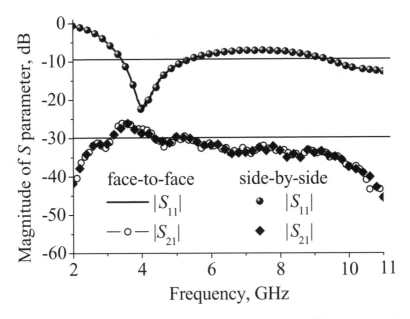

Figure 8.8 Measured $|S_{11}|$ and $|S_{21}|$ of the antenna system with difference orientations

selected as a first-order Rayleigh pulse $p_t(t) = t \exp[-(t/\sigma)^2]$ with $\sigma = 50$ ps or a duration of $T = 250$ ps ($\approx 5\sigma$) for the single-band scheme.

Figure 8.9 shows the measured waveform of the received pulse at the output of the receive antenna. The small size of the antennas prevents the nonlinear phase response. Therefore, the miniaturing design of UWB antennas is conducive to system performance and portable design of devices. However, the varying transfer response distorts the waveform of the received pulse with ringings, which is similar to a second-order Rayleigh pulse.

The system gain is one important performance parameters and defined as

$$G_s(\theta, \phi) = \frac{\int_0^\infty |V_o(t)|^2/(Z_l)dt}{\int_0^\infty P_s(\theta, \phi, t)/(4\pi r^2)dt} \text{ dBi} \tag{8.5}$$

where, Z_l and V_o stand for the loading impedance and the output voltage of a receive antenna, and P_s the source power at the input of a transmit antenna, respectively.

The system gain against the first-order Rayleigh pulse of a pulse duration σ varying from 10 ps to 110 ps is given in Figure 8.10. For the source pulse of $\sigma = 30$–110 ps, the antenna system has higher gain ranging from -35 to -32 dBi.

8.3.2 Directional Antipodal Vivaldi Antenna

The directional design is used for fixed base stations with certain coverage and point-to-point communication systems. Usually, the stable radiation performance of the antenna is a critical design issue. Horn antennas are good options for such purposes but not suitable for applications with size constraints. Alternatively, two-dimensional linearly polarised tapered slot antennas, in particular, Vivaldi antennas are

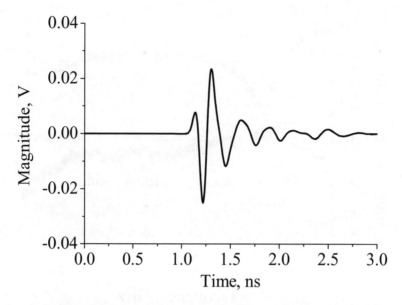

Figure 8.9 Measured waveform of received pulse in a single-band scheme

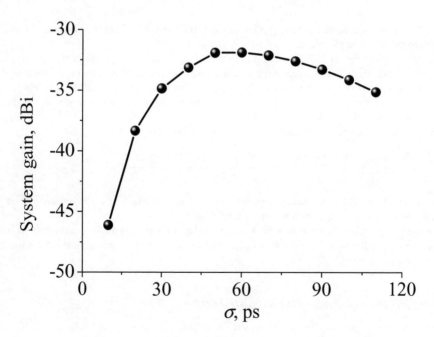

Figure 8.10 Measured system gain for a first-order Rayleigh pulse

Antenna Elements for Impulse Radio

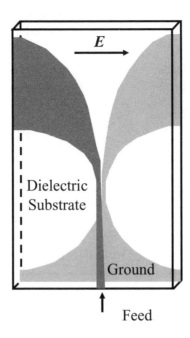

Figure 8.11 Geometry of antipodal Vivaldi antenna

widely used for UWB systems. A Vivaldi antenna is formed by elliptically tapering the inner and outer edges of the slotline conductors of the antipodal Vivaldi radiator. To simplify the feeding structure, a modified version of Vivaldi antenna, antipodal Vivaldi antenna is designed as shown in Figure 8.11. The antenna can be further enhanced by using semicircular loadings at the ends of the slotline as shown in Figure 8.12 [32, 33].

The antipodal Vivaldi antenna comprises tapered radiating slot and feeding transition. The symmetric tapered radiating slot is formed by two arms printed on opposite surfaces of a dielectric substrate. The maximum flare width of the slot determines the lower edge operating frequency. The feeding transition consists of a 50 Ω microstrip line linearly tapered to a parallel strip line to feed the tapered slot radiator, whereas the grounded trace is elliptically tapered. Two semicircular loadings are added at the ends of the slotline. Adding the curved loadings at the ends improves the radiation performance due to the reduction in the reflection at sharp ends as shown in Figure 8.11, and reduces the lower edge operating frequencies by widening the flare width.

Figure 8.13 compares the return losses and gain between conventional and modified antipodal Vivaldi antennas, which have the same configurations except for the semicircular loadings added to the latter as shown in Figures 8.11 and 8.12. The lower edge frequency is reduced from 3.05 GHz for the conventional antipodal Vivaldi antenna to 2.35 GHz for the modified one. The impedance matching of the latter antenna is also improved across 2.35 GHz to 12.0 GHz. More importantly, the modified design is capable of providing flatter gain response ranging from 4.0 dBi to 8.0 dBi than that of conventional design.

Figures 8.14 and 8.15 show the measured radiation patterns for the modified antipodal Vivaldi antenna at 7 GHz in both the E- and H-planes. The directional radiation pointing to boresight is observed. The half-power beamwidths are 42° in the E-plane and 68° in the H-plane. The measured results also show desired radiation performance with stable radiation direction, beamwidths and gain over the UWB bandwidth.

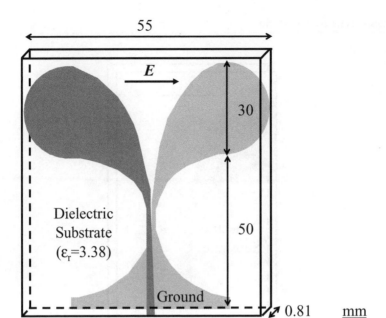

Figure 8.12 Geometry of antipodal Vivaldi antenna with semicircular loadings

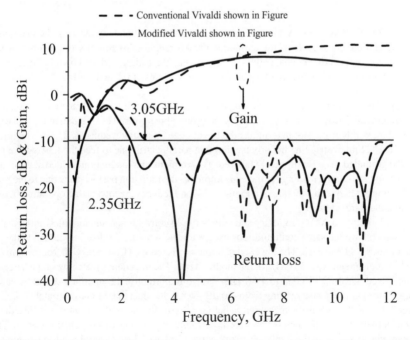

Figure 8.13 Comparison of return losses and gain between conventional and modified antipodal Vivaldi antenna with semicircular loadings

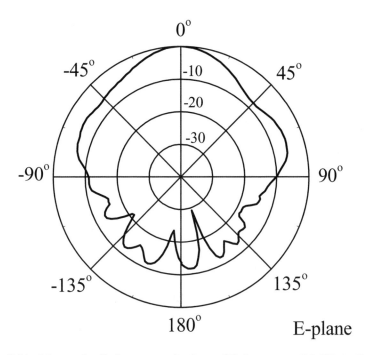

Figure 8.14 Measured radiation patterns for the modified antenna at 7.0 GHz in E-planes

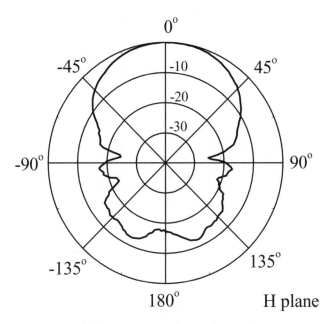

Figure 8.15 Measured radiation patterns for the modified antenna at 7.0 GHz in H-planes

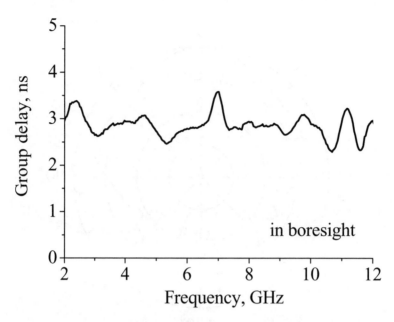

Figure 8.16 Measured group delay for the modified antenna at boresight

Moreover, group delay of the modified antipodal Vivaldi antenna is examined, which is defined as the negative derivative of the phase response with respect to frequency. Figure 8.16 shows the measured group delay at boresight due to the antenna. Across the UWB band, the group delay varies from 2.5 ns to 3.5 ns. Due to the good gain and phase (group delay) responses, the distortion in the waveform of radiated pulses by the modified antipodal Vivaldi antenna is small. For example, the fidelity of the radiated pulse with Gaussian monocycle source pulse $V_{(t)} = -\frac{t}{\sigma}e^{-(\frac{t}{\sigma})^2}$ is higher than 0.9 for $\sigma = 35\text{--}200$ ps.

8.4 Summary

In this chapter, antenna design issues for impulse radio systems were presented. First, the existing UWB antennas were classified in terms of radiation characteristics and geometry. The features of several typical antenna solutions were described briefly. Then the special design considerations for UWB antennas were highlighted with the introduction of some parameters for evaluating UWB antenna performance. Finally, the case study was carried out with two design examples. One was the omnidirectional design. A new roll monopole antenna which features broadband width and omnidirectional radiation was exemplified. The other one was the modified antipodal Vivaldi antenna, which was capable of providing good impedance and stable directional radiation responses across the UWB bandwidth.

In short, in antenna designs for impulse radio systems, the unique consideration is the phase response of the antenna. The good way to measure the performance of antenna system is to use a transfer function, which covers impedance matching, system gain, polarisation matching and phase delay (group delay) in the frequency domain. By means of inverse fast Fourier transform, the time-domain characteristics of the system can be obtained. In addition, the transfer function can readily be calculated and measured.

References

[1] Z.N. Chen, X.H. Wu, N. Yang and M.Y.W. Chia, Considerations for source pulses and antennas in UWB radio systems, *IEEE Trans. Antennas Propagat.*, **52**(7), 1739–48, 2004.
[2] X.M. Qing and Z.N. Chen, Transfer functions measurement for UWB antenna, *IEEE Int. Symp. Antennas Propagat.*, 2532–5, 2004.
[3] E.A. Theodorou, M.R. Gorman, P.R. Rigg and F.N. Kong, Broadband pulse-optimized antenna, *Proc. Inst. Elec. Eng.*, **128**(H), 124–30, 1981.
[4] M. Kanda, Transients in a resistively loaded linear antenna compared with those in a conical antenna and a TEM horn, *IEEE Trans. Antennas & Propagat.*, **28**(1), 132–6, 1980.
[5] M. Kanda, The effect of resistive loading of 'TEM' horns, *IEEE Trans. Electromagn. Compat.*, **24**, 245–55, 1982.
[6] K.L. Shlager, G.S. Smith and J.G. Maloney, Accurate analysis of TEM horn antennas for pulse radiation, *IEEE Trans. Electromagn. Compat.*, **38**(3), 414–23, 1996.
[7] L.T. Chang and W.D. Burnside, An ultrawide-bandwidth tapered resistive TEM horn antenna, *IEEE Trans. Antennas Propagat.*, **48**(12), 1848–57, 2000.
[8] R.T. Lee and G.S. Smith, On the characteristic impedance of the TEM horn antenna, *IEEE Trans. Antennas Propagat.*, **52**(1), 315–18, 2004.
[9] B. Scheers, M. Acheroy and A. Vander Vorst, Time-domain simulation and characterisation of TEM horns using a normalised impulse response, *IEE Proc. Microw., Antennas Propagat.*, **147**(6), 463–8, 2000.
[10] C.W. Harrison, Jr and C.S. Williams, Jr, Transients in wide-angle conical antennas, *IEEE Trans. Antennas Propagat.*, **13**(2), 236–46, 1965.
[11] S.S. Sandler and R.W.P. King, Compact conical antennas for wide-band coverage, *IEEE Trans. Antennas Propagat.*, **42**(3), 436–9, 1994.
[12] T.W. Hertel and G.S. Smith, On the dispersive properties of the conical spiral antenna and its use for pulsed radiation, *IEEE Trans. Antennas Propagat.*, **51**(7), 1426–33, 2003.
[13] T.T. Wu and R.W.P. King, The cylindrical antenna with nonreflecting resistive loading, *IEEE Trans. Antennas & Propagat.*, **13**, 369–73, 1965.
[14] D.L. Senguta and Y.P. Liu, Analytical investigation of waveforms radiated by a resistively loaded linear antenna excited by a Gaussian pulse, *Radio Science*, **9**, 621–30, 1974.
[15] G.H. Brown and O.M. Woodward, Experimentally determined radiation characteristics of conical and triangular antennas, *RCA Review*, **13**, 425–52, 1952.
[16] H. Meinke and F.W. Gundlach, *Taschenbuch der Hochfrequenztechnik*, Berlin, Springer-Verlag, pp. 531–5, 1968.
[17] G. Dubost and S. Zisler, *Antennas a Large Bande*, Paris, Masson, pp. 128–9, 1976.
[18] S. Honda, M. Ito, H. Seki and Y. Jinbo, A disk monopole antenna with 1:8 impedance bandwidth and omnidirectional radiation pattern, *ISAP'92*, Sapporo, Japan, 1145–8, 1992.
[19] M. Hammoud, P. Poey and F. Colombel, Matching the input impedance of a broadband disc monopole, *Electronics Letters*, **29**, 406–7, 1993.
[20] K.L. Shlager, G.S. Smith, and J.G. Maloney, Optimization of bow-tie antennas for pulse radiation, *IEEE Trans. Antennas & Propagat*, **42**, 975–82, 1994.
[21] M.J. Ammann, Square planar monopole antenna, *National Conf. Antennas & Propagat.*, York, England, 37–40, 1999.
[22] M.J. Ammann, Impedance bandwidth of the square planar monopole, *Microw. Opt. Techno. Letters*, **24**, 185–7, 2000.
[23] Z.N. Chen and M.Y.W. Chia, Impedance characteristics of trapezoidal planar monopole antenna, *Microw. Opt. Techno. Letters*, **27**, 120–2, 2000.
[24] M.J. Ammann and Z.N. Chen, Wideband monopole antennas for multi-band wireless systems, *IEEE Antennas Propagat. Magazine*, **45**(2), 146–50, 2003.
[25] Z.N. Chen, X.H. Wu, N. Yang and M.Y.W. Chia, Considerations for source pulses and antennas in UWB radio systems, *IEEE Trans. Antennas Propagat.*, **52**(7), 1739–48, 2004.
[26] R.Q. Lee and R.N. Simons, Tapered slot antennas, Chapter 9 of *Advances in Microstrip and Printed Antennas*, edited by K.F. Lee and W. Chen, New York, 1997.

[27] D. Lamensdorf and L. Susman, Baseband-pulse-antenna techniques, *IEEE Antennas Propagat. Magazine*, **36**(1), 20–30, 1994.
[28] T. Wang, Z.N. Chen and K.S. Chen, Effect of selecting antenna and template on BER performance in pulsed UWB wireless communication systems, *IEEE Intl. Workshop on Antenna Technology*, Singapore, 446–9, 7–9 March, 2005.
[29] Z.N. Chen, Broadband roll monopole, *IEEE Trans. Antennas and Propagat.*, **51**(11), 3175–7, 2003.
[30] Z.N. Chen, M.Y.W. Chia and M.J. Ammann, Optimization and comparison of broadband monopoles, *IEE Proceedings: Microw. Antennas and Propagat.*, **150**(6), 429–35, 2003.
[31] Z.N. Chen, A new bi-arm roll antenna for UWB applications, *IEEE Trans. Antennas Propagat.*, **53**(2), 672–7, 2005.
[32] E. Gazit, Improved design of the Vivaldi antenna, *IEE Proceedings: Microw. Antennas and Propagat.*, **135**, 89–92, 1988.
[33] X.M. Qing and Z.N. Chen, Antipodal Vivaldi antenna for UWB applications, *Euro Electromag.-UWB SP7*, Magdeburg, Germany, 12–16 July, 2004.

9

Planar Dipole-like Antennas for Consumer Products

Peter Massey

9.1 Introduction

Mass market consumer products require antennas that are cheap to construct. Three-dimensional structures such as bicones, discones and volcano smoke antennas have relatively good radiation properties across a very broad band. However their bulk and construction argue against use in low-cost consumer devices. Consequently types that are flat or near flat are considered here. This chapter concentrates on the properties of broadband printed dipole antennas and their practical variations.

The chapter is arranged as follows:

- Section 9.2 overviews the computer simulation methods and measurement techniques that are used in the chapter.
- In Section 9.3 the classical bicone antenna is reviewed. While this is unsuitable for low-cost applications, this design can be analysed using partly analytic techniques, and therefore is useful in introducing the general properties of broadband dipole antennas, and in outlining a simple model that describes their operation.
- Section 9.4 considers a range of flat-element broadband dipoles. This includes bowtie, circular and elliptical element and diamond dipole designs.
- Section 9.5 describes some practical designs that can be made from or mounted on printed circuit boards.
- Section 9.6 summarises the main points of the chapter, and attempts to compare the merits of the antennas studied here.

Ultra-wideband Antennas and Propagation for Communications, Radar and Imaging   Edited by B. Allen,
M. Dohler, E. E. Okon, W. Q. Malik, A. K. Brown and D. J. Edwards
© 2007 John Wiley & Sons, Ltd

9.2 Computer Modelling and Measurement Techniques

Computer modelling. The bicone modelling was carried out using a commercial finite difference time domain (FDTD) type code. Most of the other simulations were undertaken using a commercial frequency domain finite element (FE) code. The choice of software was mainly determined by what was accessible rather than by the program's particular advantages.

Many wideband antenna designers prefer FDTD codes for their ability to quickly compute antenna properties across a large frequency range from a single time-domain simulation. In the author's experience, the computational speed of time-domain codes can be severely reduced by the limitation that the time step is proportional to the smallest feature size. Consequently the faster FE codes are often as quick as and sometimes quicker than FDTD software.

To preserve accuracy, the finite element simulations were done in three parts: 1–3 GHz, 3–6 GHz and 6–10 GHz bands. For each frequency band the FE mesh was optimised to give accurate results at the central frequency of the band. Consequently the antenna return loss results are also in three parts, and the accuracy of the simulation can be in part judged by how close the S parameter values of the different parts are at their band edges.

Measurements. Measurements of reflection coefficient can be easily made and compared with the simulation results. This is done as spot checks for many of the computer predictions discussed in the following sections. Radiation pattern measurements are far more difficult to compare with theory as often they are strongly influenced by the presence of the feed cable to the antenna under test.[1] This effect is described in some detail for the commercial antennas discussed in Sections 9.5.1 and 9.5.2.

9.3 Bicone Antennas and the Lossy Transmission Line Model

The fields within the mouth of a bicone antenna (Figure 9.1) can be expressed in terms of algebraic functions of the r, θ, ϕ spherical coordinates [2]. The lowest order mode is the TEM mode. This mode can be thought of as the fundamental mode of the radial transmission line that lies between the cones, and has a characteristic impedance Z_0 which is given by:

$$Z_0 = \frac{377}{2\pi} \ln \frac{\cot(\theta_1/2)}{\cot(\theta_2/2)} \tag{9.1}$$

where θ_1 is the angle from the z axis subtended by the surface of the upper cone, and θ_2 is the angle from the z axis subtended by the surface of the lower cone. When the bicone is symmetric about the $x - y$ plane, $\theta_1 = 180° - \theta_2$. The grey curve in Figure 9.2 shows Z_0 versus θ_1 for this symmetric case.

When the bicone is transmitting, the wave in the bicone's radial transmission line leaves its terminals, travels to the outer edges of the bicone, from where it is reflected, and travels back to the antenna terminals. The wave loses energy to radiation along its path, and this radiation loss increases with increasing frequency. Hence the reflection coefficient seen at the antenna terminals is consistent with that from a transmission line of length corresponding to the distance between the antenna terminals and the edge of the antenna, which is open circuited at its far end, and which has an attenuation constant that increases with frequency. When this lossy open-ended transmission line model's behaviour is plotted on

---

[1] For a discussion of this see references [11] or [12]. The 'References' section of reference [13] serves as an extensive bibliography on this subject and also contains extended results on the effects of feed cable location and the use of ferrite and coaxial chokes to suppress the influence of the feed cable.

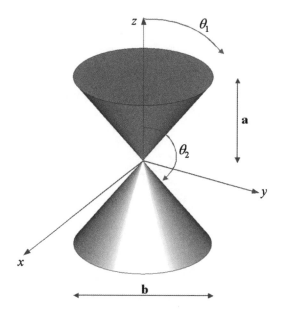

Figure 9.1 Bicone, showing flare angles of cones

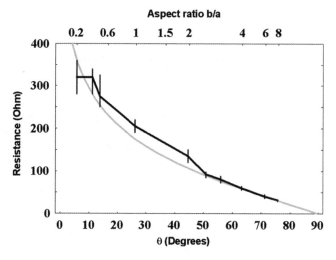

Figure 9.2 Characteristic impedance and limit impedance for symmetric bicones. Characteristic impedance shown in grey. High frequency limit impedance shown in black. Vertical lines denote error bars for limit impedance values

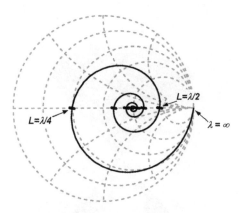

Figure 9.3 Smith chart for reflection from a transmission line with an attenuation constant that increases linearly with frequency

a Smith chart, it results in a curve such as shown in Figure 9.3. The curve starts at the low-frequency open-circuit condition, and follows a clockwise spiral inwards towards a limit that corresponds to the characteristic impedance of the transmission line.

The Smith chart plot shown in Figure 9.3 is for a the reflection coefficient of 100 ohm transmission line, with the frequency varying between 0 and the value required to make L (the line length) two wavelengths long. The attenuation coefficient α has been made proportional to frequency, with the proportionality constant set so that when the wavelength equals the line length, then the voltage of a travelling wave drops to $1/e$ of its initial value along a distance of one wavelength. The assumption that the attenuation constant is proportional to frequency means that the power decay is proportional to the frequency squared.

The reflection coefficient curve in Figure 9.3 crosses the real impedance line when the line length L is an integer number of quarter-wavelengths. When L is an odd number of quarter-wavelengths the resistance is lower than the high-frequency limit impedance. When L is an even number of quarter-wavelengths the resistance is higher than the high-frequency limit impedance. The $L = \lambda/4$ case is corresponds to the antenna being used as an approximately half-wavelength long dipole.

The Smith charts for the responses of real bicone antennas are very similar to the response shown in Figure 9.3, the major difference being in the value of the limit impedance, which depends on the cone flare angles. A number of symmetric bicone antennas of varying values of θ_1 were modelled with FDTD software and their high-frequency limit impedance values are plotted in black in Figure 9.2. The high frequency limit impedance curve follows the characteristic impedance curve to within the accuracy of the computer simulations.

As subsequent sections show, the lossy transmission line model introduced above is a fair representation for the response of many other antennas. Also it predicts the standing and travelling wave behaviour of the antenna currents described in Chapter 7. At low frequencies the attenuation is insufficient to greatly degrade the amplitude of the signal reflected from the end of the antenna, and the transmitted and reflected waves combine to give a standing wave across the antenna surface. At high frequencies the attenuation is so high that there is little reflected wave and then the currents form a travelling wave. At intermediate frequencies the attenuation is sufficient to allow the transmitted wave to dominate at the base of the antenna, giving a travelling wave distribution in this region. However, near the end of the antenna element, there is insufficient distance in which to significantly attenuate the transmitted and

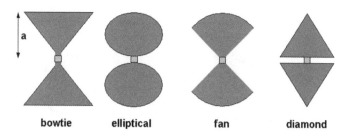

Figure 9.4 Types of wide planar dipole

reflected waves and these waves have similar amplitudes. Therefore near the top of the antenna, the current distribution is that of a standing wave.

The lossy transmission line model can break down due to a host of mechanisms. As discussed in Chapter 7, a major limitation is the influence of the finite size of the feed upon the high-frequency match. Subsection 9.5.1.3 illustrates how the antenna response is influenced by several other aspects, including the presence of a dielectric substrate and feedline.

9.4 Planar Dipoles

This section discusses the performance of a number of shapes for thick flat dipole elements. The dipole elements considered are shown in Figure 9.4. In all the cases, the element height a is kept at 25 mm. However, the width is varied as described in the individual subsections. The antennas are modelled as fed by a 50 ohm element depicted by the square in the middle of the dipoles shown in Figure 9.4. This feed element is 0.8 mm square. Where necessary to avoid shorting out the feed element, the gap between the dipole elements is maintained at 0.8 mm by trimming the elements. Also, where necessary to avoid bad contact with the feed element, a small 0.4 mm wide rectangle of conductor is welded to the antenna elements.

9.4.1 Bowtie Dipoles

The bowtie elements were modelled with a range of ratios b/a, shown in Figure 9.5. This figure shows the S parameter responses for some typical cases. The curve between 1 to 3 GHz and 6 to 10 GHz is shown in black, and for 3 to 6 GHz is shown in grey. Markers are placed at 1 GHz intervals between 2 and 9 GHz. All the Smith charts are normalised with 50 ohms at their centre.

The high frequency limit of the matching impedance at high frequency is around 200 to 250 ohms. Figure 9.6 shows the return loss if a 50 ohm feed is used. A narrow band match is achieved at around 2 GHz. When a 200 ohm feed is used, a better broadband match is achieved for frequencies above 3 GHz, as shown in Figure 9.7.

Figure 9.8 shows radiation patterns for $b/a = 1$ at a range of frequencies. Above 3 GHz the radiation pattern moves away from the classical doughnut-shaped pattern of a dipole.[2] For other values of b/a, the pattern variation is qualitatively very similar. However, the shape of the patterns at higher frequencies varies with b/a.

---

[2] This is true even for the $b/a = 0.4$ case. The 6 GHz pattern of this antenna is doughnut shaped, but is much flatter than that at 1 or 3 GHz.

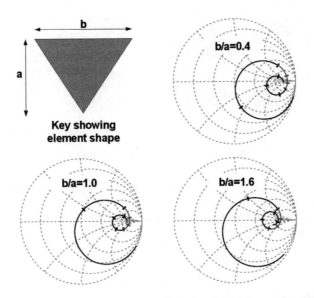

Figure 9.5 Free-standing triangular element S parameter responses for a range of aspect ratios. 1 to 3 GHz and 6 to 10 GHz results shown in black, 3 to 6 GHz results shown in grey

Phase linearity and distortion. The distortion of the transmitted signal is determined by the phase as well as the amplitude of the antenna's radiated field. Figure 9.9 shows the phase of the θ component of the electric field multiplied by e^{+jkR}, where $k = 2\pi/\lambda$ is the wavenumber (λ is the wavelength) and R is the distance from the coordinate system origin. This multiplication by e^{+jkR} makes the phase independent of distance from the origin, and dependent only upon angle. The phase is relative to the phase of the voltage of the signal incident upon the feed.

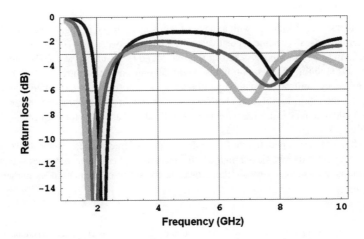

Figure 9.6 Bowtie element return loss for 50 Ω feed. $b/a = 0.4$ shown in black, $b/a = 1.0$ shown in grey, $b/a = 1.6$ shown in thick light grey. Horizontal gridlines are at -3, -6, -7 and -10 dB

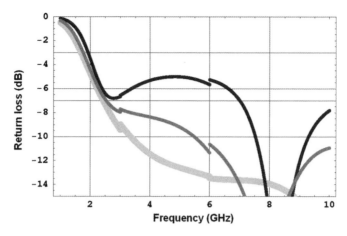

Figure 9.7 Bowtie dipole return loss for 200 Ω feed. $b/a = 0.4$ shown in black, $b/a = 1.0$ shown in grey, $b/a = 1.6$ shown in thick light grey. Horizontal gridlines are at $-3, -6, -7$ and -10 dB

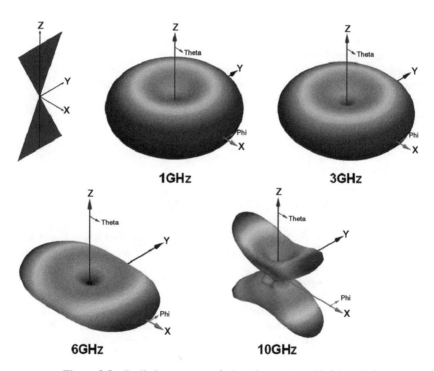

Figure 9.8 Radiation patterns of a bowtie antenna with $b/a = 1.0$

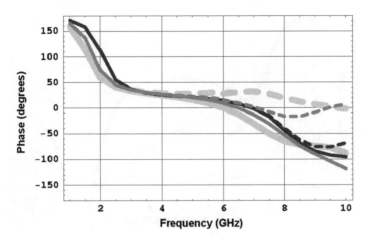

Figure 9.9 Phase of far-field for bowtie antennas for 50 Ω feed at $\theta = 90°$. Phase at $\phi = 0°$ shown in solid lines. Phase at $\phi = 90°$ shown in dashed lines. $b/a = 0.4$ shown in black, $b/a = 1.0$ shown in grey and $b/a = 1.6$ shown in thick light grey

In this chapter, the convention is that the coordinate system origin is coincident with the antenna feed. Because of the symmetries of the antennas described in Sections 9.3 and 9.4, for all of them, there is no ϕ component of electric field in the $x - y$ plane.

In order to explain the behaviour of the phase with frequency it is necessary to review the formulae for radiation. This is done below. First the relationship between current and far-field is given, and then this is related to the voltage across the antenna feed.

The formulae for electric far-field radiation **E** in terms of the current on the antenna is:

$$\mathbf{E} = -j\omega\{0, A_\theta, A_\phi\} \tag{9.2}$$

where the magnetic potential vector **A** is given by:

$$\mathbf{A} = \{A_r, A_\theta, A_\phi\} = \frac{\mu_0}{4\pi} \frac{e^{-jkR}}{R} \iiint_v \mathbf{J} e^{jk\vec{\rho}\cdot\hat{r}} dv \tag{9.3}$$

where R is the distance between the origin and the point in the far-field where **E** and **A** are being evaluated, and $\mathbf{J}(\vec{\rho})$ is the current density, $\vec{\rho}$ is the location vector of the current density, and $\hat{r}$ is the direction vector in the (θ, ϕ) far-field direction.

Assuming that the current's phase is constant across the dipole – then these formulae show that for electrically small antennas where $e^{jk\vec{\rho}\cdot\hat{r}} \approx 1$, $e^{+jkR}\mathbf{E}$ lags behind the current vector **J** by 90°.

Feed voltage/current relationship in the low frequency limit. An electrically small dipole looks like a slightly lossy capacitor. For electrically very small dipoles, the resistive part of the impedance is very small, and the current **J** leads the feed voltage **V** by 90°. Therefore $e^{+jkR}\mathbf{E}$ is in phase with **V**. This means that in the $x - y$ plane, the z component: $e^{+jkR}E_z$ is in phase with the voltage **V** across the feed. As in the $x - y$ plane E_θ points in the opposite direction to E_z, then $e^{+jkR}E_\theta$ is 180° out of phase with the voltage across the feed.

Phase variation with increasing frequency. As the frequency increases the resistive contribution to the impedance of the small dipole becomes significant. This results in a reduction of current **J**'s phase

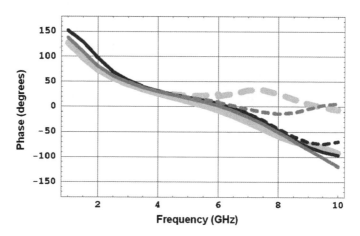

Figure 9.10 Phase of far-field for bowtie antennas for 200 Ω feed at $\theta = 90°$. Phase at $\phi = 0°$ shown in solid lines. Phase at $\phi = 90°$ shown in dashed lines. $b/a = 0.4$ shown in black, $b/a = 1.0$ shown in grey and $b/a = 1.6$ shown in thick light grey

relative to that of feed voltage **V**'s phase. Therefore as the frequency increases, then $e^{+jkR}E_\theta$ reduces from 180° towards 0°.

Phase of larger dipoles. For larger dipoles $e^{jk\bar{\rho}\bullet\hat{r}} \approx 1$ is no longer true. Then a dipole antenna's radiation pattern diverges from the classical doughnut shape. Also the phase relationship between far-field and feed voltage becomes more complex. This is seen in Figure 9.9 above around 5 GHz. As the pattern looses its z-axis symmetry and the amplitude of the far-field in the $+x$ and $+y$ directions diverges, the phase also diverges. The amplitude and phase divergence is stronger for the wider antennas.

Phase variation with feed impedance. Figure 9.9 shows the results for a feed impedance of 50 ohms, as this is a common choice for feed impedance. However the Smith chart plots (Figure 9.5) show that 200 ohms would give a better match at high frequencies. The phase variation with a 200 ohm feed is shown in Figure 9.10. Qualitatively the plots have the same features as depicted for a 50 ohm feed. There are just slight variations in the phase values.

9.4.2 Elliptical Element Dipoles

This antenna type is similar to the PulsON 200™ antenna described in Section 9.5.1, except that the PulsON antenna is built onto the dielectric sheet of a PCB, and this strongly influences the antenna impedance.

The elliptical elements were modelled with a range of ellipticities, as shown in Figure 9.11. The Smith chart key is the same as described in Section 9.4.1.

As the ellipticity increases the high-frequency response converges better and its centre value reduces slightly. For $b/a = 1.6$, the value is about 100 ohms. Figure 9.12 shows the return loss seen for a 50 ohm feed, and Figure 9.13 shows the return loss seen at a 100 ohm feed. The wider antennas have the best match.

A description of the radiation patterns of the oval element dipole is given in Section 9.5.1. The behaviour of the radiation patterns is similar to that of the other antennas, in that with increasing frequency, it moves away from the doughnut shape and becomes multi-lobed.

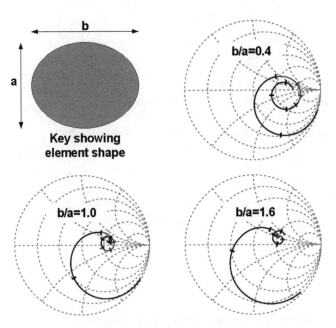

Figure 9.11 Free-standing oval element dipole S parameter responses for a range of aspect ratios. 1 to 3 GHz and 6 to 10 GHz results shown in black, 3 to 6 GHz results shown in grey

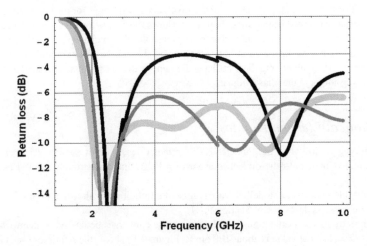

Figure 9.12 Return loss for oval element dipoles connected to 50 Ω feed. $b/a = 0.4$ shown in black, $b/a = 1.0$ shown in grey, $b/a = 1.6$ shown in thick light grey

Planar Dipole-like Antennas for Consumer Products

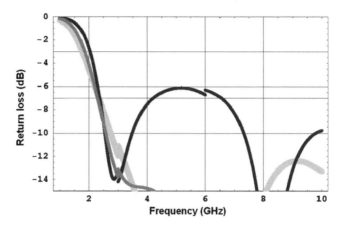

Figure 9.13 Return loss for oval element dipoles connected to 100 Ω feed. $b/a = 0.4$ shown in black, $b/a = 1.0$ shown in grey, $b/a = 1.6$ shown in thick light grey

Figures 9.14 and 9.15 show plots of the phase of the far-field against frequency for a standard 50 ohm feed impedance, and for 100 ohm feed, which is better matched at high frequencies. The general features of these plots are very similar to those for the bowtie antennas reported in Figures 9.9 and 9.10. The explanation for these features can be found in the text accompanying Figures 9.9 and 9.10.

9.4.3 Fan Element Dipoles

The fan element dipoles were modelled with a range of fan angles. Figure 9.16 shows the S parameter responses. The Smith chart key is the same as described in Section 9.4.1.

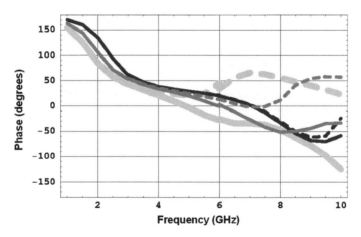

Figure 9.14 Phase of far-field for oval element dipoles for 50 Ω feed at $\theta = 90°$. Phase at $\phi = 0°$ shown in solid lines. Phase at $\phi = 90°$ shown in dashed lines. $b/a = 0.4$ shown in black, $b/a = 1.0$ shown in grey and $b/a = 1.6$ shown in thick light grey

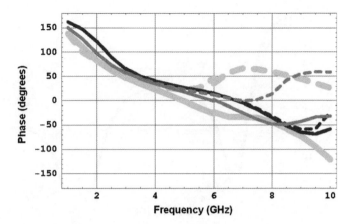

Figure 9.15 Phase of far-field for oval element dipoles for 100 Ω feed at $\theta = 90°$. Phase at $\phi = 0°$ shown in solid lines. Phase at $\phi = 90°$ shown in dashed lines. $b/a = 0.4$ shown in black, $b/a = 1.0$ shown in grey and $b/a = 1.6$ shown in thick light grey

As the fan angle increases the high-frequency limit reduces slightly from around 250 to 150 ohms. However, beyond fan angles of around 70°, the diameter of the high-frequency S parameter curve starts increasing. At a fan angle of 170° the curve crosses the real axis at about 8 GHz and less than 50 ohms. However, the curve then spans 40 to 300 ohms, which is too large for easy broadband matching.

Figures 9.17 and 9.18 show some radiation patterns for the fan element with ang $= 90°$ and 140°. For small angles both the antenna and its radiation patterns are very similar to thin bowties. The radiation patterns for ang $= 90°$ are similar to those of the bowtie with $b/a = 1$ (Figure 9.8).

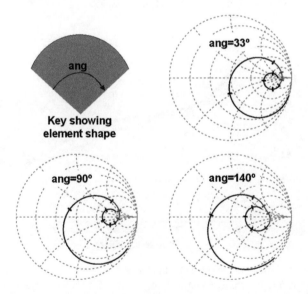

Figure 9.16 Free-standing fan element dipole responses for a range of aspect ratios 1 to 3 GHz and 6 to 10 GHz results shown in black, 3 to 6 GHz results shown in grey

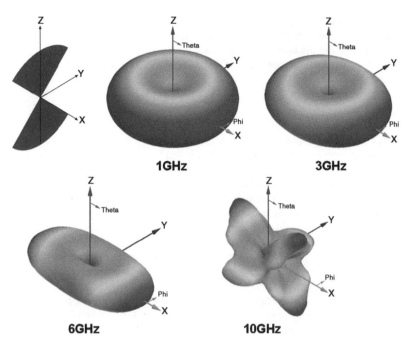

Figure 9.17 Radiation patterns of fan dipole with ang = 90°

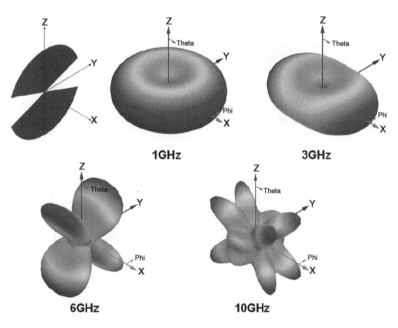

Figure 9.18 Radiation patterns of fan dipole with ang = 140°

The radiation patterns for ang = 140° change more rapidly with increasing frequency, and have similarities to those for the diamond dipole shown in Figure 9.20. These last features are probably due to the reduced width of the slot between the elements.

Phase plots for the fan dipole are not shown. They are generally similar to those for the bowtie antenna.

9.4.4 Diamond Dipoles

These antenna elements are similar to those described by Schantz and Fullerton [7]. The main difference between Schantz's dipoles and the ones discussed here is that Schantz's dipoles are printed on a dielectric substrate.

Figure 9.19 shows the element shape key and Smith charts for various element shapes. The Smith chart key is the same as described in Section 9.4.1. As promised in Schantz and Fullerton's paper, for sufficiently wide dipole elements, the antenna reduces its resonant behaviour.

In view of Schantz's claims of little pulse distortion, it is interesting to examine the radiation pattern variation with frequency. Figure 9.20 shows the radiation patterns at various frequencies for the $b/a = 1.6$ case. This case was chosen as the S11 parameter response shows little resonance for this value of b/a. While at low frequencies the radiation pattern is close to the doughnut pattern of the small dipole, at higher frequencies there are significant variations in directionality.

Figure 9.21 shows the phase of the far-field radiation in the $x - y$ plane. As for the bowtie, fan dipole and oval element dipoles, at low frequencies the phase is close to 180°, and tends to 0° with increasing frequency. The explanation for this behaviour is given in Section 9.4.1 on bowtie antennas.

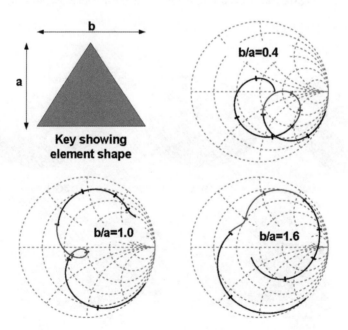

Figure 9.19 Free-standing diamond dipole S parameter responses for a range of aspect ratios. 1 to 3 GHz and 6 to 10 GHz results shown in black, 3 to 6 GHz results shown in grey

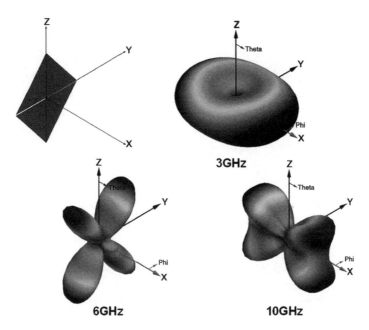

Figure 9.20 Radiation patterns of a diamond dipole with $b/a = 1$

At higher frequencies the phase goes through an inflection. The frequency where this happens is higher for narrower designs. This rapid change of phase is related to the sweeping of S11 from the top (inductive) half of the Smith chart to the bottom half of the Smith chart, resulting in a switch in the relative phases of the voltage and current of the incident signal.

As in the bowtie, fan and oval element dipoles, the phase in the +x and +y directions diverges at high frequencies, with the divergence being greater for wider aspect ratio designs.

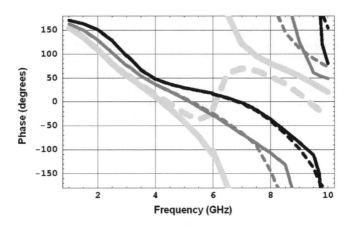

Figure 9.21 Phase of far-field at $\theta = 90°$ for diamond dipoles with 50 Ω feed

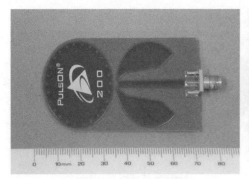 

Figure 9.22 Front and rear views of the PulsON antenna

9.5 Practical Antennas

The antennas described in Section 9.4 are not mechanically robust as their elements are only connected by the feed structure. This section discusses several antennas that are stronger as they are built on PCB materials. The first two antennas have been manufactured commercially. The first antenna is a printed elliptical element dipole, which is a development of the elliptical element dipoles of Section 9.4.2, where the dipole elements are formed from copper track patterns upon a PCB. The second antenna uses monopole with a matching line connecting the top of the monopole to the ground. This differs from other antennas discussed in the chapter in that an antenna element is formed from a low profile three-dimensional metal structure, which is mounted upon a PCB. The final antenna discussed is much larger than the others. This is a Vivaldi (or tapered slot) antenna, which offers directional radiation at the cost of increased size.

9.5.1 Printed Elliptical Dipoles

This section starts by describing a commercial example of a printed elliptical dipole: Time Domain's 'PulsON 200'.[3] Then simulations are used to identify the contributions of the features of the practical design to the electrical performance of the antenna.[4]

Figure 9.22 shows the PulsON 200 antenna. The elliptical outline of the elements is claimed to give low phase distortion across the operating band. The upper arm (to the left in the pictures) is printed on both sides of the substrate, the sides being connected by a ring of via holes at the edge of the ellipse.

The feed is a microstrip, which through a gradual narrowing of the earth plane becomes a twinline where it connects to the antenna elements. This transition has a balun action. The lower arm is fed by the hot side of the microstrip feed, and has to have a large slot (shown in the picture on the left side of Figure 9.22) to accommodate the microstrip. The upper arm is fed from the earth side of the microstrip line.

The sample antennas were intended for operation between 3.0 and 5.5 GHz. Two antennas were available for measurements, and any differences between the performances of these antennas are described below.

---

[3] 'PulsON' and 'PulsON 200' are registered trade marks of Time Domain Corporation.
[4] The simulations and the conclusions drawn from them are of approximations to practical antenna designs and do not necessarily reflect the actual performance of any commercially available antennas.

Planar Dipole-like Antennas for Consumer Products

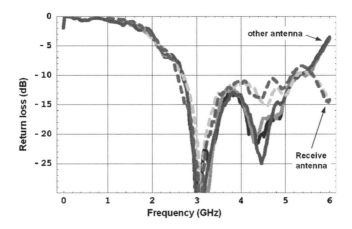

Figure 9.23 Measured PulsON 200 antenna return losses. Dashed lines denote 'receive' antenna measurements. Solid lines denote 'other' antenna measurements

9.5.1.1 Measured S Parameter Response

Two PulsON 200 antennas were available for measurement: one labelled 'receive' and the other denoted here as 'other'. They came with a range of coaxial feed arrangements. Return loss measurements of the antennas with and without these feed adaptors are shown overlaid in Figure 9.23. Their responses are very similar.

The 'other' PulsON 200 antenna came with a coaxial cable and plastic holder that clamped around the antenna's base. Its return loss below 9 GHz was measured on a wider band network analyser, and in Figure 9.24 the measurement result is shown compared with the original 0 to 6 GHz measurement. This

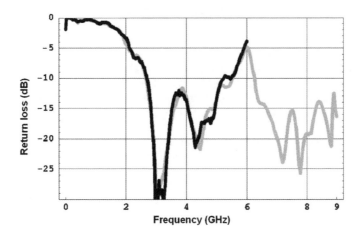

Figure 9.24 Return loss measurements of PulsON 200 antenna with feed cable and plastic holder. Black line shows measurement between 0 and 6 GHz of 'other' antenna with cable support. Grey line shows measurement between 0 and 9 GHz of same antenna

suggests that the reduced return loss on the right of Figure 9.23 is a peak, and despite being specified for operation between 3 and 5.5 GHz, the antenna continues to have good return loss up to much higher frequencies.

Smith charts of the S11 measurements are not shown here as they consist of curves that are tightly wound due to the electrical length of the feed cable and microstrip feed line.

9.5.1.2 Radiation

The radiation patterns vary dramatically with frequency. To obtain some understanding of the cause of their features, simulations were undertaken of a range of oval element based dipoles, where the features of the PulsON 200 design were progressively added.

The following simulation experiments were undertaken:

- A centrally fed dipole antenna consisting of two oval conductors.
- A centrally fed dipole antenna consisting of two oval conductors upon one side of the dielectric.
- A centrally fed dipole antenna with oval conductors printed upon both sides of the dielectric. The upper element is the same as used on the PulsON 200 antenna. The lower element is the same as the upper element.
- The PulsON 200 like antenna without the coaxial to microstrip connector or cable.
- The PulsON 200 like antenna.

The dielectric substrate used in the actual PulsON 200 antenna is Roger's 4003, which has a dielectric constant of 3.38 and loss tangent of 0.0021 at 2.5 GHz, rising to 0.0027 at 10 GHz. However, as this was not known at the time of simulation, the simulated antenna's substrate is modelled as GTEK, which has a relative dielectric constant of 3.9 and 0.012 loss tangent.

Figure 9.25 shows the radiation patterns for a dipole with oval elements the same size and shape as used on the PulsON 200 antenna. The radiation pattern at 1 GHz is the classical doughnut shape. As the frequency increases, the radiation pattern moves away from this simple shape. Above 3 GHz it goes through many different forms. At 10 GHz, the antenna is similar to back-to-back Vivaldi elements.[5]

The oval elements printed on one side of a PCB have practically the same (directivity) radiation patterns as those for oval elements printed on both sides of the PCB. Figure 9.26 shows the radiation patterns of a centre-fed structure, where the dipole elements are printed on both sides of a PCB of the same size and shape as used in upper element of the PulsON 200 antenna. The same size and number of vias as used in the upper element of the PulsON 200 antenna are used to make electrical contact between the upper dipole elements. This arrangement is reflected in the lower elements (This last detail is different from the PulsON 200 antenna, which instead has a single lower element on one side, which has been modified to accommodate the microstrip feed.)

Between 1 and 3 GHz the radiation patterns for the double-sided PCB dipole are the same as for the simple dipole. At higher frequencies the patterns differ, however, they still demonstrate increased directionality with frequency.

Figure 9.27 shows the radiation pattern for an antenna with dipole elements of the same dimensions as the PulsON 200 antenna. At 3 GHz and 6 GHz the pattern is slightly tilted out of the $x - y$ plane. The simulations above PCB suggest that this is not due to the dielectric. Rather it is due to the asymmetries in the element configuration, introduced by the feed line.

---

[5] This analogy was reported in Schantz's elliptical element paper, reference [5].

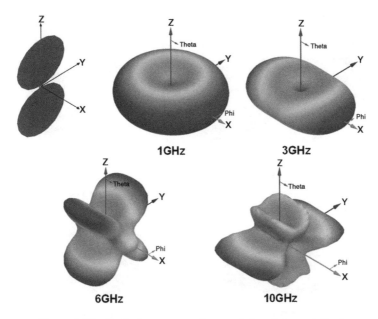

Figure 9.25 Radiation patterns of centre-fed oval element dipole

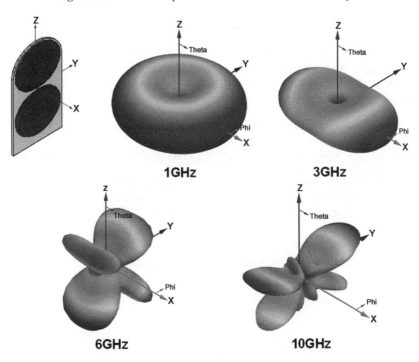

Figure 9.26 Radiation patterns of centre-fed oval elements printed on both sides of PCB

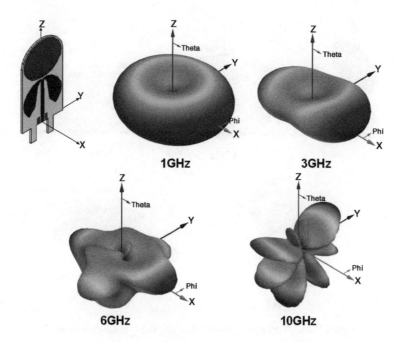

Figure 9.27 Radiation patterns of printed elliptical dipole antenna with gap source feed

Figure 9.28 shows the simulated radiation patterns for the full Pulson 200 like printed elliptical dipole antenna with coaxial to microstrip transition and feed cable. The feed cable extends down from the antenna to the edge of the modelled region.[6] The distance to the edge of the modelled region varies according to the simulation frequencies:

- For the 1 to 3 GHz simulations, the cable extends 75 mm beyond the lower edge of PCB.
- For the 3 to 6 GHz simulations, the cable extends 25 mm beyond the lower edge of the PCB.
- For the 6 to 10 GHz simulations, the cable extends ~13 mm beyond the lower edge of the PCB.

As only a small length of cable has been modelled, the simulated radiation patterns are likely to be different from those measured. Nevertheless the differences between the simulated radiation patterns shown in Figures 9.27 and 9.28 indicate that currents travel down the cable, and this cable strongly perturbs the radiation.

9.5.1.3 S Parameter Responses of Simulation Experiments

In Section 9.5.1.2, a number of progressively more complex models to investigate the contributions to the radiation patterns in the practical printed elliptical dipole antenna were discussed. This section discusses the S parameter responses of those models.

---

[6] The method of extending the feed cable to the radiation boundary of field-based simulations is discussed in reference [11].

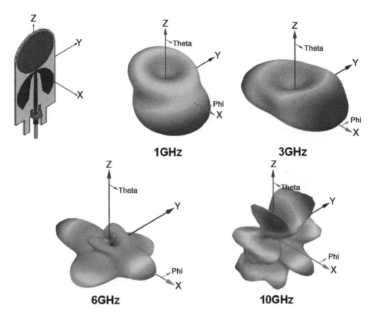

Figure 9.28 Simulated radiation patterns for printed elliptical dipole antenna with coaxial cable feed

Figure 9.29 shows Smith charts for five of the simulation models. The response for the antenna without the PCB converges on around 100 ohms for 3 GHz and above. Printing this antenna on the PCB reduces the matching impedance to around 75 ohms. Printing the oval dipole elements on both sides of the PCB results in the real part of the impedance further reducing to 50 ohms. However, this is at the cost of a greater spread in the reactance.[7]

The Smith chart in lower left corner of Figure 9.29 shows that the microstrip feed structure spreads out the 3 to 6 GHz response. The 3 GHz+ response is spread out in the full PulsON 200 like practical antenna case. Simulations of the PulsON 200 like antenna without feed cable, but with coaxial connector give a very similar response to the one shown for the cable, which suggests that the variation in S parameter response is mainly determined by the coaxial to microstrip transition.

Figure 9.30 shows the simulated return loss for the Pulson 200 like antenna fitted with a feed cable. In comparison with the measured responses of Figures 9.23 and 9.24, the simulation correctly predicts that the first return loss peak is at 3 GHz and that for higher frequencies the return loss is better than 8 dB. However, the exact locations and values of the higher frequency peaks and troughs do not coincide with either of the measured antennas.

Summarising, the printed elliptical dipole antennas studied are well matched at frequencies above around 2.5 GHz. The radiation patterns vary markedly with frequency. Radiating currents travel down the feed cable.

---

[7] A close-up view of the dipole on one and both sides of the PCB cases shows that the response curves do not quite line up. It appears that for these two cases, there are multiple resonances at around 6 GHz which to be accurately modelled require adaptive solutions at frequencies between 5 and 7 GHz.

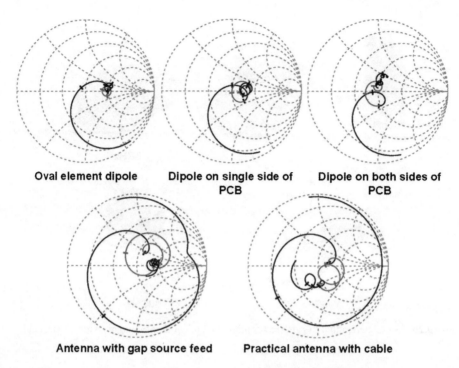

Figure 9.29 Smith charts for simulations of a free-standing and several printed elliptical dipole models. 1 to 3 GHz and 6 to 10 GHz results shown in black, 3 to 6 GHz results shown in grey

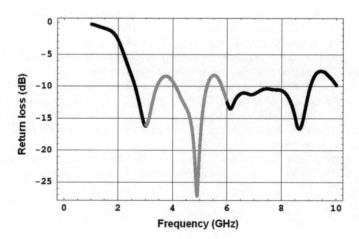

Figure 9.30 Simulated return loss for the Pulson 200 like antenna. 1 to 3 GHz and 6 to 10 GHz results shown in black, 3 to 6 GHz results shown in grey

9.5.2 Line-Matched Monopoles

The impedance match of dipoles and monopoles can be improved at certain frequencies by using an extra line. In dipoles this line connects the dipole elements directly to each other. However, such a line is more often used in monopoles, where it runs between the monopole element and the ground. Reference [9] includes a good overview of the simpler types of line-matched monopoles, where the line runs between part of the monopole that is near to the ground and the ground itself. In these cases, the line to ground improves the low frequency match of the antenna. It achieves this by introducing an inductance across the feed, which helps cancel out the capacitance of the monopole's low frequency response. The line also helps transform upwards the otherwise low radiation resistance experienced at low frequencies. At high frequencies the line's inductance increases, helping reduce its effects on the antenna impedance, which is generally satisfactory without the presence of the line.

The rest of this section concentrates on a commercial example of a line-matched monopole: the 'SMT-3TO1M', which is one of several designs manufactured by SkyCross.[8] In contrast to the elements discussed above, this antenna has a line that runs from the far end of the monopole element back to the ground. Section 9.5.2.1 describes the antenna in more detail, and in subsequent sections simulations are used to identify the contributions of the features of the practical design to the electrical performance of the antenna.[9]

9.5.2.1 Description of SMT-3TO1M Antenna

The SkyCross SMT-3TO1M antenna element is a commercially available low-cost stamped metal realisation consisting of a wideband monopole and a self-contained matching structure. Figure 9.31 shows the SMT-3TO1M antenna element in isolation (top of left photograph) and when mounted on a PCB test board (lower left of this photograph). The bare test board is shown on the right of the left photograph, and the rear test board is shown in the right-hand photograph. When mounted on the test board the element is fed by a microstrip line, which connects the element to an SMA coaxial connector. But in contrast to the Pulson and Vivaldi antennas, the monopole element itself is not printed on a PCB substrate, but instead the folded metal element is surface mounted.

The matching structure consists of a line from the middle of the underside of the open end of the monopole that runs back to near the monopole's feed and connects to the earth side of the PCB. This structure has the effect of reducing the mismatch over the antenna bandwidth and effectively lowers the low frequency cut-off frequency for a given monopole element size. Since the design is monopolar, the ground of the PCB that the antenna element is mounted upon acts as a counterpoise, carrying opposing currents from the feed.

9.5.2.2 Measured S Parameter Responses

Figures 9.32 and 9.33 show the measured return loss and S11 phase response. The 1 to 6 GHz measurements were made using a metre length of phase stable test cable. Moving a hand along this cable visibly perturbed the return loss display, indicating that the cable supported currents from the antenna.

---

[8] 'SkyCross' is a trademark of SkyCross Inc.
[9] The simulations and the conclusions drawn from them are of approximations to practical antenna designs and do not necessarily reflect the actual performance of any commercially available antennas.

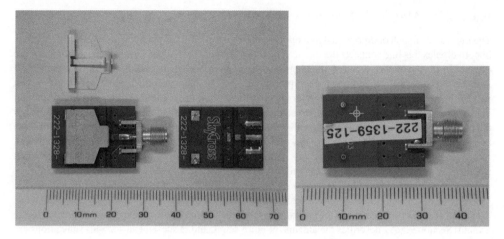

Figure 9.31 SMT-3TO1M antenna

The measurements repeated over 1 to 9 GHz were made using a network analyser that is mounted in a metal rack with a metal work surface. It was fitted with large diameter braided test cables. Consequently these measurements were more strongly affected by currents following down the test cable and coupling to the test furniture. The relatively close match to the 1 to 6 GHz measurements was only achieved by using a hand on the test cable to suppress resonances.[10]

9.5.2.3 Simulated S Parameter Responses

In view of the measurement observations of currents along the feed cable, and also in view of previous experience with similar antenna arrangements, the antenna was simulated with and without the feed cable. In the following figures, the result curves for the simulation with the feed cable are shown in black and the result curves for the simulations without the feed cable are shown in mid-grey. In addition there are some results for simulations in which both the feed cable and the SMA connector have been omitted, and these curves are portrayed in light grey.

The strip line from the middle of the underside of the monopole to the ground plane is an innovative feature of the SkyCross design that distinguishes it from the other antenna types in this chapter. Therefore simulations were undertaken of the antenna with and without this strip. These return loss curves for antennas without the strip are portrayed as dashed lines.

Figure 9.34 shows the return loss. The strip introduces extra notches in the return loss, which tend to reduce the antenna mismatch. However, as shown in Figure 9.35, the introduction of the notches also

---

[10] Current travelling from the ground down the feed cable is a very common phenomenon with monopole antenna designs. Using a ground that is several times larger than that of the test board would go some way in reducing the feed cable coupling. To achieve high levels of feed cable isolation in monopole-type designs. The use of current-choking techniques and careful placement of the feed cable connector have often been necessary. See references [11] to [13]. In commercial use the antennas are normally operated connected directly to the transceiver PCB, and therefore the feed cable coupling is not an issue.

Planar Dipole-like Antennas for Consumer Products

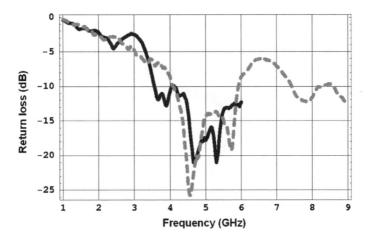

Figure 9.32 Measured return loss of SkyCross antenna. Solid line is measurement between 1 and 6 GHz. Dashed lighter line is measurement repeated between 1 and 9 GHz on a different analyser with different test cables

leads to jumps in the phase of S11. This suggests that the phase of the transmitted signal will not be linear across the band. However, as is shown in the next subsection, the radiation pattern varies across the band, and in view of this an analysis of the transmitted signal phase was not thought useful.

In Figure 9.35, the slopes of the light grey curves are lower than those of the other curves. This is because the lumped port used was nearer to the antenna feed than the coaxial port used in the other simulations.

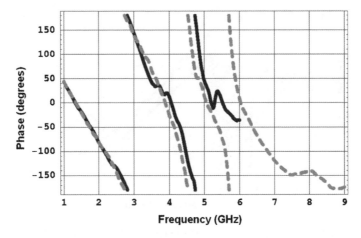

Figure 9.33 Measured phase of Skycross antenna S11 response. Solid line is measurement between 1 and 6 GHz. Dashed lighter line is measurement repeated between 1 and 9 GHz on a different analyser with different test cables

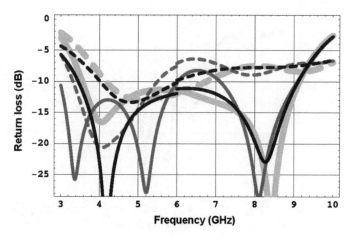

Figure 9.34 Simulated return loss curves for line-matched monopole-type antennas. Solid lines denote complete antenna. Dashed lines denote antenna without strip connecting element to ground. Black curves include connector and feed cable. Mid-grey curves omit feed cable, and light grey thicker curves omit feed cable and connector

9.5.2.4 Radiation Patterns

Figure 9.36 shows the radiation patterns of the line-matched monopole antenna modelled without a connector and feed cable. The radiation pattern starts as a near classical dipole doughnut and with increasing frequency evolves into a directional pattern. With increasing frequency, there is also a rotation about the y axis in the direction of increasing theta. Neither the omission of the line between the monopole

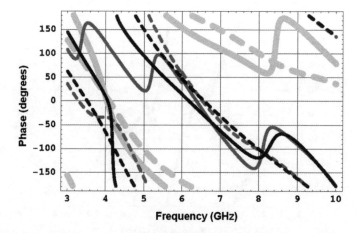

Figure 9.35 Simulated S11 phase for line-matched monopole-type antennas. Solid lines denote complete antenna. Dashed lines denote antenna without strip connecting element to ground. Black curves include connector and feed cable. Mid-grey curves omit feed cable, and light grey thicker curves omit feed cable and connector

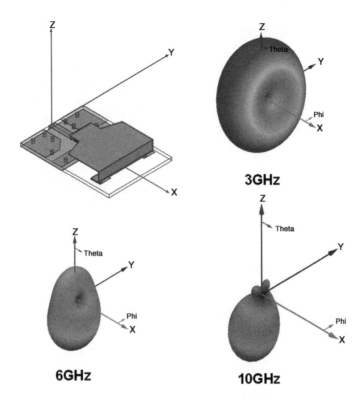

Figure 9.36 Radiation patterns of line-matched monopole antenna

edge and ground plane, nor the inclusion of the feed cable, change this qualitative behaviour. They just result in changes in the details of the radiation pattern.

Summarising, the line-matched monopole antennas described above are the smallest of the antennas studied in this chapter. The antenna pattern varies significantly with frequency across the 3 to 10 GHz band. As could be expected from the smallest antenna and its monopole design structure, there was significant coupling to the feed cable.

These antennas are usually operated without an attached feedline and exhibit performance that depends upon the actual size and configuration of the counterpoise ground of the PCB that they are mounted on.

9.5.3 Vivaldi Antenna

Vivaldi antennas are printed structures that possess a very wide bandwidth and directionality [1]. The example discussed here is one designed in the early 1980s by Peter Gibson, the originator of Vivaldi antennas, and originally intended as the feed for a dish reflector. It consists of two main conductors printed onto an alumina substrate, with a slot flaring out between the conductors, shown on the right side of Figure 9.37. The left side of Figure 9.37 shows the microstrip feed, which uses one of the main conductors as its earth, and connects to the other conductor via a conducting pin to form the feed at the narrowest part of the slot.

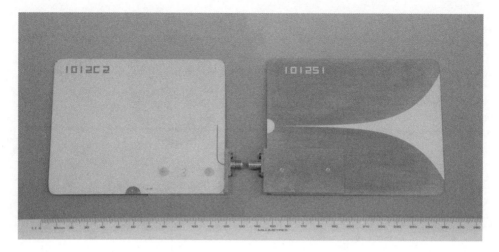

Figure 9.37 Vivaldi antenna example

The alumina substrate is 115 mm long by 95.2 mm wide by 0.55 mm thick sheet of alumina. The flare curve follows the formula $y = 0.101345e^{0.0538341x}$, where x and y are measured in millimetres, x is the distance along the slot's centreline, and the coordinate origin is at the start of the slot.

9.5.3.1 Measurement and Simulation Results

Return loss and reflection coefficient. Figures 9.38 and 9.39 show the measured return loss and S11 phase response for three Vivaldi antennas. They have very similar responses, with very good phase linearity.

Figure 9.40 shows the simulated return loss. Here the black curves represent simulations running between 1 to 3, and 6 to 10 GHz, and the grey curves represent the 3 to 6 GHz simulation. The response is very similar to that measured. The slight differences are probably due to errors in entering the antenna geometry into the simulator, and in simulating the antenna with a gap feed across the slot. The actual antennas had a 50 ohm microstrip line feeding the slot via a shorting via through the alumina substrate.

Figure 9.41 shows a Smith chart of the simulated response. Smith charts of the measured responses look similar, but the curve is more tightly wrapped about the centre. This is due to the long microstrip feed line between the antenna connector and the slot.

In common with the other Smith charts shown in this chapter, the curve in Figure 9.41 starts at 1 GHz. However, with decreasing frequency the response crosses the real axis at least twice before reaching its DC limit of an open circuit response.

Before leaving the S parameter response, it is worth making a few comments on the lossy transmission line model of Section 9.2 and the behaviour of the transmitted wave upon reaching the end of the antenna's flare. With the biconical antennas of Section 9.2 and various dipoles of subsequent sections the wave that reaches the tip of the antenna is reflected back towards the feed. However, for the Vivaldi antenna the nature of the antenna's extremities is significant. As discussed before, semicircular conductors added to the ends of the arms help reduce reflections. An alternative method described in [3] is to have the antenna's conducting sheets extending further away from the slot. For example, in the picture on the right

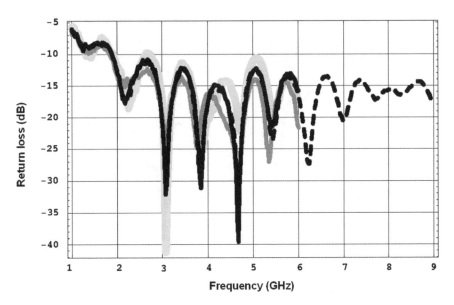

Figure 9.38 Return loss measurements of Vivaldi antennas. The measurements of the three antennas are represented by the thick light grey, grey and black curves. The first two antennas were only measured between 1 and 6 GHz. The third antenna was measured between 1 and 6 GHz (solid black curve) and remeasured between 1 and 9 GHz (dashed black curve)

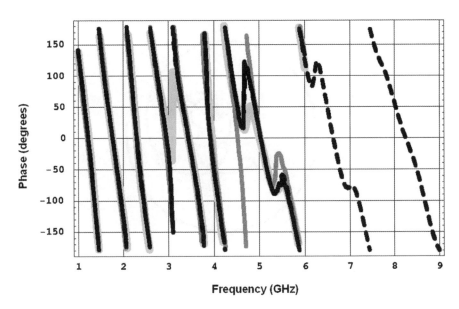

Figure 9.39 S11 phase measurements of Vivaldi antenna. The measurements of the three antennas are represented by the thick light grey, grey and black curves. The first two antennas were only measured between 1 and 6 GHz. The third antenna was measured between 1 and 6 GHz (solid black curve) and remeasured between 1 and 9 GHz (dashed black curve)

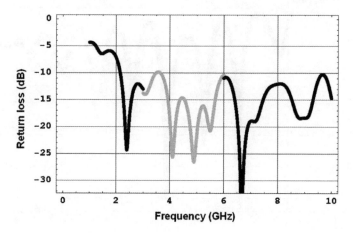

Figure 9.40 Vivaldi antenna simulated return loss. 1 to 3 GHz and 6 to 10 GHz results shown in black, 3 to 6 GHz results shown in grey

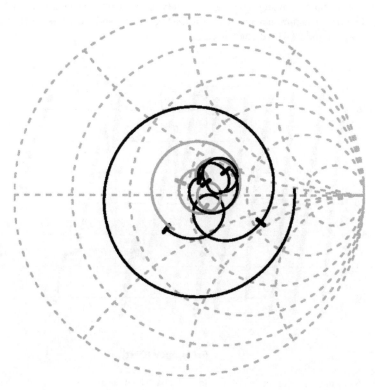

Figure 9.41 Smith chart of simulated Vivaldi antenna response. 1 to 3 GHz and 6 to 10 GHz results shown in black, 3 to 6 GHz results shown in grey

Planar Dipole-like Antennas for Consumer Products 193

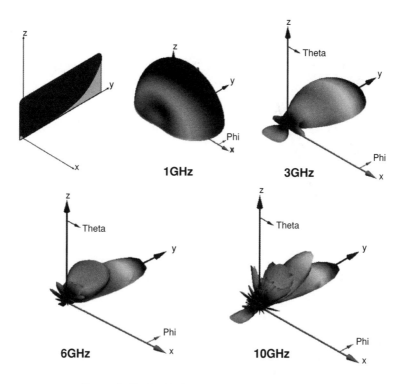

Figure 9.42 Radiation patterns of Vivaldi antenna

side of Figure 9.37 the conducting sheets extend right to the edge of the rectangular substrate. The extra conductor allows the wave to propagate around the end of the antenna flare and back along the outer edges of the antenna.

Radiation. To save computation, and for the 6 to 10 GHz simulations, to enable the model to fit within the 1.5 Gbyte memory limit, FE simulations of the antenna were done using just the top half of the antenna and a perfectly conducting mirror plane in the $x - y$ plane. Therefore the radiation patterns are only for the $+z$ direction. Figure 9.42 shows the radiation patterns. At 1 GHz the pattern looks dipole like, with a doughnut shape about the y axis. By 3 GHz the pattern has become a pencil beam, and it stays this way right through to 10 GHz, with fairly constant gain, and just changes to the beamwidth and sidelobes. Figure 9.43 shows the simulated directivity and gain values of the vivaldi antennas.

Figure 9.44 shows the phase of $\exp(jkR) * R * E_\theta$ in the boresight direction, where R is the radius, and E_θ is the θ component of the electric field computed at 1 metre from the coordinate centre, using the far-field formula. For this simulation the coordinate centre is very close to the feed across the slot. The figure shows the phase variation with frequency for the far field at any point in the boresight direction. It is very linear. This is even true down to 1 GHz, where the antenna is not launching a pencil beam.

In summary, the Vivaldi antenna is unique among the antennas described in this report, in that above 3GHz it has a relatively stable radiation pattern. This is achieved at the expense of occupying a larger area.

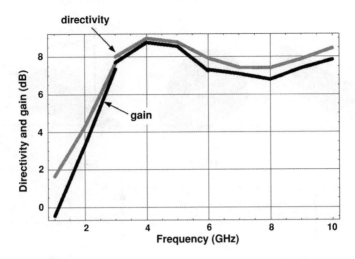

Figure 9.43 Vivaldi antenna simulated directivity and gain

9.6 Summary

Reflection coefficient. In this chapter the bicone antenna was studied as the characteristic impedance of the antenna's biconical waveguide can be solved analytically. It was shown that the bicone antenna's reflection coefficient response is very similar to that for a transmission line possessing the same characteristic impedance as the bicone's, and with frequency-dependent attenuation. The reflection coefficient converges to a high frequency limit. Similar reflection coefficient behaviour was also found in many of the planar dipole antennas studied. The value of the high-frequency limit depends on the antenna geometry and upon the dielectric constant of any substrate.

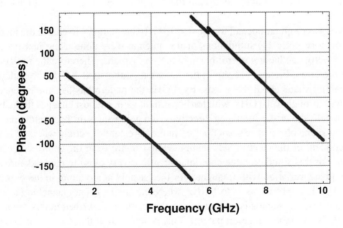

Figure 9.44 Phase of Vivaldi antenna radiation along boresight. Data extracted at 1 GHz intervals from radiation simulations

Table 9.1 Relative properties of UWB antennas

Feature	VSWR	Phase linearity	Omni-directionality	Ergonomics	Total
Bowtie and fan dipole	7	6	7	3	23
Diamond dipole	2	4	7	3	16
Printed elliptical dipole	8	6	7	5	26
Line-matched monopole	8	3	5	5	21
Vivaldi	9	8	1	2	19

Radiation. The radiation patterns of almost all dipole-like antennas studied vary with frequency. Usually, when the antenna is around half a wavelength long the pattern is the classical doughnut shape. However, with increasing frequency the radiation pattern becomes more complex. The details of the higher frequency radiation patterns depend on the antenna geometry. The presence of a feed cable can severely perturb the radiation pattern.

It is possible to construct antennas that have more constant radiation patterns with frequency. The Vivaldi antenna is an example of this, and the author believes that the constancy of its radiation patterns could be further improved. However, in order to have constant radiation patterns, such antennas have to be large enough to satisfy the rules described by Rumsey [4].

Which antenna type is best? It is difficult to come to a definite conclusion on which antenna is the most suitable for consumer UWB equipment, since the space and location available for the antenna, size and cost enter into the decision-making process. As a tentative guide, Table 9.1 shows an attempt to place numerical values on some of their relative merits based on several criteria important to UWB applications. Each of the criteria is marked on a scale of 0 to 10.

The diamond dipole can be ignored because of its appalling match. The Vivaldi antenna is probably disqualified because it has a directional pattern and is much larger than the other designs. Between the remaining designs, the free-standing elliptical, bowtie and fan dipoles and the printed elliptical dipole have similar performances. The printed design is usually easier to construct using conventional PCB lithographic techniques. The line-matched monopole antenna is a more compact design – but its radiating properties depend on the PCB it is mounted upon.

Acknowledgements

The author thanks Xiaodong Chen and his team at Queen Mary University of London for the very useful dialogue that led to the model for explaining the high-frequency impedance asymptotes of broadband dipoles. Adrian Jennings of Time Domain helped with the section on printed elliptical dipole antennas, and Frank M. Caimi and his team at SkyCross helped with the section on the SMT3-TO-1M antenna.

References

[1] P.J. Gibson, The Vivaldi aerial, *Proceedings of 1979 European Microwave Conference*. pp. 101 onwards. Paper reprinted in P.J.B. Clarricoats, *Advanced Antenna Technology*, Microwave Exhibitions and Publishers Limited, 200–4, 1981.
[2] R.F. Harrington, *Time Harmonic Electromagnetic Fields*, McGraw-Hill, section 6.5, 1961.
[3] P.J. Massey, Tapered slot antenna analysis using FDTD simulation, *IEE Colloquium on 'Millimetre-wave and quasi-optical antennas'*, London, 14 June 1990.

[4] V.H. Rumsey, *Frequency Independent Antennas*, Academic Press, 1966.
[5] H.G. Schantz, Planar elliptical ultra-wideband dipole antennas, *IEEE AP Symposium,* June, Vol. 3, 16–21, 2002.
[6] H.G. Schantz, Ultra wideband technology gains a boost from new antennas, *Antenna Systems and Technology*, **4**(1), 2001.
[7] H.G. Schantz and L. Fullerton, The diamond dipole: a Gaussian impulse antenna, *IEEE AP Symposium*, July, Vol. 4, 8–13, 2001.
[8] H.G. Schantz, *The Art and Science of Ultrawideband Antennas*, Artech House, 2005.
[9] Z. Wu, P. Sevret and M.J. Ammann, Broadband dual-plate monopole antennas, *IEE International Conference on Antennas and Propagation* (ICAP), 31 March to 3 April, Exeter, UK. Vol. 2, 493–5, 2003.
[10] K.Y. Yazdandoost and R. Kohno, Ultra wideband printed bow-tie antenna, CRL submission to IEEE P802.15 Working Group for Wireless Personal Area Networks (WPANS), September 2003.
[11] P.J. Massey and K.R. Boyle, Controlling the effects of feed cable in small antenna measurements, *Proceedings of 12th International Conference on Antennas and Propagation*, 31 March to 3 April, Exeter, UK. Vol. 2, 561–4, 2003.
[12] B.S. Collins and S.A. Saario, The use of baluns for measurements on antennas mounted on small groundplanes, *Proceedings of IEEE International Workshop on Antenna Technology: Small Antennas and Novel Metamaterials*, Singapore, 7 to 9 March, 2005.
[13] S.A. Saario, *FDTD Modelling for Wireless Communications: Antennas and Materials*, PhD thesis, Griffith University, Australia, September 2002.

10

UWB Antenna Elements for Consumer Electronic Applications

Dirk Manteuffel

10.1 Introduction

Following the discussion on antenna elements for UWB consumer applications in Chapter 9 we will now focus on the integration aspect of the antenna in the application. As a first step a method to extract the spatio-temporal UWB antenna characterisation from an FDTD simulation is given. Thereafter, the shape of a planar monopole is optimised to provide broadband matching. This is followed by the integration of the latter antenna into a model of a DVD player and a mobile device such as a PDA or a mobile phone. The impact of this integration on the antenna performance is evaluated. Finally, the transfer function of the complete system is extracted and used for indoor propagation modelling in an exemplary home environment. The results show that the antenna integration into the DVD chassis results in a directive radiation pattern that shows a significant frequency dependency. When this antenna is used for the propagation modelling, a single frequency ray-tracing simulation shows significant variation in the radiated power distribution in the room as a result of the directive pattern. When the received power is averaged over a larger bandwidth, the coverage becomes smoother mainly due to the frequency dependency of the radiation pattern and frequency-dependent propagation effects. Indeed, when comparing the results of the integrated antenna to the case of the ideal isotropic radiator, no major disadvantages can be discerned. In the final step the complete system including the antennas integrated into different applications as well as the indoor channel is characterised by means of the transfer function and the impulse response for different links scenarios. The calculated delay spread of the system is significantly more affected by the channel than by the antenna elements.

The use of ultra-wideband (UWB) systems, e.g. for wireless multimedia data communication between different home entertainment systems (DVD player, flat screen, Internet PC and so on) has become very appealing since the FCC released the spectrum from 3.1 GHz to 10.6 GHz for unlicensed low-power use [1]. At the moment in PANs (personal area networks) both wired and unwired links are used. While the link between the modem and the laptop computer is nowadays typically achieved by WLAN (wireless

---

Ultra-wideband Antennas and Propagation for Communications, Radar and Imaging Edited by B. Allen,
M. Dohler, E. E. Okon, W. Q. Malik, A. K. Brown and D. J. Edwards
© 2007 John Wiley & Sons, Ltd

Figure 10.1 Prospective of a future WPAN in a typical living room. DVD, screen and PDA equipped with UWB modules use a wireless high data rate link

local area network) and the connection between the laptop and the printer is also unwired by Bluetooth, high data rate links, e. g. the transmission of the video signal between the DVD player and the flat screen is still established as a wired connection. It is the idea of major companies working in the definition and the design of consumer applications to unwire any wired connection, even the high data rate video links. Such a concept could be called *wireless* USB (universal serial bus) similar to the well-established *wired* USB interconnections. An interesting prospective on systems and applications is given in [2].

We can assume that there is a big market for this kind of UWB application and forecast that UWB modules in terms of wireless USB will be used in nearly every consumer electronic application in the future, as is already the case for the wired USB module.

Figure 10.1 illustrates a typical scenario of a living room with different consumer electronic applications linked to each other by such a UWB connection.

UWB modules for the integration into consumer electronic products have to fulfil technical and nontechnical requirements. The general technical requirement is that the module has to be able to establish a link with the required data rate. Nontechnical requirements are low cost, small size and the ability to be integrated into the device in order to prevent distortion of the aesthetic design. It is unlikely that a design-driven product will use an external protruding antenna. For the antenna, especially the integration into the chassis of the application can be a serious problem, because it will interact with all components nearby. The performance parameters of the antenna such as matching, radiation pattern and ringing will be affected by the specific integration. Moreover it will be necessary to adapt and tune the antenna to a certain integration scenario in a specific application.

Especially for the design of systems entailing different antenna integration scenarios, it is essential to have fast and easy access to the antenna characteristics by simple numerical simulations. This is important in the definition phase of a new product in order to predict the performance when prototypes are not yet available. In this respect, the FDTD method is well suited for such computations because of the fact that a large frequency spectrum can be investigated in one simulation run. In the first part of this chapter the methodology to model the antenna by using the FDTD method will be described. In particular it will be discussed how the UWB characteristic of the antenna can be extracted from the full-wave simulation for later post-processing of all relevant antenna parameters.

A parametric antenna model with the ability to retune the antenna according to different integration situations will be derived in the next subsection. Various integration scenarios for different typical applications like DVD players or mobile phones will be investigated and the impact of the integration on the prior defined antenna parameters will be discussed.

In order to derive the overall performance of a complete system a ray-tracing model of a typical living room environment is set up. The prior assessed UWB characteristics of different applications is used to evaluate realistically the link between different applications. This hybrid investigation enables us in the final step to evaluate the overall link performance taking into account realistic antenna integration scenarios as well as a typical indoor channel.

10.2 Numerical Modelling and Extraction of the UWB Characterisation

10.2.1 FDTD Modelling

Numerical simulations are essential for the analysis and design of modern radio frequency (RF) systems. Especially for these future applications where prototypes are not yet available it is important to get realistic insight into all physical aspects of the desired product.

Among the numerical techniques for electromagnetic (EM) simulations the FDTD method is chosen in this work for the following reasons:

- The FDTD method directly applies Maxwell's curl equations, and has therefore little restrictions with respect to EM problems. Especially, it is capable of handling various types of material distribution.
- FDTD is a time-domain method; it perfectly complies to the ultra-wideband behaviour of the desired antennas and enables investigation of a large frequency spectrum within one simulation run.
- Due to the fact that no linear equation systems have to be resolved within an FDTD simulation, the memory consumption is not that demanding as in case of MoM (method of moments) or FEM (finite element method), and therefore enables fast and efficient computation of even complicated antennas using standard PCs.

The FDTD software used for this work is the 3D-field solver EMPIRE™. The FDTD method typically applies an orthogonal grid. This means that the original structure is divided into cubical pieces for the simulation. One can assume that this complicates the simulation of arbitrarily curved objects, as they are often used in UWB antennas. In order to lower this threat the following example shows the simulation of a curve-shaped UWB antenna element similar to the ones discussed in Chapter 9 by the use of different discretisation. Figure 10.2 shows the calculated matching of the antenna modelled using different dense discretisation. The first model uses a uniform coarse discretisation of 1 mm. It is clearly visible how the discretisation modifies the shape by a so-called *staircase* approximation. Especially the smooth taper near the antenna feeding point turns to a straight horizontal line. The next model also uses a homogeneous mesh, but the cell size is only 0.25 mm. Therefore the mesh is four times denser than the one use in the first model. Although the shape of the antenna can never be smooth by the use of an orthogonal grid it

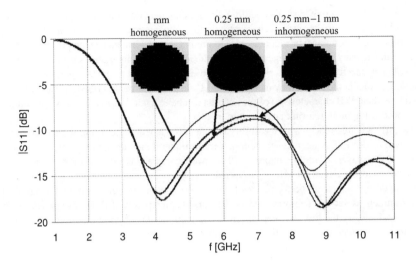

Figure 10.2 Staircasing: the influence of the discretisation on the calculated matching of the antenna [3]

resembles the real antenna shape much better. Due to this there is a significant difference in the calculated matching of the antennas using different discretisation. It is obvious that the dense discretisation results in a longer simulation time. The coarse model needs less than 1 minute on a standard desktop computer while the dense model needs around 6 minutes. Although this is not quite long to achieve broadband results, a better trade off can be made by applying an inhomogeneous mesh. With this regards the third model uses a dense mesh only in the lower part of the antenna in order to resolve the smooth taper near the feeding of the antenna. The simulation time of this model is around 2 minutes and the results are quite close to the dense case. As a rule of thumb we can state that the staircasing by the orthogonal grid does not necessarily mean that the results are incorrect. If we look for global quantities like matching or far-field, the values are close to the real smooth structure. There is only the minor influence of the staircasing. Of course this is not the case if we want to observe e.g. the current density on the edge of the antenna. This microscopic measure is directly affected by the staircasing.

Numerical modelling is a difficult task and requires insight in the method used as well as some experience. This chapter only gives an outline on numerical modelling. The interested reader is referred to more specialised literature in this field, e.g. [3].

For the antenna characterisation the FDTD simulation is only used to calculate the near-field problem. The far-field is later determined by a near-field-to-far-field-transformation (Figure 10.3). The technique applied here is based on Huygens' principle.

The computational domain consists of the antenna and some 'air-filled' space (typically less than $\lambda_{max}/8$) to the boundaries which are chosen to be a six-layer PML (perfectly matched layer) [3] in order to ensure free-space conditions. Between the boundaries and the antenna, a closed surface (a box) enclosing the antenna is defined on which the near-field is recorded during the simulation run. Based on the recorded tangential electric and magnetic field values on the surface, equivalent electric and magnetic sources can be derived:

$$S = \hat{n} \times H \quad (10.1)$$
$$M = -\hat{n} \times E \quad (10.2)$$

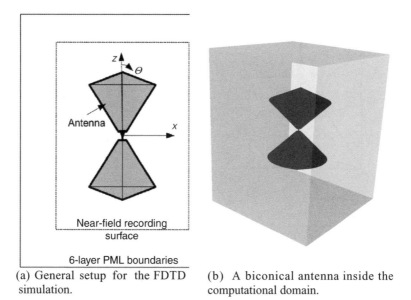

Figure 10.3 Illustration of the computational domain in the FDTD model showing the antenna and a box to record the near-field for the later near-field-to-far-field-transformation

In Equations (10.1) and (10.2) **E** and **H** denote the electric and the magnetic field strength at a certain point on the surface, and **S** and **M** are the equivalent unit source elements. The outward directed unit vector is denoted $\hat{\mathbf{n}}$ in the above equations.

Based on the equivalent sources on the enclosing surface the far-field is expended by standard techniques describes in many places in the literature [3].

10.2.2 UWB Antenna Characterisation by Spatio-Temporal Transfer Functions

From a signal-processing point of view the antenna can be considered an LTI (linear time-invariant) system which can be fully characterised by its transfer function [5]. This can be expressed by

$$\frac{\mathbf{E}_2(\mathbf{r}_2, \omega)}{\sqrt{Z_{F0}}} = \frac{U_{1,in}(\omega)}{\sqrt{Z_L}} \mathbf{H}_{TX}(\hat{\mathbf{r}}_{12}, \omega) \frac{e^{-jk_0 r_{12}}}{\sqrt{4\pi r_{12}}} \tag{10.3}$$

In Equation (10.3) $\mathbf{E}_2(\mathbf{r},\omega)$ denotes the electric field strength at a point P_2 in the far-field of the antenna, positioned at location P_1, which is excited by an incoming voltage $U_{1,in}(\omega)$ at the antenna port (see Figure 10.4). While $e^{-jk_0 r_{12}}/\sqrt{4\pi r_{12}}$ describes the propagation of the wave from the antenna to the observation point in the direction $\mathbf{r}_{12}$, $\mathbf{H}_{TX}(\mathbf{r}_{12},\omega)$ represents the transmit transfer function of the antenna. In Equation (10.3) Z_{F0} and Z_L are the free-space and feed-line impedances, respectively, and $\hat{\mathbf{r}}_{12} = \mathbf{r}_{12}/r_{12}$ is the unit vector pointing from the antenna to the observation point (far-field conditions assumed). Consequently, $\mathbf{H}_{TX}(\mathbf{r}_{12},\omega)$ is independent from the distance between the antenna and the observation point but one has always to take into account that the definition of the transfer function

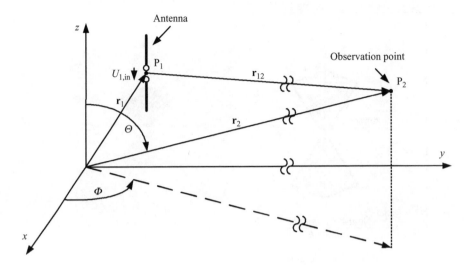

Figure 10.4 Definition of the coordinate system (transmit antenna) [3]

according to Equation (10.3) requires local plane wave propagation and thus is only valid if the far-field conditions apply.

On the other hand, following [4] the reception of the antenna from an incident plane wave can be expressed by

$$\frac{U_{2,out}(\omega)}{\sqrt{Z_L}} = \sqrt{4\pi} \frac{E_{1,inc}}{\sqrt{Z_{F,0}}} H_{RX}\left(\hat{k}, \omega\right) \qquad (10.4)$$

In Equation (10.4) $U_{2,out}(\omega)$ denotes the voltage at the antenna port connected to the receiving system when the antenna is exposed to a plane wave. Note that $\mathbf{E}_{1,inc}$ is the electric field strength of the incident plane wave, i.e., the field at the location of the receiving antenna in absence of this antenna. According to this definition $\mathbf{H}_{RX}(\mathbf{k},\omega)$ can be considered as the receive transfer function of the antenna.

Figures 10.4 and 10.5 illustrate the above definitions.

Both transmit and receive transfer functions are related to each other by Lorentz' theorem of reciprocity. An expression that takes into account the ultra-wideband properties of the system has been derived in [5]:

$$2j\omega \mathbf{H}_{RX}\left(-\hat{k}, \omega\right) = c_0 \mathbf{H}_{TX}\left(\hat{k}, \omega\right) \qquad (10.5)$$

10.2.3 Calculation of Typical UWB Antenna Measures from the Transfer Function of the Antenna

The above-defined transfer functions provide a complete characterisation of the far-field properties of the antenna. They can be used to treat the antenna as a module in system simulations or as the basis to calculate any other useful antenna measure. For ultra-wideband antennas specialised measures have been derived. Some of them, discussed in the following subsection, are also reported in related literature, e.g. [7].

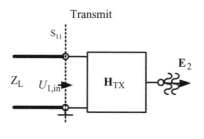
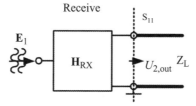

Figure 10.5 Representation of the antenna as an LTI system for transmit and receive mode [3]

Gain. For UWB antennas it is not sufficient to define only angle-dependent gain at a certain centre frequency as is typically done for narrowband antennas. The radiation characteristic usually shows significant frequency dependence in the UWB frequency range. Therefore the effective continuous-wave (CW) gain pattern G_{eff} can be derived from the transfer function as follows:

$$G_{\text{eff}}(\omega, \Phi, \Theta) = \frac{\omega^2}{\pi c_0^2} |\mathbf{H}_{\text{TX}}(\omega, \Phi, \Theta)|^2 \tag{10.6}$$

However, the IEEE definition of the gain is slightly different as it takes into account the matching of the antenna:

$$G(\omega, \Phi, \Theta) = \frac{G_{\text{eff}}(\omega, \Phi, \Theta)}{1 - |s_{11}(\omega)|^2} \tag{10.7}$$

Group delay. As defined above the transfer function of an antenna is a complex measure and can be written in the following form:

$$\mathbf{H}_{\text{TX}}(\omega) = |\mathbf{H}_{\text{TX}}(\omega)| e^{j\varphi(\omega)} \tag{10.8}$$

The transfer function describes how a signal's amplitude and phase are distorted by the antenna. A standard measure in filter theory is the group delay:

$$\tau_g(\omega) = -\frac{d\varphi(\omega)}{d\omega} \tag{10.9}$$

which is defined as the negative derivative of the transfer functions phase. The group delay gives the average time delay the input signal suffers at each frequency, thus it is related to the dispersive nature of the antenna.

In order to investigate the distorting effects a relative group delay can be defined:

$$\tau_{g,rel}(\omega) = \tau_g(\omega) - \tau_{g,mean}(\omega) \qquad (10.10)$$

Finally the standard derivative of the relative group delay provides a single number that characterises the dispersive behaviour:

$$\tau_{g,RMS} = \sqrt{\frac{1}{\omega_2 - \omega_1} \int_{\omega_1}^{\omega_2} \tau_{g,rel}^2(\omega) d\omega} \qquad (10.11)$$

Impulse response. By inverse Fourier transformation of the transfer function of the antenna the impulse response can be calculated. For simplicity we use the equivalent baseband signal in this work. Furthermore, due to the fact that we calculated a limited frequency spectrum, we observe a truncation error if we apply the inverse Fourier transform directly. Therefore we multiply the frequency domain transfer function with a standard window function (Kaiser window) first, before applying the inverse Fourier transformation. Detailed formulations are given in [7].

Ringing and delay spread. A plot of the impulse response shows the ringing of the antenna, which is a quite intuitive measure. The ringing indicates that an impulse is spread by the antenna. A measure for the delay spread is derived from the impulse response as follows:

$$\tau_{DS}(\Phi, \Theta) = \sqrt{\frac{\int_{-\infty}^{\infty} (t - \tau_{DS,mean}(\Phi, \Theta))^2 |\mathbf{h}_{TX}(t, \Phi, \Theta)|^2 dt}{\int_{-\infty}^{\infty} |\mathbf{h}_{TX}(t, \Phi, \Theta)|^2 dt}} \qquad (10.12)$$

with

$$\tau_{DS,mean}(\Phi, \Theta) = \frac{\int_{-\infty}^{\infty} t |\mathbf{h}_n(t, \Phi, \Theta)|^2 dt}{\int_{-\infty}^{\infty} |\mathbf{h}_n(t, \Phi, \Theta)|^2 dt} \qquad (10.13)$$

10.2.4 Example

In order to prove the above-derived method, the biconical antenna shown in Figure 10.3 has been analysed. The antenna is designed to operate in the frequency range above 3.1 GHz. For the FDTD simulation the antenna is modelled with all necessary details. The distance to the PML boundaries is less than $\lambda/8$ of a wavelength at the lowest frequency of interest thus resulting in a time- and memory-efficient simulation. The antenna is excited by a broadband Gaussian pulse centred at 0 Hz, and having a half-bandwidth of 20 GHz with reference to a signal decrease of 20 dB. The near-field of the antenna is recorded at every 200 MHz intervals between 1 GHz and 20 GHz on a Huygens' surface enclosing the antenna. The EMPIRE™ field solver uses this near-field data to derive equivalent electric and magnetic sources on the surface and extrapolate the near-field into the far-field. The total simulation time, including the post-processing of the near-field data, takes only a few minutes on a standard 2 GHz PC. The results from

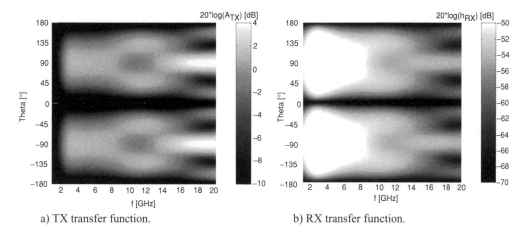

Figure 10.6 Calculated transfer functions of the biconical antenna in the E-plane [3]

this simulation are used to determine the transmit and receive transfer functions of the antenna according to the above-mentioned method.

Figure 10.6 shows the calculated transfer functions of the biconical antenna in the E-plane. The TX transfer function has been directly calculated from the FDTD simulation using Equation (10.3). The RX transfer function is derived from the TX transfer function applying Lorentz' reciprocity principle according to Equation (10.5). Figure 10.6(a) illustrates the TX transfer function of the biconical antenna in the E-plane. It can be observed that the antenna is matched above 3 GHz. The characteristics are similar to those of a standard first-order dipole up to a frequency of 8 GHz. For higher frequencies, the characteristic changes showing sidelobes and gain deviations.

The impulse response of the biconical antenna is calculated from the frequency domain transmit transfer function of the antenna. Figure 10.7 shows the transfer function for a single position. In order to achieve a smooth curve a Kaiser window has also been used.

It can be noted from Figure 10.7 that the ringing of the antenna is quite small. The delay spread is calculated to be:

$$\tau_{DS}(\Phi = 0, \Theta = 90°) = 0.01417 \text{ ns}$$

10.3 Antenna Design and Integration

The above method is used in the following section to design a broadband miniaturised antenna and integrate it into different devices of consumer electronic applications such as a model of a DVD player or a PDA. The problem is divided into two parts. First a miniaturised UWB antenna is designed and optimised. Afterwards this antenna is integrated into the chassis of the application and the influence is investigated. In a second phase the antenna is retuned and optimised to work in the specific integration scenario. After this the complete antenna characterisation is extracted from the FDTD simulation for later use with other modelling tools.

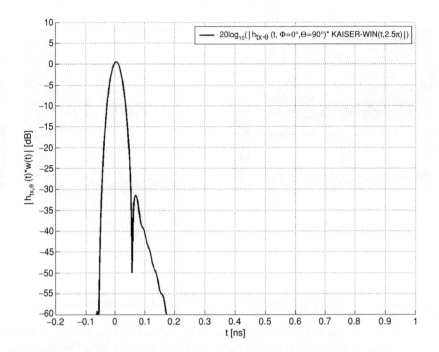

Figure 10.7 Impulse response of the biconical antenna ($\Phi = 0°$, $\Theta = 90°$)

10.3.1 Antenna Element Design and Optimisation

From the EM modelling point of view a suitable antenna has to be designed in the first step. Because of the intended application, the following basic requirements for the antenna element can be listed:

- large bandwidth (e. g.: $s_{11} \leq -10$ dB $\forall$ f $\in$ {3.1 GHz, 10.6 GHz});
- small size (integration into device);
- low-cost technology.

Additionally often other quality criteria like:

- omnidirectional radiation pattern;
- frequency stability (radiation pattern);
- low ringing

are associated with and requested for UWB antennas [6]. However, the following investigation will also evaluate if the last-mentioned criteria are of great importance when we focus on a UWB module integrated into consumer electronic products.

It has been shown in Chapter 9 that a broadband planar monopole antenna can be good candidate. The antenna shape differs from a basic wire monopole in the way that the element provides a smooth taper relative to the groundplane. By optimising this taper broadband impedance matching can be obtained quite easily.

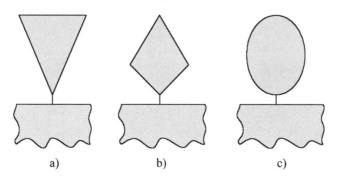

Figure 10.8 Different basic shapes of planar broadband monopoles on a groundplane [3]

Figure 10.8 shows some examples of basic shapes for planar broadband monopoles.

In order to apply an automated optimisation process for the simulation modelling it is advantageous to describe the antenna by an analytic mathematical function [7]. This can be e.g. polynomial, elliptical, exponential or any other kind of analytical function which provides a smooth taper. In this paper the following parametric polynomial approach is used because it offers maximum flexibility in the generation of various shapes:

$$x = \frac{w}{2} \sin(\pi \alpha) \qquad (10.14)$$

$$y = \frac{h}{2} \cos(\pi (k |\alpha|^{p_1} - 1)) \qquad (10.15)$$

w is the maximum width of the antenna and h is its maximum height, α ranges within $-1 \leq \alpha \leq 1$. In Equation (10.15) the factor k is

$$k = \frac{\cos^{-1}(2p_2 - 1)}{\pi} + 1 \qquad (10.16)$$

These continuous Euclidic functions depend on the power p_1 (see Figure 10.9(a)) and the parameter p_2 (see Figure 10.9(b)), which enable a continuous transition into manifold antenna shapes (circle, ellipse, triangle-like, etc.).

The shape is first optimised to offer good matching when the antenna is located on a groundplane. The variables p_1 and p_2 can now be assessed using the GUI of the simulator. This provides an easy way to generate a large number of different shapes for a series of simulations or the use of an optimiser.

When an optimiser is available within the simulation software, different variations are generated automatically in order to approximate a certain goal function which has to be defined in advance. For our example the goal function could be:

$$\text{Goal}(p_1, p_2) = s_{11} \leq -10 \text{ dB } \forall \text{ f} \in \{3.1 \text{ GHz}, 10.6 \text{ GHz}\} \qquad (10.17)$$

Starting from a certain initial state, the parameters are now varied and after each step the approximation of the simulation result to the goal function is evaluated. The next evolution of the shape is now chosen on the basis of an estimation by a gradient approach or a genetic algorithm, depending on the optimiser. Using a prior optimised grid a single simulation takes less than a minute on a state-of-the-art PC. Figure 10.10

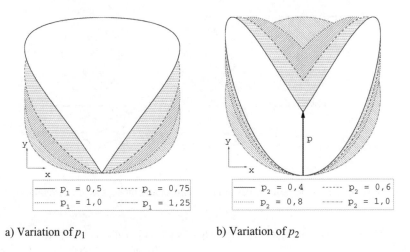

a) Variation of p_1 b) Variation of p_2

Figure 10.9 Parametric generation of various shapes [3]

shows the results of two different shapes that fulfil the goal function in comparison to the initial design of the monopole.

10.3.2 Antenna Integration into a DVD Player

It is obvious that all antenna parameters are affected when the antenna is transferred from the groundplane to a specific situation inside a (partly) metal chassis. For a detailed analysis three different integration scenarios are observed. They are illustrated in the photograph of the DVD player shown in Figure 10.11.

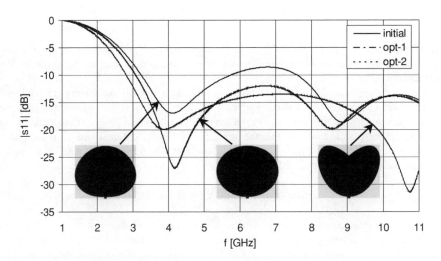

Figure 10.10 Results from the parametric optimisation [3]

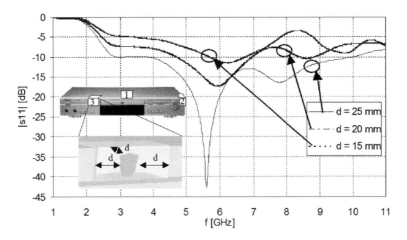

Figure 10.11 Antenna integration in the front side of the DVD player: influence of the integration volume on the antenna matching [3]

In scenario no. 1 the antenna is mounted on the top edge of the DVD player. This is the most similar situation compared to the groundplane case that may have at least some practical relevance. A more likely case may be scenario no. 2 in which the antenna is integrated within a certain volume at the corner of the device. In scenario no. 3 the antenna is integrated within a volume inside the front face of the DVD player. It shows the influence of the integration volume on the matching of the antenna for scenario no. 3. It can be observed that allowing a distance of 25 mm to the metal walls of the surrounding structure, the antenna still fulfils the requirements in terms of matching. Note: the assumption that the DVD player is completely composed of metal is quite pessimistic. It may be possible to find a smaller volume with less interference in a more realistic scenario. Additionally, as mentioned earlier, in an industrial development the antenna element would be optimised for this specific integration scenario.

In the next step the influence of the different integration scenarios on the radiation pattern of the antenna is investigated. Figure 10.12 shows the simulation models for the different integration scenarios together with the far-field pattern at $f = 7$ GHz (as an example):

Figure 10.12 shows clearly that the pattern depends very much on the specific integration scenario and is in any case no longer quasi omnidirectional as it was mounted on the groundplane. Figure 10.13 shows two planes of the radiation pattern of the model in Figure 10.12(c) as a function of frequency. It can be observed that the radiation pattern shows a strong frequency dependency. The system provides a single main beam up to 6 GHz, and sidelobes occurring at higher frequencies. Furthermore, it is clearly illustrated that there are shadow areas located at the backside of the DVD player.

In general we can conclude that even if the assumption of a metal chassis with only a small integration volume for the antenna may be quite pessimistic, it is clear that we cannot reach a quasi omnidirectional radiation which is stable over the frequency range when the antenna is integrated in (or even mounted on) a realistic device. Therefore, the demand for omnidirectionality or frequency stability that is often requested is surely not feasible for low-cost UWB antennas in consumer applications. However, at this stage it is an open question if we finally need such strict antenna demands to achieve sufficient network coverage in a home environment. In order to answer these questions the broadband antenna characterisation can be extracted from the FDTD simulation and used as input data for propagation modelling.

(a) Antenna on top at front side.

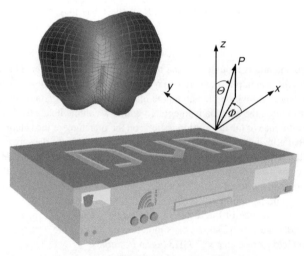

(b) Antenna integrated in corner.

Figure 10.12 Influence of different integration scenarios on the radiation pattern (e.g. at $f = 7$ GHz)

The impulse response of the antenna in the integration scenario in Figure 10.12(c) is calculated from the frequency domain transmit transfer function of the antenna. Figure 10.14 shows the transfer function for a single angle ($\Phi = 270°$, $\Theta = 90°$). In order to achieve a smooth curve a Kaiser window has also been used.

It can be noted from Figure 10.14 that the ringing of the antenna integrated in the DVD player is significantly higher compared to the biconical antenna in free space. The delay spread is calculated to be:

$$\tau_{DS}(\Phi = 270°, \Theta = 90°) = 0.027 \text{ ns}$$

and is therefore also significantly higher compared to the biconical antenna in free space.

UWB Antenna Elements for Consumer Electronic Applications

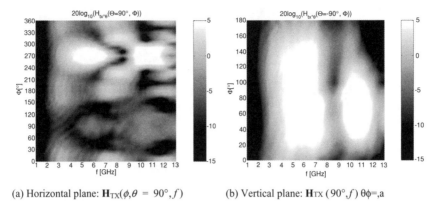

(a) Horizontal plane: $\mathbf{H}_{TX}(\phi, \theta = 90°, f)$

(b) Vertical plane: $\mathbf{H}_{TX}(90°, f)\, \theta\phi=,a$

Figure 10.13 Horizontal and vertical plane of the frequency-dependent transmit transfer function of the scenario in Figure 10.12(c)

10.3.3 Antenna Integration into a Mobile Device

As a second realistic example the planar antenna is adapted to be integrated into a mobile device, such as a mobile phone or a PDA.

In a mobile device space is even more limited than in a larger device like a DVD player. If we consider e.g. a mobile phone, it turns out that the integrated antenna is in close vicinity to other components and

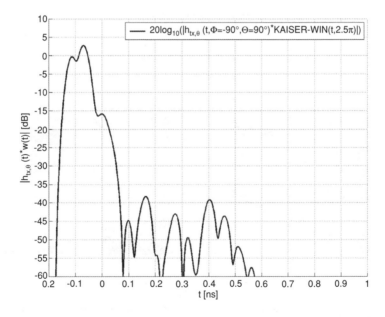

Figure 10.14 Impulse response of the biconical antenna ($\Phi = 270°$, $\Theta = 90°$)

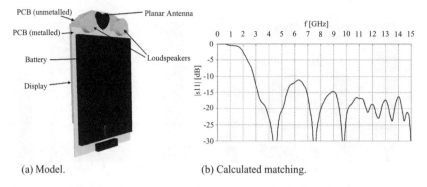

(a) Model. (b) Calculated matching.

Figure 10.15 Mobile device equipped with integrated planar UWB antenna

even has to share its volume with some of them. The model in Figure 10.15 forecasts how it could look like for a mobile device with a UWB antenna. The planar antenna element is located at top of the PCB. Other components, e.g. loudspeakers, are located next to the antenna. The battery comes close to the antenna on the one side and the display partly overlaps on the other side. In order to apply the antenna concept we have to recess the metallisation of the PCB under the antenna. The antenna is adapted to the specific integration scenario by the same optimisation procedure as the DVD player in the previous section.

Figure 10.15 shows that it is possible to achieve a good broadband matching by this procedure.

The transfer function is displayed for the Θ-component only, however, due to the radiation from the edge of the PCB and the other components the antenna radiation contains also certain part of the Φ-component.

Figure 10.16(a) shows that the omnidirectionality holds only for the lower frequencies due to the planar antenna element on the one hand and the current distribution on the chassis of the mobile on the other hand.

It can be noted from Figure 10.17 that the ringing of the antenna integrated into the mobile device is also significantly higher compared to the biconical antenna in free space. This is a result of the realistic

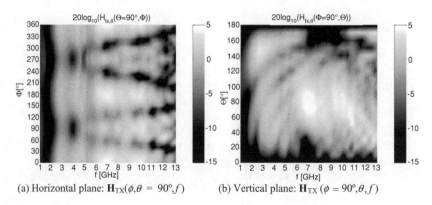

(a) Horizontal plane: $\mathbf{H}_{TX}(\phi, \theta = 90°, f)$ (b) Vertical plane: $\mathbf{H}_{TX}(\phi = 90°, \theta, f)$

Figure 10.16 Horizontal and vertical plane of the frequency-dependent transmit transfer function of the mobile device with integrated UWB antenna

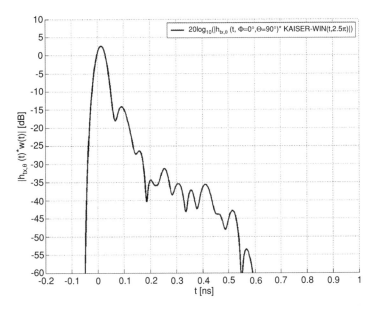

Figure 10.17 Impulse response of the mobile device with integrated UWB antenna ($\Phi = 90°$, $\Theta = 90°$)

integration situation of the antenna which leads to additional currents on the chassis and thus results in a much larger antenna which is composed out of the antenna element and the chassis of the mobile itself. The delay spread is calculated to be: $\tau_{DS}(\Phi = 90°, \Theta = 90°) = 0.0217$ ns and is therefore also significantly higher compared to the biconical antenna in free space.

10.3.4 Conclusion

The results presented in this section allow different findings:

- It is possible to integrate small antennas which provide UWB matching into devices of home entertainment systems and mobile devices.
- The antenna has to be tuned individually to operate in the specific integration scenario. This results in an entire system that works as the antenna which is composed of the antenna element itself and the chassis of the device the antenna element is integrated in.
- Therefore the specific integration scenario significantly affects the antenna performance resulting in:
  - non-omnidirectional radiation even if the antenna element is omnidirectional;
  - additional ringing due to currents which provide radiation from the chassis;
  - frequency-dependent radiation pattern, even if the radiation pattern of the antenna element is nearly stable in the desired frequency range.

Furthermore these *non-ideal* antenna properties cannot be prevented because they result from the physical principle of any antenna.

Therefore, as a general conclusion, an antenna has to be designed for a specific application scenario and the performance is a result of the antenna in the application and not the antenna element alone.

10.4 Propagation Modelling

UWB for consumer equipment is a new technology where there is very little experience in the design and performance of the complete system. One main fear of the system designer who decides to replace the cable by a UWB module is that certain areas in the room may not be covered. Because of this often omnidirectional antennas are requested by the system designers. However, we have seen that this demand is not feasible when the antenna is integrated into the chassis of the application. In addition very often people think of UWB in terms of impulse transmission. With this regard ringing and frequency-dependent radiation patterns lead to deformations of the radiated pulse by the antenna and the channel. Therefore frequency independent antennas in combination with low ringing are also often requested by the system designer. However, we have seen from the above investigations that this is also a non-feasible demand if we think of antenna integration.

Nevertheless it is not that simple to answer if all of these requirements are really necessary or what is the impact on the performance at all if they cannot be fulfilled completely. For example pulse deformation should be less a problem for OFDM-based systems as it is for impulse radio. All in all, the question can only be answered by a complete investigation of the system including propagation modelling and knowledge about the system conception. The final decision on the system conception has not yet been taken. At the moment there are still different proposals based on OFDM or DS-CDMA proposed by different interest groups. Therefore the final step in our investigation is a general investigation of the communication link.

In order to investigate the network coverage in a home environment the antenna characteristic is transferred to ray-tracing software for propagation modelling. An artificial living room is set up. The DVD player is placed in a shelf at one side of the room. Figure 10.18 shows a picture of the room including an interior.

The room has been modelled using the 3D ray-tracing software *Wireless Insite*. A script for the transition from the antenna characteristics calculated using EMPIRE™ to *Wireless Insite* is written within *MatLab*.

Figure 10.18 shows a map of the power distribution in a plane of 1 m constant height from the floor in the living room when the DVD player with the antenna integrated according to scenario no. 3 (see Figure 10.12(c)) is used. In order to show the effect of the directive pattern a single frequency of $f = 7$ GHz

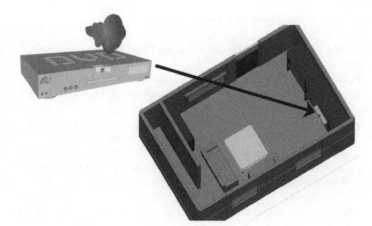

Figure 10.18 Typical living room modelled using the ray-tracing tool *Wireless Insite*. The DVD player is positioned on a shelf at the right side of the room

UWB Antenna Elements for Consumer Electronic Applications   215

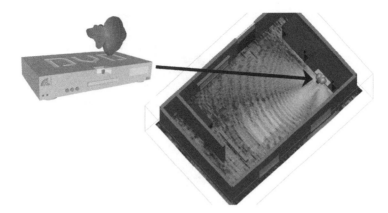

Figure 10.19 Map of the power distribution in the room at $f = 7$ GHz using the planar monopole antenna integrated in the DVD player (scenario 3, see Figure 10.12(c))

is used. It can be observed from Figure 10.13(a) that the antenna patterns split into several beams in this frequency range. The effect of these separated beams leads to a significant variation on the received power at different locations in the room as shown in Figure 10.19. As a result some directions benefit from the higher gain, some areas suffer from being in the direction of a null in the pattern.

Additionally the coverage is significantly affected by fading as it can be observed from the standing wave modulation on the pattern of the power map. At this stage another advantage of UWB propagation can be mentioned: it has been reported in [8] that fading becomes smaller when the bandwidth of the transmitted pulse becomes broader. In this respect a second propagation investigation has been performed. First propagation simulations at single frequencies between 3.1 GHz and 10.6 GHz are performed. Later the received power of all simulations is averaged over the frequency range. (Note: the usage of the complete bandwidth between 3.1 GHz and 10.6 GHz is of course not intended for early systems. More realistic bandwidth will be smaller for the first generation of systems.) The results of this investigation are displayed in Figure 10.20 first for the case of the DVD player applying again integration scenario 3 (see Figure 10.20(a)) and later for an *artificial* isotropic antenna for the sake of comparison (see Figure 10.20(b)). It can be observed that the map of the power distribution becomes smoother due to the averaging. In the case of the DVD player this is not only because of the beneficial effect of the multipath propagation but also the frequency-dependent radiation pattern (that was deemed to be unwanted in the beginning). The comparison to the 'ideal' isotropic radiator shows that only small areas at the outer regions of the room are covered less using the realistic model while the largest part of the overall room shows even better signal.

10.5 System Analysis

As mentioned earlier, depending on the system conception, network coverage by smooth power distribution may not be the only aspect. It might also be interesting to investigate how the signal will be distorted either by the antennas or the channel. In particular, because of the fact that the channel is given (living room environment), the designer can only affect (at least in principle) the distortion of the pulse by the choice of the antenna.

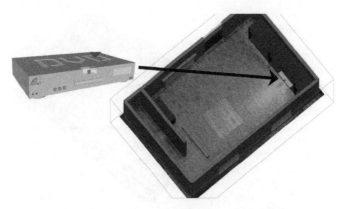

(a) Monopole antenna integrated in the frontside of the DVD player.

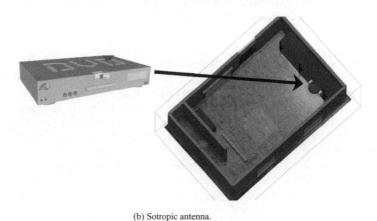

(b) Sotropic antenna.

Figure 10.20 Ray-tracing simulation of DVD player with integrated UWB antenna in an artificial living room environment. Map of the received power averaged over a frequency span from 3.1 GHz < f < 10.6 GHz. Comparison between the integrated antenna (a) and an isotropic radiator (b)

In order to investigate the related aspects we again make use of the derived antenna characterisations including the specific integration scenario in combination with the ray-tracing model of the living room. As an example we can investigate the link between the DVD player (TX1) and the mobile device on the table (RX1) or any other combination shown in Figure 10.21.

The entire link can be described in terms of the block diagram in Figure 10.22. It includes the transmit transfer function of the transmit application, the transfer function of the channel and the receive transfer function of the receive application. Transmit and receive antenna transfer functions are taken from the FDTD modelling. The transfer function of the channel is calculated by the use of the ray-tracing software. In combination we have a transfer function of the link system which results in the product of the single transfer functions in the frequency domain or the convolution in the time domain, respectively. The notation chosen in Figure 10.22 indicates that the transfer function of the link can also be interpreted in terms of scattering parameters, which is a familiar measure to RF engineers.

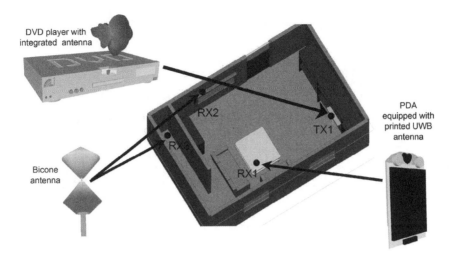

Figure 10.21 Ray-tracing model of a living room. The UWB module inside the DVD player acts as a transmitter (TX1). Other applications (PDA on table (RX1), flat screen on wall (RX2), modem in corridor) using different antennas are positioned at certain locations in the room and act as receivers for this example

Figure 10.23 shows the transfer function of the channel from the link between TX1 and RX1. It can be observed that the exponential decrease in frequency is superimposed by the effect of the multipath propagation in this indoor environment.

By including the transfer functions of the transmit and the receive application we are able to calculate the impulse response of the entire link including realistic antenna integration scenarios by means of the inverse Fourier transform. In order to cope with the truncation of the spectrum we again make use of a

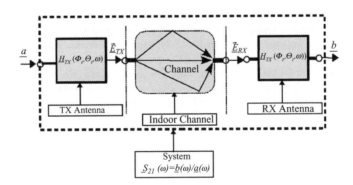

Figure 10.22 Block diagram of the entire system including the antennas integrated in the applications and the indoor channel. The links between two devices is characterised by the transfer function of the system, which can be expressed in terms of s_{21}

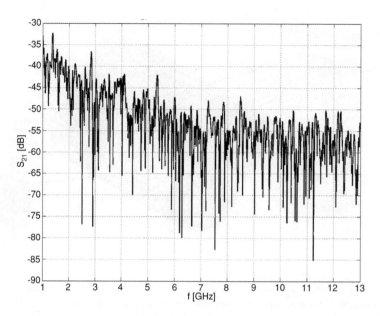

Figure 10.23 Transfer function of the channel between TX1 and RX1 in the living room using ideal isotropic radiators

window function. Figure 10.24 shows the impulse response of the link between the DVD player (TX1) on the shelf and the mobile device and the mobile device on the table (RX1) according to Figure 10.21:

It can be observed from the impulse response in Figure 10.24 that due to the LoS channel between TX1 and RX1 the first part of the signal arrives after 10 ns corresponding to a direct distance of 3 metres. The direct signal is followed by a number of reflections that are smaller in amplitude and are delayed according to the distance travelled. All in all the impulse excited in the transmit system is spread over a significant time.

If we replace the antenna transfer functions of the realistic integration scenarios (antennas integrated into DVD player of mobile device) by the transfer functions of ideal isotropic radiators in free space, we are able to compare the above result to a *quasi ideal* case with respect to the antenna. With this regard Figure 10.25 shows the impulse response of link TX1 (isotropic)-channel-RX1(isotropic) neglecting realistic antenna integration. It can be observed that the overall spread of the signal for the *ideal* antennas is in the same order as that for the case of the realistic *non-ideal* antenna integration. This is mainly due to the fact that the ringing imposed by the channel is significantly longer than the ringing of the antennas when they are integrated into the chassis of the application. Moreover, with respect to the specific link investigated there are even more echoes when the omnidirectional antennas are used compared to the directive chassis integration.

10.6 Conclusions

Planar monopole antennas are good candidates for the cost- and size-efficient integration into the UWB module for devices of consumer equipment. The FDTD method is ideally suited for the numerical analysis and design of the entire antenna system. Two different realistic antenna integration scenarios

UWB Antenna Elements for Consumer Electronic Applications

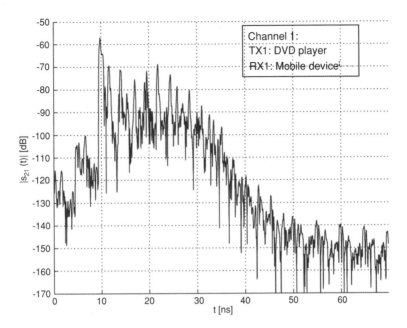

Figure 10.24 Impulse response of the link TX1 (DVD)-channel-RX1 (mobile) taking into account the realistic antenna integration

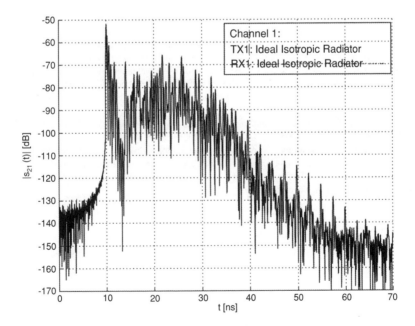

Figure 10.25 Impulse response of the link TX1 (isotropic)-channel-RX1 (isotropic) neglecting the realistic antenna integration

have been investigated – a UWB module integrated into the chassis of a DVD player and a planar UWB antenna integrated into a mobile device. As a result the specific integration scenario drastically affects all antenna parameters. While the matching of the antenna can be retuned by variation of the shape, enhanced frequency dependence and directional radiation patterns occur as a result of the integration.

The antenna characterisation of entire application (antenna integrated in the chassis) can be extracted and used either to calculate typical UWB antenna measures or as a basis for propagation modelling. With respect to typical UWB antenna measures we can observe that the ringing of the antenna is significantly enlarged by the antenna integration into the chassis of the device. All in all, the antenna performance tends to be less ideal with respect to traditional UWB antenna measures when the antenna is integrated into the chassis. In order to investigate the influence of these unideal properties the transfer function of the antennas is used to investigate the entire link of the system in an indoor environment by means of propagation modelling.

As a result it seams that the impact of the directive and frequency-dependent radiation pattern of the integrated antenna is not as strong as one would expect. Moreover, the frequency-dependent radiation pattern results in a smooth coverage in the room when the received power is averaged over a certain frequency span. In addition, when investigating the entire link, it can be observed that the ringing imposed by the channel is significantly longer than the ringing due to the antenna integration. Therefore, no major disadvantage of the realistic antenna integration can be observed when we model the entire link and not the antenna alone.

References

[1] Federal Communications Commission (FCC): Revision of Part 15 of the communication's rules regarding ultra wideband transmission systems. First report and order, ET Docket 98-153, FCC 02-48; Adopted: 14 February 2002; Released: 22 April 2002.
[2] Intel:Wireless USB – The first high-speed personal wireless interconnect. White Paper. http://www.intel.com/technology/comms/wusb/download/wirelessUSB.pdf.
[3] F. Gustrau and D. Manteuffel, *EM Modelling of Antennas and RF Components for Wireless Communication Systems – A Practical Guide. Series: Signals and Communication Technology*, Springer, 2006.
[4] D. Manteuffel and J. Kunisch, Efficient characterization of UWB antennas using the FDTD method. *AP-S – International Symposium on Antennas and Propagation*, Monterey, CA, USA, June 2004.
[5] J. Kunisch and J. Pamp, UWB radio channel modelling considerations. *Proc. of ICEAA'03*, Turin, September 2003.
[6] W. Sörgel and W. Wiesbeck, Influence of antennas on the ultra-wideband transmission, *Eurasip Journal on Applied Signal Processing*, **3**, 296–305, 2005.
[7] D. Manteuffel, Considerations for the design of UWB antennas for mobile and consumer equipment. *ICT 2005 – 12th International Conference on Telecommunications*, Cape Town, South Africa, May 2005.
[8] J. Romme and B. Kull, On the relation between bandwidth and robustness of indoor UWB communication. *UWBST Conference*, Reston, November 2003.

11

Ultra-wideband Arrays

Ernest E. Okon

11.1 Introduction

This chapter examines UWB arrays for communication systems. Antenna arrays are often applied in communication and sensor systems to realise increased sensitivity and range extension. In UWB systems, low power levels will enable coexistence of UWB devices with more traditional communication systems in the same frequency bands. UWB arrays will be used to extend the existing range without necessarily increasing power levels. Emphasis will be placed on compact and cost-effective designs in realising UWB arrays for communication systems.

In this chapter, simple design parameters for pattern synthesis and array scanning are presented. Design parameters for the avoidance of array grating lobes are given. We then examine antenna elements applicable to UWB array design. Wideband array design parameters are presented. Modelling techniques for UWB arrays are also examined. Subsequently, various wideband arrays types are studied under passive and active configurations. Finally, design considerations for UWB arrays are highlighted.

11.2 Linear Arrays

The linear array is often used as the basic foundation for the design of practical array systems. We will adopt this strategy in order to establish basic design parameters for practical arrays. Consider a linear array of N isotropic elements as depicted in Figure 11.1.

The elements are separated by an equal distance d and we assume a phase reference at the first element $n = 1$. Thus the radiated field at a large distance away for the array is given as

$$AF = \sum_{n=1}^{N} A_{n-1} e^{j(n-1)\psi} \tag{11.1}$$

and

$$\psi = kd \cos\phi + \beta \tag{11.2}$$

Ultra-wideband Antennas and Propagation for Communications, Radar and Imaging Edited by B. Allen,
M. Dohler, E. E. Okon, W. Q. Malik, A. K. Brown and D. J. Edwards
© 2007 John Wiley & Sons, Ltd

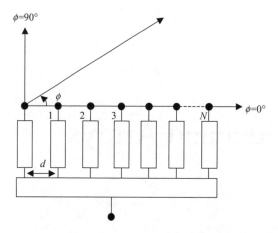

Figure 11.1 Linear array configuration

where $k = (2\pi/\lambda)$ is the wave number, λ is the wavelength in free space, ϕ denotes the direction angle of the far-field measured from the horizontal axis and β is the excitation phase for each element. If the phase reference is chosen at the centre of the array and the amplitudes are uniform and equal to unity, the radiated field can be written in normalised form as

$$AF_n = \frac{1}{n}\frac{(1-e^{jn\psi})}{(1-e^{j\psi})} = \frac{1}{n}\frac{\sin(n\psi/2)}{\sin(\psi/2)} \tag{11.3}$$

We now consider some array pattern characteristics, such as broadside arrays, ordinary end-fire arrays and highly directive end-fire arrays.

11.2.1 Broadside Array

In a broadside array of N isotropic point sources, the radiation pattern is perpendicular to the plane of the array and the amplitude and phase of the elements are equal. Thus, the phase term of Equation (11.2) reduces to

$$\psi = kd\cos\phi \tag{11.4}$$

The maximum value of the field occurs when $\psi = 0$, this occurs when $\phi = (2m+1)\pi/2$, where m is an integer value (i.e. $m = 0, 1, 2, 3\ldots$). Thus, the field is a maximum when $\phi = \pi/2$ and $\phi = 3\pi/2$. In Figure 11.2(a) and (b), the normalised field patterns for broadside arrays of eight elements with spacing $d = \lambda/2$ and $d = \lambda$ respectively are shown.

We observe from Figure 11.2(a) and (b) that the patterns have main lobes broadside to the plane of the array with sidelobes. As the spacing increases, the amplitude of the sidelobes becomes larger. Thus in Figure 11.2(b) the sidelobe amplitude is of the same magnitude as the main lobe. These lobes are known as grating lobes.

11.2.2 End-fire Array

In an end-fire array, the main lobes are in the direction of the array. We require the phase angle between adjacent elements that will make the array pattern a maximum. In this regard, we substitute $\psi = 0$ and

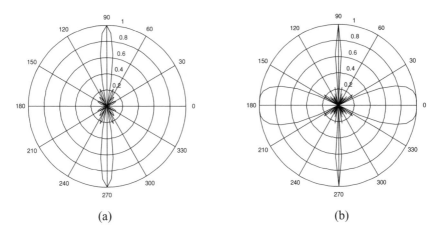

Figure 11.2 Broadside array field patterns for an eight-element array with (a) spacing $d = \lambda/2$ and (b) spacing $d = \lambda$

$\phi = 0$ into Equation (11.2). This yields

$$\beta = -kd \qquad (11.5)$$

Thus, to obtain an end-fire pattern with a linear array, the phase difference between successive elements must be progressively retarded as stated in Equation (11.5). The phase retardation is equal to the distance between elements expressed in radians. Depicted in Figure 11.3(a) and (b) are the normalised field patterns for an end-fire array of eight elements with element spacing of $d = \lambda/2$ and $d = \lambda/4$ respectively.

When the spacing between elements is $d = \lambda/2$, $\beta = -\pi$ and the resulting field pattern is bidirectional as depicted in Figure 11.3(a). The same field pattern is obtained if $\beta = +\pi$. Also, when the element spacing is $d = \lambda/4$, $\beta = -\pi/2$ and the resulting field pattern is unidirectional with the maximum radiation

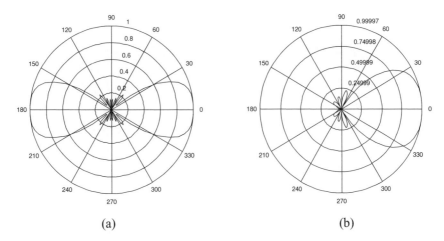

Figure 11.3 End-fire array field pattern for an eight-element array with (a) spacing $d = \lambda/2$ and (b) spacing $d = \lambda/4$

along $\phi = 0°$ as observed in Figure 11.3(b). However, when $\beta = +\pi/2$ the maximum radiation is in the direction $\phi = 180°$.

11.2.3 End-fire Array with Increased Directivity

The phase formula for an end-fire array in the previous section produced a field pattern with maximum radiation in the end-fire direction. However, the array does not produce maximum directivity (higher selectivity) in that direction. The directivity can be maximised by employing the formula given by Hansen and Woodward [1]. This states that a larger directivity is obtained if the phase differential between sources is

$$\beta = -\left(kd + \frac{\pi}{n}\right) \quad (11.6)$$

where n is the number of elements, d is the spacing between elements and all other expressions are as previously defined.

Figure 11.4(a) and (b) shows the field pattern of an eight-element array using Equation (11.6) for the phase differential as defined in the previous section for Figure 11.3(a) and (b), with element spacing of $d = \lambda/2$ and $d = \lambda/4$ respectively. We observe from Figure 11.4 that the patterns indicate a higher directivity for the main lobes compared to Figure 11.3. However, this is at the expense of higher sidelobe levels elsewhere.

11.2.4 Scanning Arrays

Scanning arrays permit the field patterns to have the maximum in some arbitrary direction ϕ_o. In this regard, we write Equation (11.2) as

$$0 = kd \cos \phi_o + \beta \quad (11.7)$$

Thus by specifying the element spacing d, the phase shift β is obtained. Alternatively we can specify β and deduce d. Figure 11.5 depicts the field pattern for an eight-element array with the maximum at an arbitrary angle $\phi = 30°$ with spacing $d = \lambda/4$.

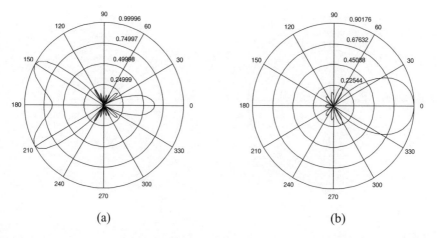

Figure 11.4 Directive end-fire array field pattern for an eight-element array with (a) spacing $d = \lambda/2$ and (b) spacing $d = \lambda/4$

Ultra-wideband Arrays

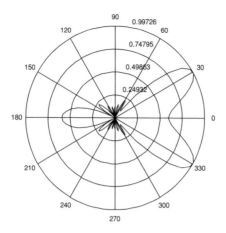

Figure 11.5 Field pattern for an eight-element array with spacing $d = \lambda/4$ and maximum radiation at 30°

We shall return to the scanning array later on in discussing phased arrays and formulating design equations for UWB arrays.

11.3 Null and Maximum Directions for Uniform Arrays

In the design of arrays, it is often useful to form a null in the array pattern. This could be used to mitigate noise or interference sources. Maximum directions are also important as they maximise signal-to-noise ratios in specific directions. Schelkunoff advanced pioneering work in determining the null and maximum directions for uniform arrays [3]. We shall now describe the approach.

11.3.1 Null Directions

The null directions for uniform arrays can be obtained by examining Equation (11.3). We observe that the expression goes to zero when

$$e^{jn\psi} = 1 \tag{11.8}$$

provided the denominator of Equation (11.3) is not zero. Thus, to satisfy Equation (11.8) we require

$$n\psi = \pm 2N\pi, \quad N = 1, 2, 3 \ldots \tag{11.9}$$

Using the expression for ψ in Equation (11.2), we can obtain

$$kd \cos \phi_o + \beta = \frac{\pm 2N\pi}{n} \tag{11.10}$$

We then deduce

$$\phi_o = \cos^{-1}\left[\left(\pm \frac{2N\pi}{n} - \beta\right)\frac{1}{kd}\right] \tag{11.11}$$

Thus we can obtain the angle ϕ_o for a uniform array using Equation (11.11). We note that values must be excluded for which $N = mn$, where $m = 1, 2, 3\ldots$ since Equation (11.9) reduces to $\psi = \pm 2m\pi$ and the denominator in Equation (11.3) goes to zero. Thus, the condition for the numerator being zero becomes insufficient.

If we consider a broadside array, $\beta = 0$, thus Equation (11.11) reduces to

$$\phi_o = \cos^{-1}\left[\left(\pm\frac{2N\pi}{n}\right)\frac{1}{kd}\right] \tag{11.12}$$

The same principle can be applied to end-fire arrays. In a similar manner, for an ordinary end-fire array, $\beta = -kd$, thus Equation (11.11) reduces to

$$\phi_o = \cos^{-1}\left[\left(\pm\frac{2N\pi}{nkd} + 1\right)\right] \tag{11.13}$$

Similarly, for a an end-fire array of high directivity as proposed by Hansen and Woodward [1],

$$\beta = -\left(kd + \frac{\pi}{n}\right),$$

Substituting this in Equation (11.11) yields

$$\phi_o = \cos^{-1}\left[\left(\pm\frac{2N\pi}{n} + kd + \frac{\pi}{n}\right)\frac{1}{kd}\right] \tag{11.14}$$

Equations (11.12)–(11.14) can be used to determine the null directions for uniform broadside, ordinary end-fire and high directivity end-fire arrays respectively.

As an example, we consider the broadside array of Figure 11.2(a) with eight elements ($n = 8$), $d = \lambda/2$. Thus, Equation (11.12) becomes

$$\phi_o = \cos^{-1}\left(\pm\frac{N}{4}\right) \quad N = 1, 2, 3\ldots \tag{11.15}$$

We can now deduce the null of the array, for $N = 1$, $\phi_o = \pm 75.5°$ and $\pm 104.5°$, for $N = 2$, $\phi_o = \pm 60°$ and $\pm 120°$, for $N = 3$, $\phi_o = \pm 41.4°$ and $\pm 138.6°$ and for $N = 4$, $\phi_o = \pm 0°$ and $\pm 180°$. These are the 14 nulls for the array as observed from Figure 11.2(a).

11.3.2 Maximum Directions

The maximum directions of the radiated field for the uniform array can be obtained by examining Equation (11.3). The main lobe maximum occurs when $\psi = 0$ for the broadside and ordinary end-fire arrays. For the broadside array, the maximum occurs in the direction $\phi = 90°$ and $\phi = 270°$. This is easily seen from Figure 11.2. For the ordinary end-fire array, the maximum occurs at $\phi = 0°$ and $\phi = 180°$ or both. For the end-fire array with increased directivity, the maximum occurs at $\psi = \pm \pi/n$ with the main lobe in the direction $\phi = 0°$ or $\phi = 180°$.

We now consider the maxima for the minor lobes. These are situated between the first and high-order nulls and occur whenever the numerator is a maximum as noted by Schelkunoff [3]. Thus, for the minor lobes, using Equation (11.3) the maximum directions occur when

$$\sin\frac{n\psi}{2} = 1 \tag{11.16}$$

This implies that

$$\frac{n\psi}{2} = \pm(2N+1)\frac{\pi}{2} \tag{11.17}$$

Thus, substituting the value of ψ from Equation (11.2) we obtain

$$kd \cos\phi_m + \beta = \frac{\pm(2N+1)\pi}{n} \tag{11.18}$$

Rearranging Equation (11.18) yields

$$\phi_m = \cos^{-1}\left[\frac{\pm(2N+1)\pi}{n} - \beta\right]\frac{1}{kd} \tag{11.19}$$

where ϕ_m denotes the direction of the minor-lobe maxima.

We now specialise Equation (11.19) for the case of a broadside array and end-fire arrays. In this regard, for a broadside array $\beta = 0$, thus the minor-lobe maxima becomes

$$\phi_{mb} = \cos^{-1}\left[\frac{\pm(2N+1)\pi}{n}\right]\frac{1}{kd} \tag{11.20}$$

For the ordinary end-fire array $\beta = -kd$, thus the minor-lobe maxima becomes

$$\phi_{me} = \cos^{-1}\left[\frac{\pm(2N+1)\pi}{nkd} + 1\right] \tag{11.21}$$

For the high directivity end-fire array $\beta = -\left(kd + \frac{\pi}{n}\right)$, thus the expression for the minor-lobe maxima is

$$\phi_{med} = \cos^{-1}\left[\{\pm(2N+1)+1\}\frac{\pi}{nkd} + 1\right] \tag{11.22}$$

where ϕ_{mb}, ϕ_{me} and ϕ_{med} represent the minor-lobe maxima for the broadside, ordinary end-fire and highly directive end-fire arrays, respectively.

As an example, we consider the field pattern of Figure 11.2(a), ($n = 8$, $d = \lambda/2$, $\beta = 0$). We obtain the directions of the minor-lobe maxima from Equation (11.19), which becomes

$$\phi_{mb} = \cos^{-1}\left[\frac{\pm(2N+1)\pi}{8}\right] \tag{11.23}$$

For $N = 1$, $\phi_{mb} = \pm 67.97°$ and $\pm 112.03°$, also for $N = 2$, $\phi_{mb} = \pm 51.32°$ and $\pm 128.68°$, and for $N = 3$, $\phi_{mb} = \pm 28.96°$ and $\pm 151.04°$. These are the 12 nulls of the pattern.

It has also been shown by Schelkunoff that the relative amplitudes of the minor lobes of the patterns can be obtained by observing that since the numerator of Equation (11.3) is approximately unity at the maximum of the minor lobe, the relative amplitude of a minor lobe maximum is given as

$$AF_{ML} = \frac{1}{n\sin(\psi/2)} \tag{11.24}$$

Substituting the expression for ψ from Equation (11.17) into Equation (11.24) we obtain

$$AF_{ML} = \frac{1}{n\sin[(2N+1)\pi/2n]} \tag{11.25}$$

As an example, we consider Figure 11.2(a) once more for $n = 8$, substituting values for $N = 1,2,3$ in Equation (11.18) we obtain the relative amplitudes as 0.22, 0.15 and 0.13, where the main lobe has a unit amplitude. The values can be verified from the plot of the field pattern.

11.3.3 Circle Representations

We now examine an alternative pictorial representation for obtaining the maxima and minima of the field patterns for a uniform array. This is due to the original work of Schelkunoff. In this representation, we recall Equation (11.1) for the uniform linear array with unit amplitudes as

$$AF = 1 + e^{j\psi} + e^{j2\psi} + e^{j3\psi} + \ldots e^{j(n-1)\psi}$$

If we substitute the expression $z = e^{j\psi}$, then the array factor can be expressed as

$$AF = 1 + z + z^2 + z^3 + \ldots z^{n-1} \tag{11.26}$$

The expression above is a polynomial equation of the $(n-1)$th order. Schelkunoff has shown that every linear array can be expressed by a polynomial and the converse that every polynomial can be interpreted as a linear array is valid. Also, since $\psi = kd \cos \phi + \beta$ is always real, the absolute value of z equals unity and z is always on the circumference of the unit circle. As ϕ increases from 0° (which is in the direction of the line of sources) to 180° (which is in the opposite direction), ψ decreases and z moves in the clockwise direction as shown in Figure 11.6. When $\phi = 0°$, $\psi = kd + \beta$ and when $\phi = 180°$, $\psi = -kd + \beta$. Hence, the range described by z is

$$\overline{\psi} = 2kd \tag{11.27}$$

When the separation d between the successive elements is equal to half a wavelength, the range of $z = 2\pi$ and as ϕ varies from 0° to 180°, z completes a full cycle and returns to its original position. There is a one-to-one correspondence between the points of the circumference of the unit circle and the conical surfaces coaxial with the line of sources. Such conical surfaces (radiation cones) are loci of directions in which the radiation amplitudes are equal. Also, if the separation between the elements is less than $\lambda/2$, then the range of z is smaller than 2π and z describes a portion of the unit circle. Finally, if $d > \lambda/2$ then the path of z overlaps itself.

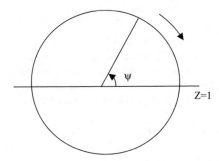

Figure 11.6 The unit complex circle with centre at origin represents a typical direction in space. As the angle ϕ, made by a typical direction with the line of sources, increases from 0° to 180°, point z moves clockwise

The use of algebraic representation such as Equation (11.26) is useful for pattern synthesis using arrays. Thus, polynomial expressions can be multiplied to yield arrays with desired radiation patterns. In addition, the maxima and null points of the array can be deduced by recasting Equation (11.26) in the form

$$AF = (z - t_1)(z - t_2)(z - t_3)\ldots(z - t_{n-1}) \tag{11.28}$$

where the fundamental theorem of algebra which states that a polynomial of degree $(n - 1)$ has $(n - 1)$ zeros and can be factored into $(n - 1)$ binomials.

Hence, multiplying Equation (11.26) by $e^{j\psi}$ and subtracting the result from Equation (11.26) we obtain a representation expressed as

$$AF = \frac{1 - z^n}{1 - z} \tag{11.29}$$

The null points of the array are then the nth roots of unity, excluding $z = 1$. Since z is a unit complex number, its higher powers are also unit complex numbers. In addition, each multiplication by $z = e^{j\psi}$ represents a displacement through an arc of ψ radians. Hence the nth roots of unity divide the circle into n equal parts. This is expressed analytically as

$$z^{n-1} = 0, t_m = e^{-\frac{j2m\pi}{n}}, m = 1, 2, 3, \ldots n - 1 \tag{11.30}$$

$$\psi_m = -\frac{2m\pi}{n}, \cos\phi = -\frac{2m\pi}{n} - \frac{\beta}{kd} \tag{11.31}$$

Based on Equations (11.30) and (11.31) we deduce the radiation pattern nulls and maxima as illustrated on a unit circle in Figure 11.7. This corresponds to the expressions derived in the ϕ directions in Sections 11.3.1 and 11.3.2.

We have considered uniform arrays in this presentation so as to convey the concept of pattern synthesis and design. However, nonuniform arrays using binomial, Taylor and Dolph–Tschebyscheff distributions can be applied to reduce sidelobe levels in the array pattern [1], [2], [4].

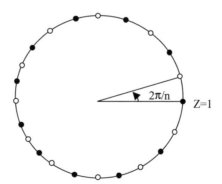

Figure 11.7 The null and maximum points of a uniform linear array and the point $z = 1$ representing the direction of greatest radiation divides the circle into equal parts. Maximum radiation points are represented by the solid circles while null points are represented by the hollow circles

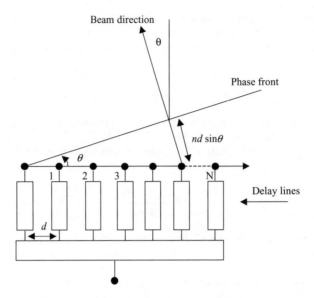

Figure 11.8 Basic linear phased array configuration

11.4 Phased Arrays

The phased or scanning array allows the radiated or received pattern of the array to be scanned in any preferred direction permitted by the array geometry. This is particularly useful for UWB systems, where the array can be scanned to preferred directions and enables range extension without the need to increase the radiated power of the system. We consider a linear array of N elements with spacing d to describe the principle of the phased array as shown in Figure 11.8.

We assume that the direction of a received plane wave makes the angle θ with the array normal and the phase front of the received beam has the same angle as shown in Figure 11.8. The response of the array or the array factor can be expressed as

$$AF(\theta) = \sum_{n=1}^{N} A_{n-1} e^{j(n-1)\psi} \quad (11.32)$$

and

$$\psi = kd\sin\theta + \beta \quad (11.33)$$

Where $k = (2\pi/\lambda)$ is the wave number, λ is the wavelength in free space, θ denotes the angle the received beam makes with the array normal and β is the excitation phase for each element. If we assume that the excitation phase for each element is zero, then there is a progressive phase shift between successive elements in the array given by $\Delta\psi = kd\sin\theta$. This phase shift corresponds to the difference in the time of arrival τ of the phase front of $\tau = d/c\sin\theta$, where c denotes the speed of light. In a UWB system, a pulse may be received by the array. Due to the difference in time of arrival across the array, the pulse will appear distorted. In order for the pulse to be received undistorted, phase shifters or delay lines will need to be used behind the array elements to obtained a uniform phase front when summing the array signal. In UWB systems, true time delay lines are normally used as these are not frequency dependent and are valid at all frequencies.

In Equation (11.32), the coefficients A_{n-1}'s are known as the array amplitude taper, while ψ_{n-1}'s are known as the phase taper. If we desire to combine the received signal from all elements to produce a maximum response in the direction θ_o, then from Equation (11.33), $\psi = 0$ and we deduce the expression for β as

$$\beta = -kd \sin \theta_o \qquad (11.34)$$

Thus, when the excitation phase of the elements are set according to Equation (11.34), a main beam will be produced in the θ_o direction on transmit or receive mode for the array. Hence, the array factor can be expressed as

$$AF(\theta) = \sum_{n=1}^{N} A_{n-1} e^{j(n-1)kd(\sin \theta - \sin \theta_o)} \qquad (11.35)$$

If we assume the special case of a uniform array with unit amplitudes (i.e. $A_{n-1} = 1$ for all n), the array factor for the N-element array becomes

$$AF(\theta) = \frac{\sin\left[N\pi \frac{d}{\lambda}(\sin \theta - \sin \theta_o)\right]}{N \sin\left[\pi \frac{d}{\lambda}(\sin \theta - \sin \theta_o)\right]} \qquad (11.36)$$

11.4.1 Element Spacing Required to Avoid Grating Lobes

In the design of phased arrays, it is imperative that grating lobes are eliminated as they reduce the power in the main beam and thus reduce the effective antenna gain. An example of grating lobes is shown in Figure 11.2(b) and these lobes occur when the spacing between the array elements is beyond a certain limit. We need to know the limits for element spacing in the array so to avoid the grating lobes. In this regard, using the variable $\nu = \sin \theta$, the array factor in Equation (11.32) can be expressed as

$$AF(\nu) = \sum_{n=1}^{N} A_{n-1} e^{j(n-1)kd(\nu - \nu_o)} \qquad (11.37)$$

where $\nu_o = \sin \theta_o$, denotes the beam-pointing direction and the differential excitation phase between elements in the array is $\Delta \phi = -kd\nu_o \sin \theta$. Equations (11.35) and (11.37) are related by a one-to-one mapping in the region $|\nu| \leq 1$, this is often referred to as the visible space corresponding to the angle of θ [6]. We observe that $AF(\nu)$ is a Fourier-series representation and that its maxima occurs whenever the argument of Equation (11.37) is a multiple of 2π. Thus, $kd(\nu - \nu_o) = 2n\pi$ where $n = \pm1, \pm2, \pm3, \ldots$ this can be further expressed as

$$\nu - \nu_o = \frac{n}{d/\lambda} \qquad (11.38)$$

The principal or main lobe coincides with the maximum of Equation (11.37) and this occurs when $\nu = \nu_o$ or when $n = 0$. The other maxima occur as a result of grating lobes. Thus, when the beam is scanned to ν_o, the closest grating lobe is located at $\nu = \nu_o - \lambda/d$. This occurs when $\nu = -1$ which is just at the end of visible space. Thus,

$$\frac{d}{\lambda} = \frac{1}{1 + \sin |\theta_o|} \qquad (11.39)$$

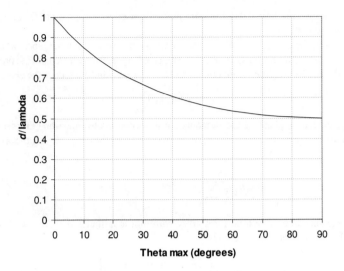

Figure 11.9 Allowable element spacing versus maximum scan angle

To avoid grating lobes, the element spacing criterion can be expressed as

$$\frac{d}{\lambda} < \frac{1}{1 + \sin |\theta_{max}|} \tag{11.40}$$

where θ_{max} denotes the maximum scan angle. In a planar array, the scan angle varies from boresight to 90°; the variation of d/λ over this scan range is depicted in Figure 11.9. We also observe that for a maximum scan angle of 90°, the spacing must be less than $\lambda/2$.

11.5 Elements for UWB Array Design

In the design of UWB arrays, it is imperative to consider the type of elements to be employed in the design of the array. Emphasis will be on low cost and compact elements that are easily integrated into array configurations. The general bandwidth requirement for UWB antennas is for elements that have a bandwidth greater than 25% of the centre frequency. In UWB systems for communications, bandwidths will be in the range of 3.1–10.6 GHz [7]. The bandwidth can also be expressed as a ratio of the upper to the lower frequency of operation. Thus, the percentage bandwidth can be expressed as

$$B_{pt} = \frac{f_U - f_L}{f_C} \times 100 \tag{11.41}$$

where f_U and f_L denote upper and lower band frequencies respectively and f_C is the centre frequency with the relation $f_C = (f_L + f_U)/2$. The bandwidth ratio is expressed as

$$B_{rt} = \frac{f_U}{f_L} \times 100 \tag{11.42}$$

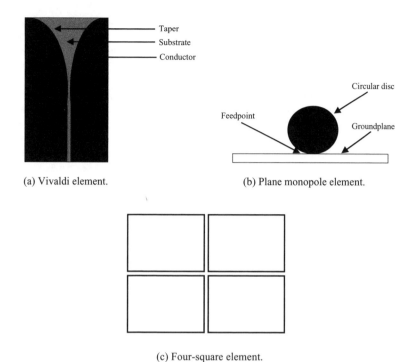

(a) Vivaldi element. (b) Plane monopole element.

(c) Four-square element.

Figure 11.10 Wideband antenna elements

Also, from Equations (11.41) and (11.42) the percentage bandwidth can be related to the bandwidth ratio by the expression

$$B_{pt} = 2\frac{B_{rt} - 1}{B_{rt} + 1} \times 100 \tag{11.43}$$

We now review a number of wideband elements that are applicable in UWB array design [7]. Emphasis is placed on elements that are compact, relatively low cost and easy to manufacture. The tapered notch antenna is typical for wideband array design. Vivaldi antennas are a class of tapered notch antennas with a typical bandwidth of 18:1. They consist of a microstrip or stripline to slotline transition with the slot flared in an exponential taper as depicted in Figure 11.10(a). The radiation pattern is fairly constant over a fairly wide frequency range and the element is linearly polarised. Arrays of UWB elements have also been shown to yield bandwidth of about 5:1 [8], [9]. The antipodal, travelling wave and bunny ear elements are variants of the Vivaldi element and provide improvements in bandwidth, compactness or ease of manufacture.

Another applicable element is the planar bowtie antenna. The antenna is based on the principle of the biconical horn and has bandwidths in the range of 1.5:1 (40%). The transverse electromagnetic (TEM) horn is a very wideband element and is applicable for UWB systems. The microstrip quasi-horn is based on the principle of the TEM horn and is applicable for UWB arrays. It consists of a printed upper conductor on a tapered dielectric structure backed by a groundplane. A bandwidth ratio of over 30:1 has been reported for the antenna [11].

The planar disc monopole is another element with wideband characteristics [12]. The element consists of a planar metallic disc on a groundplane as depicted in Figure 11.10(b). The circular and elliptical disc monopoles are known to yield very wide bandwidths. A typical bandwidth of 10.7:1 has been observed for this antenna type. However, the antenna is sensitive to the position of feed and height above the groundplane.

The spiral antenna is also applicable in UWB arrays. The Archimedean and equiangular spirals are well-known wideband elements producing circularly polarised radiation patterns. A typical isolated bandwidth for the equiangular spiral is 10:1 while that for an Archimedean spiral is 18:1. A variant of the spiral antenna is the sinuous antenna which is superficially similar to the spiral but is linearly polarised with a typical bandwidth of 9:1. However, the spiral antenna tends to be relatively large in the lateral dimension compared to other antenna elements. A typical value of 0.5λ is used where λ is the lowest operating frequency.

The four-square antenna depicted in Figure 11.10(c) is a relatively broadband antenna with a typical bandwidth of 1.8:1. It is a low profile antenna and consists of four patches that could be excited individually or through a feed network. The four-square elements are considered to be a good choice for wideband elements as they can be made quite compact. They also exhibit nearly equal beam widths in the E- and H-planes, which remain fairly constant over their bandwidth [10].

The antenna elements listed above are by no means an exhaustive list of wideband elements. However, they broadly depict the types of antenna elements applicable for cost-effective UWB array systems.

11.6 Modelling Considerations

The prediction of approximate radiation patterns of UWB arrays can be effected through the use of the equations expressed in Sections 11.1 to 11.4. However, the analysis of mutual coupling effects and input impedance values is specific to the geometry of the array and this requires more detailed modelling of the elements in the array. There are a number of full-wave techniques that can be applied to obtain a more rigorous model for the array. The method of moments technique can be applied to model UWB arrays and variants of this technique using mixed potential integral equations and incorporating exact Green's functions are applicable [17] [18]. This approach is a frequency-domain technique and is best suited for structures that are perfectly electrically conducting or for structures where the exact Green's functions can be derived and computed.

Another full-wave approach is the time-based technique where Maxwell's equations are stepped through in space and time. These include methods such as the finite element method (FEM) and finite difference time domain (FDTD) techniques [19] [20]. The approach is best suited for UWB systems as the analysis in the time domain can be obtained and structures of arbitrary geometry can be modelled. In the FDTD approach, the structure is meshed using a Cartesian grid approach. The mesh size can be reduced to account for fine details in the structure and various excitation signals can be analysed.

The FEM approach is a variational method where the energy in the system is stationary with respect to a variation in potential. This leads to the derivation of potential values and thus the field and current characteristics of the structure. Time-domain solvers can also deal with infinite array systems by the use of periodic boundary conditions.

11.7 Feed Configurations

In the design of UWB arrays, it is pertinent to consider the nature of the feed network. The feed configuration has an effect on the practical bandwidth of the array, the array size and the radiation pattern.

There are two main types of feed configurations, namely the active and passive array configurations. The effective realised gain of a planar phased array of n elements is given by the expression [5]

$$G_A = \frac{4\pi n A_e}{\lambda^2} \left(1 - \Gamma_{mn}^2\right) \cos\theta \tag{11.44}$$

where A_e denotes the effective element area, the pattern is assumed to vary as cos $\cos\theta$ and the wavelength is denoted by λ. Also, the reflection coefficient Γ_{mn} accounts for the effects of mutual coupling between the elements in the array. Equation (11.44) is assumed for the array when no grating lobes are present. The effect of grating lobes on the array will further reduce the effective gain.

11.7.1 Active Array

An active array consists of elements that are excited individually by separate generators. In the active array, the refection coefficient in Equation (11.44) accounts for the effect of mutual coupling due to proximity of the array elements. It accounts for the power returned to the generators and is dependent on the impedance value of the generators. Figure 11.11 depicts a section of an infinite active array of aperture elements (e.g. horn antennas). The gain of each element is assumed to be identical and can be obtained by measuring the immersed element pattern with the element excited and all other elements terminated in matched loads. The effective gain of the array is obtained by measuring the array pattern when all elements in the array are excited simultaneously.

11.7.2 Passive Array

The passive array consists of a single generator and excites all the elements through a network of power dividers and phase shifters. The passive array does not have the identical performance for every element that the infinite active array has. In addition, reflections from the elements do not necessarily return to the generator but may be absorbed in terminations or reradiated in other directions, depending on the particular network used. Thus, the active array is more complicated to analyse than the active array. In

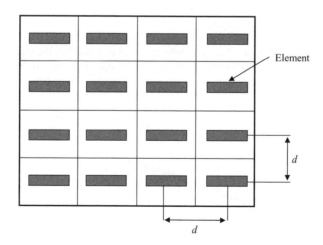

Figure 11.11 Active array of aperture antennas

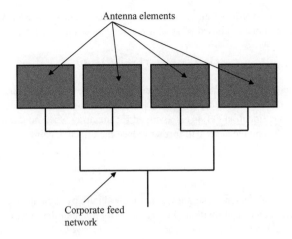

Figure 11.12 Array with corporate feed

modelling passive arrays, it is typical to use full-wave electromagnetic solvers based on the method of moments, finite elements or finite difference time domain methods.

Figure 11.12 depicts a corporate feed arrangement for a patch array antenna. The antenna is fed from a single generator and a power splitter is used to feed each element with the same excitation phase. Phase shifters may be included before each element to provide a phase gradient across the array. The array arrangement in Figure 11.12 provides a beam at boresight. Variations in the feed network such as notches or different line lengths may be employed to improve the bandwidth or alter the excitation phase. In a corporate feed network, the number of levels of power splitters required is equivalent to $\text{Log}_2 n$ where n is a number obtained by raising 2 to a certain power 2^P and P is a positive natural number. The bandwidth of the array is also dependent on the bandwidth of the antenna elements in the array. Thus, for a microstrip patch antenna printed on dielectric substrate, it is imperative to adopt low permittivity substrates and relatively thick substrate heights in order to improve the bandwidth of the array [13], [14]. However, in making this choice, care must be taken not to excite surface waves as this could lead to scan blindness in the array leading to pattern nulls for specific scan angles of the array.

Another passive array type of interest is the series fed array. The series array consists of an array of elements each feeding the other in succession and terminated in a matched load. The array is fed from a single generator. Figure 11.13 illustrates a microstrip patch series fed array for broadband millimetre wave application. The elements are printed on a substrate and bandwidths of greater than 11% have been

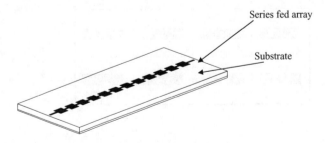

Figure 11.13 Series fed patch array. Source [15] (Reproduced by permission of © 2002 IET)

Ultra-wideband Arrays 237

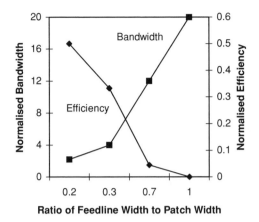

Figure 11.14 Bandwidth and efficiency plots for loaded two-port microstrip patch. Substrate permittivity 2.2, height 1.5 mm. Normalised to values for an unloaded patch. Source [15] (Reproduced by permission of © 2002 IET)

reported with this array [15]. The array bandwidth is dependent on the ratio of the feedline width to the patch width of the elements.

In addition, the amplitude distribution of the array is tapers with distance from the generator. The patch dimensions are approximately half a wavelength in the substrate although this dimension could be varied to provide different shaped beam patterns for the array. The series fed array offers the advantage of lower feedline losses compared to the corporate fed array. This becomes more pronounced as the number of array elements increase. Figure 11.14 shows the variation of bandwidth and efficiency of the array with the ratio of feedline width to patch width. The values have been normalised to that for an unloaded patch antenna with a single feed. A dielectric substrate of permittivity 2.2 and height 1.5 mm has been assumed.

Figure 11.15 shows the return loss for a five-element series array at millimetre wave frequencies. The computed results have been obtained using a full-wave method of moments approach with exact Green's functions expressed in the form of Sommerfeld type integrals [15]. In the design of series fed arrays, the

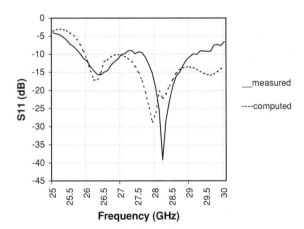

Figure 11.15 Return loss plot for a five-element series array. Source [15] (Reproduced by permission of © 2002 IET)

radiation pattern is obtained by a combination of simple formulation for the array factor and the full-wave technique.

The angle of the main lobe of the microstrip series array varies with frequency and can be designed to squint at a specific angle from the normal. The squint angle is obtained approximately in a simple manner, by modelling the array as a series of M point sources with array factor given as

$$AF = \sum_{m=1}^{M} I_{m1} e^{j(m-1)(kd_x \sin\theta \cos\phi + \psi_x)} \quad (11.45)$$

θ and ϕ being the angles of elevation and azimuth respectively. I_{m1} is the excitation coefficient of the mth element m, k denotes the wave number, d_x is the spacing and ψ_x represents the progressive phase shift between elements respectively. The array is assumed orientated in the x-direction. Thus, it is easily deduced that the value of ψ_x for maximum radiation intensity is given by

$$\psi_x = -kd_x \sin\theta_o \cos\phi_o \quad (11.46)$$

where we assume that the beam is squinted at the angle θ_o, ϕ_o.

We note other feed networks such as Butler matrices and Blass networks that may be used to effect switched beam configurations. Details of these networks can be referred to elsewhere [6].

11.8 Design Considerations

The design of UWB arrays is dependent on three main factors, namely, the array element, the array architecture and the feed network. Although several elements are available as noted in Section 11.7, the element must be appropriate for the array architecture. The array architecture is determined by requirements on the array bandwidth, pattern and the scan range. The feed network is also important and must be designed for the elements and the architecture to satisfy system performance requirements.

In the design process, the array architecture is investigated to determine the array grid size, next the element specifications are determined from the selected array grid and system requirements. This is followed by a feed network for the selected element and array architecture. The limiting factor for wideband arrays is most often thought to be the physical size of the element. We have assumed a uniform array throughout most of this presentation, however, nonuniform arrays can be designed using elements of different sizes and a less dense array grid geometry [10].

The scan angle range determines the grid dimensions and it is often appropriate to design array systems to prevent the emergence of grating lobes during the scan. This has been discussed in Section 11.4. We note that the formation of nulls in the pattern can also limit the scan range more than the appearance of grating lobes. This may be due to mutual coupling or the effect of surface waves on the array structure. The physical element size limits the performance of wideband elements. Thus, for wide bandwidth the elements tend to be electrically and physically large to radiate effectively at the lowest operating frequency. The lowest operating frequency is usually fixed by the physical size of the element. This can be expressed as

$$D = \alpha \lambda_L \quad (11.47)$$

where D is the array grid size, α is a constant term and λ_L is the wavelength at the lowest operating frequency. We assume D is the same in both dimensions for the grid. If we assume that the scan limit is given by Equation (11.40), and $D = d$, we obtain

$$D = \frac{\lambda_U}{1 + \sin\theta_{max}} \quad (11.48)$$

where λ_U is the wavelength at the upper operating frequency. Thus, the bandwidth is determined by the element size at the low end of the operating frequency range and by the scan performance at the upper end of the frequency range. Thus, the array bandwidth is given from Equation (11.42) as

$$B_{rt} = \frac{f_U}{f_L} = \frac{\lambda_L}{\lambda_U} \quad (11.49)$$

Substituting Equations (11.47) and (11.48) into Equation (11.49), we obtain the array bandwidth as

$$B_{rt} = \frac{1}{\alpha(1 + \sin\theta_{max})} \quad (11.50)$$

Equation (11.50) indicates that the array bandwidth is determined by the maximum scan angle and the size of α. Hence the smaller the scan range and the smaller the size of α the larger the bandwidth.

There is a direct relation between the volumetric size of an element or its Q and the bandwidth achievable with the element. This relationship gives the designer an estimate of a trade-off between size and desired bandwidth. The relation is given as

$$Q_{radiation} = \frac{1 + 2(ka)^2}{(ka)^3(1 + (ka)^2)} \quad (11.51)$$

where k is the wave number and a is the radius of the sphere. Thus, this relationship must be considered in constraining the element size. The model is a sphere that completely encloses the antenna under test. By computing the ratio of the radiating fields to the reactive fields (the evanescent mode) for the antenna and then transferring this understanding into an equivalent two-port ladder network, Q can be computed. No practical antenna can achieve this ideal limit; a real antenna can only approach it [21] [22].

In addition, to achieve scan angles in planes, broad beamwidths are required in the E- and H-planes. A narrow beamwidth will lead to scan loss due to the loss of gain at large scan angles

The challenge in UWB array design is to satisfy the impedance and pattern bandwidths, scan angle, array size, element size and polarisation requirements. This must be done while taking into consideration the coupling environment, the effect of scan loss due to element pattern effects, grating lobes and null formation due to blind spots in the array.

11.9 Summary

This chapter has examined UWB arrays for communication systems. A review of basic formulations for array pattern synthesis has been presented. This has been based on the linear array in order to outline basic array concepts. Basic array design equations to avoid grating lobes were highlighted. Subsequently, antenna elements for UWB arrays were presented. Emphasis has been placed on low-cost, compactness and ease of manufacture. Active and passive array feed configurations have been examined and various array types based on these configurations were presented. Finally, design considerations for UWB arrays were discussed.

References

[1] J.D. Kraus, *Antennas*, McGraw-Hill, New York, 1988.
[2] C.A. Balanis, *Antenna Theory*, John Wiley & Sons, Inc, New York, 1997.
[3] S.A. Schelkunoff, A mathematical theory of linear arrays, *Bell System Technical Journal*, **22**, 80–107, 1943.
[4] C.L. Dolph, A current distribution for broadside arrays which optimises the relationship between beamwidth and side-lobe level, *Proc. IRE Waves and Electrons*, **34**(6), 335–48, 1946.
[5] P.W. Hannan, The element-gain paradox for a phased array antenna, *IEEE Trans. Antennas Propagat.*, **AP-12**, 423–33, 1964.
[6] R.C. Johnson and H. Jasik, *Antenna Engineering Handbook*. McGraw-Hill, New York, 1981.
[7] E.E. Okon, J.C.G. Matthews and R.A. Lewis, Antenna elements for UWB array design, *Proc. Loughborough Antennas and Propagation Conference*, 2005.
[8] H. Holter, T. Chio and D.H. Schaubert, Exponential results of 144-element dual polarised endfire tapered-slot phased arrays, *IEEE Trans. Antennas Propagat.*, **48**(11), 1707–18, 2000.
[9] J.D.S. Langley, P.S. Hall and P. Newham, Balanced antipodal Vivaldi antenna for wide bandwidth phased arrays, *IEE Proc.-Microw. Antennas Propagat.*, **143**(2), 97–102, 1996.
[10] W.L. Stutzman and C.G. Buxton, Radiating elements for wideband phased arrays, *Microwave Journal*, 1999.
[11] C. Nguyen, J.S. Lee and J.S. Park, Ultra-wideband microstrip quasi-horn antenna, *Electron. Lett.*, **37**(12), 731–2, 2001.
[12] N.P. Agrawall, G. Kummar and K.P. Ray, Wide-band planar monopole antennas, *IEEE Trans. Antennas Propagat.*, **46**(2), 294–5, 1998.
[13] R.J. Mailloux, J. McIlevenna and N. Kerweis, Microstrip array technology, *IEEE Trans. Antennas Propagat.*, **AP-29**(1), 25–38, 1981.
[14] D.M. Pozar and D.H. Schaubert (eds), *Microstrip Antennas: The Analysis and Design of Microstrip Antennas and Arrays*, IEEE Press, 1995.
[15] E.E. Okon and C.W. Turner, Design of broadband microstrip series array for mm-wave applications, *IEE Electron. Lett.*, **38**(18), 1036–7, 2002.
[16] R.F. Harrington, *Field Computations by Moment Methods*. IEEE Press, 1993.
[17] J.R. Mosig and F.E. Gardiol, General integral equation formulation for microstrip antennas and scatterers, *Proc. Inst. Elec. Eng.*, **132**(H), 424–32, 1985.
[18] J.R. Mosig, Arbitrary shaped microstrip structures and their analysis with a mixed potential integral equation, *IEEE Trans. Microwave Theory Tech.*, **36**(2), 314–23, 1988.
[19] A. Ishimaru, *Electromagnetic Wave Propagation, Radiation, and Scattering*, Prentice-Hall, 1991.
[20] A. Taflove and S.C. Hagness, *Computational Electrodynamics: The Finite Difference Time Domain Method*, Artech House, 2000.
[21] L.J. Chu, Physical limitations of omni-directional antennas, *Journal of Applied Physics*, **19**, 1163, 1948.
[22] R.F. Harrington, Effect of antenna size on gain, bandwidth, and efficiency, *Journal of Research of the National Bureau of Standards*, **64-D**, 1, 1960.
[23] L.Y. Astanin and A.A. Kostylev, *Ultra-wideband Radar Measurements Analysis and Processing*, IEE, 1997.

12

UWB Beamforming

Mohammad Ghavami and Kaveh Heidary

A major challenge of UWB systems is the equalisation of the channel impulse response with a corresponding delay spread in indoor environments. Not only is the channel estimation a real challenge, but also the equaliser may require a large amount of calculating power, which is a contradiction to the main objective of low-cost UWB systems. A further major limitation of a UWB system is its restricted range due to the extremely low transmit power. To overcome these difficulties and to further increase the maximum number of simultaneous users, multiple antenna techniques seem to be a promising approach.

The benefits of multiple antennas in UWB systems are threefold. First, the delay spread can be reduced and, consequently, the number of significant paths is decreased, which facilitates the equalisation. Second, grating lobes are negligible for UWB systems with sufficient antenna spacing so that undesired users or interferers can be suppressed. Third, focusing on a particular user enlarges the antenna gain and therefore the coverage of a UWB system is extended. However, from the technology point of view, numerous challenges resist. For instance, digital beamforming seems to be prevented due to the extremely high sampling rate. In turn, analogue beamforming requires adjustable true time delays. Such analogue components are not only simple to implement, but they also exhibit noticeable tolerance and therefore less precision. In conclusion, even if the principal approach seems to be very attractive, numerous barriers have to be overcome before a high data rate UWB system can be realised.

12.1 Introduction

Array signal processing or formation of beams involves the manipulation of signals induced on the elements of an array system or transmitted from the elements and received at a distant point from the array. In a narrowband beamformer, in which the signal bandwidth is much smaller than the central frequency, signals corresponding to each sensor element are multiplied by a complex weight and then summed to form the array output. As the signal bandwidth compared to the centre frequency increases,

---

Ultra-wideband Antennas and Propagation for Communications, Radar and Imaging Edited by B. Allen,
M. Dohler, E. E. Okon, W. Q. Malik, A. K. Brown and D. J. Edwards
© 2007 John Wiley & Sons, Ltd

the performance of the narrowband beamformer starts to deteriorate because the phase provided for each element and the desired angle is not a function of frequency and, hence, will change for the different frequency components of the communication wave.

For processing broadband signals a tapped delay line (TDL) structure can normally be used on each branch of the array antenna. The TDL structure allows each element of the array to have a phase response that varies with frequency. In this way, we can achieve a sort of compensation for the fact that lower frequency signal components have less phase shift for a given propagation distance, whereas higher frequency signal components have greater phase shift as they travel the same length. This structure can also be considered to be an equaliser that makes the response of the array the same across different frequencies. In addition to UWB signals, inherent baseband signals, such as audio and seismic signals, are examples of wideband or broadband signals. In sensor array processing applications, such as spread spectrum communications and passive sonar, there is also growing interest in the analysis of broadband sources and data.

12.2 Basic Concept

Figure 12.1 shows the general structure of an ultra-wideband array antenna system using delay lines in receiving mode. A uniform linear array is positioned symmetrically along an axis of the Cartesian coordinate system. The TDL network permits adjustment of gain and phase as desired at a number of frequencies over the wideband of interest. The far-field wideband signal is received by N antenna elements. Each element is connected to $M - 1$ delay lines, with the time delay of T seconds. The delayed input signal of each element is then multiplied by a real weight C_{nm}, where $1 \leq n \leq N$ and $1 \leq m \leq M$. If the input signals at N antenna elements are denoted by $x_1(t), x_2(t), \ldots x_N(t)$, the output signal that is the sum of all intermediate signals can be written as

$$y(t) = \sum_{n=1}^{N} \sum_{m=1}^{M} C_{nm} x_n(t - (m-1)T) \tag{12.1}$$

In a linear and uniform array, such as Figure 12.1, the signals $x_n(t)$ are related according to the angle of arrival and the distance between elements. Figure 12.2 shows that a time delay τ_n exists between the signals received at element n and at reference element $n = 1$. This amount of delay can be found as follows:

$$\tau_n = (n-1)\frac{d}{c}\sin\theta \tag{12.2}$$

where d is inter-element spacing and c is propagation speed. It is assumed that the incoming signal is spatially concentrated around the angle θ. Using the time delays corresponding to the antenna elements we can now write $x_n(t)$ with respect to $x_1(t)$ as follows:

$$x_n(t) = x_1(t - \tau_n) = x_1\left(t - (n-1)\frac{d}{c}\sin\theta\right) \tag{12.3}$$

Different methods and structures can be utilised for computation of the adjustable weights of a wideband beamforming network. In the following section we will explain one of them.

UWB Beamforming

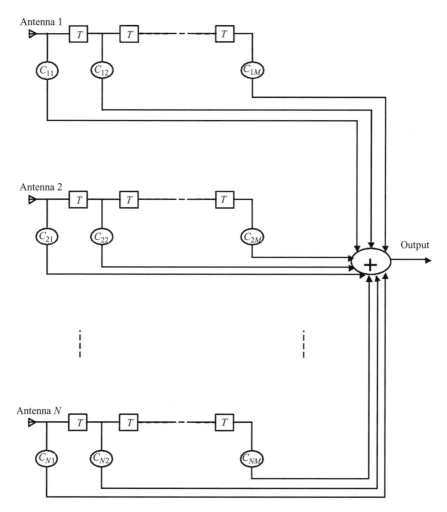

Figure 12.1 General structure of a TDL wideband array antenna using N antenna elements, $M - 1$ delay elements, and M multipliers

12.3 A Simple Delay-line Transmitter Wideband Array

The problem of designing a uniformly spaced array of sensors for operation at a narrowband frequency domain is well understood and widely investigated in literature. However, when a single frequency design is used over a wide or ultra-wide bandwidth the array performance degrades significantly. At lower frequencies the beamwidth increases, resulting in a reduced sharpness of beams and damaged spatial resolution; at frequencies above the narrowband frequency the beamwidth decreases and grating lobes may be introduced into the array beam pattern.

Figure 12.3 shows the basic structure of a delay-line wideband transmitter array system. The adjustable delays T_n, $1 \leq n \leq N$, where N is the number of antennas, are controlled by the desired angle of the

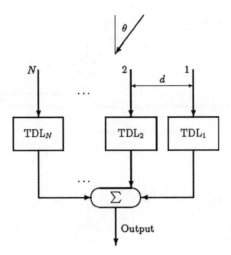

Figure 12.2 The incoming signal arrives at the antenna array with angle θ

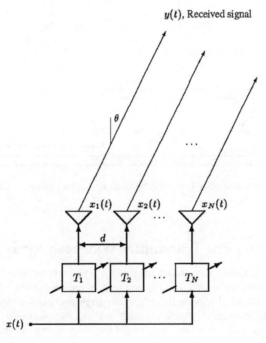

Figure 12.3 Beam formation using adjustable delay lines in the transmit mode

UWB Beamforming

main lobe of the directional beam pattern θ_0 as follows:

$$T_n = T_0 + (n-1)\frac{d}{c}\sin\theta_0 \tag{12.4}$$

where d is the inter-element spacing. The constant delay $T_0 \geq (N-1)d/c$ is required because without T_0 a negative delay will be obtained for negative values of θ_0 and cannot be implemented in practice. The signal received at the far field in the direction of $\theta(-90° < \theta < +90°)$ is equal to

$$y(t) = A(\theta)\sum_{n=1}^{N} x_n(t - \tau_n)$$

$$= A(\theta)\sum_{n=1}^{N} x(t - T_n - \tau_n) \tag{12.5}$$

where $x_n(t)$ indicates the transmitted signal from transducer n, τ_n is the delay due to the different distances between the elements and the receiver, and $A(\theta)$ is the overall gain of the elements and the path. The time delay τ_n regarding Figure 12.3 is equal to

$$\tau_n = \tau_0 - (n-1)\frac{d}{c}\sin\theta \tag{12.6}$$

where τ_0 is the constant transmission delay of the first element and is independent of θ. The gain $A(\theta)$ can be decomposed into two components as follows:

$$A(\theta) = A_1(\theta)A_2 \tag{12.7}$$

where $A_1(\theta)$ is the angle-dependent gain of the elements and A_2 is attenuation due to the distance. Substituting Equations (12.7) and (12.6) into Equation (12.5) yields:

$$y(t) = A_1(\theta)A_2 \sum_{n=1}^{N} x(t - \alpha_0 - (n-1)\frac{d}{c}(\sin\theta_0 - \sin\theta)) \tag{12.8}$$

where $\alpha_0 = T_0 + \tau_0$. In the frequency domain we may write

$$H(f,\theta) = \frac{Y(f,\theta)}{X(f)}$$

$$= A_1(\theta)A_2 e^{-j2\pi f\alpha_0} \sum_{n=1}^{N} e^{-j2\pi f(n-1)\frac{d}{c}(\sin\theta_0 - \sin\theta)}$$

$$= A_1(\theta)A_2 e^{-j2\pi f\alpha_0} e^{-j2\pi f(N-1)\frac{d}{c}(\sin\theta_0 - \sin\theta)} \frac{\sin[\pi f N\frac{d}{c}(\sin\theta_0 - \sin\theta)]}{\sin[\pi f\frac{d}{c}(\sin\theta_0 - \sin\theta)]} \tag{12.9}$$

From this equation we can derive several properties of a wideband delay beamformer. An important characteristic of the beamformer is the directional patterns for different frequencies.

Example 12.1
Consider a UWB signal with a centre frequency of 4.5 GHz and a bandwidth of 1 GHz. Calculate and sketch the normalised amplitude of (13.9) for $\theta_0 = -10°$, $-90° < \theta < +90°$, $N = 10$, $d = 3$ cm, $c = 3 \times 10^8$ m/s, and perfect antennas (i.e., $A(\theta) = A_2$).

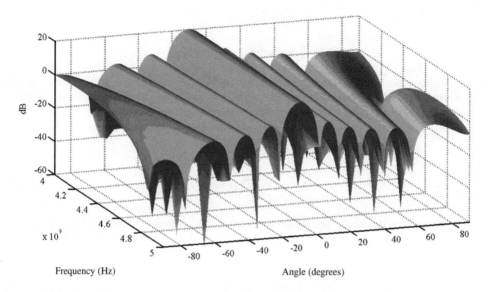

Figure 12.4 Directional patterns of a delay beamformer for frequencies uniformly distributed from 4 GHz to 5 GHz

Solution
The result is plotted in Figure 12.4. We observe that at $\theta = \theta_0$ frequency independence is perfect, but as we move away from this angle the dependence increases. Nevertheless, the beamformer is considered wideband with a fractional bandwidth of 1/4.5 or 22%.

Increasing the inter-element spacing has positive and negative consequences. As we will shortly see, it will produce a sharper beam and it is clearly more practical. On the other hand, this increase will result in some extra main lobes in the same region of interest (i.e., $-90° < \theta < +90°$).

12.3.1 Angles of Grating Lobes

We can now derive the angles of grating lobes and conditions for their existence. Assuming perfect antennas (i.e., $A(\theta) = A_2$) we can write from Equation (12.9) for $\theta = \theta_0$ as follows:

$$|y(f, \theta_0)| = A_2 N \tag{12.10}$$

This situation can happen for some other angles, denoted by θ_g. To calculate θ_g, it follows from Equation (12.9) that

$$|H(f, \theta_g)| = A_2 N = A_2 \frac{\sin[\pi f N \frac{d}{c}(\sin\theta_0 - \sin\theta_g)]}{\sin[\pi f \frac{d}{c}(\sin\theta_0 - \sin\theta_g)]} \tag{12.11}$$

Now, Equation (12.11) should be solved for θ_g:

$$\sin[\pi f \frac{d}{c}(\sin\theta_0 - \sin\theta_g)] = 0$$

or

$$\pi f \frac{d}{c}(\sin\theta_0 - \sin\theta_g) = m\pi \qquad (12.12)$$

where $m = \pm 1, \pm 2, \ldots$. The result is

$$\theta_g = \sin^{-1}\left(\sin\theta_0 - m\frac{c}{fd}\right) \qquad (12.13)$$

The first grating lobes are given for $m = \pm 1$. The necessary condition for having no grating lobe for a beamformer is that θ_g does not exist for any values of $-90° < \theta_0 < +90°$). The worst case happens for $\theta_0 = +90°$ and the condition of no grating lobe can be inferred from Equation (12.13) as

$$\frac{c}{fd} \geq 2 \quad or \quad d \leq \frac{c}{fd} = \frac{\lambda}{2} \qquad (12.14)$$

where λ indicates the wavelength. It is interesting to note that θ_g is not a function of N, but is very dependent on d. To show this more adequately, Figure 12.4 is replotted for $d = 12$ cm in Figure 12.5. We observe that the nearest grating lobes for $f = 4$ GHz are at $-53°$ and $+27°$, and for $f = 5$ GHz they are at $-42°$ and $+19°$. The frequency dependence of the beam patterns increases as we move away from the desired angle.

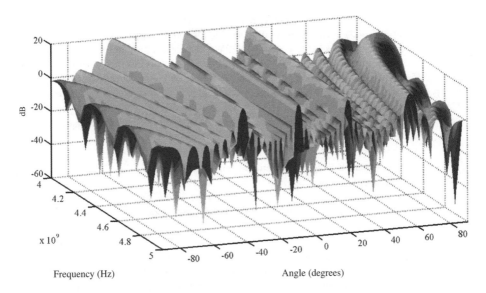

Figure 12.5 Grating lobes appear as a result of the increase of spacing between antennas

12.3.2 Inter-null Beamwidth

Comparing Figures 12.4 and 12.5 reveals that the main beamwidth of Figure 12.5 is less than that of Figure 12.4. The inter-null beamwidth (INBW) is defined as the difference between the nearest two nulls around the desired angle. The corresponding equation can be derived easily. Starting from Equation (12.9) and equating it to zero gives the following:

$$\pi f N \frac{d}{c}(\sin\theta_0 - \sin\theta) = m\pi \tag{12.15}$$

where $m = \pm 1, \pm 2, \ldots$. The first two angles around θ_0 are denoted by θ_1 and θ_2 and are computed from Equation (12.15) for $m = +1$ and $m = -1$, respectively

$$\theta_1 = \sin^{-1}\left(\sin\theta_0 - \frac{c}{fdN}\right) \tag{12.16}$$

$$\theta_2 = \sin^{-1}\left(\sin\theta_0 + \frac{c}{fdN}\right) \tag{12.17}$$

Hence, the INBW, $\Delta\theta = \theta_2 - \theta_1$, is written as

$$\text{INBW} = \sin^{-1}\left(\sin\theta_0 + \frac{c}{fdN}\right) - \sin^{-1}\left(\sin\theta_0 - \frac{c}{fdN}\right) \tag{12.18}$$

It is clear that for $|\sin\theta_0 \pm \frac{c}{fdN}| > 1$, there exists no null on the left or right side of the main angle θ_0. As a special case, for $\theta_0 = 0$ we have

$$\text{INBW} = 2\sin^{-1}\left(\frac{c}{fdN}\right) \tag{12.19}$$

that is, increasing d lowers the INBW and produces sharper beams. It is easy to test Equation (12.18) for values of the first and second cases, which are illustrated in Figures 12.4 and 12.5.

As is obvious from Equation (12.18) or (12.19), INBW is a function of frequency f. To observe the effect of frequency variations on the beam pattern of the delay-line beamforming network we repeat Example 12.1 for a wide frequency range from 4 GHz to 8 GHz and inter-element spacing of 1.87 cm. The results are shown in Figure 12.6 for frequencies of 4, 5, 6, 7 and 8 GHz. The computed values of INBW for these frequencies are 47.3, 37.4, 31, 26.5 and 23.1 degrees, respectively.

From the previous discussion we can conclude that pure delay-line wideband antenna arrays have the following properties:

1. A relatively simple structure using only a variable delay element.
2. No multiplier in the form of amplification or attenuation.
3. A perfect frequency independence characteristic only for the desired angle of the array.
4. Their INBW and sidelobe characteristics vary considerably with frequency of operation.

Because of the existence of some distinctive differences between conventional and UWB antenna arrays the well-known conventional concepts of phased array antennas have to be modified appropriately to accommodate UWB signals. One significant difference from narrowband theory is that frequency domain analysis alone is insufficient to treat UWB arrays. In fact, the time domain may be a more natural setting for understanding and analysing the radiation of UWB signals.

UWB Beamforming

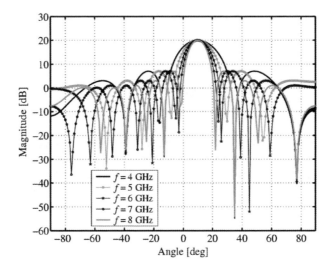

Figure 12.6 Directional patterns of the delay beamformer for five different frequencies show that the beamwidth is very sensitive to frequency

12.4 UWB Mono-pulse Arrays

In this section, the radiation characteristics of UWB mono-pulse arrays are discussed. The UWB arrays considered here consist of ultra-wideband sources radiating coded sequences of ultra-narrow Gaussian mono-pulses with very low duty cycle. Pseudo-random (PN) time-hopping spread spectrum technique is utilised for further spreading the signal spectrum and providing multiple access, interference rejection and secure communication. Pulse position modulation (PPM) at the rate of many pulses per data bit is used for data modulation. At the receiver correlation is applied and the PN code is used to discriminate between the desired and interfering signals. A model for the general UWB mono-pulse volume array has been developed. Mathematical formulas for parameters of interest have been derived and computed results have been presented.

12.4.1 Problem Formulation

The antenna array considered here consists of N discrete, isotropic (omnidirectional) source elements distributed over the array volume. Element locations are designated as $(x_n, y_n, z_n; 1 \leq n \leq N)$, and the receiver is positioned at (r, θ, Φ) where Cartesian and spherical coordinates are used for the radiating element and receiver locations, respectively. The radiated mono-pulse originating from the nth element, $p_n(t)$, is given below

$$p_n(t) = \alpha_n p(t - \beta_n) \qquad (12.20)$$

$$p(t) = 6\sqrt{\frac{e\pi}{3} \frac{t-\tau}{\tau}} e^{-6\pi(\frac{t-\tau}{\tau})^2} \qquad (12.21)$$

where $p_n(t)$ is the normalised Gaussian mono-pulse of width τ and α_n and β_n denote the magnitude and time delay (phase) of the nth element, respectively. Superposition is applied to determine the composite signal at the receiver input (output of the receiver's antenna)

$$s(t) = \frac{1}{r}\sum_{n=1}^{N} p_n\left(t - \frac{r - d_n}{c}\right) \tag{12.22}$$

$$d_n = x_n \sin\theta \cos\phi + y_n \sin\theta \sin\phi + z_n \cos\theta \tag{12.23}$$

In the above c is the speed of light, and it is assumed that the receiver is located in the far-field region of the array with $r \gg D$, where D is the largest array dimension. Time origin in Equation (12.22), $t = 0$, refers to the nominal pulse time, determined by the time-hopping PN code and the data bit, which causes an additional time lag or lead. The signal at the output of the receiver's antenna is cross-correlated with the mono-pulse (template) in order to generate the decision parameter

$$f(t_0) = M \int_{t=0}^{\infty} s(t) p(t - t_0) dt \tag{12.24}$$

In the above M is the number of pulses per data bit, and t_0 is chosen to maximise the decision parameter. In order to evaluate the integral in Equation (12.24) the energy and the auto-correlation function of the Gaussian pulse were derived as below:

$$E = \int_{t=0}^{\infty} p^2(t) dt \tag{12.25}$$

$$R(t_0) = \int_{t=0}^{\infty} p(t) p(t - t_0) dt \tag{12.26}$$

Substituting Equation (12.21) in (12.25), change of variable and utilising the symmetric nature of the resulting integrand leads to

$$E = 12e\pi\tau \left[\int_{0}^{\infty} x^2 e^{-12\pi x^2} dx + \int_{0}^{1} x^2 e^{-12\pi x^2} dx\right] \tag{12.27}$$

Applying change of variables to the above integrals and simplifying one arrives at

$$E = \frac{e\tau}{\sqrt{12\pi}} \left[\int_{0}^{\infty} y^2 e^{-y^2} dy + \int_{0}^{\sqrt{12\pi}} y^2 e^{-y^2} dy\right] \tag{12.28}$$

Noting that $\int_{0}^{\sqrt{12\pi}} y^2 e^{-y^2} dy \cong \int_{0}^{\infty} y^2 e^{-y^2} dy$, and substituting for the integrals above, one obtains

$$E = \frac{e\tau}{4\sqrt{3}} \tag{12.29}$$

Substitution from Equation (12.21) in (12.26), change of variable and simplification leads to

$$R(t_0) = 12e\pi\tau e^{-3\pi\alpha^2} \left[\int_{-1}^{\infty} (x^2 - \alpha x) e^{-(\sqrt{12\pi}x - \sqrt{3\pi}\alpha)^2} dx \right] \quad (12.30)$$

where $x = \frac{t-\tau}{\tau}$, and $\alpha = \frac{t_0}{\tau}$. Applying change of variable, subsequent to extensive algebraic manipulations and using the following equalities

$$\int_0^{(\sqrt{12\pi}+\sqrt{3\pi}\alpha)} y^2 e^{-y^2} dy \cong \int_0^{\infty} y^2 e^{-y^2} dy; \quad \alpha \geq 0 \quad (12.31a)$$

$$\int_{-(\sqrt{12\pi}+\sqrt{3\pi}\alpha)}^{\infty} y e^{-y^2} dy \cong 0 \quad (12.31b)$$

one obtains the autocorrelation function

$$R(t_0) = e\pi\tau e^{-3\pi(t_0/\tau)^2} \left[\frac{1}{4\sqrt{3\pi}} - \frac{\sqrt{3}(t_0/\tau)^2}{2} \right] \quad (12.32)$$

Substituting $t_0 = 0$ in Equation (12.32), in light of Equation (12.29), one observes $R(0) = E$ as expected.

Substituting from Equation (12.21), (12.22) in (12.24), using Equation (12.32), and extensive mathematical manipulations leads to

$$g(\xi) = \frac{Me\pi\tau}{r} \sum_{n=1}^{N} \alpha_n e^{-3\pi(\gamma_n+\xi)^2} \left[\frac{1}{4\sqrt{3\pi}} - \frac{\sqrt{3}(\gamma_n+\xi)^2}{2} \right] \quad (12.33)$$

$$\xi = \frac{\frac{r}{c} - t_0}{\tau}, \quad \gamma_n = \frac{\beta_n - \frac{d_n}{c}}{\tau} \quad (12.34)$$

The array factor describes the decision parameter or the receiver output as a function of the receiver's position as it traverses the spherical surface of radius r.

$$G(\theta, \phi) = \sum_{n=1}^{N} \alpha_n e^{-3\pi(\gamma_n+\xi)^2} \left[\frac{1}{4\sqrt{3\pi}} - \frac{\sqrt{3}(\gamma_n+\xi)^2}{2} \right] \quad (12.35)$$

Where t_0 in Equation (12.32) is chosen to maximise G and all other parameters are as before. It is noted that the array factor is determined by the element positions and their relative magnitudes and time delays.

12.4.2 Computed Results

Simulations of UWB Gaussian mono-pulse array systems based on the mathematical formulations developed above are presented here. Examples of Figures 12.1 through 12.9 pertain to a uniform linear array positioned symmetrically along the x-axis of the Cartesian coordinate system, and the receiver is positioned in the xz-plane ($\Phi = 0$). Number of elements is assumed to be 21 and the mono-pulse width is 1 ns. In Figures 12.7 and 12.8 all elements are in-phase and element spacing is 5 cm. These figures

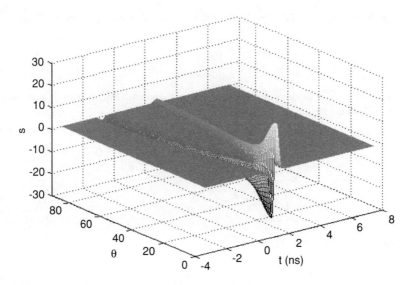

Figure 12.7 Radiated signal for a 21-element linear array, inter-element spacing = 5 cm, $\tau = 1$ ns

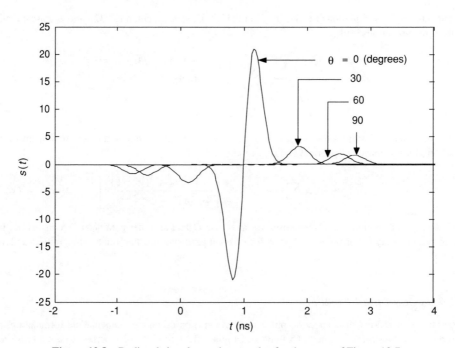

Figure 12.8 Radiated signal at various angles for the array of Figure 12.7

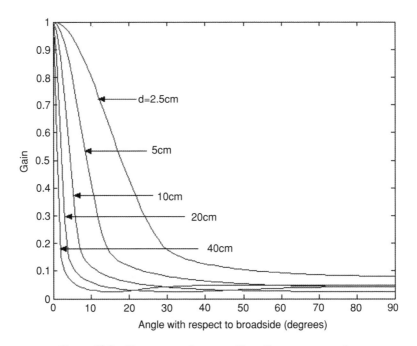

Figure 12.9 Twenty-one-element uniform linear array $\tau = 1$ ns

plot the composite signal at the receiver. It is noted that at broadside ($\theta = 0$) the signal is a mono-pulse similar to the transmitted signal. Along other directions, however, due to asynchronism of signals at the receiver, the signal consists of two half mono-pulses with diminished amplitudes. The time separation between the two half mono-pulses is proportional to the path length differential from the two extreme ends of the array to the receiver.

Figure 12.9 plots the array system gain for different values of element spacing with fixed number of elements. As expected, the gain function becomes narrower with widening the array system, which results from increased inter-element spacing. Increased inter-element spacing does not result in sidelobes.

Figure 12.10 plots the effect of element spacing on the beamwidth of a broadside linear array (synchronous elements) with 11, 21 and 41 elements.

Figure 12.11 plots the gain function of a 21-element linear uniform phased array. The beam maximum is along the broadside direction ($\theta = 0$) when all elements are fed synchronously (inter-element-delay = 0). Direction of the beam maximum is steered by adjusting inter-element time delay. It is seen that as the beam is steered away from the broadside direction it becomes wider, as expected. The beam steering, however, does not cause beam splitting. For an array with inter-element spacing of 5 cm, a linear time delay of 166.7 ps (delay between adjacent elements) results in an endfire array.

Figure 12.12 plots the beamwidth as function of scan angle for three arrays with inter-element spacing of 10 cm.

Figure 12.13 plots gain functions of a 21-element linear broadside array for uniform and binomial excitations. In the binomial array the element excitations (amplitudes) are adjusted according to the binomial coefficients. Figures 12.14 and 12.15 show the array beam polar plots for a uniform linear broadside (Figure 12.14) and scanned (Figure 12.15) array. Figure 12.16 plots the beam pattern of an

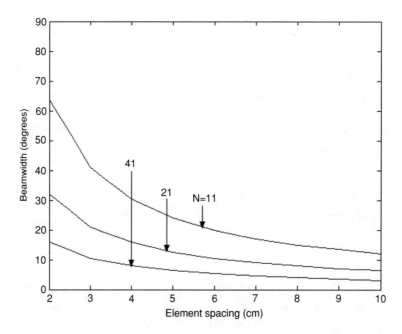

Figure 12.10 Beamwidth of uniform broadside array ($\tau = 1$ ns). N is number of array elements

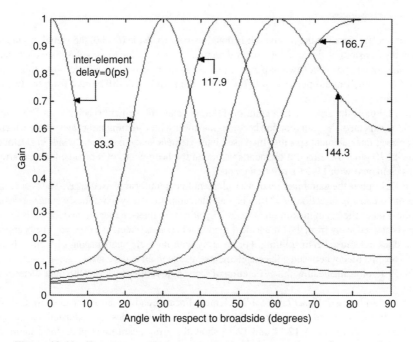

Figure 12.11 Twenty-one element array, inter-element spacing = 5 cm, $\tau = 1$ ns

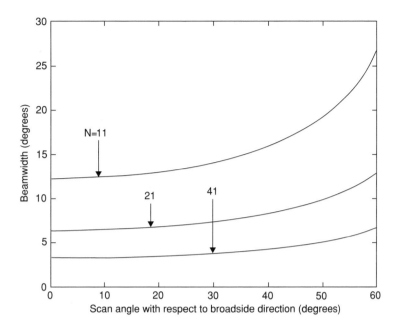

Figure 12.12 Beamwidth versus scan angle for a linear array. Inter-element spacing = 10 cm, $\tau = 1$ ns, N is number of array elements

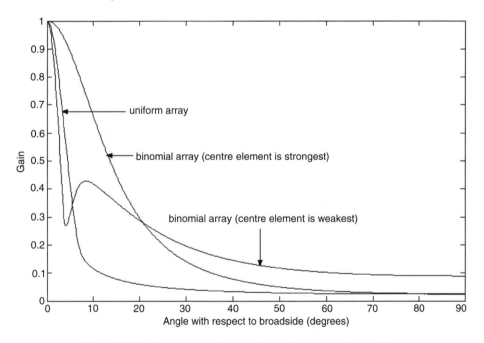

Figure 12.13 Twenty-one elements. Inter-element spacing = 10 cm

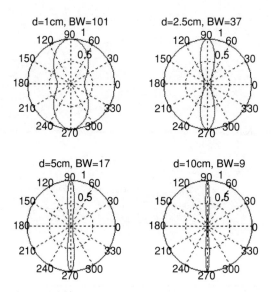

Figure 12.14 Twenty-one element uniform linear array. $\tau = 1$ ns. Angle is with respect to the array axis (90 is broadside direction)

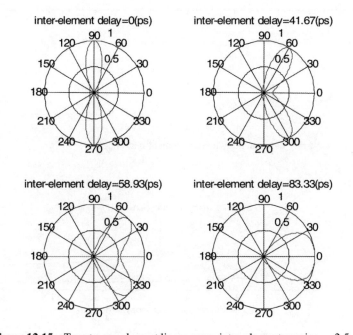

Figure 12.15 Twenty-one element linear array, inter-element spacing $= 2.5$ cm

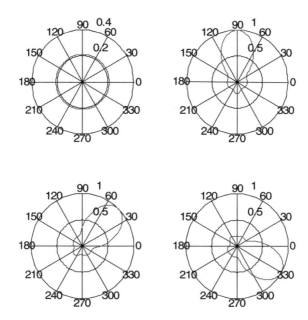

Figure 12.16 Eight radiators on a circle of radius 10 cm. Top left all sources are in phase. Top right three sources lag ($\Phi = 45, 135\,(235\,ps)$, $\Phi = 90\,(333.33\text{ ps})$) and three sources lead w/r to sources at $\Phi = 0, 180$. Bottom left sources at $\Phi = 0, 90$ (lag 235 ps), source at $\phi = 0$ (lag 333.33 ps) w/r to sources at $\Phi = -45, 135$. Bottom right lag and lead times for sources (1 at $\Phi = 0$ through 8 at $\Phi = 335$) are 289, 86, -167, -322, -289, -86, 167, 322 ps, respectively

eight-element uniform (amplitude) circular array. It is seen that by proper timing of element feed functions the single-lobe beam can be steered along desired directions.

The examples given here elucidate some of the salient features of UWB mono-pulse array systems. These systems have the following properties:

1. Mono-pulse UWB antenna arrays can be used for spatial focusing of radiated EM energy using simple time-shift structures.
2. Mono-pulse UWB antennas can be used for spatial focusing of received UWB signals using simple time-shift and sum array structures.
3. Beam patterns of UWB arrays do not contain sidelobes.
4. Sparse UWB arrays with large inter-element spacing and/or inactive elements do not give rise to grating lobes.
5. Spatial scanning of the UWB array beams does not result in grating lobes.

12.5 Summary

In this chapter we studied two kinds of antenna array systems for UWB signals. The TDL structure employs adjustable delay lines and profits from the simplicity and a straightforward design technique. On the other hand, a quite different concept of UWB mono-pulse array consists of ultra-wideband sources radiating coded sequences of ultra-narrow Gaussian mono-pulses with very low duty cycle. The characteristics of both methods were illustrated by several examples.

References

[1] F. Anderson, W. Christensen, L. Fullerton and B. Kortegaard, Ultra-wideband beamforming in sparse arrays, *IEEE Proceedings-H*, **138**(4), 1991.
[2] R.A. Scholtz, Multiple access with time-hopping impulse modulation (invited paper), *MILCOM 93*, Bedford, MA, 11–14 October, 1993.
[3] R.J. Fontana, A novel ultra wideband (UWB) communications system, *Proc. MILCOM 97*, Monterey, CA, 2–5 November, 1997.
[4] P. Withington, *In-building Propagation of Ultra-wideband RF Signals*, Time Domain Corporation Publication, June 1999.
[5] J. Forester, E. Green, S. Somayazulu and D. Leeper, Ultra-wideband technology for short- or medium-range wireless communications, *Intel Technology Journal*, **Q2**, 2001.
[6] K. Heidary, 'Ultra-Wideband Antenna Arrays', *IEEE International Symposium on Antennas and Propagation*, July 2001, Boston MA.
[7] K. Heidary, 'A Physical Scattering Model for the Ultra-Wideband (UWB) Propagation Channel', *IEEE International Symposium on Antennas and Propagation*, July 2005, Washington DC.

Part III

Propagation Measurements and Modelling for UWB Communications

Introduction to Part III

Mischa Dohler and Ben Allen

Ultra-wideband (UWB) is predicted to emerge as a fundamental technology for the home entertainment, security, tracking and high data rate wireless transmission markets. It will be a viable part of future wireless personal area networks (WPANs), wireless body area networks (WBANs) and sensor networks, among others.

Sometimes referred to as impulse radio, baseband or zero-carrier technology, UWB pulse radio systems operate by spreading energy across a wide range of frequencies relative to the centre frequency. UWB communication is hence mainly achieved by transmitting pulses of short duration, the radiated energy of which therefore occupies a very large bandwidth and overlays frequencies already partially allocated to other services. It is hence mandatory that the power emitted by UWB devices be sufficiently small to be able to coexist with other devices without causing harmful interference to them. This characteristic allows UWB transmissions to be almost completely hidden in the noise floor of co-channel narrowband receivers and gives additional security and robustness to intentional detection or jamming.

Interference, be it from a UWB transmitter to a coexisting system, or vice versa, has provided the basis of much discussion surrounding UWB systems. Proponents of UWB claim that the interference caused by UWB systems to other systems is, as mentioned above, negligible; they also claim that the interference from narrowband systems, which are known to saturate the UWB receiver input, will be insignificant. Sceptics of UWB claim that low data rate UWB systems might cause negligible interference, but high data rate systems cannot be neglected and cause significant interference to other narrowband systems; they also claim that the high power narrowband interference into the UWB receiver will require special receiver solutions, thereby making a UWB device more expensive than envisaged. Opponents of UWB claim that interference onto, e.g., cellular systems, will cost the operators significant revenue loss and any deployment of UWB has to be strictly forbidden. Only time will tell who, if anyone, is right. We expect, however, that the deployment of UWB will go ahead and that solutions to many of the challenges will be found.

Whether the objective is to model, analyse or optimise UWB data transmission and associated interference, accurate characterisation of UWB signal propagation over the ultra-wideband propagation channel is fundamental. The behaviour and statistics of pathloss, shadowing and multipath propagation are of interest; which can be measured and/or modelled.

Ultra-wideband Antennas and Propagation for Communications, Radar and Imaging   Edited by B. Allen, M. Dohler, E. E. Okon, W. Q. Malik, A. K. Brown and D. J. Edwards
© 2007 John Wiley & Sons, Ltd

Measuring the UWB propagation channel will enable the behaviour of a UWB transceiver in the measured environment to be observed; however, unfortunately, it is of no reliable use in environments different from the measured one. It has thus become common to utilise measurements to corroborate theoretical models or to characterise the electromagnetic specific environments and/or objects.

Modelling the UWB propagation channel using analytical techniques is very complex. To tackle the complexity problem, the UWB channel is traditionally modelled:

- in a deterministic fashion, thereby neglecting any randomness in the channel (e.g. ray tracing);
- in a stochastic/random fashion, thereby mapping some characteristic environments to some probabilistic functions (e.g. IEEE 802.15 channel models).

This part of the book is dedicated to both approaches, where we will detail measurements and models of the UWB propagation channel:

- Chapter 13 is dedicated to signal attenuation measurements for typical building materials;
- Chapter 14 to the modelling of pathloss and shadowing;
- Chapter 15 to the modelling of multipath fading;
- Chapter 16 to the measurements and modelling of UWB WBAN systems; and
- Chapter 17 to the modelling of the spatial properties of the UWB channel.

In particular, Chapter 13 deals with an issue rarely found in any open literature – the measurement of the electromagnetic properties of materials typically occurring in residential and office buildings. The measurement results are very valuable because they are tailored to environments in which UWB systems are most likely to operate. They have been conducted in an anechoic chamber over UWB bandwidths. Accurate attenuation values can be critical in the case of multiple systems operating in the same frequency bands since a sufficiently large attenuation would guarantee peaceful coexistence between systems. The measurements were conducted for 12 key materials, i.e., red brick, cinder block, melamine, plywood, pine wood, plasterboard, medium density fibreboard, glass, double-tinted glass, Venetian blinds, thin metal and thick metal.

Chapter 14 deals with pathloss and shadowing for UWB signals. At first this may appear a trivial issue, since it has been studied in great depth for narrowband systems. The extreme wideband nature of UWB signals, however, introduces entirely new problems. For instance, while it was easy to decouple the behaviour of antenna and channel for narrowband systems, this is not the case for UWB systems anymore; in fact, depending on the antennas at the transmitting and receiving ends, the UWB channel exhibits different dispersive properties which need to be considered for the UWB system design. Also, the electromagnetic properties of materials (permittivity, permeability and conductivity) vary with frequency, which leads to frequency-dependent reflection, transmission, diffraction and scattering coefficients. We will hence dwell on how to account for the above effects in a deterministic and stochastic fashion, thereby developing a baseline for pathloss and shadowing models. These models are corroborated by a set of pathloss and shadowing channel measurements. This chapter concludes with a description of the two stochastic IEEE 802.15 reference pathloss models, which have been developed by the IEEE standardisation groups to model low and high data rate UWB communication channels.

Chapter 15 extends the developments of Chapter 14 to the small-scale multipath fading effects in the UWB channel. The large bandwidth of the UWB signal facilitates the signal to be captured by means of the multiple, delayed signal copies. A proper modelling of this effect allows one to obtain a realistic picture on the number of required signal copies, and hence influences the hardware design of UWB receivers. Furthermore, each delayed signal copy obeys certain distributions, which have a direct influence on the error and outage performance of UWB receivers. We will therefore dwell on the modelling of these effects and show, for example, why the small-scale fading margin for UWB systems is so much smaller than for narrowband systems. The developed theory is again corroborated by a set of channel measurements. The chapter concludes with a description of the small-scale behaviour of the two stochastic IEEE 802.15 reference channel models.

Chapter 16 rigorously deals with a topic barely touched upon until today – wireless body area networks. WBANs consist of a number of wireless nodes placed on or in close proximity to the human body. The topic of WBANs will certainly gain in importance over the forthcoming years and the research described here will be of fundamental importance to the design of UWB systems in such environments. A major driver behind using WBANs is that its wired counterpart is very inconvenient to wear because either cables have to be used, or special clothing worn, or the unreliable channel via the body's tissue to be used. Using the wireless channel through and/or around the human's body, however, poses new problems and hence challenges. Do our far-field assumptions still hold true? What are the electromagnetic properties of the body, i.e., does the receiver have to rely on signal components coming through the body, or being diffracted around the body, or being reflected off some objects in the person's vicinity? These and other issues will be tackled in Chapter 16, leading to a set of WBAN models that are corroborated by a comprehensive measurement campaign.

Finally, Chapter 17 is dedicated to a much-disputed topic in the community of UWB – the spatial properties of the UWB channel and its offered gains. While narrowband systems clearly benefit from a proper multiple-input multiple-output (MIMO) channel, this is not necessarily the case for UWB. First of all, due to its large bandwidth, the definition of spatial correlation has to be altered, which impacts the performance and evaluation of MIMO techniques. The diversity offered by a single-antenna UWB system is already substantial – does the use of multiple antennas still yield gains, particularly in the light of increased complexity? Can we improve the outage rate or the capacity using MIMO, or both? Can this be achieved by means of spatial multiplexing? We endeavour to answer some of these and other questions in the last chapter of this part of the book, where the behaviour is again corroborated by results from an extensive measurement campaign.

13

Analysis of UWB Signal Attenuation Through Typical Building Materials

Domenico Porcino

This chapter introduces a study of path attenuation due to UWB radio wave propagation through common building materials. While typical propagation models generically consider an average long-term signal variation (pathloss) statistically derived from measurements in some arbitrary conditions, very few documents have analysed in detail the characteristics of signal strength reduction due to direct penetration through common furniture material.

The topic of signal attenuation in common environments is important for automated indoor wireless installation/optimisation systems but fundamental also for coexistence analysis in overlay conditions. In the case of multiple systems operating in the same frequency bands, the attenuation from materials can in fact represent a critical mitigation element for the peaceful coexistence of wireless operations.

Our work introduces a simple methodology for attenuation estimation and a set of results from measurements, which lead to empirical attenuation factors for 12 different construction materials. The results can be directly applied to link budgets or detailed coexistence analysis.

13.1 Introduction

Ultra-wideband (UWB) is set to become a fundamental technology for the home entertainment, security, tracking and high-data rate transmission markets via wireless personal area networks (WPANs). Sometimes referred to as impulse radio, baseband or zero-carrier technology, UWB pulse radio systems operate by spreading tiny quantities of energy (typically less than 0.5 mW) across a wide range of frequencies relative to the centre frequency. UWB communications are mainly achieved by transmitting pulses of short duration (typically 100–5000 picoseconds), whose radiated energy therefore occupies a

---

Ultra-wideband Antennas and Propagation for Communications, Radar and Imaging Edited by B. Allen, M. Dohler, E. E. Okon, W. Q. Malik, A. K. Brown and D. J. Edwards
© 2007 John Wiley & Sons, Ltd

very large bandwidth (typically 0.2–10 GHz) overlaying frequencies already partially allocated to other services. The power emitted by UWB devices has of course to be made sufficiently small to be able to coexist with other devices without causing harmful interference to them. This characteristic allows UWB transmissions to be almost completely hidden in the noise floor of an adjacent operating narrowband receiver and gives additional security and robustness to intentional detection or jamming. While initial UWB technologies, often referred to as baseband UWB, were implemented in the lower frequency bands with 3 dB spectrum masks starting at 500 MHz and going up to 1.5 GHz, more recent definitions have been derived from the first Report and Order (Feb 2002 [1]) by the Federal Communications Commission (FCC). According to the FCC definition, this technology has established commercial viability for communications devices in the band 3.1–10.6 GHz (see Chapter 1).

An accurate characterisation of the UWB signal propagation over air is fundamental for a good estimate of the received UWB signals after a given transmission path. The goal of appropriate UWB propagation channel models is to capture and simulate both long-term (pathloss) and short-term (multipath) propagation effects of typical environments where UWB devices are expected to operate. The reason for this accurate scientific analysis is twofold: on one hand, a channel model is generally a necessary tool for wireless system analysis and accurate receiver simulation and design. On the other hand – in the case of overlay UWB signals – it is important to understand how coexistence with different primary radio services can be guaranteed in normal operating conditions (i.e. in rooms with real people, real furniture, walls, etc.). The exposed measurements hence differ from the channel models developed in Chapter 14.

UWB signals are fundamentally electromagnetic waves travelling at the speed of light from a source (transmitter) to a sink (receiver). Along their propagation, these radio waves will interact with different physical matters (objects, people) that will alter the strength, phase and direction of the original rays. When a UWB signal hits a material, some of the power will be reflected at the surface and some of the power will be transmitted into (refracted) and possibly through the material. Common objects and materials found in the daily environment could therefore be mostly radio-transparent (i.e. allowing high penetration characteristics) or radio-absorbent (i.e. act as blockages and hence reflecting off most of the incoming energy and causing an impediment to the main transmission path). Most conductors (such as metal, aluminium, gold, silver) typically act as radio obstacles reflecting a great part of the energy away, while most dielectrics (such as paper, plastics, teflon, glass) let some part of the energy pass through them. The simple description of the phenomenon of propagation through materials is further complicated by the fact that the attenuation and distortion characteristics of each material are different at different frequencies and therefore not totally uniform over the very large bandwidths used by UWB devices.

Due to the rapid evolution of mainstream UWB technology from baseband to higher frequencies and the rather experimental nature of UWB systems for communications, current literature models are not extensive with regard to indoor material attenuation.

The most significant contribution to the scientific community comes from the research of William C. Stone and the team at the Building and Fire Research Laboratory [2] that produced a comprehensive campaign of measurements and analysis of attenuation coefficients in the 0.5–2 GHz and 3–8 GHz frequency band, using a modified ultra-wideband synthetic aperture radar. Their work is a useful and very interesting reference with a large variety of materials under investigation and an extensive set of results of signal attenuation as function of frequency. Their use of 2 MHz continuous-wave (CW) gated steps as transmission sources makes the results very suitable for narrowband wireless communication, but not immediately applicable to modern UWB communications.

A 2001 study by Estes et al. [3], while of some relevance, concentrates on in-hull RF propagation for naval ships with early prototype systems emitting UWB pulses spread from 0.8 to 2.5 GHz. The results of their campaign are very interesting even if they are not directly applicable to modern home/office environments both in terms of material types and frequencies adopted for the trials.

Interesting, also, are the results of the study conducted by Ray-Rong Lao et al. [4], which applied frequency domain analysis to characterise four typical office materials (plasterboard, Ca-Si board, chipwood and tempered glass) finding their respective transmission coefficients and the expected variation in frequency.

Attention to propagation models and attenuation characteristics has also been given by the IEEE 802.15.3a working group [5], but having emphasis on the mean pathloss characteristics in generic environments, their final models are not sufficiently detailed to derive any real conclusion on UWB absorption characteristics; the models are detailed in Chapters 14 and 15.

The work presented in the remaining part of this chapter will try to fill this gap in the literature concentrating on attenuation given to UWB communication systems by common indoor materials used in real buildings today. The results of experiments and measurements will give empirical indications of signal strength behaviour and useful parameters for system pathloss and coexistence analysis. The document is organised as follows: the next section gives a short overview of the channel characteristics, highlighting the parts most analysed in literature and the parts most relevant to the characterisation of building materials. Then, a full description of the materials under test and their geometrical characteristics is given. A description of the experimental campaign is then presented with details of the main results of signal strength attenuation caused by all the materials previously mentioned. The chapter will end with some concluding remarks and useful hints for wireless system developers.

13.2 A Brief Overview of Channel Characteristics

In general terms, a radio channel model is a complex mathematical attempt to describe the propagation phenomena through free space, physical and biological objects. The interaction among electromagnetic waves and the real world is extremely difficult to predict reliably and is clearly strongly dependent on the precise details of the propagation scenario chosen to represent the typical case under analysis.

In the scientific world, there are many approaches attempting to define a radio channel with a high degree of confidence. Two major branches can be identified [6], i.e. statistical and deterministic models. While a statistical model is generally derived from measurements conducted in different environments and then characterised statistically, a deterministic model tries to predict the exact characteristics of electromagnetic waves through the use of geometry, Maxwell's equations and models of the boundary conditions of the media.

Given the very stringent requirements in terms of computational power and the dependency on extremely precise boundary conditions, deterministic models are not very suitable for coexistence studies, where the conditions of propagation are often unknown or only roughly defined.

Statistical models are instead more readily available in the literature, simpler to implement and regularly updated with new measurement results as technology evolves or new techniques and/or frequencies are used.

The multipath radio channel has to physically represent the sum of all the effects of loss and distortion that the signals undergo during their propagation from a transmitter to a receiver. In the case of studies of UWB coexistence with other services, we will be interested in knowing how the UWB signals will propagate through free space and other media and how this might affect the link budget of other systems.

The main effects that a radio wave encounters during its propagation can be divided into:

- **Large-scale (pathloss)** characteristics. The large-scale (or long-term) characteristics of the radio mobile channel describe how the *mean* signal will behave as a function of the distance at a given frequency. The loss is gradual, with received power decreasing almost as an exponential decay in logarithm scale.

- **Medium-scale (shadowing)** characteristics. Shadowing exhibits time-varying factors caused by diffraction of travelling waves around large objects such as large obstacles or buildings. It is added on top of the pathloss and represented as a random fluctuation with a log-normal distribution, with a standard deviation, σ, dependent on propagation conditions.
- **Small-scale (multipath or fast-fading)** characteristics. The small-scale (or short-term) characteristics describe the sudden variations of the received signal strength due to multipath and reflections coming from objects. This factor is superimposed on the large- and medium-scale variations and will typically happen for very small movements of the mobile antenna. Even less than a wavelength spacing could be enough to cause fast fading: as an example, at 4 GHz the wavelength is just 7.5 cm, hence a movement of a few centimetres will cause fluctuations in the received signal.

These three effects will of course be summed together and are not easily discernible under normal conditions.

A classical way to represent the propagation phenomena independently from the transmitter and receiver characteristics is to give an appropriate definition of the channel impulse response, $h(t)$, between a source signal, $x(t)$, and a received signal, $y(t)$, as:

$$h(t) = \sum_{i=1}^{N} E_i(t) \cdot \delta(t - \tau_i(t)) \tag{13.1}$$

where $\delta(\cdot)$ is the Dirac delta function, $E_i(t)$ the ith scattered path arriving at the receiver with delay τ_i.

The channel impulse response can be described as the sum of N scatterers $E_i(t)$ arriving at the receiver (with N typically considered between 6 and 20). Each scatter will be in itself the summation of numerous partial waves.

Each single scatterer E_i is the result of the sum of N_{waves} waves (theoretically infinite, but in typical simulation models limited to 100). The fast-fading component of each of these scatterers (i) can be characterised by amplitude a_i, phase ϕ_i, and angle of incidence (relative to the vector movement of the user) α_i, as described in Equation (13.2).

$$E_{iFF}(t) = \sum_{k=0}^{N_{waves}} a_{ik}(t) \cdot e^{j(\varphi_{ik} + \frac{2\pi}{\lambda} \cdot v \cdot t \cdot \cos \alpha_{ik})} \tag{13.2}$$

The summation of these N_{waves} waves is at each instant a good representation of the short-term characteristics. Added on top of these fading effects, we have to consider also the long- and medium-scale variations of the signal strength at a given distance, represented by the attenuation At_i (including pathloss and shadowing) of each single scatter, thus:

$$E_i(t) = At_i(t) \cdot E_{iFF}(t) \tag{13.3}$$

The simple analysis often used in coexistence studies limits the propagation characteristics to the large scale of the signal at given distances (pathloss). In mathematical terms, the mean received power (around which there will still be shadowing and multipath) will vary with distance with an exponential law.

The total pathloss at a distance, d, will then be $L(d)$, often modelled as:

$$L(d) = L_o + 10 \cdot n \cdot \log_{10}\left(\frac{d}{d_0}\right) \tag{13.4}$$

with L_o = loss at the reference distance d_0, for example $d_0 = 1$ m. Choosing d_0 to be very close to the transmitter, it can be safely assumed that the line-of-sight signal will dominate over any other component and therefore it is reasonable to model this parameter L_o using the free-space propagation law as:

$$L_0 = 20 \cdot \log_{10}\left(\frac{4 \cdot \pi \cdot d_0 \cdot f'_c}{c}\right) \quad (13.5)$$

and

$$f'_c = \sqrt{f_{min} f_{max}} \quad (13.6)$$

is the geometric centre frequency of the UWB waveform (with f_{min} and f_{max} being the -10 dB edges of the waveform spectrum). The parameter n is the pathloss exponent, obtained from measurement campaigns and propagation analysis. The value of L_o is easily measurable or modelled assuming free-space conditions. As an example, choosing $f'_c = 5$ GHz, with $n = 2$ at 1 m, L_o results in 46.42 dB of attenuation at 1 m distance.

In particular, for the pathloss analysis, we can define three different conditions:

- **LOS (line of sight)**: a signal that reaches the receiver directly from the transmitter (no obstacles are present in the path).
- **OBS (obstructed line of sight)**, or **soft NLOS (soft non-line of sight)**: a situation in which the power coming from reflections is dominant with respect to the power coming from the direct path (energy from reflected path > energy from direct path).
- **Hard NLOS**: a situation in which the direct path between the transmitter and receiver is totally obstructed (energy from reflected path > 90% of total).

For a detailed discussion of pathloss models for UWB wireless systems, interested readers are invited to the consult the chapters of this book which detail this particular aspect of propagation. The above summary has, however, been included to highlight some high-level characteristics and background to keep in mind when reading results extrapolated from a 'simple' signal strength attenuation analysis over UWB bandwidths.

For the purpose of the attenuation measurements, we only considered ideal LOS conditions, as reproduced in an anechoic chamber in order to isolate the effects of the attenuation given by the materials themselves from other propagation effects caused by objects or human interactions with the electromagnetic waves. While these operating conditions require much more effort than measurements in living open environments as in [2], [3], [4], the resulting attenuation factors will be more realistic and less conditioned by potential multipath effects at the receiver.

In order to fully characterise the behaviour of different materials, we also need to remember that electromagnetic waves will be perfectly reflected only by perfect conductors, while all the typical surfaces show certain values of reflection and transmission, which are directly influencing the degree of signal penetration.

Assuming initially that the reflecting surfaces are large compared with the wavelength of the radiation (i.e. neglecting the effects of possible diffraction) and flat enough to be considered smooth, we can consider the ideal situation depicted in Figure 13.1.

In this situation, we have an incident wave that hits a flat surface (which may be a partition or a wall and is assumed to be of large thickness) and the resultant electromagnetic wave is divided in reflected and transmitted ray. With reference to Figure 13.1, we can define the angle of incidence θ_i (also known

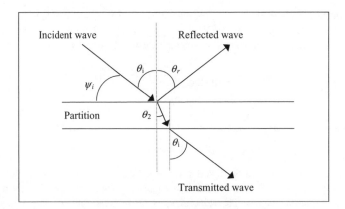

Figure 13.1 Simplified view of signal propagation trough a partition/material

as the optical angle), the angle of reflection θ_r, and the grazing angle ψ_i. According to the Snell's law, the angle of reflection is equal to the angle of incidence.

We can also define the complex *transmission coefficient* as the ratio of the transmitted to the incident electric field strengths and the *reflection coefficient* as the ratio of the reflected to the incident electric field strengths at a given wavelength. Mathematical equations describing models for these angles in both horizontal and vertical polarisation conditions can be found in [6].

The degree of signal attenuation strictly depends on the complex transmission and reflection coefficients of the surfaces; they, in turn, are influenced by the material properties, by the angle of arrival, the relative polarisation and the frequency of the transmitted signal. The results presented later in this chapter attempt to give an empirical measure to the UWB signal attenuation in the presence of multiple different attenuating surfaces under the ideal conditions where the incident wave is perfectly perpendicular to the reflecting object or partition and no diffraction is affecting the propagation.

13.3 The Materials Under Test

The propagation characteristics described in previous paragraphs apply to a variety of environments and radio operating conditions. In order to enable a more detailed analysis of propagation and services, such as NLOS communications or through-wall surveying or overlay multi-radio systems, a deeper understanding of the signal characterisation through common materials is necessary.

Given the variety of building practises around the world and the noticeable differences in material structures among nations and even towns, the analysis has tried to cover a wide range of common materials as usually found in offices and homes. During the experiments, 12 materials were analysed: red brick, cinder block, melamine, plywood, pine wood, plasterboard, Medium Density Fibreboard (MDF), glass, double-tinted glass, Venetian blinds, thin metal and thick metal.

The samples under analysis were obtained from a local (UK-based) site management company and represent the typical materials used inside buildings and homes in the western world. A short description of each of the materials used and some of their characteristics is given below:

Red brick – A brick is a moulded rectangular block of clay baked by the sun or in a kiln until hard and resistant. It is widely used as a building and paving material across Europe. Red bricks are often used in the construction of buildings as an aesthetically pleasing outer layer over a cinder block wall. In constructions, the bricks would usually be joined together (laid) using cement. For practical reasons the experiments in the anechoic chamber did not include this additional joining layer, and the bricks were simply laid out on top of each other and joined with simple bluetag inter-frames. A 'wall' of six levels of five red bricks each was built resulting in an obstacle of 1 m $\times$ 60 cm between transmitter and receiver.

Cinder block – Cinder (or breeze) blocks are rough light concrete building blocks made with cinder aggregate. They are often used for the inner layer of an external wall (with the outer layer being red bricks). The blocks used in this experiment were solid throughout with no holes passing through the block. Blocks would normally be laid using cement. As with bricks, this was not deemed to be feasible in the anechoic chamber and the blocks were simply stacked one on top of another. The stack was 1 m high by 1.58 m large dividing the transmitter and receiver which were placed at 10 cm distance from the blocks.

Melamine chipboard – Chipboard is a common woodwork product that is made from resin-coated particles ('chips') of softwood. The particles are evenly spread over a flat plate and heat bonded together under high pressure or 'glued' together with resin. The sample considered here was coated (on two faces) with a shiny melamine resin that gives the material a wipe-clean, water-resistant, surface. Often referred to simply as melamine board, this material is popular for work surfaces and, particularly in a home environment, for cheap furniture (such as coffee tables and desks). For the experiments in the anechoic chamber two boards were joined side by side forming a barrier of approximately 1.95 m height by 1.25 m width.

Plywood – Plywood is a traditionally man-made wooden surface, which consists of a number of thin sheets of wood veneer, called plies, laminated together with the direction of each ply's grain differing from its neighbours by 90° (cross-banding). The plies are usually bonded under heat and pressure with strong adhesives, often phenol formaldehyde resin, making plywood a typical composite material. Plywood can be used for furniture such as cabinets, as well as sheathing of walls, floors and roofs or used in combination with decorative wood faces such as hardwood (red oak, maple, birch, mahogany). In the sample under analysis, the sheet of plywood had five plies joined together for a height of 1.70 m and a width of 1.22 m.

Softwood (pine) – Softwood is a wooden surface coming from conifers. The difference from hardwood is in the internal microscopic structure of the wood, which lacks the vessel elements for water transport and is more uniform than hardwood. The softwood family includes woods like pine, redwood, spruce, cedar and cypress. It is used primarily in construction work, for example dry walls may consist of plasterboard screwed to softwood joists. It is also found in other products such as millwork (doors, mouldings, windows) and home furniture. The softwood used in this experiment was dry pine in long panels stacked side by side to form a barrier 1.88 m high and 0.98 m long.

Plasterboard (sheetrock) – Plasterboard, also known as gypsum board or drywall or sheetrock, is low-cost material made primarily from gypsum rock (hydrous calcium sulphate) and formed by sandwiching a core of wet gypsum between two sheets of heavy paper and drying it out. It is used for constructing dry internal walls and also as the inner layer of external walls. Plasterboard (as the name suggests) comes in the form of large boards. In this experiment the penetration loss in single and double thickness plasterboard was measured. A plasterboard panel was 2.66 m high and 1.20 m wide.

Medium Density Fibreboard – MDF is a manufactured wooden board formed by breaking down softwood into its wood fibres, combining the result with wax and resin and then forming the panels by applying high temperature and pressure to the mixture. This material is used for furniture, such as cupboards and coffee tables. It has become increasingly popular within the UK home for low-budget projects due to its extensive use on home improvement television shows. It is also used to finish walls in business types of environments.

Single pane of glass – Glass is a transparent, strong and biologically inactive material (mainly formed of amorphous silicon dioxide), which can be shaped in very smooth surfaces. These characteristics make this material ideal as a liquid container or as a front surface for windows or style doors. Glass in its natural form is, however, brittle and will break into sharp shards if a strong pressure is exercised on it. For this reason it can be commercially modified and hardened with the addition of other compounds or heat treatment. The pane of glass used in this experiment is typical of that used in single glazed windows which are common throughout the UK. The glass was not tinted and with a single layer measuring 65 × 79 cm and with a thickness of 0.5 cm.

Double glazing unit (tinted glass) – Made of multiple layers of glass, double glazing units are widely used in windows as they offer additional benefits to conventional, single pane, windows such as better acoustic and better thermal insulation. They consist of two sheets of glass that sandwich an air gap that may sometimes be filled with argon or krypton. In the case of the particular unit used for this experiment, the two layers of glass were tinted. Such tinting is common for office buildings but rarely used in domestic situations. The sample used for UWB signal attenuation measurements was 97 cm high, 153 cm wide and 2.5 cm thick.

Venetian (metal) blind – Venetian blinds are window coverings made of horizontal thin metal (or vinyl) slats whose angle can be adjusted around the horizontal axis. These blinds are very common in office buildings to hide from sight (thus 'blinding' the viewer and the name) or to reduce sunlight. For this experiment the penetration loss was measured in three different operating conditions: with the slats angled towards the transmitter, with the slats angled towards the receiving antenna and with the slats set horizontally. In the experiments, the blinds' barrier was 175 cm high and 139 cm wide.

Thin metal – A metal is a strong and electropositive element with a shiny surface and generally good conducting properties both for heat and electricity. It can be melted or fused, hammered into thin sheets, or drawn into wires. Well-known natural metals are gold, aluminium, iron, silver, copper, titanium, zinc, uranium. For the UWB experiments, we used a sheet of thin aluminium metal from a storage box. This thin material can be used in storage compartments, or for conduits for electrical cables and filing cabinets.

Thick metal – The sample of metal used for the thick metal experiment was a sheet of aluminium, which was destined to be used as cladding on the outside of an industrial building. This material, which is expensive and requires specialised working skills, has not many other uses in constructions. The layer of aluminium used for this experiment was 0.5 cm thick, 92 cm high and 125 cm wide.

Table 13.1 is a schematic and visual description of the materials used for experiments, with their main characteristics. And Figures 13.2–13.4 show some photographs of typical configurations of material under test in an anechoic chamber.

13.4 Experimental Campaign

Numerous experiments have been performed in order to establish the signal attenuation of typical building materials when UWB energy is impinging upon them.

The penetration loss (also referred to as partition loss) is dependent on the characteristics of the material being analysed (in particular the dielectric constant and conductivity), the thickness of the sample of material analysed and the frequency of the signal for which the characterisation of the material is carried out. This important loss parameter is also dependent upon the moisture content of the materials or other covering or colouring surfaces that might be added on top of them. In this chapter, we assume that the materials under investigation have negligible moisture content and the resulting signal loss is entirely due to the material under analysis.

Analysis of UWB Signal Attenuation Through Typical Building Materials

Table 13.1 Characteristics of the 12 materials used in the penetration loss experiments

Material	Width	Height	Depth	Side view	Front view
Red brick Experimental setup: built a brick wall		10 cm	5 cm		
Cinder block Experimental setup: built a cinder block wall	39 cm	19 cm	9 cm		
Melamine chipboard Experimental setup: joined two blocks together	123 cm	189 cm	1.5 cm		
Plywood	122 cm	170 cm	1.8 cm		
Pine wood (Softwood) Experimental setup: used 7 planks side by side (as in a door)	14 cm	188 cm	3.8 cm		
Plasterboard (gypsum board, sheetrock)	120 cm	266 cm	1 cm		

(continued)

Table 13.1 (*Continued*)

Material	Width	Height	Depth	Side view	Front view
Medium Density Fibreboard	132 cm	183 cm	2 cm		
Single glass	65 cm	79 cm	0.5 cm		
Double glazing glass	97 cm	153 cm	2.5 cm		
Venetian blinds	139 cm	175 cm	2.5 cm		
Thin metal	63.5 cm	94 cm	0.2 cm		
Thick metal	92 cm	125 cm	0.5 cm		

Figure 13.2 Example of measurements of UWB propagation through pine wood (softwood) planks

13.4.1 Equipment Configuration

In the case of the UWB experiments, we used an FCC compliant UWB radio transmitter emitting pulses with a frequency spectrum of between 3.1 and 5 GHz. The energy sent over air was around −43 dBm/MHz with a uniform noise-like spectrum as depicted in Figure 13.5.

This UWB signal generator was coupled to an HP83481 Digital Analyser with the wideband HP83481A plug-in module. The materials were inserted between the transmitter and receiving antennas. A detailed schematic of the test set-up is shown in Figure 13.6.

The antennas used for the experiments were commercial wideband thick elliptical shaped dipoles. The return loss of these antennas was below −10 dB between 2.5 GHz and 5.5 GHz with a nominal gain of 0–1 dB across the covered band.

The penetration loss experiments were carried out in the Philips Research Labs, Redhill, UK, main primary anechoic chamber, the schematic of which is represented in Figure 13.7.

Figure 13.3 Example of measurements of UWB propagation through plywood

Over 130 experiments were conducted in the anechoic chamber to ensure a correct characterisation of the penetration loss.

Analysis of each material was carried out in the anechoic chamber in order to reduce the effects of reflections from the surrounding environment. The use of the chamber, which is contained within a Faraday cage, also reduced the probability of external noise affecting the measurements.

A barrier made of each material was constructed in the chamber. The barrier was constructed in approximately the same position, near the centre of the chamber, for each material. A platform was constructed on each side of the material barrier using overturned anechoic wall covering panels: the transmitter was placed on one of these platforms and the receiving antenna on the other. The transmitting and receiving antennas were directly opposite each other. A toy train set was used to carry the receiving antenna – this train set was designed to facilitate accurate placement of the receiving antenna during the experiment.

The transmitter was placed at a distance of 0.5 m from the barrier, which ensured it was in the far-field of the antenna for all considered frequencies. Measurements of the minimum and maximum signal voltage received at transmitter-receiver (Tx-Rx) separation distances of 0.6 m, 1.1 m and 1.7 m were made. Measurements with identical Tx-Rx separation distances, but with no barrier in place, were also made to allow the penetration loss calculation, i.e. to enable free-space propagation loss to be calibrated out of the subsequent calculation of penetration loss.

Figure 13.4 Example of measurements of UWB propagation through Venetian blinds

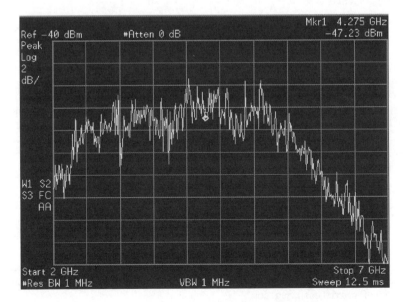

Figure 13.5 Spectrum of the reference UWB signal

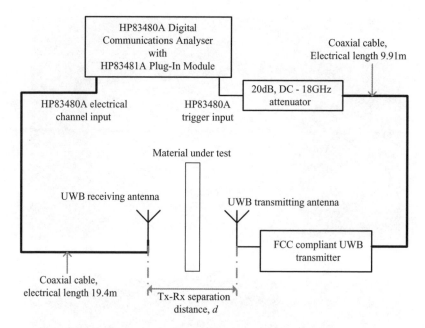

Figure 13.6 Schematic of the measurement test set-up

The penetration loss was calculated for each of the points using the below equation:

$$PenetrationLoss = -0.5^* \left(10^* \log \left(\frac{V_{max}^2}{V_{max\,Free}^2} \right) + 10^* \log \left(\frac{V_{min}^2}{V_{min\,Free}^2} \right) \right), \tag{13.7}$$

where V_{max} is the maximum received signal voltage measured with the barrier in place, $V_{maxFree}$ is the maximum received signal voltage measured without the barrier in place, V_{min} is the minimum received signal voltage measured with the barrier in place and $V_{minFree}$ is the minimum received signal voltage without the barrier in place.

Effectively, Equation (13.7) gives the average penetration loss calculated from two easily distinguishable points (maximum voltage and minimum voltage) of the received waveform. This calculation is carried out for each of the three measurement points (0.6 m, 1.1 m and 1.7 m) and an average value for penetration loss (the sum of the three penetration losses divided by three) is calculated. This average value is presented along with the standard deviation between the penetration loss measurements made at each of the three points.

13.4.2 Results

The results of the UWB penetration loss measurements are reported in Table 13.2. This summarises all of the material penetration losses calculated during the measurement campaign. The standard deviation between the measurement points for each material, in decibels and as a percentage, is also presented in the same table. Figure 13.8 presents the same information in the form of a bar chart, useful as reference for comparison of material characteristics.

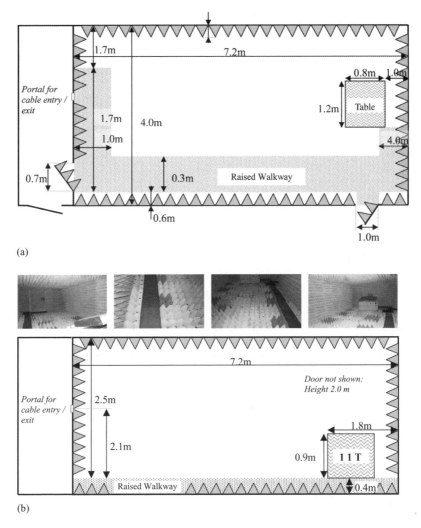

Figure 13.7 Philips Research Laboratories, Redhill, UK – Main anechoic chamber. (a) Plan view. (b) Side view

The experimental campaign presented in this chapter concentrated on the effects of material penetration loss, but interesting experiments have also been reported in the literature concerning the effects off biological beings (i.e. humans) on UWB signal propagation [8]. Even if in different operating conditions than the ones observed by our experiments, the measurements reported by Welch et al. indicate that the human body can attenuate UWB signals up to 23 dB when obstructing the full LOS pulse propagation path. This level of attenuation would be comparatively higher than many of the materials under test in our laboratories and would indicate another important factor to be considered in system analysis and simulations.

Table 13.2 Results of UWB penetration loss (Philips Research labs, UK, anechoic chamber experiments)

Material name	Penetration loss/dB	Standard deviation between measurement points/dB	Standard deviation between measurement points/%
Plasterboard (single)	8.2	0.4	4.8
Plasterboard (double)	8.5	0.5	5.5
Red brick	5.9	0.6	9.9
Cinder block	15.9	1.3	8.1
Soft wood (pine)	6.6	2.1	32.1
Plywood	2.5	0.7	27.4
Chipboard (melamine board)	2.2	0.6	26.0
Medium Density Fibreboard (MDF)	2.0	0.8	40.9
Single glass	3.9	0.4	11.4
Double glazing unit	6.1	0.9	15.1
Venetian blinds closed towards transmitter	5.6	0.6	10.4
Venetian blinds closed towards receiving antenna	6.1	1.0	15.7
Venetian blinds with slats horizontal	1.2	0.3	22.9
Thick metal	37.5	NA	NA
Thin metal	26.6	6.3	23.6

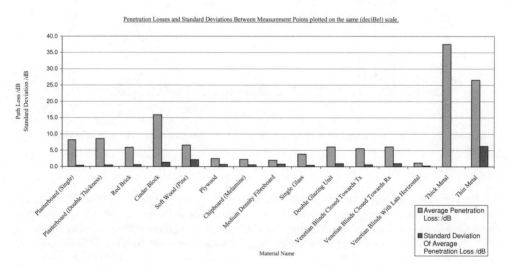

Figure 13.8 Summary of penetration loss through common building materials

13.5 Conclusions

This chapter has highlighted the importance of an accurate model of propagation for the analysis of the attenuation caused by typical materials between the transmitter and the receiver. Many studies have reported on the potential characteristics of pathloss and multipath for the typical channels between 3 and 10 GHz used by UWB communications devices. A large gap existed in the literature in the analysis of the attenuation characteristics of building materials, which are a fundamental part of a detailed link budget for performance assessment or evaluation of coexistence of UWB with other radio services.

The experimental campaign presented in this chapter has analysed the amplitude variations of UWB impulse radio signals resulting from the penetration of materials commonly found in homes or offices. Typical deployment scenarios of WLAN/WPAN systems include many of the elements taken in consideration by our measurement campaign (wooden partitions, plasterboards, glass windows, blinds, etc.) and the knowledge of their attenuation characteristics is beneficial in the system analysis.

A set of measurements in an anechoic chamber shows that UWB signals are attenuated differently by different building materials. Strong attenuation (when compared with other materials) has been registered with cinder blocks (very widespread in southern Europe) and metal blocks (often used as ornamental features on the outside part of a building). In this case, the attenuation coefficient has been calculated as 16 and 26–35 dB, respectively. Plasterboard, often used as a partition between adjacent offices, has registered an attenuation coefficient of around 8–9 dB. The effect of attenuation due to blinds (attenuation between 5.5 and 6 dB with closed slats) and double glazing and tinting (about double the attenuation than single glass transparent windows) has also been observed and should be taken into account in realistic simulation models.

Comparative results of attenuation from biological beings (humans) reported in the scientific literature have also been introduced to highlight their potential strong effect on link budgets, even if they are often neglected or not fully characterised.

These results offer a practical reference for any coexistence study and indicate that the attenuation in the path between a transmitter and a receiver unit is extremely variable depending on the exact objects the UWB radiowaves will have to penetrate. A detailed analysis of the scenarios coupled with the use of measured attenuation factors will help the system designers to understand the real coexistence potential of UWB signals with other radio services and the need for extra care in planning communication links through more reflective surfaces.

References

[1] Federal Communications Commission (FCC), Revision of Part 15 of the Commission's Rules regarding ultra-wideband transmission systems, First Report & Order, ET Docket 98-153, FCC 02-48; Adopted: 14 February, 2002; Released: 22 April, 2002.
[2] William C. Stone, Electromagnetic signal attenuation in construction materials, *NIST Construction Automation Program Report No. 3*, Building and Fire Research Laboratory, Gaithersburg, Maryland, October 1997.
[3] Daniel R.J. Estes, Thad. B. Welch, Antal A. Sarkady and Henry Whitese, Shipboard radio frequency propagation measurements for wireless networks, *MILCOM 2001 – IEEE Military Communications Conference*, no. 1, October 2001.
[4] Ray-Rong Lao, Jenn-Hwan Tarng and Chiuder Hsiao, Transmission coefficients measurement of building materials for UWB systems in 3–10 GHz, *IEEE Semiannual Vehicular Technology Conference*, 2003.

[5] J.R. Foester *et al.*, Channel modeling sub-committee report final, technical report, IEEE P802.15 Wireless Personal Area Networks, P802.15-02/490r1P802-15_SG3a, February 2003.
[6] D. Porcino, Simulation of an indoor radio channel at mm-wave frequencies, Final Year Project, Dott. Ing. Thesis, Politecnico di Torino, Italy, 1997.
[7] J.D. Parson, *The Mobile Radio Propagation Channel*, Pentech Press, 1992.
[8] Thad. B. Welch, Randall L. Musselman, Bomono A. Emessiene, Phillip D. Gift, Daniel K. Choudhury, Derek N. Cassadine and Scott M. Yano, The effects of the human body on UWB signal propagation in an indoor environment, *IEEE Journal on Selected Areas in Communications*, **20**(9), 2002.

14

Large- and Medium-scale Propagation Modelling

Mischa Dohler, Junsheng Liu, R. Michael Buehrer,
Swaroop Venkatesh and Ben Allen

14.1 Introduction

As a transmitted signal propagates towards the intended receiver, it loses in power due to three multiplicative processes:

- the free-space propagation loss governed by Friis' formula (see Section 3.3.9);
- the loss of power occurring at all reflection, diffraction and scattering processes on the way; and
- the partial or full cancelation of two or more unresolvable signals.

Averaged over time and a sufficiently large spatial area with dimensions typically greater than one thousand times the largest wavelength present in the signal, the average loss in power is referred to as *large-scale fading or average pathloss*. The random variations, usually observed when averaging the received signal over time and a spatial area with dimensions typically around 40 times the largest wavelength present in the signal, is referred to as *medium-scale fading or shadowing*. To design a proper UWB transceiver system with satisfactory power budget, it is of importance to predict the loss in power due to pathloss and shadowing with a satisfactory statistical precision.

Channel modelling for wireless systems is a well-investigated topic and is the subject of several textbooks [1, 2, 3, 4]; it is well understood that, in the absence of clutter, a pulse loses in average 20 dB in power per decade distance when propagating in the far-field. In the presence of clutter, however, the average number of reflections off clutter surfaces increases as the receiver moves away, thereby also increasing the average loss in reflected power; for that reason pulses lose more than 20 dB in power per decade distance. The only exception are waveguide-like corridors, hallways or tunnels, which yield losses below 20 dB per decade distance.

Ultra-wideband Antennas and Propagation for Communications, Radar and Imaging Edited by B. Allen,
M. Dohler, E. E. Okon, W. Q. Malik, A. K. Brown and D. J. Edwards
© 2007 John Wiley & Sons, Ltd

It is, however, important to explain the need to revisit channel modelling for UWB. The obvious reason is the extremely large bandwidth associated with UWB signals. Traditional channel models for average pathloss, for instance, assume that diffraction coefficients, the attenuation due to materials and other propagation effects are constant over the frequency band of interest. Such an assumption makes sense when the fractional bandwidth of the signal is 0.1 % or less, whereas with UWB signals, the fractional bandwidth can be in excess of 20 %. In such a case, these assumptions may no longer be valid. Additionally, narrowband models often incorporate antenna effects, such as the effective aperture, into the pathloss. Again, this is acceptable when the variation in these antenna effects is negligible over the signal bandwidth. Once again, this assumption may not be valid for a UWB system.

Fortunately, these effects have been measured, quantified and modelled for different communication scenarios and UWB frequency bands. However, because of above-mentioned reasons, UWB pathloss modelling can be very different to that employed for conventional narrowband signals, where we generally distinguish a deterministic and stochastic approach. Both approaches will be investigated in this chapter in the context of UWB pathloss modelling. The stochastic approach has led to two IEEE802.15 reference pathloss models, which are described at the end of this chapter.

14.2 Deterministic Models

If all the boundary conditions for a propagating pulse are known *a priori*, then Maxwell's equations and their simplifications can be invoked to predict the average pathloss. Typical boundary conditions are the exact location of transmitter and receiver, the bandwidth and centre frequency, the exact location and electromagnetic properties of reflective surfaces, etc. Typical simplifications of Maxwell's equations are Friis' free-space pathloss, Fresnel's reflection, Kirchhoff's scattering, and Heugen's diffraction formulas. Since UWB (impulse) radio signal propagation is inherently different from a narrowband signal propagation, it is worth revising even the simplest of all propagation environments, i.e. free space.

Traditionally, the free-space pathloss for narrowband signals is examined using the Friis transmission formula which provides a means for predicting the received signal power. For a given transmit frequency, the Friis transmission formula predicts that the received signal power falls off with the square of the distance between the transmitter and receiver. Additionally, the formula predicts that the received signal power will decrease with the square of increasing frequency due to the assumption of constant gain antennas, which has little effect on narrowband systems. However, the large bandwidth of UWB signals, coupled with this definition of pathloss, would tend to suggest that the channel results in a pathloss that varies over the entire signal bandwidth. This clearly would cause frequency-dependent attenuation and thus distort the pulse shape while propagating through the wireless channel. Therefore, the Friis transmission formula needs to be examined more closely in order to justify its application to UWB.

As shown in the subsequent sections, the channel *per se* is not frequency selective; however, when incorporating the behaviour of transmit and receive antennas at either end, the loss in power may very well be strongly frequency dependent.

14.2.1 Free-space Pathloss – Excluding the Effect of Antennas

The basis for the Friis transmission formula is the flux density of a transmitting source. The flux density ω is defined as [5]

$$\omega = \frac{EIRP}{4\pi r^2} \qquad (14.1)$$

where *EIRP* is the effective isotropic radiated power, which assumes that the transmit power is radiated equally in all directions by the transmitter, and r is the radius of the sphere for which the flux density is being calculated, *i.e. the distance between transmitter and receiver.*

Equation (14.1) shows that the flux density assumes no frequency dependence and that with a doubling of distance the flux decreases by a factor of four. This flux density can then be used to determine received power, P_{Rx}, by multiplying with A_e, the effective aperture of the receive antenna, resulting in

$$P_{Rx} = \frac{EIRP}{4\pi r^2} A_e \quad (14.2)$$

The Friis equation is typically stated in terms of the gains of the antennas where the gain is related to the effective aperture of the antenna, A_e, by [5]

$$G = \frac{4\pi}{\lambda^2} A_e \quad (14.3)$$

where λ is the wavelength. Rearranging (14.3) to solve for A_e, and substituting the result into (14.2) gives

$$P_{Rx} = \frac{EIRP}{4\pi r^2} \frac{\lambda^2}{4\pi} G_{Rx} \quad (14.4)$$

where G_{Rx} is the gain of the receive antenna. Further, we can write $EIRP = P_{Tx} G_{Tx}$, where P_{Tx} is the transmit power and G_{Tx} is transmit antenna gain. This results in the standard Friis transmission formula:

$$P_{Rx} = P_{Tx} \frac{G_{Tx} G_{Rx} \lambda^2}{4\pi r^2} \quad (14.5)$$

The term $\lambda^2/4\pi r^2$ is typically defined as the pathloss.[1] The existence of λ in the pathloss equation is interpreted as frequency-dependent pathloss. However, this term is *explicitly* an antenna effect. To make this more obvious, it is instructive to consider another type of antenna, a constant aperture antenna. A constant aperture antenna has a flux density which is a function of wavelength given by

$$\omega = P_{Tx} \frac{4\pi A_{e,Tx}}{\lambda^2} \frac{1}{4\pi r^2} = \frac{P_{Tx} A_{e,Tx}}{\lambda^2 r^2} \quad (14.6)$$

This flux density can be used in the same manner as above to give the expected received power:

$$P_{Rx} = P_{Tx} \frac{A_{e,Tx} A_{e,Rx}}{\lambda^2 r^2} \quad (14.7)$$

This result again shows frequency dependence, but here the received power increases with frequency. For systems with a constant gain antenna on one end of the link and a constant aperture antenna on the other end of the link, it can be easily shown that the received power is independent of frequency.

The key point being made here is that, while the received power may be dependent on frequency, this is due to the antennas, not the path taken by the signal *per se*. The pathloss (or more accurately the spreading loss) in free space is not frequency dependent. This can be seen by examining line-of-sight (LOS) measurements using UWB signals at various distances [6]. As an example, consider a 200 ps Gaussian pulse transmitted and received with wideband biconical antennas. Figure 14.1 plots the received

---

[1] Note that due to the multiplication of $\lambda^2/4\pi r^2$ with P_{Tx} to obtain P_{Rx}, the coefficient should correctly be referred to as *pathgain coefficient*. We will, however, stay with convention and use *pathloss coefficient*.

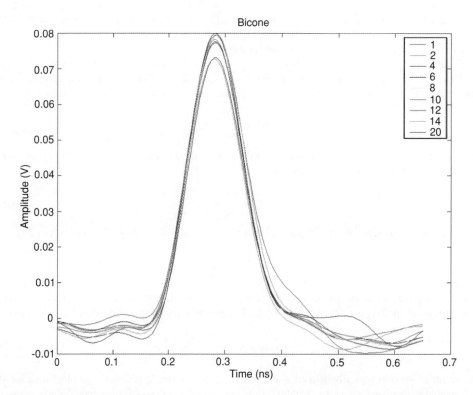

Figure 14.1 LOS received pulses normalised according to their respective distances using Bicone antennas (source: R.M. Buehrer, A.Safaai-Jazi, W.A. Davis, and D. Sweeney, "Characterization of the UWB Channel," in Proceedings of *IEEE Conference on Ultra-Wideband Systems and Technologies*, pp. 26–31, Reston, VA, Nov. 2003; reproduced by permission of © 2003 IEEE)

waveforms, which are normalised by distance[2] for distances from 1 m to 20 m. Figure 14.1 shows that, within measurement error, the pulses maintain the same shape for distances between 1 m and 20 m, indicating that there is no frequency dependency in the path *per se*.

It must be noted that the received pulse is not necessarily the same as the generated pulse. This is due to the fact that the antenna response is frequency dependent, and different antennas can have very different frequency responses. However, the impact of the antennas on the received signal is similar, regardless of distance. Thus, it is not necessary to accommodate frequency dependence into the pathloss for UWB signals, at least not in LOS environments. The received signal power will clearly vary with frequency due to the antenna response. Additionally, any individual measurement may exhibit frequency dependence for a specific measurement due to multipath. All antennas regardless of their characteristics or frequency of operation have their power flux density varying as $1/r^2$ in free space, which is a direct result of the linearity of Maxwell's equations. This can be made explicit by defining the received power at some reference distance r_0 as P_0. The average received power at any distance r in a line-of-sight environment

---

[2] Since power spreading loss is relative to the square of distance, the loss in voltage is expected to be relative to distance. Thus, normalising the received signal values with respect to distance should result in similar voltage levels.

can then be calculated as

$$P_{Rx} = P_0 \left(\frac{r_0}{r}\right)^2 \qquad (14.8)$$

where the frequency dependence due to antenna effects is captured entirely in the reference measurement, P_0.

14.2.2 Free-space Pathloss – Considering the Effect of Antennas

In the previous section, we demonstrated that the free-space propagation channel is not frequency dependent as such. Due to the frequency dependency of UWB antenna response, thereby stipulating the design of appropriate receiver structures, it is, however, customary to jointly consider the distance dependency of the propagation channel and the frequency selectivity of transmit and receive antennas. Although the effects of channel and antennas can be factored into multiplicative terms, we will keep them together when quantifying the pathloss here.

We resolve a UWB pulse into its spectral components and consider the effect of loss in power for each of these components. For this we invoke the Friis transmission formula which, in contrast to Sections 3.3.9 and 14.2.1, we now apply per spectral component. Using the identity

$$S_{Rx}(r, f) = \frac{\partial P_{Rx}(r, f)}{\partial f} = \frac{\partial P_{Rx}(r, f)}{\partial P_{Tx}(f)} \cdot \frac{\partial P_{Tx}(f)}{\partial f} = \frac{\partial P_{Rx}(r, f)}{\partial P_{Tx}(f)} \cdot S_{Tx}(f) \qquad (14.9)$$

we obtain

$$S_{Rx}(r, f) = \frac{A_{e,Tx}(f) \cdot A_{e,Rx}(f)}{\lambda^2 \cdot r^2} \cdot S_{Tx}(f) \qquad (14.10)$$

where r is the distance between transmitter and receiver, f is the frequency, $S_{Tx}(f)$ and $S_{Rx}(r, f)$ are the transmitted and distance-dependent received power spectral densities respectively, $P_{Tx}(f)$ and $P_{Rx}(r, f)$ are the transmitted and distance dependent received power respectively, $A_{e,Tx}(f)$ and $A_{e,Rx}(f)$ are the effective areas of the transmitting and receiving antennas per spectral component respectively, λ is the wavelength and r the distance between transmitter and receiver. Furthermore, as said in the previous section, the antenna gain, $G(f)$, and the effective aperture, $A_e(f)$, are generally frequency dependent, and they relate via

$$G(f) = \frac{4\pi}{\lambda^2} A_e(f) = \frac{4\pi f^2}{c^2} A_e(f), \qquad (14.11)$$

where f is the frequency, c the speed of light, and $c = \lambda f$. This allows (14.10) to be rewritten, leading to Friis transmission formula in the spectral domain as

$$S_{Rx}(r, f) = S_{Tx}(f) \cdot G_{Tx}(f) G_{Rx}(f) \left(\frac{c}{4\pi}\right)^2 \cdot \frac{1}{r^2} \cdot \frac{1}{f^2} \qquad (14.12)$$

where $G_{Tx}(f)$ and $G_{Rx}(f)$ are the gains of the transmitting and receiving antenna elements respectively. Remember that the above equation holds assuming that the transmitting and receiving antennas are in each other's far-field, and that each frequency component at the transmitting and receiving antenna is impedance matched to the associated circuitry. Generally, perfect antenna matching is difficult to achieve across the entire UWB bandwidth, which is the reason why matching efficiency factors are introduced in more realistic pathloss characterisations, as considered in the IEEE802.15 channel modelling sub-committee discussions [7]. In this chapter, however, we will neglect the mismatch (if required, it can be absorbed into the frequency-dependent antenna gains).

The strong frequency dependency of the received pulse assuming, for instance, a constant-gain antenna with $G_{Tx}(f) = G_{Tx}$ and $G_{Rx}(f) = G_{Rx}$ has serious implications for the UWB transceiver design. This is because to deliver the received power to the detection process, the pulse has to be received by a correlation receiver which needs to be matched to the received pulse. Due to the strong frequency dependency, however, the received pulse is clearly dispersed in time. This requires special attention during the correlation receiver design (assuming coherent detection is employed). The frequency dependency can be demonstrated by means of a Gaussian pulse [8], which in the time domain can be expressed as[3]

$$s_{Tx}(t) = \frac{t}{\tau} \exp\left(-\frac{1}{2}\left(\frac{t}{\tau}\right)^2\right), \qquad (14.13)$$

where τ is some variable responsible for the temporal spread of the Gaussian pulse. Such a time representation leads to a power spectral density of

$$S_{Tx}(f) = (2\pi)^3 \left(\tau^2 f\right)^2 \exp\left(-(2\pi\tau f)^2\right). \qquad (14.14)$$

Inserting (14.14) into (14.12) yields the power spectral density of the received pulse. Applying the inverse Fourier transform to the square-root of the power spectral density yields the received pulse in the time domain. Assuming constant-gain antennas, this gives

$$s_{Rx}(t) = \frac{\pi c \tau}{r} \cdot \sqrt{G_{Tx} G_{Rx}} \exp\left(-\frac{1}{2}\left(\frac{t}{\tau}\right)^2\right). \qquad (14.15)$$

The transmitted and received Gaussian pulse in the time and frequency domains are depicted in Figures 14.2–14.5, where we assumed $\tau = 0.0265$ ns and frequency independent antenna gains. From Figure 14.4, it is clear that here the channel and antennas jointly act as a low-pass filter, and from Figure 14.5 it is clear that the resulting pulse is dispersed in time and has been attenuated.

It is hence desirable to develop a pathloss model which incorporates the distance and frequency dependency of the UWB pulse propagation through antenna and channel. This can be accomplished by commencing with the pathloss[4] definitions as follows [4]

$$L(r) \triangleq \frac{P_{Rx}(r)}{P_{Tx}} \qquad (14.16)$$

$$= \frac{\int_W S_{Rx}(r,f) df}{P_{Tx}} \qquad (14.17)$$

$$= \int_W \frac{1}{r^2} \left(\frac{c}{4\pi}\right)^2 \frac{1}{P_{Tx}} \frac{S_{Tx}(f) G_{Tx}(f) G_{Rx}(f)}{f^2} df \qquad (14.18)$$

$$= \int_W \frac{1}{r^2} \left(\frac{c}{4\pi}\right)^2 \frac{f_0}{P_{Tx}} \frac{S_{Tx}(f) G_{Tx}(f) G_{Rx}(f)}{f^2} d(f/f_0) \qquad (14.19)$$

$$\triangleq \int_W \overline{L}(r,f) d(f/f_0) = \int_W \overline{L}(r)\overline{L}(f) d(f/f_0) \qquad (14.20)$$

---

[3] Although the used Gaussian pulse does not exactly fit the power spectral density mask set by the FCC nor is it a transmittable pulse due to the non-existence of the derivative at $t = 0$ s, it yields sufficient insight into the behaviour of UWB pulse propagation. Also, the given pulse shape is before it decouples from the transmit antenna, since the decoupling process distorts the pulse shape.

[4] Again, we will stay with the term *pathloss*, instead of *pathgain*. Note, also, that $L(r)$ should be smaller than 1 in linear scale or below 0 dB in decibels.

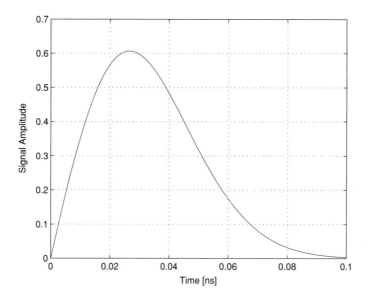

Figure 14.2 Time domain signal behaviour of transmitted pulse

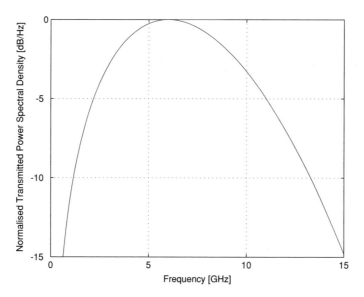

Figure 14.3 Power spectral density of the transmitted pulse

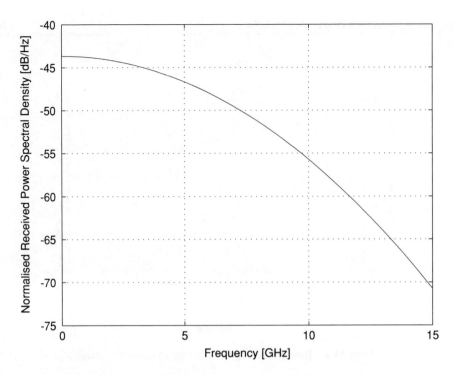

Figure 14.4 Power spectral density of received pulse at 1 m distance

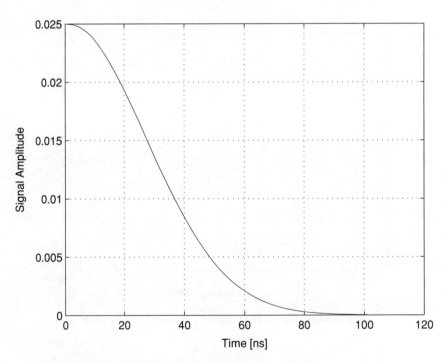

Figure 14.5 Time domain signal behaviour of received pulse at 1 m distance

Here, f_0 is some reference frequency and, comparing (14.19) with (14.20), we impose

$$\overline{L}(r, f) = \overline{L}(r)\overline{L}(f) \equiv \frac{1}{r^2} \left(\frac{c}{4\pi}\right)^2 \frac{f_0}{P_{Tx}} \frac{S_{Tx}(f)G_{Tx}(f)G_{Rx}(f)}{f^2}, \qquad (14.21)$$

which leads to the unitless distance and frequency-dependent pathloss, $\overline{L}(d, f)$. Note that here the distance and frequency dependencies are independent, i.e. $\overline{L}(r, f) = \overline{L}(r)\overline{L}(f)$, which is a direct consequence of the linearity of Maxwell's equations. Furthermore, if the physical values are properly distributed between $\overline{L}(d)$ and $\overline{L}(f)$, they are also unitless.

The above equations resemble the pathloss behaviour of a narrowband signal, however, taking the dispersive character of the antennas into account. For instance, if one deploys a UWB system with constant-gain antennas and the spectral mask of the signal being constant, i.e. $S_{Tx}(f) = S_{Tx} =$ constant over the UWB signal frequency band stretching from $f_c - W/2$ to $f_c + W/2$, then the pathloss equation simplifies to

$$L(r) = \frac{1}{r^2} \cdot \left(\frac{c}{4\pi}\right)^2 \frac{\int_{f_c-W/2}^{f_c+W/2} S_{Tx}(f).G_{Tx}(f)G_{Rx}(f)f^{-2}df}{\int_{f_c-W/2}^{f_c+W/2} S_{Tx}(f)df} \qquad (14.22)$$

$$= \frac{1}{r^2} \cdot \left(\frac{c}{4\pi}\right)^2 \frac{S_{Tx}(f_c)G_{Tx}(f_c)G_{Rx}(f_c) \int_{f_c-W/2}^{f_c+W/2} f^{-2}df}{S_{Tx}(f_c) \int_{f_c-W/2}^{f_c+W/2} df} \qquad (14.23)$$

$$= \frac{1}{r^2} \cdot \frac{1}{f_c^2} \cdot \left(\frac{c}{4\pi}\right)^2 \cdot G_{Tx}(f_c)G_{Rx}(f_c) \cdot \left[\frac{1}{1 - \left(\frac{W}{2f_c}\right)^2}\right] \qquad (14.24)$$

which is equivalent to the narrowband pathloss behaviour up to a fairly small multiplicative factor of $\left[1 - (W/2f_c)^2\right]^{-1}$; as $W \to 0$, the above equation reduces completely to the narrowband pathloss formula. Note that, invoking the first mean value theorem for integration, antennas and signals with nonconstant spectral behaviour will lead to a similar expression, however, with a different multiplicative factor.

14.2.3 Breakpoint Model

We will now quantify the deterministic pathloss behaviour for one or several reflected UWB pulses. Due to the close proximity between transmitter and receiver, the loss in power exhibits a breakpoint behaviour similar to the one observed in cellular systems, albeit the underlying reasons are totally different. We will hence deviate from frequency-dependent behaviour of the UWB system, but rather concentrate on its distance-dependent tendencies [9, 10, 11].

As mentioned before, of prime importance for the performance of a UWB system is the amount of power captured by the UWB receiver pulse correlator (assuming coherent detection). A perfect correlation is extremely difficult to obtain; however, we assume that the correlation receiver is capable of resolving M multipath components, which allows us to use the distance-dependent pathloss in (14.18) to[5]

$$L(d) \approx K \cdot \sum_{i=1}^{M} \frac{1}{d_i^2} \qquad (14.25)$$

---

[5] We have purposely changed the notation of the distance from r to d to emphasise that we do not deal with simple point-to-point free-space propagation anymore.

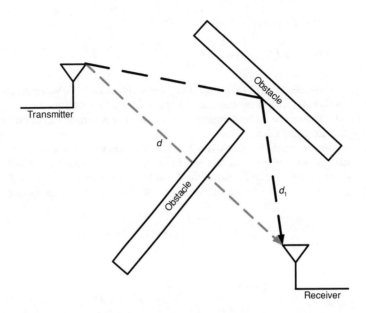

Figure 14.6 Spatial UWB transmitter, receiver and obstacle arrangement

where the distance traversed by the ith multipath component is denoted by $d_i = d + \Delta d_i$. Here, Δd_i is the incremental distance travelled with respect to the LOS distance, d. In (14.25), the distance dependencies have been decoupled from the frequency dependencies which have been associated with the constant K. Also, to facilitate the analytical framework, the average reflection coefficients are absorbed by K. Further analysis hence concentrates first on the case of a single resolvable multipath component. It is then shown that the same analysis approximately holds for arbitrary values of M.

It is assumed first that a single multipath component is resolvable at the receiver under NLOS conditions, i.e. the LOS path is obstructed, as depicted in Figure 14.6. Under the given conditions, Equation (14.25) simplifies to

$$L(d) \propto \frac{1}{(d + \Delta d_1)^2} \qquad (14.26)$$

which is given in decibels as

$$L(d) \propto -20\log_{10}(d) - 20\log_{10}(1 + \Delta d_1/d) \qquad (14.27)$$

The dependency of the received power versus distance is depicted in Figure 14.7, parameterised on various realisations of Δd_1. The distance corresponds to the physical distance, d, between UWB transmitter and receiver. The power is normalised with respect to the power received at 1 m distance from the transmitter. The frequency effects of transmitting and receiving antennas, as well as the wireless channel, have hence been eliminated.

The LOS communication link corresponds to $\Delta d_1 = 0$ m, which is shown to exhibit the typical pathloss behaviour of -20 dB/dec. However, for $\Delta d_1 > 0$ m, the pathloss does not follow the traditional -20 dB/dec over the anticipated communication range from 1 m to 10 m. The effect is particularly predominant in the vicinity of the transmitter.

The cases for $\Delta d_1 = 1$ m and $\Delta d_1 = 5$ m are depicted, which levels off as the receiver approaches the transmitter. The difference between a traditionally assumed pathloss of -20 dB/dec and the real

Large- and Medium-scale Propagation Modelling

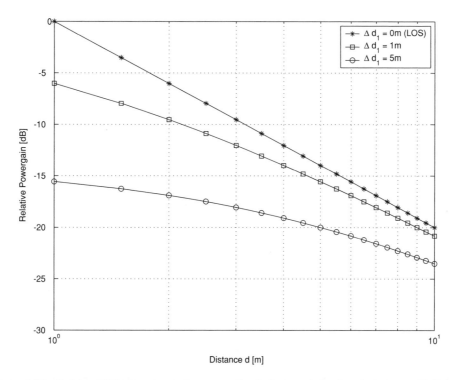

Figure 14.7 Pathloss [dB] versus transceiver separation [m] parameterised on the incremental distance Δd assuming one multipath component (source: [9] reproduced by permission of © 2006 IET)

occurring loss can be considerable, as assessed below in more detail. This is the reason why a dual-slope pathloss model is suggested, which yields a closer fit to the theoretical pathloss behaviour. The derived model results in tighter link budgets, and thus in a better performance and/or power savings.

Single-slope powerloss models. The single-slope pathloss model is only dealt with for reference. Here, a free-space behaviour with a loss of -20 dB/dec is assumed which allows the pathloss to be expressed as

$$L(d) = L(d_0) - 20 \log_{10}(d) \tag{14.28}$$

where $L(d_0)$ is the measured or calculated power at reference distance d_0 in free space; this is a value typically below 0 dB. Figure 14.8 analyses the pathloss behaviour for $\Delta d_1 = 1$ m in more detail. The original pathloss curve (thick line) is depicted, as well as two single-slope pathloss models with a loss of -20 dB/dec. The first single-slope model (stars) is too optimistic, as it predicts less pathloss than that occurring in theory. This is because the reference point of $d_0 \to \infty$ is in the region where the pathloss slope is indeed -20 dB/dec.

The second single-slope model (pentagons) is too pessimistic, as it predicts more loss than that occurring in theory. This is because the reference point was chosen to be $d_0 = 1$ m; the pathloss slope is far below -20 dB/dec.

Therefore, the optimistic model may neglect and the pessimistic model may add losses of up to 5 dB, as observed from Figure 14.8 for the respective asymptotic cases.

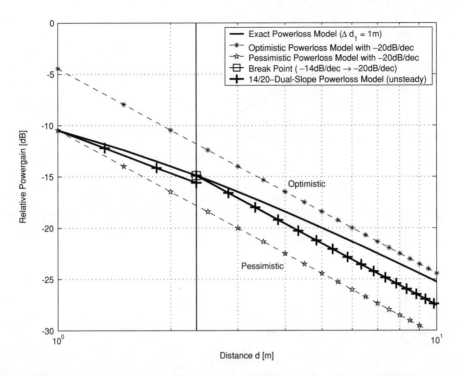

Figure 14.8 Traditional pathloss models assuming -20dB/dec and novel dual-slope pathloss model for the one multipath component case (source: [9] reproduced by permission of © 2006 IET)

Dual-slope powerloss model. Deviations of 5 dB in the communication link budget are considerable, particular for anticipated low-power UWB communication systems. A novel dual-slope model is hence suggested, which attempts to predict the pathloss in a more precise fashion. The well-known dual-slope model for micro and macro cells [12] is characterised by a breakpoint, d_{bp}, which separates the regions of -20 dB/dec ($d < d_{bp}$) and -40 dB/dec ($d \geq d_{bp}$).

In total analogy, it is suggested here that the imposed breakpoint separates the regions of $-\Omega$ dB/dec ($d < d_{bp}$) and -20 dB/dec ($d \geq d_{bp}$), which is equivalent to a change in slope by $(20 - \Omega)$ dB/dec. It hence remains to deduce the location of this breakpoint; as mentioned earlier, the reasons for this breakpoint are very different from the one described in [12]. To obtain the breakpoint, (14.27) is differentiated with respect to $\log_{10}(d)$ to arrive at

$$\frac{\partial L(d)}{\partial \log_{10}(d)} = \frac{\partial}{\partial \log_{10}(d)} \left[-20 \log_{10}(d) - 20 \log_{10}(1 + \Delta d_1/d) \right] \qquad (14.29)$$

$$= -20 - 20 \frac{\partial \log_{10}(1 + \Delta d_1/d)}{\partial \log_{10}(d)} \qquad (14.30)$$

$$= -20 - 20 \left[\frac{\partial \log_{10}(1 + \Delta d_1/d)}{\partial d} \right] \left[\frac{\partial \log_{10}(d)}{\partial d} \right]^{-1} \qquad (14.31)$$

$$= -20 + 20 \frac{\Delta d_1/d}{1 + \Delta d_1/d}. \qquad (14.32)$$

Large- and Medium-scale Propagation Modelling

The breakpoint which separates the $-\Omega$ dB/dec from the -20 dB/dec regions is easily obtained by equating the above-mentioned derivative to $-\Omega$ dB/dec. After some elementary manipulations this allows to write the dual-slope powerloss components:

$$L(d) = L(1\text{m}) - \Omega \log_{10}(d) \tag{14.33}$$

for $1 \text{ m} \leq d < d_{bp}(\Delta d_1, \Omega)$

$$L(d) = L(d_{bp}) - 20 \log_{10}(d) \tag{14.34}$$

for $d_{bp}(\Delta d_1, \Omega) \leq d < 10$ m. Since the change from an arbitrary gradient of $-\Omega$ dB/dec to -20 dB/dec, the dual-slope model is henceforth referred to as the UWB $\Omega/20$-dual-slope pathloss model. Note that the modelled power at the breakpoint may or may not be equal either side.

For a *steady* breakpoint, (14.33) and (14.34) need to equate at the breakpoint. For an *unsteady* breakpoint, the slope of -20 dB/dec may commence from the original pathloss curve.

The respective $L(d_{bp})$ can be related to $L(1\text{m})$ by

$$L(d_{bp}) = L(1 \text{ m}) - 20 \log_{10}(1 + \Delta d_1/d_{bp}) \tag{14.35}$$

for an unsteady breakpoint, and

$$L(d_{bp}) = L(1 \text{ m}) + (20 - \Omega) \log_{10}(d_{bp}) \tag{14.36}$$

for steady breakpoint.

Example UWB 14/20-dual-slope pathloss model. In Figure 14.8, the unsteady UWB 14/20-dual-slope pathloss model is also depicted, where we assumed a gradient change of 6 dB so as to yield good fitting results for a broad class of communication configurations. The model mismatch has been reduced to a maximum of 2 dB. The link budget is thus enhanced by 3 dB compared to the case where a free-space pathloss model is employed. This corroborates the usefulness of this dual-slope pathloss model.

Multiple impinging multipath components. In a realistic UWB receiver, more multipath components will be resolved. In this case, M will be larger than one, which allows (14.25) to be expressed as

$$L(d) \propto -20\log_{10}(d) + 10 \log_{10} \sum_{i=1}^{M}(1 + \Delta d_i/d)^{-2} \tag{14.37}$$

Differentiating (14.37) with respect to $\log_{10}(d)$ leads to

$$\frac{\partial L(d)}{\partial \log_{10}(d)} = -20 \cdot \frac{\sum_{i=1}^{M}(1 + \Delta d_i/d)^{-3}}{\sum_{i=1}^{M}(1 + \Delta d_i/d)^{-2}} \tag{14.38}$$

To derive a general breakpoint analogous to the case where $M = 1$ is unfortunately not possible. However, assuming that d is sufficiently larger than Δd_i for any i, the approximation

$$(1+x)^\alpha \approx 1 + \alpha \cdot x \tag{14.39}$$

can be invoked. Applying (14.39) to (14.38), the breakpoint is easily derived to be

$$d_{bp} \approx \frac{3 - 2\Omega/20}{1 - \Omega/20} \cdot \frac{\sum_{i=1}^{M} \Delta d_i}{M} \tag{14.40}$$

which can be utilised to estimate the location where the pathloss curve separates the region of $-\Omega$ dB/dec from the -20 dB/dec region.

UWB channel measurements are available to corroborate this dual-slope behaviour; see, for example, reference [13] or Figure 13.12 in Chapter 13. Note, however, that they were not (yet) analysed from the point of view as done here.

14.2.4 Ray-tracing and FDTD Approaches

If more deterministic boundary conditions, such as walls, tables, etc., are introduced, an analytical approach becomes cumbersome and hence of little use. For this reason, other tools have been developed which allow Maxwell's equations to be solved numerically for a large number of deterministic boundary conditions.

Ray tracing or ray launching is one of such a deterministic modelling approach, where narrowband rays are either traced from transmitter to receiver or launched from the transmitter and then followed. Clearly, such a narrowband ray can propagate infinitely and therefore requires some termination criteria, e.g. the ray's average power or the number of undergone reflections/diffractions. Ray-tracing tools have successfully been deployed in narrow and wideband cellular and WLAN planning tools [14, 15]. Although the error of the input to the ray-tracing tool in terms of clutter location can be very high, it yields sufficient insight into the general power behaviour in a given geographical area. Here, if we feed the ray-tracer with the location of walls and doors only, the measured power in a very specific point in the room may vary from the predicted value, but the average power measured in that room will sufficiently coincide with the value produced by the ray-tracer.

Ray-tracing or ray-launching tools are not easily applicable to the UWB propagation case [16, 17], since the UWB pulse clearly does not fulfil the narrowband assumption of the tools. A straightforward approach is hence to combine the results of the ray-tracer at different frequencies. Ideally, the sampling step in frequency should be within the coherence bandwidth of the UWB pulse [18], which includes the frequency-dependent behaviour of free-space propagation, the reflection coefficients, etc. A further approach to refine the prediction is to introduce contributions of randomly distributed clutter, thereby accounting for diffusely scattered UWB components [19]. It has further been observed that the outdoor environment, be it used in the context of UWB or not, requires the inclusion of a direct and ground reflected wave into the ray-tracing tools [20]. Indoor propagation environments, however, were observed to require about three reflections to encompass the major loss in power [19].

Finite-difference time-domain (FDTD) methods discretise the time and the space domain and calculate the propagation of the pulse timestep by timestep [21]. FDTD has been successfully used in the field strength prediction and design process of fairly complex and intricate radiation structures, e.g. specifically formed antennas. Spatial and temporal discretisation, however, need to satisfy certain constraints so as to guarantee good numerical stability and no artificial pulse dispersion (sampling is known to act as a filter). Furthermore, standard FDTD assumes permittivities, conductivities and permeabilities to be constant, which is clearly not the case over most UWB frequency bands [22, 23]. Enhancements have hence been developed, which incorporate the frequency dependency into the FDTD tools. Generally, because the spatial boundary conditions are discretised, only fairly small areas can be predicted.

14.3 Statistical-Empirical Models

As outlined above, a deterministic modelling approach is fairly confined to predefined scenarios with fixed boundary conditions. The radio propagation environment, however, is known to be random due to

- unpredictable precise location of all the clutter;
- unpredictable movement of transmitter and/or receiver; and
- unpredictable movement of clutter.

Statistics is an appropriate mathematical tool to describe the propagation process and is hence applied to UWB pathloss modelling.[6]

A typical approach in pathloss and channel modelling is to propose a given statistical distribution and then use real-world channel measurements to justify such a distribution and/or associated parameters. For instance, physical reasoning behind the multiplicative effect of reflection processes, allowed the presupposition that the effect of shadowing exists and obeys a log-normal distribution [4]; measurements have then corroborated this assumption and empirical propagation environment specific shadowing standard deviations have been extracted from the measurements.

The hybrid approach of statistical and empirical modelling is henceforth referred to as *statistical-empirical modelling*, and dealt with in subsequent paragraphs.

14.3.1 Pathloss Coefficient

The location-dependent pathloss, describing the average of the received power over time, frequency and a sufficiently large area, can be expressed as

$$L(d) = \frac{E_{\sigma(d),t}[P_{Rx}(d, t, W)]}{P_{Tx}}, \quad (14.41)$$

where the instantaneous received power, $P_{Rx}(d, t, W)$, is dependent on the location d, the time moment t, and the UWB frequency band W. Its average, $E_{\sigma(d),t}[P_{Rx}(d, t, W)]$, is obtained by averaging over

- a spatial area of size σ at location d, which is sufficiently large so as to get rid of the spatially random effects of shadowing and fading;
- and time t, so as to get rid of the temporally random effects of shadowing and fading.

---

[6] Strictly speaking, due to the temporal character of the random processes, the theory of stochastic processes needs to be involved; however, since the pathloss modelling process results in any temporal behaviour being averaged out, a statistical approach suffices.

Invoking the power spectral density, the above equation can be redrafted as

$$L(d) = \frac{\mathrm{E}_{\sigma(d),t}[\int_W S_{Rx}(d,t,f)\mathrm{d}f]}{P_{Tx}} \tag{14.42}$$

$$= \frac{\int_W \mathrm{E}_{\sigma(d),t}[S_{Rx}(d,t,f)]\mathrm{d}f}{P_{Tx}} \tag{14.43}$$

$$= \frac{\int_W \overline{S}_{Rx}(d,f)\mathrm{d}f}{P_{Tx}}, \tag{14.44}$$

where $\overline{S}_{Rx}(d,f) = \mathrm{E}_{\sigma(d),t}[S_{Rx}(d,t,f)]$ is the Fourier transform of the autocorrelation function of the stochastic pathloss realisations averaged over a spatial area $\sigma(d)$. Due to the linearity of Maxwell's equations and a presumed linear dependency of dielectric constants, one can show that the spectral density can be resolved into a frequency- and distance-dependent part, i.e.

$$\overline{S}_{Rx}(d,f) = \overline{S}_{Rx}(d) \cdot \overline{S}_{Rx}(f), \tag{14.45}$$

which is similar to (14.12). Note that $\overline{S}_{Rx}(d)$ and $\overline{S}_{Rx}(f)$ on their own do not carry the units of power spectral density. With this factorisation in mind, we can write for the pathloss

$$L(d) = P_{Tx}^{-1} \int_W \overline{S}_{Rx}(d) \cdot \overline{S}_{Rx}(f)\mathrm{d}f \tag{14.46}$$

which, for compliance with traditional pathloss analysis, we express as

$$L(d) = \int_W \overline{L}(d) \cdot \overline{L}(f) \, \mathrm{d}(f/f_0) \tag{14.47}$$

$$= \int_W \overline{L}(d,f) \, \mathrm{d}(f/f_0) \tag{14.48}$$

Here, we define $\overline{L}(d,f) \triangleq \overline{L}(d) \cdot \overline{L}(f)$, f_0 is a reference frequency and we impose $f_0^{-1} \cdot \overline{L}(d) \cdot \overline{L}(f) \equiv P_{Tx}^{-1} \cdot \overline{S}_{Rx}(d) \cdot \overline{S}_{Rx}(f)$, which leads to the unitless average location and frequency-dependent pathlosses $\overline{L}(d)$ and $\overline{L}(f)$. This allows us to reuse known narrowband concepts for $\overline{L}(d)$ but requires novel approaches to $\overline{L}(f)$.

Distance dependency. When modelling the distance dependency of the pathloss, we eliminate any frequency dependency by defining pathloss relative to a reference spectral density or reference power; the antenna effects are hence subsumed into this reference measurement. This approach facilitates the narrowband concept of pathloss exponent or pathloss coefficient, n, over distance to be used to express the UWB pathloss in dB as

$$\overline{L}(d) \propto -10n \log_{10}\left(\frac{d}{d_0}\right) \tag{14.49}$$

or equally

$$L(d) \propto -10n \log_{10}\left(\frac{d}{d_0}\right) \tag{14.50}$$

where d_0 is some reference distance; also note that the proportionality factors are different for both quantities. The pathloss coefficient, n, clearly depends on the communication environment; in particular, whether line-of-sight (LOS) conditions exist between the transmitter and receiver or not (i.e. NLOS).

Table 14.1 Parameters for the pathloss model of Ghassemzadeh et al. [32].

	Mean (LOS)	Std (LOS)	Mean (NLOS)	Std (NLOS)
$L(d_0)$ [dB]	−47	—	−51	—
n	1.7	0.3	3.5	0.97
σ_S [dB]	1.6	0.5	2.7	0.98

Some published results relating to UWB pathloss characterisation differentiate even further between LOS, 'obstructed/soft LOS' (OLOS), and 'hard NLOS'.

As demonstrated by means of measurements, the LOS coefficients in indoor environments range from 1.0 in a narrow corridor [29], to 1.2 in an industrial environment [24], to 1.5–2 in office and residential environments [25, 26, 27, 28, 29]. NLOS coefficients typically range from 2 to 2.5 in industrial and outdoor environments [24, 27], 3 to 4 for soft-NLOS in office and residential environments [25, 26, 30], and from 4 to 7 for hard-NLOS in indoor environments [29]. Other papers use a breakpoint model so that the propagation exponent attains a coefficient n_1 up to a breakpoint distance, and then n_2 beyond that; an analytical framework for this had been developed in Section 14.2.3.

An interesting refinement of pathloss modelling was introduced by [31] and applied to UWB systems in [32]. Due to the proximity between transmitter and receiver and the large differences in clutter environments in which UWB transceivers operate, the pathloss coefficient has been found to be a random variable which changes from location to location. It has been observed that the probability density function of this random variable can be approximated by a Gaussian distribution. For the indoor environment, the means and standard deviations of n for LOS and NLOS situations are given in Table 14.1. This model has been proposed with random shadowing effects and is hence detailed in a subsequent paragraph.

On a more general note, remember that the reference measurement is important when determining the pathloss model: it can be defined to include multipath effects, or to be the received power in free space. Both are commonly used in conventional narrow/wideband pathloss models. Defining the reference measurement to be free space has the advantage that it is possible to directly calculate the reference point eliminating the need for a reference measurement. This is somewhat more complicated in UWB systems than in narrowband systems. When defining the reference measurement to include environmental factors (e.g., including multipath) the reference point becomes environment specific. In such a case, statistical characterisation of the reference may be useful.

Note that if free-space attenuation is to be used for the reference point in UWB, care must be taken in determining this value. It may be calculated, provided that sufficient information concerning the antennas is available. It may also be obtained via measurement provided that all multipath effects are eliminated. For example, in [33, 34] pathloss was calculated with respect to free-space pathloss at 1 m. The reference measurement was taken at $d_0 = 1$ m and time gating was used to obtain only the LOS component and thus eliminate multipath effects. This makes the pathloss calculations relative to free space at 1 m. Pathloss can be calculated for any subsequent measurement taken at a distance d, with received power $P_{Rx}(d)$ using $L(d) = P_{Rx}(d_0)/P_{Rx}(d)$. The pathloss exponent, n, can be determined using the ensemble average of the pathloss, at various distances d. This is typically done performing a linear regression to obtain a least squares fit to data from a large number of measurements taken in the environment of interest. An example is given in Figure 14.9 for four scenarios (LOS, NLOS, bicone antennas and TEM horn antennas) [34].

Frequency dependency. As already observed in the previous sections, a frequency dependency of the pathloss enters the equation mainly through the antennas. The second place where frequency dependency

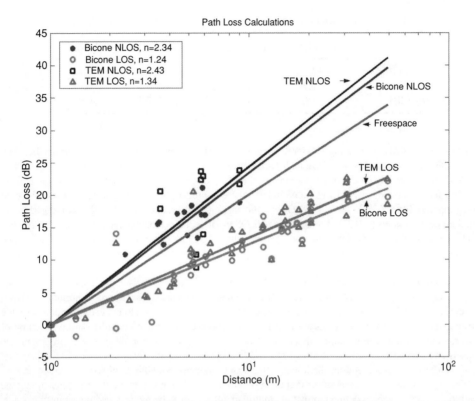

Figure 14.9 Averaged Path Loss Results Directional and Omni-Directional Antennas for LOS and NLOS environments (source: B. Donlan and R.M. Buehrer, "The Indoor UWB Channel," *Proc. IEEE Veh. Technol. Conf.*, Milan, Italy, May 2004; reproduced by permission of © 2004 IEEE)

can enter the equation is in the pathloss exponent, n. Due to the fact that diffraction, material penetration and other effects are frequency dependent [22, 23], the pathloss exponent could change with frequency. While this may not be the case for LOS scenarios, the NLOS environment allows for substantial pulse interaction, possible frequency selectivity and the introduction of frequency dependence into the channel.

Note that if we consider the frequency-domain transfer function of a single channel realisation, we will clearly observe frequency selective behaviour. However, since we are interested in the statistical modelling of the channel, we are concerned with the ensemble average of transfer functions in a particular environment at various distances. If the average transfer function shows frequency dependence apart from antenna effects, then this must be accommodated in the model.

There is currently no consensus on this effect. While some measurements did not exhibit frequency dependence, other measurements have shown some frequency dependence in pathloss. Specifically, [19] discusses the frequency-dependent decay of channel gain with increasing frequency. It was demonstrated from the measurements that the UWB channel shows, apart from a frequency selective fading pattern, a general decaying trend with increasing frequency. It is also shown that for the NLOS case, the decay is slightly steeper at higher frequencies than the LOS case; specifically, the pathloss at a given distance increases exponentially with frequency.

Analytically, the frequency dependence of UWB propagation channel, and hence up to a multiplicative constant also of the integral kernel of (14.47), has been observed to obey in dB [35, 36]

$$\overline{L}(f) \propto -20\chi \log_{10}\left(\frac{f}{f_0}\right) \tag{14.51}$$

where f_0 is some reference frequency and χ is some frequency decay coefficient. Including antenna effects, [19] found χ to be bound between 0.8 and 1.4. Excluding antenna effects, [47] found χ between -1.4 in industrial environments and $+1.5$ in residential areas. With reference to the strong frequency dependency of the antennas and the propagation channel, it is not surprise that χ can take positive and negative values in dependency whether the environment exhibits a stronger low or high pass filter behaviour.

In [37], it has been shown that the frequency-dependent pathloss is well approximated by considering the pathloss at an effective 'centre frequency' given by the geometric mean of the lower and upper band edge frequencies. Furthermore, reference [38] also observed some frequency-dependent behaviour in pathloss. Specifically, they proposed modelling the average channel transfer function as being frequency dependent according to (in dB)

$$\overline{L}(f) \propto e^{-\delta f} \tag{14.52}$$

where δ is between 1 for LOS environments and 1.36 for NLOS environments with significant blockage of the path.

Finally, it shall be noted that an alternative modelling of the frequency dependence of the pathloss is introduced in [45], which proposes a frequency-dependent pathloss coefficient, $n(f)$. However, it was found in [33] that for short-range indoor measurements frequency-dependent behaviour was not observable in the pathloss exponent. This can be seen in Figure 14.10 (taken from [33, 39]) which plots the mean and standard deviation of the pathloss exponent versus frequency over the range 1–10 GHz. The figure contains LOS and NLOS environments using bicone and TEM horn antennas. It can be seen that the pathloss exponent does not vary with frequency. The standard deviation of the pathloss exponent over the band was found to be less than 0.1. It should be noted, however, that the measurements were taken at relatively short distances (generally less than 10 m). It is possible that larger distances may reveal frequency dependencies for NLOS channels due to the frequency dependency of many materials.

14.3.2 Shadowing

The random variation of the received power averaged over a small spatial area is referred to as *shadowing* and denoted by S. The averaging over the small area, usually around $40\lambda_{max}$, is required so as to eliminate small-scale fading effects due to constructive/destructive superposition of pulses; however, the area should not to be too large so as not to average out the effects of shadowing. It is hence a random effect between the pathloss behaviour and small-scale fading. Excluding small-scale fading, the total loss in power due to pathloss *and* shadowing can be expressed in dB as

$$PL(d) = \left[L(d_0) - 10n \log_{10}\left(\frac{d}{d_0}\right)\right] + S, \tag{14.53}$$

where $L(d_0)$ is the average pathloss in dB at reference location d_0; this value is typically less than 0 dB. As can be seen from (14.53), there are two main terms to determine from the environment, the pathloss exponent and the standard deviation of the shadowing. There are two common approaches to obtaining

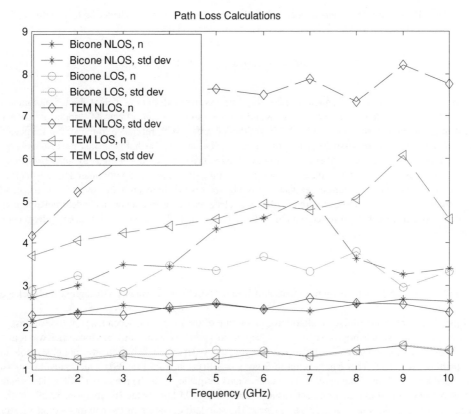

Figure 14.10 Path Loss Exponent and Standard Deviation for Different Frequencies (source: [39] reproduced by permission of © 2006 IEEE)

these values. The first is to simply combine all measurements from similar environments (e.g., office NLOS) and determine the best fit to the data. This is the more common approach [33, 40, 41, 6]. A second approach is to determine the best fit for each specific building or room and create a statistical model for the pathloss exponent and shadowing standard deviation [42, 43].

Generally, due to the multiplicative but random realisations of reflections along a pulse's propagation trajectory, the statistics of the shadowing, S, can be shown to obey a log-normal distribution [44, 3, 6]. The mean of this distribution is usually absorbed into the pathloss behaviour (explaining why pathloss coefficients of $n > 2$ exist), hence the mean of shadowing is taken to be 0 dB. The standard deviation, σ_S, is determined empirically from sets of measurements.

The strength, i.e. the variance or standard deviation, of shadowing strongly depends on the propagation conditions, i.e. the distance between the transmitter and receiver and also whether there is LOS or NLOS. Compared to cellular systems, UWB systems are operating at much closer distances, hence invoking for smaller shadowing variations. For the standard deviation, values of 1–2 dB have been observed in UWB LOS conditions and 2–6 dB in NLOS [24, 27, 30, 28, 26].

S. Ghassemzadeh et al., have performed an extensive UWB indoor measurement campaign that took place in several representative homes at a centre frequency of 5 GHz and a bandwidth of 1.25 GHz [32].

The following key observations have been made (see the summary in Table 14.1):

- The pathloss exponent, n, changes from one home to another; it is a Gaussian random variable $\mathcal{N}(\mu_n, \sigma_n)$ with mean μ_n and standard deviation σ_n, where data reduction yielded $\mathcal{N}(1.7, 0.3)$ for LOS and $\mathcal{N}(3.5, 0.97)$ for NLOS conditions.
- Shadowing, S, is a zero-mean Gaussian random variable (hence log-normal in a linear scale) with standard deviation, σ_S, that changes from one home to another; this standard deviation also obeys a Gaussian distribution $\mathcal{N}(\mu_\sigma, \sigma_\sigma)$ with mean μ_σ and standard deviation σ_σ, where data reduction yielded $\mathcal{N}(1.6, 0.5)$ for LOS and $\mathcal{N}(2.7, 0.98)$ for NLOS conditions.

We can hence express the random pathloss coefficient as $n = \mu_n + \sigma_n \cdot \mathcal{N}_1$, the shadowing process as $S = \sigma_S \cdot \mathcal{N}_2$ and its standard deviation as $\sigma_S = \mu_\sigma + \sigma_\sigma \cdot \mathcal{N}_3$, where $\mathcal{N}_{i=1,2,3}$ are independent unit variance Gaussian random processes. Note that $\mathcal{N}_1$ varies from one home to another, while $\mathcal{N}_2$ and $\mathcal{N}_3$ vary from one location to another within each home.

With the given substitutions, we can express (14.53) as

$$PL(d) = \left[L(d_0) - 10\mu_n \log_{10}\left(\frac{d}{d_0}\right) \right] + \left[-10\sigma_n \log_{10}\left(\frac{d}{d_0}\right) \cdot \mathcal{N}_1 + \mu_\sigma \cdot \mathcal{N}_2 + \sigma_\sigma \cdot \mathcal{N}_2 \cdot \mathcal{N}_3 \right], \quad (14.54)$$

where the first bracketed term expresses the median pathloss and the second bracketed term the random variations about the median pathloss. To exclude impossible realisations of the total loss in power, it has been recommended in [32] to truncate $\mathcal{N}_{i=1,2,3}$ to $\mathcal{N}_1 \in [-0.75, +0.75]$ and $\mathcal{N}_{i=2,3} \in [-2.0, +2.0]$.

It shall be noted here that although the measured bandwidth of 1.25 GHz formally qualifies as an UWB bandwidth, the deductions on pathloss and shadowing may turn out to be very different if a bandwidth of 7 GHz is used; it can be anticipated that the random variations decrease. Finally, the suggested approach can also be used for other UWB propagation environments, where data reduction has to be performed on the empirical measurements to obtain the required mean values and standard deviations.

14.4 Standardised Reference Models

The aim of standardised reference models is not to describe the UWB channel as precisely and generically as possible, but to provide some 'close-to-reality' reference UWB channel models which allow a fair comparison between various UWB transceiver proposals. UWB systems have been considered in two antipodal working groups of the IEEE 802.15 wireless personal area network (WPAN) networking group: IEEE 802.15.3a and IEEE 802.15.4a.

The IEEE 802.15.3 High Rate (HR) Task Group (TG3) for wireless WPANs has been chartered to draft and publish a standard for high-rate (20 Mbit/s or greater) WPANs. Besides a high data rate, the standard should provide for low-power, low-cost solutions addressing the needs of portable consumer digital imaging and multimedia applications. Its extension, the IEEE 802.15.3a High Rate Alternative PHY Task Group (TG3a) for WPANs is working to define a project to provide a higher speed physical layer enhancement to IEEE 802.15.3 for applications which involve imaging and multimedia. UWB is hence a good candidate due to its ability to deliver high data rates.

The IEEE 802.15.4 Low Rate (LR) Task Group (TG4) was chartered to investigate a low data rate solution with multi-month to multi-year battery life and very low complexity. It is targeting operation in an unlicensed, international frequency band. Potential applications are sensors, interactive toys, smart badges, remote controls and home automation. The IEEE 802.15.4a Low Rate Alternative PHY Task Group (TG4a) for WPANs has defined a project for an amendment to IEEE 802.15.4 for an alternative physical layer, and is thereby working on the provision of communications with high precision location capabilities (1 m accuracy and better), high aggregate throughput, and ultra low power; as well as adding scalability to data rates, longer range and lower power consumption and cost. These additional capabilities over the existing 802.15.4 standard are expected to enable significant new applications and market opportunities. The low power consumption and small fading margins of UWB make it a good candidate for the envisaged system.

The system design of both working groups clearly required a suitable channel model, part of which is the pathloss and shadowing model. These aspects are briefly summarised in the sections below.

14.4.1 IEEE 802.15.3a

IEEE 802.15.3a is all about high data rate PANs deployed in residential and office environments, where the wireless links are constrained to distances of maximum 10 m. The model has been compiled by measurements conducted by S. Ghassemzadeh *et al.*, M. Pendergrass, J. Foerster *et al.*, J. Kunisch *et al.*, A.F. Molisch *et al.*, G. Shor *et al.* [46]. The model distinguishes four radio environments, i.e. LOS with a distance between TX and RX of up to 4 m (CM1), NLOS for a distance of up to 4 m (CM2), NLOS for a distance of 4–10 m (CM3), and 'heavy multipath' environments (CM4). The main aim of the model was to account for the multipath and shadowing behaviour, and less of the distance and frequency dependent pathloss behaviour.

As such, the frequency selectivity of the channel has not been considered at all and the distance dependency was initially modelled by the narrowband free-space pathloss formula, where the frequency was chosen as the geometrical mean of the upper and lower 10 dB cutoff frequencies. While this approach allowed a fair comparison between various proposals, it was clearly far off reality. The final report of IEEE 802.15.3a hence recommended to use the Ghassemzadeh pathloss model as introduced in Section 14.3.2 without the shadowing, i.e.

$$L(d) = L(d_0) - 10\mu_n \log_{10}\left(\frac{d}{d_0}\right) - 10\sigma_n \log_{10}\left(\frac{d}{d_0}\right) \quad (14.55)$$

where the values provided in [32] (see Table 14.1) should be used.

Shadowing has been modelled as a multiplicative effect to pathloss and small-scale fading. It is independently applied to each cluster of arriving multipath components, where a log-normal process with standard deviation (of the underlying Gaussian process) of $\sigma_S = 3.4$ dB has been extracted.

14.4.2 IEEE 802.15.4a

IEEE 802.15.4a is all about communication with extremely low power consumptions [7], thereby not excluding larger communication distances, specific data rates or frequency bands. The variety of different operating scenarios has hence triggered the development of channel models for different environments at low (100–960 MHz) and high (3–10 GHz) frequency bands. The developed models have overcome the shortcomings of the IEEE 802.15.3a channel models, thereby including frequency and distance dependency in a more sophisticated manner. Also, the channel modelling subworking group of IEEE 802.15.4a has managed to developed a generic pathloss model which is only parameterised on the different scenarios. Of interest to the system designer is clearly the frequency and distance dependency, as well as the shadowing behaviour.

Distance and frequency dependency. This dependency has been formulised as

$$\overline{L}(d, f) = \overline{L}(d) \cdot \overline{L}(f) = \frac{1}{2} \cdot K_0 \cdot \eta_{Tx}(f) \cdot \eta_{Rx}(f) \cdot \frac{c^2}{(4\pi d_0 f_0)^2} \cdot \frac{(f/f_0)^{-2(k+1)}}{(d/d_0)^n} \qquad (14.56)$$

where

- the factor of $1/2$ has been included to account for average attenuations caused by people close to the antennas;
- K_0 is a normalisation constant which has to be chosen so that at reference distance, d_0, and reference frequency, f_0, the value $\overline{L}(d, f)$ is equal to the tabled parameter $\overline{L}(d_0)$;
- $\eta_{Tx}(f)$ and $\eta_{Rx}(f)$ are the frequency-dependent transmit and receive antenna efficiencies and have to be provided by the system designer;
- d_0 is the reference distance, which, in subsequent parameterisations, is equal to 1 m;
- f_0 is the reference frequency, which, in subsequent parameterisations, is equal to 5 GHz (no frequency dependency in the lower bands has been reported so far);
- k is the frequency decay factor; and
- $c \approx 3 \cdot 10^8$ m/s is the speed of light.

For example, if we assume for simplicity that the antennas are ideal isotropic radiators, then the constant K_0 can easily be determined as [47]

$$K_0 = (4\pi d_0 f_0)^2 / c^2 \cdot \overline{L}(d_0), \qquad (14.57)$$

which reduces (14.56) to

$$\overline{L}(d, f) = \overline{L}(d) \cdot \overline{L}(f) = \frac{1}{2} \cdot \overline{L}(d_0) \cdot \eta_{Tx}(f) \cdot \eta_{Rx}(f) \cdot \frac{(f/f_0)^{-2(k+1)}}{(d/d_0)^n} \qquad (14.58)$$

which can be expressed in dB as

$$\overline{L}(d, f) = -3dB + \overline{L}(d_0) + \eta_{Tx}(f) + \eta_{Rx}(f) - 10n \log_{10}\left(\frac{d}{d_0}\right) - 20(k+1) \log_{10}\left(\frac{f}{f_0}\right)$$

assuming that all values have been converted to decibels. Different parameterisations for typical scenarios have been summarised in Table 14.2.

Shadowing. This dependency has simply been formulised as

$$PL(d) = \left[L(d_0) - 10n \log_{10}\left(\frac{d}{d_0}\right)\right] + S, \qquad (14.59)$$

where

- $L(d_0)$ is the pathloss measured at reference distance d_0; and
- S is a Gaussian distributed random variable with zero mean and standard deviation σ_S.

Different parameterisations for typical scenarios have been summarised in Table 14.2.

Finally, it should be mentioned that IEEE 802.15.4a also standardised a body area network (BAN) pathloss model, which significantly differs from the above models. This is mainly due to the fact that

Table 14.2 IEEE 802.15.4a pathloss parameterisation for various scenarios [47]

	$\overline{L}(d_0)$ [dB]	n	k	σ_S [dB]
Residential, LOS, high frequency	−43.9	1.79	1.12 ± 0.12	2.22
Residential, NLOS, high frequency	−48.7	4.58	1.53 ± 0.32	3.51
Office, LOS, high frequency	−35.4	1.63	0.03	1.90
Office, NLOS, high frequency	−57.9	3.07	0.71	3.90
Industrial, LOS, high frequency	−56.7	1.20	−1.10	6.00
Industrial, NLOS, high frequency	−56.7	2.15	−1.43	6.00
Outdoors, LOS, high frequency	−45.6	1.76	0.12	0.83
Outdoors, NLOS, high frequency	−73.0	2.50	0.13	2.00
Open outdoors, LOS, high frequency	−49.0	1.58	0.00	3.96
Indoors, NLOS, low frequency	n.a.	2.40	0.00	5.90

transmit and receive antennas operate in each other's near-field. Far-field simplifications are hence not applicable anymore, and the more sophisticated modelling approach can be obtained from [47] or a subsequent chapter on channel modelling for BANs.

14.5 Conclusions

In this chapter, we have introduced different modelling approaches to UWB large-and medium-scale fading. Novel models are required because the very large bandwidth of the UWB channel exhibits a different behaviour than its narrowband counterpart. We have shown that the free-space LOS pathloss is not frequency dependent as such; however, frequency-dependent behaviours can be observed due to the transmit and receive antennas, as well as the NLOS propagation channel. We have revised deterministic and stochastic modelling approaches, identifying weaknesses, strengths and prior art for either of the two approaches. This allowed the introduction and understanding of the pathloss and shadowing models adopted by the IEEE 802.15.3a and 802.15.4a standardisation bodies. The exposed pathloss and shadowing modelling is complemented by the small-scale fading modelling in the subsequent chapter, both of which lead to a complete UWB channel characterisation.

References

[1] H.L. Bertoni, *Radio Propagation for Modern Wireless Systems*, Prentice Hall, 2000.
[2] J.D. Parsons, *The Mobile Radio Propagation Channel*, Second Edition, John Wiley and Sons, Inc., 2000.
[3] T.S. Rappaport, *Wireless Communications: Principles and Practice*, Second Edition, Prentice Hall, 2002.
[4] R. Vaughan and J.B. Andersen, *Channels, Propagation and Antennas for Mobile Communications*, IEE, UK, 2003.
[5] W. Stutzman and G. Thiele, *Antenna Theory and Design*, John Wiley and Sons, Inc., 1981.
[6] R.M. Buehrer, W.A. Davis, A. Safaai-Jazi and D. Sweeney, Ultra-wideband propagation measurements and modeling, *DARPA NETEX Program Final Report*, 31 January, 2004.
[7] IEEE 802.15 Working Group for WPAN, http://www.ieee802.org/15/.
[8] B. Allen, A.S. Ghorashi and M. Ghavami, A review of pulse design for impulse radio, *IEE Ultra Wideband Workshop*, London, June, 2004.
[9] J. Liu, M. Dohler, B. Allen and M. Ghavami, Power-loss modeling of short-range ultra wideband pulse transmissions, *IEE Proceedings on Communications*, vol. 153, no. 1, pp. 143–153, February 2006.

[10] M. Dohler, B. Allen, A. Armogida, S. McGregor, M. Ghavami and A.H. Aghvami, A new twist on UWB pathloss modelling, *IEEE VTC Spring*, Milan, Italy, May 2004, Conference CD-ROM.
[11] M. Dohler, B. Allen, A. Armogida, S. McGregor, M. Ghavami and A.H. Aghvami, A novel powerloss model for short range UWB transmissions, *IWUWBS*, Koyoto, Japan, May 2004, Conference CD-ROM.
[12] H. Bertoni, W. Honcharenko, L.R. Maciel and H. Xia, UHF propagation prediction for wireless personal communications, *Proceedings of the IEEE*, **82** (9), 1333–59, 1994.
[13] A. Fort, C. Desset, J. Ryckaert, P. de Doncker, L. van Biesen and S. Donnay, Ultra wide-band body area channel model, *IEEE International Conference on UWB*, Zurich, Switzerland, 2005.
[14] A. Toscano, F. Bilotti and L. Vegni, Fast ray-tracing technique for electromagnetic field prediction in mobile communications, *IEEE Transactions on Magnetics*, **39**(3), 1238–41, 2003.
[15] A. Falsafi, K. Pahlavan and G. Yang, Transmission techniques for radio LANs – a comparative performance evaluation using ray tracing, *IEEE Journal on Selected Areas in Communications*, **14**(3), pp. 477–91, 1996.
[16] S. Emami, C.A. Corral and G. Rasor, Ultra-wideband outdoor channel modeling using ray tracing techniques, *Second IEEE Consumer Communications and Networking Conference CCNC*, 3–6 January, 466–70, 2005.
[17] Y. Zhang and A.K. Brown, Ultra-wide bandwidth communication channel analysis using 3-D ray tracing, *1st International Symposium on Wireless Communication Systems*, 20–22 September, 443–7, 2004.
[18] H. Sugahara, Y. Watanabe, T. Ono, K. Okanoue and S. Yarnazaki, Development and experimental evaluations of rs-2000 - a propagation simulator for UWB systems, *Proc. IEEE UWBST 04*, 76–80, 2004.
[19] J. Kunisch and J. Pamp, Measurement results and modeling aspects for the UWB radio channel, *IEEE Conference on Ultra Wideband Systems and Technologies Digest of Technical Papers*, 19–23, 2002.
[20] A. Domazetovic, L.J. Greenstein, N.B. Mandayam and I. Seskar, A new modeling approach for wireless channels with predictable path geometries, *Proc. VTC 2003 fall*, 454–8, 2003.
[21] N.V. Venkatarayalu, C.-C. Chen, F.L. Teixeira and R. Lee, Numerical modeling of ultrawide-band dielectric horn antennas using FDTD, *IEEE Transactions on Antennas and Propagation*, **52**(5), 1318–23, 2004.
[22] K.Y. Yazdandoost and R. Kohno, Complex permittivity determination of material for indoor propagation in ultra-wideband communication frequency, *IEEE International Symposium on Communications and Information Technology, ISCIT*, **2**, 26–29 October, 1224–27, 2004.
[23] R.C. Qiu, A study of the ultra-wideband wireless propagation channel and optimum UWB receiver design, *IEEE Journal on Selected Areas in Communications*, **20**(9), 2002.
[24] J. Karedal, S. Wyne, P. Almers, F. Tufvesson and A.F. Molisch, Statistical analysis of the UWB channel in an industrial environment, *Proc. VTC fall 2004*, 2004.
[25] D. Cassioli, W. Ciccognani and A. Durantini, D3.1 - UWB channel model report, *Tech. Rep. IST-2001-35189-ULTRAWAVES*, November, 2003.
[26] A. Durantini, W. Ciccognani and D. Cassioli, UWB propagation measurements by PN sequence channel sounding, *Proc. IEEE Int. Conf. on Commun.*, Paris, France, June, 2004.
[27] B. Kannan et al., UWB channel characterization in office environments, *Tech. Rep. Document IEEE 802.15-04-0439-00-004a*, 2004.
[28] Y.K.C.C. Chong and S.S. Lee, A modified s-v clustering channel model for the UWB indoor residential environment, *Proc. IEEE VTC spring 05*, 2005.
[29] J. Keignart, J.-B. Pierrot, N. Daniele, A. Alvarez, M. Lobeira, J.L. Garcia, G. Valera and R.P. Torres, UCAN report on UWB basic transmission loss, *Tech. Rep. IST-2001-32710 UCAN*, March, 2003.
[30] B. Kannan et al., UWB channel characterization in outdoor environments, *Tech. Rep. Document IEEE 802.15-04-0440-00-004a*, 2004.
[31] V. Erceg et al., An empirically based path loss model for wireless channels in suburban environments, *IEEE. J. Sel. Areas Commun.*, 1205–11, 1999.
[32] S. Ghassemzadeh, R. Jana, C. Rice, W. Turin and V. Tarokh, Measurement and modeling of an ultra-wide bandwidth indoor channel, *IEEE Trans. on Commun.*, 1786–96, 2004.
[33] R.M. Buehrer, A. Safaai-Jazi, W.A. Davis and D. Sweeney, Characterization of the UWB channel, *Proceedings of IEEE Conference on Ultra-Wideband Systems and Technologies*, 26–31, Reston, VA, November, 2003.
[34] B. Donland and R.M. Buehrer, The indoor UWB channel, *Proceedings of the Spring 2004 Vehicular Technology Conference*.
[35] R. Qiu and I.-T. Lu, Wideband wireless multipath channel modeling with path frequency dependence, *IEEE International Conference on Communications (ICC96)*, 1996.

[36] R.C. Qiu and I. Lu, Multipath resolving with frequency dependence for broadband wireless channel modeling, *IEEE Trans. Veh. Tech.*, 1999.
[37] V.S. Somayazulu, J.R. Foerster and S. Roy, Design challenges for very high data rate UWB systems, *Conference Record of the Thirty-Sixth Asilomar Conference on Signals, Systems and Computers*, **1**, 3–6 November, 717–21, 2002.
[38] A. Alvarez, G. Valera, M. Lobeira, R. Torres and J.L. Garcia, New channel impulse response model for UWB indoor system simulations, *Proceedings of the Spring 2003 Vehicular Technology Conference*, 1–5, 2003.
[39] B. Donlan and R.M. Buehrer, Large and small scale channel modeling for indoor UWB channels, submitted to *IEEE Trans. Wireless Commun.*, in press.
[40] J. Foerster, Channel, modeling sub-committee report final (doc.: IEEE 802-15-02/490r1-SG3a), submitted to IEEE P802.15 Working Group for Wireless Personal Area Networks (WPANs), February, 2002. Available: http://grouper.ieee.org/groups/802/15/pub/2002/Nov02/.
[41] C. Prettie, D. Cheung, L. Rusch and M. Ho, Spatial correlation of UWB signals in a home environment, *Proceedings of IEEE Conference on Ultra Wideband Systems and Technology*, Baltimore, MD, 2002.
[42] S.S. Ghassemzadeh, R. Jana, C.W. Rice, W. Turin and V. Tarokh, A statistical path loss model for in-home UWB channels, *Proceedings of IEEE Conference on Ultra Wideband Systems and Technology*, Baltimore, MD, 2002.
[43] S.S. Ghassemzadeh and V. Tarokh, UWB path loss characterization in residential environments, *Proceedings of IEEE Radio Frequency Integrated Circuits Symposium*, 2003.
[44] S.S. Ghassemzadeh, L.J. Greenstein and V. Tarokh, The ultra-wideband indoor multipath loss model (doc: IEEE P802.15-02/282-SG3a and IEEE P802.15-02/283-SG3a), submitted to IEEE P802.15 Working Group for Wireless Personal Area Networks (WPANs), June, 2002. Available: http://grouper.ieee.org/groups/802/15/pub/2002/Ju102/.
[45] D. Cassioli, A. Durantini and W. Ciccognani, The role of path loss on the selection of the operating bands of UWB systems, *Proc. IEEE Int. Symp. on Personal, Indoor and Mobile Radio Communications*, September, Barcelona, Spain, 2004.
[46] A.F. Molisch, J.R. Foerster and M. Pendergrass, Channel models for ultrawideband personal area networks, *IEEE Personal Communications Magazine*, **10**, 1421, 2003.
[47] Andreas F. Molisch, Kannan Balakrishnan, Dajana Cassioli, Chia-Chin Chong, Shahriar Emami, Andrew Fort, Johan Karedal, Juergen Kunisch, Hans Schantz, Ulrich Schuster and Kai Siwiak, *IEEE 802.15.4a channel model – Final report*, IEEE802.15.4a working group, 2005.

15

Small-scale Ultra-wideband Propagation Modelling

Swaroop Venkatesh, R. Michael Buehrer, Junsheng Liu and Mischa Dohler

15.1 Introduction

In a communication system, the received signal is an attenuated, delayed and possibly distorted version of the transmitted signal plus noise. The relationship between the received signal and the transmitted signal (ignoring the noise) is called the 'channel'. In general, there are two ways to model electromagnetic (EM) wave propagation. The first type of modelling is often termed 'site-specific' modelling or 'deterministic' modelling. This technique attempts to model the exact interaction of the EM wave with the specific environment of interest. The second type, called 'statistical modelling', attempts to model the relevant statistics of the received signal. These statistics are based on the type of environment assumed and the bandwidth of the signal used. Statistical modelling is particularly useful in communication system development where the system is required to work in a wide variety of environments, and it is this type of modelling which we discuss in this chapter.

In the modelling of the UWB channel, we typically divide the effects of propagation into two categories: (a) large-scale effects and (b) small-scale effects. The phrase 'large scale' typically refers to the impact that the channel has on the transmit signal over large distances and generally includes attenuation effects due to distance and objects which are in the propagation path. These aspects of the channel were discussed in the previous chapter. The small-scale characteristics include the small-scale fading in a local environment, as well as the distortion of the transmitted waveform due to multipath. Since the channel modelling discussed in this chapter is applicable to short-range ($d < 10$ m) high data rate (HDR) applications and long range low data rate (LDR) applications, we will typically restrict the term 'small scale' to refer to signal variations within a 1 m^2 area. Large-scale fading will refer to variation of the received signal over distances greater than 1 m.

This chapter is organised as follows: Section 15.2 discusses the small-scale modelling of the UWB channel in the time domain and several traditional models for channel impulse responses are characterised,

---

Ultra-wideband Antennas and Propagation for Communications, Radar and Imaging Edited by B. Allen, M. Dohler, E. E. Okon, W. Q. Malik, A. K. Brown and D. J. Edwards
© 2007 John Wiley & Sons, Ltd

with special attention given to the Saleh–Valenzuela [24] model. A discussion of the spatial variation of UWB signals in indoor scenarios is presented in Section 15.3. We present an overview of the small-scale modelling components of the IEEE 802.15.3a and IEEE 802.15.4a standard channel models in Sections 15.4 and 15.5 respectively, based on the discussed concepts. The contents of the chapter are summarised in Section 15.6.

15.2 Small-scale Channel Modelling

In Chapter 14, we discussed the statistical modelling of the *large-scale* characteristics of the received signal. This is primarily concerned with the loss in received signal power versus distance (and possibly frequency) between the transmitter and receiver. It was shown that UWB presents some interesting challenges and care must be taken when applying a model to a UWB communications system.

While the statistical properties of the large-scale characteristics are necessary for proper link budget design, the statistical properties of the small-scale characteristics are needed for efficient receiver design. Specifically, understanding small-scale fading and signal correlation over a small area aids in the evaluation of the receiver, diversity mechanisms and potential multiple antenna applications. Additionally, understanding the waveform distortion suffered by a UWB signal is vital to designing the modulation scheme, the signal demodulator and detector as well as for evaluating various receiver architectures. These properties are typically analysed by characterising the *channel impulse response* of the propagation environment.

It must be noted that if the angular dependencies of the antennas (and thus the vector nature of the problem) are subsumed into the effects of the channel, the impulse response that is calculated is not strictly only due to the channel. Typically, the antenna effects are not easily disentangled from the received measurements. As a result, very often, some antenna effects are inherently part of the channel model. For example, the measured channel impulse response for directional antennas (such as TEM horn antennas) would not be influenced significantly by reflectors/diffractors behind the antennas. However, the channel impulse response for omnidirectional (in one plane only) antennas would be affected to a greater extent by reflectors or diffractors in the environment than the channel impulse response for directional antennas. Often, only the antenna response at boresite is considered, which assumes that the pulse distortion due to the antenna is the same in all transmit/receive directions. While this is a strong assumption, it is justified by the fact that generally the strongest paths are those that are transmitted and received via boresite. Broadly speaking, all omnidirectional antennas behave in a similar fashion, and all directional antennas display similar patterns of behaviour. Therefore, the characterisation of the 'channel' for antennas such as TEM horns and bicones should be performed separately, and may be representative of many other antenna structures.

15.2.1 Statistical Characterisation of the Channel Impulse Response

The small-scale channel is most often modelled as a time-varying linear filter, where the received signal is given by [25]:

$$r(t) = h(t, \tau) \otimes s(t) + n(t)$$

In the above equation, $s(t)$ is the transmitted signal, which is assumed to be the received LOS pulse in the case of UWB (due to our inability to extricate the actual transmit pulse from the antenna response), $h(t,\tau)$ is the time-varying channel impulse response, $n(t)$ is additive noise, and $\otimes$ denotes convolution. The propagation delay can be incorporated into the channel impulse response. The impulse response can change as a function of time (or as a function of spatial variation) due to the motion of the transmitter or receiver and/or changes in the channel itself. The channel impulse response is typically modelled using a tap-delay line approach [25, 31]. Thus, the channel model can be written as

$$h(t, \tau) = \sum_{k=1}^{L(t)} \beta_k(t) \delta(\tau - \tau_k(t))$$

where $\beta_k(t)$ and $\tau_k(t)$ are the time-varying amplitude and delay of the kth path respectively, t is the observation time and τ is the application time of the impulse. $L(t)$ is the number of multipath components, which is also, in general, a function of time. Note that for narrowband (or wideband) bandpass channel models, an additional phase term $e^{j\theta(t)}$ is introduced, whereas in UWB modelling, the phase is typically ignored, since it is assumed that baseband pulses are being sent. If the pulses are up-converted or down-converted, a complex baseband model may be appropriate and the phase term should be reintroduced. If the channel is assumed to be static over the time interval of interest, the time-invariant model of the channel can be used:

$$h(\tau) = \sum_{k=1}^{L} \beta_k \delta(\tau - \tau_k). \tag{15.1}$$

The main goal of small-scale channel models is to statistically characterise the amplitudes, delays and polarities of the multipath components of the channel. Besides specific statistical characterisation of the multipath parameters, the channel can also be characterised by coarse statistics such as mean excess delay, RMS delay spread and maximum excess delay that describe the time-dispersive properties of the channel. These are useful as single-number descriptions of the channel to estimate the performance and potential for intersymbol interference. These values tend to increase with greater transmitter/receiver separation [4]. The mean excess delay of the channel $h(\tau)$ defined in (15.1) is defined as [25]:

$$\tau_m = \frac{\sum_k \beta_k^2 \tau_k}{\sum_k \beta_k^2} \tag{15.2}$$

The RMS delay spread of $h(\tau)$ is defined as

$$\tau_{RMS} = \sqrt{\frac{\sum_k \beta_k^2 \tau_k^2}{\sum_k \beta_k^2} - \tau_m^2} \tag{15.3}$$

The number of significant multipath components that form the channel is another metric that can be used to characterise the time-dispersion of the channel. The number of paths, $N_P(S)$, is defined as the number of multipath components with amplitudes that are greater than $-S$ dB relative to the strongest multipath component. Another important metric is the number of paths $N_E(m)$ that capture m % of the total multipath energy.

The key to any model, which assumes the discrete channel model presented above, is the joint statistical characterisation of the path amplitudes and delays. This first step is therefore to extract the discrete channel impulse response $h(\tau)$, given the received signal and the transmit pulse. This is briefly discussed in the next section. For a more detailed treatment, the reader is referred to [35]. Various wireless statistical channel models are discussed in subsequent sections. Although these models were originally developed to characterise indoor wideband channels, it has been found that the general approaches are also useful for UWB channels.

15.2.2 Deconvolution Methods and the Clean Algorithm

As mentioned previously, the small-scale variations of the channel are typically characterised through the channel impulse response. Channel modelling depends on how the channel impulse response and other channel parameters are obtained from the measurements. Deconvolution is the process of separating two signals that have been combined by convolution, and can be used to extract the channel impulse response from the received signal given the transmitted pulse shape. Several deconvolution techniques have been developed [45], often for specific types of signals or for use with a specific application. Deconvolution can be performed in the time domain or the frequency domain. Different techniques emphasise different aspects of the deconvolved signal and can offer different advantages depending on what further analysis is desired.

A time-domain deconvolution technique often considered in UWB measurement analysis is the CLEAN algorithm [35, 48, 53], which does provide an impulse response similar to Equation (15.1). The algorithm is important in channel modelling because it assumes that the impulse response being found is not band-limited, but is rather a sum of scaled and time-delayed impulses, which is consistent with classic small-scale channel models.

15.2.3 The Saleh-Valenzuela Model

The most common statistical model for the discrete indoor channel impulse response is the Saleh–Valenzuela model [33]. The physical intuition behind this model is that groups or 'clusters' of multipath components arrive at the receiver due to the interaction of the transmitted signal with scatterers in the propagation environment.[1] This model for multipath consists of two processes: a cluster-arrival process and a ray-arrival process. The cluster-arrival process defines the arrival times and the amplitudes of the clusters of multipath components, and the ray-arrival process determines the arrival times and amplitudes of multipath components or 'rays' within a cluster. In this sense, the model is essentially a 'clustered-point process' as shown in Figure 15.1.

Mathematically, the Saleh–Valenzuela model is given by

$$h(t) = \sum_{l=0}^{L} \sum_{k=0}^{K} \beta_{k,l} \delta\left(t - T_l - \tau_{k,l}\right)$$

where T_l is the arrival time of lth cluster, $\beta_{k,l}$ and $\tau_{k,l}$ are respectively the amplitude and arrival time of kth ray within the lth cluster, L is the number of clusters and K is the maximum number of rays within a cluster. The cluster arrivals are described by a Poisson process, and thus the cluster inter-arrival times, T_l, are described by exponential random variables:

$$p\left(T_l | T_{l-1}\right) = \Lambda \exp\left[-\Lambda\left(T_l - T_{l-1}\right)\right], \quad l > 0$$

where Λ is the mean cluster arrival rate. Within the lth cluster, the ray arrivals are also described by a Poisson process, so that the distribution of the interarrival times is described using

$$p\left(\tau_{k,l} | \tau_{(k-1),l}\right) = \lambda \exp\left[-\lambda\left(\tau_{k,l} - \tau_{(k-1),l}\right)\right], \quad k > 0$$

where λ is the mean ray arrival rate. The average power of both the clusters and the rays within the clusters are assumed to decay exponentially, such that the average power of a multipath component at a

---

[1] It should be noted that although the Saleh–Valenzuela model was developed for NLOS channels, it has also been applied to LOS channels where it is less valid [13].

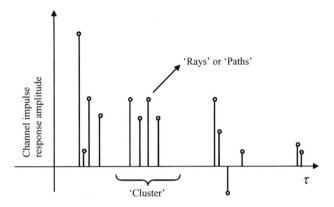

Figure 15.1 Illustration of the Saleh–Valenzuela model

given delay, $T_l + \tau_{k,l}$, is given by

$$\langle \beta_{k,l}^2 \rangle = \langle \beta_{0,0}^2 \rangle e^{-\frac{T_l}{\Gamma}} e^{-\frac{\tau_{k,l}}{\gamma}} \tag{15.4}$$

where $\langle \beta_{k,l}^2 \rangle$ is the expected value of the power of the first arriving multipath component, Γ is the decay exponent of the clusters and γ is the decay exponent of the rays within a cluster. This expression defines the power-delay profile (PDP) of the model. In addition to the average power decay, a common practice with small-scale channel models is to normalise the channel impulse responses, such that the *average power* is unity.

The amplitude of each path is assumed to be a random variable about an exponential mean. Several distributions have been proposed for the amplitudes, although the original model assumed a Rayleigh distribution. The polarity is assumed to be a binary random variable, taking on the values ± 1 with equal probability. When a random phase is used, the phase is assumed to be uniformly distributed. The Rayleigh-distributed amplitudes and random phase model come from the assumption that several paths arrive at delays that are not resolvable to the measurement system used. This is a more valid assumption for wideband channels, but is questionable for UWB channels. As a result, many researchers have found that log-normal or Nakagami distributions provide a better fit [8, 53, 59].

The use of log-normal random variables is especially convenient when they are also used for the cluster amplitudes. This implies two independent log-normal variables to represent the amplitude variations of the clusters (σ_1) and rays (σ_2). However, these random variables can be combined as a single log-normal random variable with standard deviation σ such that $\sigma^2 = \sigma_1^2 + \sigma_2^2$. The polarity of the path is represented as an equiprobable binary random variable, $p_{k,l}$, taking on the values ± 1. For log-normal amplitudes, the path amplitudes (and polarity) are given by

$$\beta_{k,l} = p_{k,l} 10^{(\mu_{k,l} + X_{\sigma,k,l})/20}, \tag{15.5}$$

where

$$\mu_{k,l} = \frac{10 \log \left(\langle \beta_{0,0}^2 \rangle \right) - \frac{10 T_l}{\Gamma} - \frac{10 \tau_k}{\gamma}}{\log 10} - \frac{\sigma^2 \log 10}{20}$$

and

$$X_{\sigma,k,l} = N(0, \sigma^2) \quad (\sigma \text{ is in dB}).$$

To summarise, this model is described by five parameters:

- Λ the mean cluster arrival rate;
- λ is the mean ray arrival rate;
- Γ is the cluster exponential decay factor;
- γ is the ray exponential decay factor;
- σ is the standard deviation of the log-normal distributed path powers (combining the log-normal variations of the clusters (σ_1) and rays (σ_2)).

We examine the impact that the parameters of the Saleh–Valenzuela model have on the statistics of the RMS delay spread, the mean excess delay and the number of significant multipath components $N_P(10)$. For the Saleh–Valenzuela model, the parameters that may be varied are $\{\lambda, \Lambda, \gamma, \Gamma, \sigma\}$. Figures 15.2, 15.3 and 15.4 show the effect of doubling the parameters $\{\gamma, \Gamma, \sigma\}$ individually from the 'original' parameters given in Table 15.1.

We observe the following trends:

1. Increasing Γ (time-constant of exponential energy decay of clusters) increases the RMS delay spread, mean excess delay and the number of paths. It is also decreases the slopes of the cumulative density

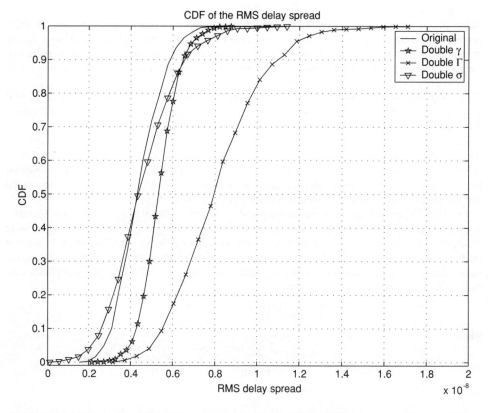

Figure 15.2 Effect of doubling SV parameters $\{\gamma, \Gamma, \sigma\}$ on CIR RMS delay spread (source: [53])

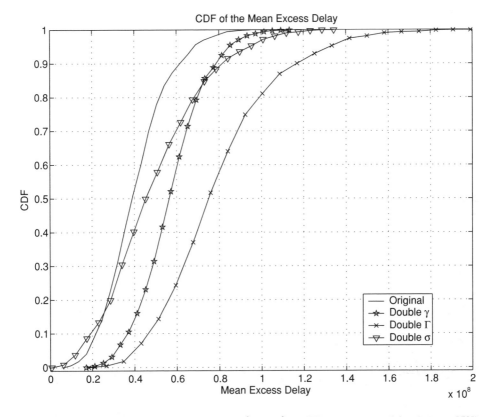

Figure 15.3 Effect of doubling SV parameters $\{\gamma, \Gamma, \sigma\}$ on CIR mean excess delay (source: [53])

functions (CDFs) of the RMS delay spread and mean excess delay. This can be interpreted as a larger region of support for the PDF, i.e. a larger range of values.

2. Increasing γ (time-constant of exponential energy decay of rays) increases the RMS delay spread, mean excess delay and the number of paths, since it takes a longer time for the paths to die out.
3. Increasing σ (proportional to the standard deviation of small-scale variation) decreases the slope of the CDF curves of the RMS delay spread and the mean excess delay. It also decreases the number of paths. This is because due to a large variance in amplitude, a larger number of multipath components fall below the 20 dB threshold set for this example.

Figures 15.5, 15.6 and 15.7 show the effect of doubling the parameters $\{\lambda, \Lambda\}$ individually from the original parameters shown in Table 15.1. We observe the following trends:

1. Increasing Λ (cluster arrival rate) does not significantly impact the RMS delay spread, but increases the mean excess delay and the number of paths.
2. Increasing λ (arrival rate of rays) does not significantly impact either the RMS delay spread or the mean excess delay but increases the number of paths. As mentioned previously, changing the arrival rate does not affect the energy distribution of paths, which impacts the RMS delay and the mean excess delay.

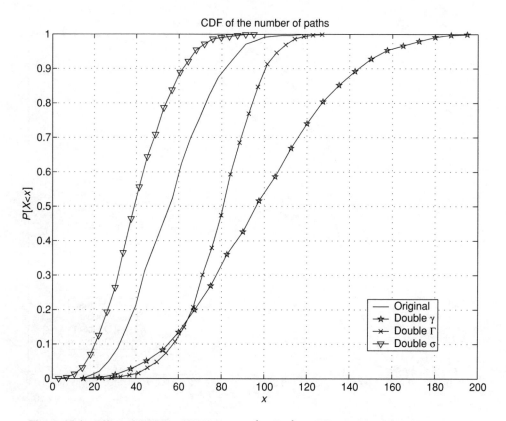

Figure 15.4 Effect of doubling SV parameters $\{\gamma, \Gamma, \sigma\}$ on CIR number of paths (source: [53])

15.2.4 Other Temporal Models

The Saleh–Valenzuela model is one of the most popular temporal models used for indoor wireless communications. However, there are some other models that have been applied to these situations, and these are discussed below.

Table 15.1 'Original' Saleh–Valenzuela model parameters for examining the impact of parameter variation

Parameter	Value
$\frac{1}{\Lambda}$	5 nsec
$\frac{1}{\lambda}$	1 nsec
Γ	5 nsec
γ	2 nsec
σ	4

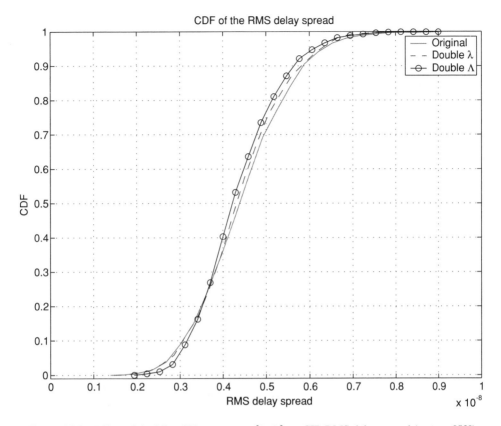

Figure 15.5 Effect of doubling SV parameters $\{\lambda, \Lambda\}$ on CIR RMS delay spread (source: [53])

15.2.4.1 Δ-K Model

The Δ-K model [18,31], like the Saleh–Valenzuela model, is based on the assumption that multipath components arrive in clusters. The probability that a path arrives at any given delay is higher by a factor of K if a path has arrived within the past Δ seconds. By increasing the arrival rate when a path has recently arrived, the paths tend to arrive in clusters. The arrival times thus follow a modified two-state Poisson process and the interarrival times follow an exponential distribution where the arrival rate is based on the state. When in the first state (S1), the mean arrival rate is given by λ. The transition to the second state (S2) is triggered when a path occurs. In (S2), the mean arrival rate is given by $K\lambda$. After Δ seconds, if a path has not arrived, a transition back to (S1) occurs.

Exponential energy decay is assumed in the PDP to describe the expected value of the energy in a path at a given delay. The polarity of paths is assumed to be ± 1 with equal likelihood, and amplitude fading is assumed to be log-normal such that the amplitude of a path is given by Equation (15.5). Here,

$$\mu_k = \frac{20 \ln\left(\overline{|\beta_0|}\right) - 10\tau_k/\gamma}{\ln(10)} - \frac{\sigma^2 \ln(10)}{20}$$

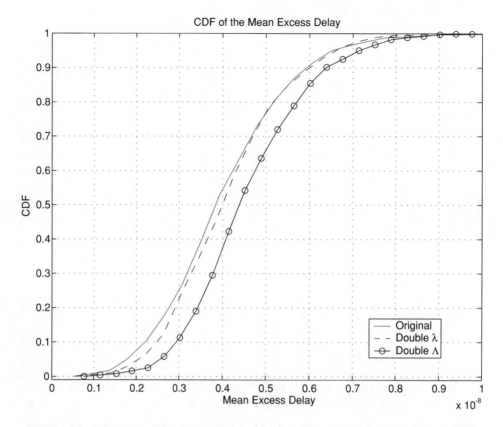

Figure 15.6 Effect of doubling SV parameters $\{\lambda, \Lambda\}$ on CIR mean excess delay (source: [53])

and

$$X_{\sigma,k} = N(0, \sigma^2) \quad (\sigma \text{ is in dB})$$

In the discrete version of the above model, the time axis is divided into bins, and the probability of a path arriving in a given bin is based on whether a path arrived in the previous bin (probability being higher by a factor of K if a path was present).

15.2.4.2 Single Poisson Model

This is a simplified version of both of the previously introduced models and assumes that only one cluster is present in the impulse response, indicative of the lack of clustering of multipath components. The arrivals of paths are treated as a Poisson process with arrival rate λ. Also, the decay of the paths is assumed to be exponential, with a decay time-constant γ. The amplitudes of the paths are modelled using a log-normal random variable with parameter σ. This simple model proves to be useful in environments containing a small number of scatterers, such as outdoor environments.

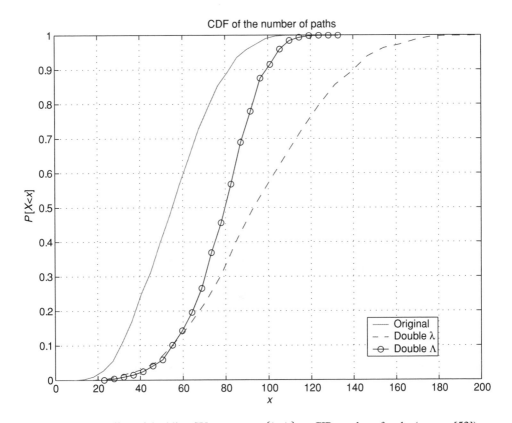

Figure 15.7 Effect of doubling SV parameters $\{\lambda, \Lambda\}$ on CIR number of paths (source: [53])

15.2.4.3 Two-Cluster or Split-Poisson Model

The Saleh–Valenzuela model is based on the generation of multiple exponentially decaying clusters. However, a sizeable portion of the measurement data indicates very few clusters may exist, perhaps due to the limited range used for UWB systems and measurements. An example taken from [53] is shown in Figure 15.8. As can be seen, the data seems to indicate that the average channel impulse response (CIR) consists of two clusters, the first short cluster containing several strong paths which decay quickly, and the longer second cluster containing slowly decaying paths.

Based on these observations, the split-Poisson or two-cluster model was proposed in [53, 56, 57]. The model assumes two clusters of Poisson arrivals, one delayed by t_1 relative to the other. The first cluster is generated using a set of parameters, i.e., $\lambda_1, \sigma_1, \gamma_1$, while the second cluster is generated using a separate set of parameters, i.e., $\lambda_2, \sigma_2, \gamma_2$. The overall CIR is created by adding a delayed version of the second cluster to the first cluster. Also, in order to maintain continuity in the decay of energy in the overall CIR, the first cluster is weighted higher than the second cluster by a factor α. This is shown in Figure 15.9. Each of these parameters is estimated from the data as described in [56, 57]. The two-cluster model was found to be a good match to measurement data [57] in terms of Rake-receiver energy capture, compared to other models. Figure 15.10 shows a comparison of the average Rake receiver energy capture

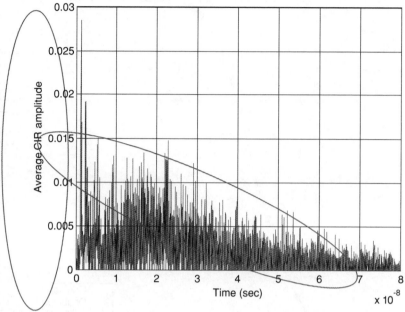

Figure 15.8 Average CIR amplitude over one location from example measurement – two distinct clusters are visible (source: [56] reproduced by permission of © 2004 IEEE)

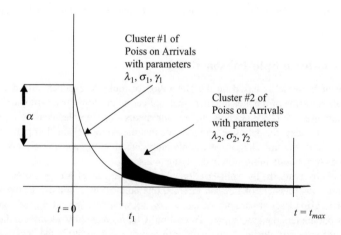

Figure 15.9 Illustration of the two-cluster model (source: [56] reproduced by permission of © 2004 IEEE)

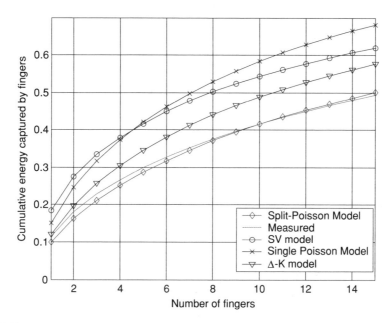

Figure 15.10 Average cumulative Rake finger energy capture for measured indoor bicone NLOS data and various models for different numbers of RAKE fingers (source: [56] reproduced by permission of © 2004 IEEE)

versus the number of Rake correlators or 'fingers' for the measured and model-based received signals in an indoor/office environment. We see that the two-cluster model closely matches the measured data in this respect. The details of this model can be found in [56]. This figure also demonstrates a typical problem with receiver design for UWB channels: UWB channels exhibit extreme energy capture. In the plot, we see that a Rake receiver with 15 fingers only captures approximately 50 % of the energy in this NLOS scenario. This is not uncommon and points to a challenge associated with the use of time-domain receivers for UWB signals, namely limited energy capture.

15.3 Spatial Modelling

The previous sections have described the small-scale time-domain variations of the UWB channel. Specifically, the discussion focused on the time-domain statistics of the received signal and models that can mimic those statistics. Another important aspect of wireless channels is spatial variation: it is important to know how the received signal varies in the local area, both in time and space. These characteristics are important for understanding applications which attempt to exploit multiple antennas [64] as well as for understanding the channel behaviour in a mobile scenario.

In the time domain, the short duration of the transmitted pulses considerably reduces the interaction between multipath components when compared to narrowband signals. This substantially reduces multipath fading in the time domain. Spatial fading (sometimes also called *local* fading) refers to the variation of the received power over a local area. The spatial fading characteristics (i.e. the fading variation in a local area) of UWB signals were examined in [10–12] and [53]. In order to quantify spatial fading, we compare the entire received signal energies over a local area, which is equivalent to comparing received

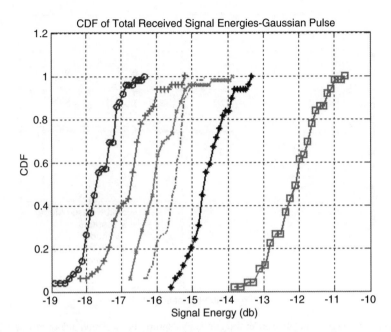

Figure 15.11 Estimated CDFs for the total received signal energy at six measurement locations (source: [53])

signal powers for a fixed observation interval. Figure 15.11 shows a typical set of CDFs of the received signal energies over a 1 m² measurement grid at various measurement locations. Biconical antennas were used, and the 10 dB bandwidth of the transmitted pulse was 3.5 GHz. Each measurement location had a different distance between the transmit antenna and the centre of the receiver location grid. Further details of this experiment can be found in [53]. Figure 15.11 shows that the received signal energy over the entire measurement grid does not vary by more than 4 dB. This result is typical of UWB signals and is in dramatic contrast to spatial fading observed in narrowband signals [25]. However, it must be noted that most practical receivers cannot capture the entire received signal [63] energy due to limitations in computational complexity, thus tempering this benefit to some degree.

A classic receiver structure for channels with resolvable multipath is the Rake receiver [38]. Since a Rake receiver has a limited number of correlators and thus only captures a fraction of the energy, the fading seen by a Rake receiver may be very different from the fading observed over the entire signal. While the total received signal power may exhibit little variation, this may be irrelevant if a receiver cannot capture the entire received signal energy. Figure 15.12 shows the variation of the received signal energy captured over a measurement grid for different numbers of Rake fingers [53]. We observe that for a Rake receiver with a single finger, the variation in the captured energy over the entire measurement grid can be quite large, when compared with a receiver which captures all the received signal energy shown in Figure 15.11. Therefore, UWB communication links are likely to be robust to spatial multipath fading when a significant fraction of the total received signal is captured by the receiver.

A Rake receiver with a finite number of correlators is likely to capture the dominant multipath components in the multipath profile. For LOS scenarios, this would correspond to the earliest arriving paths

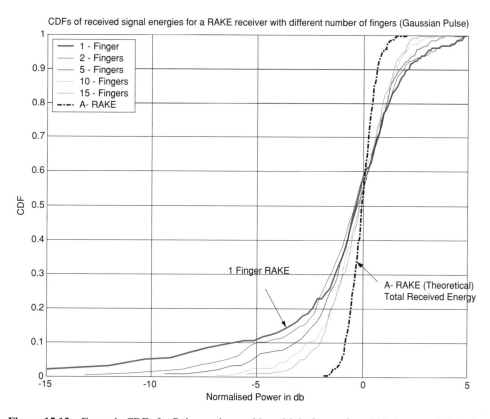

Figure 15.12 Example CDFs for Rake receivers with multiple fingers for a 200 picosecond Gaussian pulse; measurements for each pulse are normalised to unit average energy (source: [53])

at the receiver. Figure 15.13 shows the correlation [53] between different portions of the CIR measured over a grid versus distance. The correlation $\xi(d)$ between measured signals at different locations as a function of distance is defined as:

$$\xi(d) = \int_{\tau_1}^{\tau_2} r(\tau, \mathbf{x}_j) r(\tau, \mathbf{x}_j) dt$$

where $r(\tau, \mathbf{x}_i)$ and $r(\tau, \mathbf{x}_j)$ are respectively, the received signals at grid points $\mathbf{x}_i$ and $\mathbf{x}_j$, and $d = \|\mathbf{x}_i - \mathbf{x}_j\|$.

We observe that the dominant multipath components, which are likely to be contained in the initial portion of the CIR, are highly correlated, whereas the subsequent diffuse multipath shows very little correlation. This result has important implications for indoor multi-antenna applications: low spatial correlation is observed when we consider the entire received signal. However, if the receiver captures only the early, dominant components, a much larger antenna separation is required to obtain low spatial correlation.

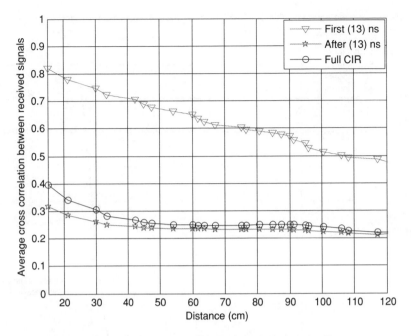

Figure 15.13 The variation of spatial correlation versus distance

15.4 IEEE 802.15.3a Standard Model

The IEEE 802.15.3a task group evaluated a number of indoor channel models to determine which model best fits the important characteristics from realistic channel measurements using UWB waveforms designed for short-range high data rate (HDR) applications. The goal of the channel model is to capture the multipath characteristics of typical environments where IEEE 802.15.3a devices are expected to operate. Another requirement was that the model should be relatively simple to use in order to allow easy evaluation of different proposals in terms of their performance in typical operational environments.

As shown previously, it may be impossible to create a single model that reflects all of the possible channel environments and characteristics. The standard model was formulated using several sets of measurement data [13] which represented a wide variety of operational scenarios, from residential to office-type environments in both LOS and NLOS conditions. The IEEE 802.15.3a committee categorised these measurements into the following four general groups, each representing different UWB WPAN operational environments:

1. LOS, 0–4 m (CM1);
2. NLOS, 0–4 m (CM2);
3. NLOS, 4–10 m (CM3);
4. Very high multipath NLOS channel (CM4).

The attempt was then to match the following primary characteristics of the multipath channel:

- RMS delay spread τ_{RMS} (defined in Equation (15.3));
- mean excess delay τ_m (defined in Equation (15.2));

- mean number of paths that capture 85 % of the energy in the channel $N_E(85\%)$;
- number of multipath components $N_P(10)$ (defined as the number of multipath arrivals that are within 10 dB of the strongest multipath arrival).

Three main indoor channel models were considered: the tap-delay line Rayleigh fading model [13], the Saleh–Valenzuela model, and the Δ-K model described in [36], as well as several novel modifications to these approaches that better matched the measurement characteristics. Each channel model was parameterised in order to best fit the important channel characteristics described above. Although many good models were contributed to the group, the model finally adopted was based on a modified Saleh–Valenzuela model that seemed to best fit the channel measurements.

The Saleh–Valenzuela model was modified for the IEEE 802.15.3a model by prescribing a log-normal amplitude distribution for the cluster and multipath amplitudes. The model also includes a shadowing term to account for total received multipath energy variation which results from the blockage of the LOS path. The small-scale channel model is defined by the impulse response given by:

$$h(t) = X \sum_{l=1}^{L} \sum_{k=1}^{K} \beta_{k,l} \delta(t - T_l - \tau_{k,l}), \tag{15.6}$$

where the random variable X is called the *shadowing factor*. The standard deviation, σ_X, of the shadowing factor is termed the *shadowing depth*. The shadowing depth σ_X was set to 3 dB, which implies that $20\log_{10} X = \sigma_X Y = 3Y$, where Y is a Gaussian random variable with zero mean and unit variance. It must be noted, since X captures the lognormal shadowing of the total multipath energy, the total energy of the paths $\beta_{k,l}$ is normalised to unity for each channel realisation.

Table 15.2 compares the target channel characteristics and the results from the proposed model for several different channel types. The channel characteristics and corresponding parameter matching results in Table 15.2 correspond to a time resolution of 167 ps (corresponding to the 6 GHz bandwidth of the underlying measurements). Comparing the target channel characteristics with the model characteristics in Table 15.2, we see a good fit between the model and measurement characteristics.

15.5 IEEE 802.15.4a Standard Model

The main goal of the IEEE 802.15.4a working group is to define an amendment to 802.15.4 for an alternative PHY [60, 61], designed for long-range (tens of metres), low data rate (1 kbps to several Mbps) (LDR) applications. The other requirements include high precision ranging, location capability (within 1 metre accuracy), high aggregate throughput, ultra low power and adding scalability to data rates, longer range and lower power consumption and cost. The possible application environments for the WPANs defined in 802.15.4a include:

- indoor residential;
- indoor office;
- industrial environment;
- body area network (BAN);
- outdoor;
- agricultural areas/farms.

A fair comparison of different proposals submitted to the 802.15.4a working group necessitates a standard channel model. It is important to note that no specific technology or frequency range is mandated

Table 15.2 Example multipath channel characteristics and corresponding model parameters

Target channel characteristics	CM 1[1]	CM 2[2]	CM 3[3]	CM 4[4]
Mean excess delay (nsec) (τ_m)	5.05	10.38	14.18	
RMS delay (nsec) (τ_{ms})	5.28	8.03	14.28	25
$N_P(10)$			35	
N_E (85%)	24	36.1	61.54	
Model parameters				
Λ (1/nsec)	0.0233	0.4	0.0667	0.0667
λ (1/nsec)	3.5	1	3	3
Γ	7.1	5.2	14.00	24.00
γ	5	6.5067	8.5	12
$\sigma_1(dB)$	3.3941	3.3941	3.3941	3.3941
$\Sigma_2(dB)$	3.3941	3.3941	3.3941	3.3941
Model characteristics				
Mean excess delay (nsec) (τ_m)	5.5	9.2 ns	14.9	26.3
RMS delay (nsec) (τ_{ms})	6	8	14	25
$N_P(10)$	14.9	22.0	31.7	43.8
$N_E(85\%)$	23.4	35.7	60.8	115.5

[1] This model is based on LOS (0–4 m) channel measurements reported in [49].
[2] This model is based on NLOS (0–4 m) channel measurements reported in [49].
[3] This model is based on NLOS (4–10 m) channel measurements reported in [49], and NLOS measurements reported in [14].
[4] This model was generated to fit a 25 nsec RMS delay spread to represent an extreme NLOS multipath channel.

by the 802.15.4a working group. However, UWB technology remains a strong candidate and thus the channel model recommended in the final report [60] for the IEEE 802.15.4a channel model is, almost exclusively, an ultra-wideband channel model.

The pathloss and shadowing components of the general model proposed in [60] were discussed in Chapter 14. We present a brief overview of the small-scale aspects of the model. The impulse response defined in the IEEE 802.15.4a channel is, like the 802.15.3a model, based on the Saleh–Valenzuela model. The number of the clusters of multipath components, L, obeys the Poisson distribution with the PDF:

$$p_L(L) = \frac{(\bar{L})^L \cdot \exp(-\bar{L})}{L!},$$

where $\bar{L}$ is the mean of L. The cluster arrival times, T_l, are given by a Poisson process, exactly as in 802.15.3a channel model. The ray arrival times are modelled by a mixture of two Poisson processes as follows:

$$p\left(\tau_{k,l}|\tau_{(k-1),l}\right) = \alpha\lambda_1 \exp\left[-\lambda_1\left(\tau_{k,l} - \tau_{(k-1),l}\right)\right] + (\alpha - 1)\lambda_2 \exp\left[-\lambda_2\left(\tau_{k,l} - \tau_{(k-1),l}\right)\right] \quad k > 0$$

The PDP is defined using the cluster decay constant, Γ, and the ray decay constant, γ, of the lth cluster. The double exponentially decaying power profile of the 15.4a channel model is once again similar to that of the 802.15.3a model defined in Equation (15.6). The Nakagami distribution is chosen to describe the amplitudes of the multipath components in the 802.15.4a channel mode with PDF $p(x, m, \Omega)$ defined by,

$$p(x, m, \Omega) = \frac{2}{\Gamma(m)} \left(\frac{m}{\Omega}\right)^m x^{2m-1} \exp\left(-\frac{m}{\Omega} x^2\right)$$

where m is the Nakagami m-factor, and Ω is the mean-square value of the random variable. The parameter m is modelled as a log-normal random variable, i.e., $\log(m) \in N(\mu_m, \sigma_m)$. Other important parameters describing the 802.15.4a channel model include excess delay and RMS delay, which are exactly the same as those of the 802.15.3a channel model.

15.6 Summary

In this chapter, we have discussed the statistical small-scale modelling of UWB channels. A discussion of the fundamental techniques and models used in this type of modelling for indoor channels was presented. Particular attention was given to the Saleh–Valenzuela model, as this model constitutes an integral part of the short-range HDR IEEE 802.15.3a and the long-range LDR IEEE 802.15.4a channel models. We briefly reviewed these standard channel models in the light of the discussed small-scale modelling approaches. A discussion of the spatial fading in indoor UWB channels was also presented that is particularly useful in UWB multiple-antenna and mobile applications.

References

[1] H.L. Bertoni, *Radio Propagation for Modern Wireless Systems*. Upper Saddle River, NJ: Prentice Hall, 2000.
[2] J.D. Parsons, *The Mobile Radio Propagation Channel*, Second Edition. New York: John Wiley and Sons, Inc, 2000.
[3] W. Stutzman and G. Thiele, *Antenna Theory and Design*. New York: John Wiley and Sons, Inc, 1981.
[4] R.M. Buehrer, A.Safaai-Jazi, W.A. Davis and D. Sweeney, Characterisation of the UWB channel, *Proceedings of IEEE Conference on Ultra-Wideband Systems and Technologies*, 26–31,Reston, VA, November 2003.
[5] B. Donlan and R.M. Buehrer, The indoor UWB channel, *Proceedings of the Spring 2004 Vehicular Technology Conference*.
[6] D.M. McKinstry and R.M. Buehrer, UWB small-scale channel modelling and system performance, *Proceedings of the Fall 2003 Vehicular Technology Conference*, Orlando, FL, September 2003.
[7] A. Alvarez, G. Valera, M. Lobeira, R. Torres, J.L. Garcia, New channel impulse response model for UWB indoor system simulations, *Proceedings of the Spring 2003 Vehicular Technology Conference*, 1–5, 2003.
[8] D. Cassioli, M.Z. Win and A.F. Molisch, A statistical model for the UWB indoor channel, *Proceedings of Spring 2001 Vehicular Technology Conference*, **2**, 1159–63, 2001.
[9] D. Cassioli, M.Z. Win and A.F. Molisch, The ultra-wide bandwidth indoor channel: from statistical study to simulations, *IEEE Journal on Selected Areas in Communications*, **20**(6), 1247–57, 2002.
[10] R.J.-M. Cramer, R.A. Scholtz and M.Z. Win, Spatio-temporal diversity in ultra-wideband radio, *Proceedings of IEEE Wireless Communications and Networking Conference*, **2**, 888–92, 1999.
[11] R.J.-M. Cramer, R.A. Scholtz and M.Z. Win, Evaluation of an ultra-wideband propagation channel, *IEEE Transactions on Antennas and Propagation*, **50**(5), 561–70, 2002.
[12] R.J.-M. Cramer, R.A. Scholtz and M.Z. Win, Evaluation of an indoor ultra-wideband propagation channel (doc: IEEE P802.15-02/286-SG3a and IEEE P802.15-02/325-SG3a), submitted to IEEE P802.15 Working Group for Wireless Personal Area Networks (WPANs), June 2002. Available: http://grouper.ieee.org/groups/802/15/pub/2002/Jul02/

[13] J. Foerster, Channel Modelling Sub-committee Report Final (doc.: IEEE 802-15-02/490r1-SG3a), submitted to IEEE P802.15 Working Group for Wireless Personal Area Networks (WPANs), February 2002. Available: http://grouper.ieee.org/groups/802/15/pub/2002/Nov02/.

[14] J. Foerster and Q. Li, UWB channel modelling contribution from Intel (doc: IEEE P802.15-02/279-SG3a), submitted to IEEE P802.15 Working Group for Wireless Personal Area Networks (WPANs), June 2002. Available: http://grouper.ieee.org/groups/802/15/pub/2002/Jul02/.

[15] S.S. Ghassemzadeh, L.J. Greenstein and V. Tarokh, The ultra-wideband indoor multipath loss model (doc: IEEE P802.15-02/282-SG3a and IEEE P802.15-02/283-SG3a), submitted to IEEE P802.15 Working Group for Wireless Personal Area Networks (WPANs), June 2002. Available: http://grouper.ieee.org/groups/802/15/pub/2002/Jul02/.

[16] S..S. Ghassemzadeh, R. Jana, C.W. Rice, W. Turin and V. Tarokh, A statistical path loss model for in-home UWB channels, *Proceedings of IEEE Conference on Ultra Wideband Systems and Technology*, Baltimore, MD, 2002.

[17] S.S. Ghassemzadeh and V. Tarokh, UWB path loss characterisation in residential environments, *Proceedings of IEEE Radio Frequency Integrated Circuits Symposium*, 2003.

[18] H. Hashemi, The indoor radio propagation channel, *Proceedings of the IEEE*, **81**(7), 943–68, 1993.

[19] V. Hovinen et al., A proposal for a selection of indoor UWB path loss model, http://grouper.ieee.org/groups/802/15/pub/2002/Jul02, 02280r1P802.15.

[20] J. Keignart and N. Daniele, Subnanosecond UWB channel sounding in frequency and temporal domain, *IEEE Conference on Ultra Wideband Systems and Technology*, Baltimore, MD, 2002.

[21] J. Keignart, J.B. Pierrot, N. Daniele and P. Rouzet, UWB channel modelling contribution from CEA-LETI and STMicroelectronics (doc: IEEE P802.15-02/444-SG3a), submitted to IEEE P802.15 Working Group for Wireless Personal Area Networks (WPANs), October 2002. Available: http://grouper.ieee.org/groups/802/15/pub/2002/Nov02/.

[22] J. Keignart and N. Daniele, Channel sounding and modelling for indoor UWB communications, *Proceedings of 2003 International Workshop on Ultra Wideband Systems*, 2003.

[23] P. Pagani, P. Pajusco and S. Voinot, A study of the ultra-wideband indoor channel: propagation experiment and measurement results, in COST273, TD(030)060, January 2003.

[24] C. Prettie, D. Cheung, L. Rusch and M. Ho, Spatial correlation of UWB signals in a home environment, *Proceedings of IEEE Conference on Ultra Wideband Systems and Technology*, Baltimore, MD, 2002.

[25] T.S. Rappaport, *Wireless Communications: Principles and Practice*, Second Edition. Upper Saddle River, NJ: Prentice Hall, 2002.

[26] L. Rusch, C. Prettie, D. Cheung, Q. Li and M. Ho, Characterisation of UWB propagation from 2 to 8 GHz in a residential environment, *IEEE Journal on Selected Areas in Communications*, submitted for publication.

[27] J.A. Hogbom. Aperture synthesis with a non-regular distribution of interferometer baselines, *Astron. and Astrophys. Suppl. Ser.*, **15**, 1974.

[28] R.J.-M. Cramer, An evaluation of ultrawideband propagation channels, PhD dissertation, Dept. of Electrical and Computer Engineering, University of Southern California, December 2000.

[29] R.A. Scholtz, M.Z. Win and J.M. Cramer, Evaluation of the characteristics of the ultra-wideband propagation channel, *Proceedings of Antennas and Propagation Society International Symposium*, **2**, 626–30, 1998.

[30] P. Withington, R. Reinhardt and R. Stanley, Preliminary results of an ultra-wideband (impulse) scanning receiver, *Proceedings of IEEE Military Communications Conference*, **2**, 1186–90, 1999.

[31] G.L. Turin, F.D. Clapp, T.L. Johnston, S.B. Fine and D. Lavry, A statistical model of urban multipath propagation, *IEEE Transactions on Vehicular Technology*, **VT-21**, 1–9, 1972.

[32] J. Foerster, Channel Modelling Sub-committee Report Final (doc: IEEE 802-15-02/490r1-SG3a), submitted to IEEE P802.15 Working Group for Wireless Personal Area Networks (WPANs), February 2002. Available: http://grouper.ieee.org/groups/802/15/pub/2002/Nov02/.

[33] A.A. Saleh and R.A. Valenzuela, A statistical model for indoor multipath propagation, *IEEE Journal on Selected Areas in Communications*, **SAC-5**(2), 128–37, 1987.

[34] H. Suzuki, A statistical model for urban radio propagation, *IEEE Transactions on Communications*, **COM-25**(7), 673–80, 1977.

[35] D. McKinstry, Ultra-wideband small-scale channel modelling and its application to receiver design. Master's Thesis, Dept. of Electrical and Computer Engineering, Virginia Tech, 2002.

[36] H. Hashemi, Impulse response modelling of indoor radio propagation channels, *IEEE Journal on Selected Areas in Communications,* **11**(7), 1993.
[37] M.Z. Win and Robert A. Scholtz, Characterisation of ultra-wide bandwidth wireless indoor channels: a communication–theoretic view, *IEEE Journal on Selected Areas in Communications,* **20**(9), 2002.
[38] R. Price and P. Green, A communication technique for multipath channels, *Proc. IRE,* **46**, 555–70, 1958.
[39] R.J.-M. Cramer, M.Z. Win and R.A. Scholtz, Impulse radio multipath characteristics and diversity reception, in press.
[40] R.J.-M. Cramer, M.Z. Win and R.A. Scholtz, Evaluation of the multipath characteristics of the impulse radio channel, *Proc. PIMRC'98,* **2**, 864–8, 1998.
[41] M.Z. Win, F. Ramirez-Mireles, R.A. Scholtz and M.A. Barnes, Ultra-wide bandwidth (UWB) signal propagation for outdoor wireless communications, *IEEE 47th Vehicular Technology Conference,* **1**, 251–5, 1997.
[42] R.A. Scholtz and M.Z. Win, Impulse radio, *Personal Indoor Mobile Radio Conference.* September 1997.
[43] M.Z. Win, R.A. Scholtz and M.A. Barnes, Ultra-wide bandwidth signal propagation for indoor wireless communications, *IEEE International Conference on Communications: Towards the Knowledge Millennium,* **1**, 56–60, 1997.
[44] J. Kunisch and J. Pamp, Measurement results and modelling aspects for the UWB radio channel, *IEEE Conference on Ultra Wideband Systems and Technology,* 2002.
[45] S.M. Riad, The deconvolution problem, an overview, *Proceedings of the IEEE,* **74**(1), 82–5, 1986.
[46] R.G. Vaughan and N.L. Scott, Super-resolution of pulsed multipath channels for delay spread characterisation, *IEEE Transactions on Communications,* **47**(3), 343–7, 1999.
[47] A. Bennia and S.M. Riad, Filtering capabilities and convergence of the Van-Cittert deconvolution technique, *IEEE Transactions on Instrumentation and Measurement,* **41**(2), 246–50, 1992.
[48] S.M. Yano, Investigating the ultra-wideband indoor wireless channel, *IEEE VTS 55th Vehicular Technology Conference,* **3**, 1200–4, 2002.
[49] M. Pendergrass and W. Beeler, Empirically based statistical ultra-wideband (UWB) channel model (doc.: IEEE 802-15-02/240SG3a),' presented to IEEE P802.15 Working Group for Wireless Personal Area Networks (WPANs), June 2002. Available: http://grouper.ieee.org/groups/802/15/pub/2002/Jul02/.
[50] W. Turin, R. Jana, S.S. Ghassemzadeh, C.W. Rice and V. Tarokh, Autoregressive modelling of an indoor UWB channel, *IEEE Conference on Ultra Wideband Systems and Technologies,* 71–4, 2002.
[51] V.S. Somayazulu, J.R. Foerster and S. Roy, Design challenges for very high data rate UWB systems, in *Conference Record of the Thirty-Sixth Asilomar Conference on Signals, Systems and Computers,* **1**, 717—21, 2002.
[52] R.C. Qiu, A study of the ultra-wideband wireless propagation channel and optimum UWB receiver design, *IEEE Journal on Selected Areas in Communications,* **20**(9), 2002.
[53] R.M. Buehrer, W.A. Davis, A. Safaai-Jazi and D. Sweeney, Ultra-wideband propagation measurements and modelling, *DARPA NETEX Program Final Report,* 31 January, 2004.
[54] B. Donlan and R.M. Buehrer, Large and small-scale channel modelling for indoor UWB channels, submitted to *IEEE Transactions on Wireless Communications.*
[55] J.A. Dabin, N. Ni, A.M. Haimovich, E. Niver and H. Grebel, The effects of antenna directivity on path loss and multipath propagation in UWB indoor wireless channels, *IEEE Conference on Ultra Wideband Systems and Technologies,* 2003.
[56] S. Venkatesh, J. Ibrahim and R.M. Buehrer, A new two-cluster model for indoor UWB channel measurements, *2004 IEEE International Symposium on Antennas and Propagation,* **1**, 946–9, 2004.
[57] S. Venkatesh, J. Ibrahim and R.M. Buehrer, A new two-cluster model for indoor UWB channel measurements, submitted to *IEEE Transactions on Communications,* June 2004.
[58] A. Molisch, Status of models for UWB propagation channels, unpublished work, 2004.
[59] A.F. Molisch, J.R. Foerster and M. Pendergrass, Channel models for ultra-wideband personal area networks, *IEEE Wireless Communications Magazine,* **10**(6), 14–21, 2003.
[60] A.F. Molisch, K. Balakrishnan, C.-C. Chong, S. Emami, A. Fort, J. Karedal, J. Kunisch, H. Schantz, U. Schuster and K. Siwiak, IEEE 802.15.4a Channel Model – Final Report.
[61] Part 15.4: Wireless Medium Access Control (MAC) and Physical Layer (PHY) Specifications for Low-Rate Wireless Personal Area Networks (LR-WPANs), IEEE Std 802.15.4 TM, 2003.

[62] S.R. Saunders, *Antennas and Propagation for Wireless Commmunication Systems*, John Wiley and Sons, Ltd, 1999.
[63] W.Q. Malik, D.J. Edwards and C.J. Stevens, Experimental evaluation of RAKE receiver performance in a line-of-sight ultra-wideband channel, *2004 International Workshop on Ultra Wideband Systems*, 217–20, 18–21 May 2004.
[64] H.A. Khan, W.Q. Malik, D.J. Edwards and C.J. Stevens, Ultra wideband multiple-input multiple-output radar, *2005 IEEE International Radar Conference*, 900–4, 9–12 May, 2005.

16

Antenna Design and Propagation Measurements and Modelling for UWB Wireless BAN

Yang Hao, Akram Alomainy and Yan Zhao

16.1 Introduction

Current communication systems are driven by the concept of being connected anywhere at anytime. An essential part of this concept is a user-centric approach in which services are constantly available and systems provide reconfigurability, unobtrusiveness and true extension of the human's mind. Body area networks (BANs) consist of a number of nodes and units placed on the human body or in close proximity, such as on everyday clothing [1]. Currently, they are used to receive or transmit simple information which requires very low processing capabilities, e.g. patient monitoring systems that transmit low data rate information (heart rate, blood pressure, etc.). However, some high performance and complex units are needed in the future to provide the facilities for powerful computational processing with high data rates for applications such as video streaming and heavy data communications.

A major drawback of current body-worn systems is the wired communication which is often undesirable because of the inconvenience for the user. Other connection methods have been proposed for solving this problem, including the use of smart textiles and communication by the currents on the user's body [2], [3]. Smart clothes imply the need for a special garment to be worn, which may conflict with the user's personal preferences. Similarly, the body current communication is limited because it has a relatively low capacity and is hence not suitable for real-time video transfer, where very high data rates will be required. Wireless body-centric networks present an apparent suitable alternative and since low power transmission is required for body-worn devices, the human body can be used as a communication channel between wireless wearable devices. The wireless body-centric network has special properties and requirements in comparison to other available wireless networks that is due to the rapid changes in communication channel behaviour on the body during the network operation. This raises some important issues regarding the

---

Ultra-wideband Antennas and Propagation for Communications, Radar and Imaging Edited by B. Allen,
M. Dohler, E. E. Okon, W. Q. Malik, A. K. Brown and D. J. Edwards
© 2007 John Wiley & Sons, Ltd

propagation channel characteristics, radio systems compatibility with such environments and the effect that it has on the human.

Ultra-wideband (UWB) communication is a low-power, high data rate technology with large bandwidth signals that provide robustness to jamming and have low probability of detection [4]. UWB low transmit power requirements allow longer battery life for body-worn units. This leads to UWB being a potential candidate for BAN. The possibility of transmitting data with various requirements in short-range communication with low-power consumption offered by UWB introduces an attractive solution for wireless BAN (WBAN) and implant radio system designers.

The chapter describes the main characteristics of UWB antennas used for WBAN and the effects they have on the reliability and efficiency of such systems. Measurements of UWB on-body radio propagation applying frequency-domain techniques to deduce important channel characteristics for both large-scale and small-scale analyses are presented with details and also with respect to different antennas used for the measurements. The application of well-known modelling techniques and modified methods to suit the requirement of WBAN systems is studied and investigated with emphasis on potential solutions, such as sub-band FDTD (finite-difference time-domain) method. System-level modelling of common UWB impulse radio transceivers are computationally examined to give an overview of on-body channel effects on system performance from a bit error rate (BER) prospective.

16.2 Propagation Channel Measurements and Characteristics

Accurate prediction of radio propagation behaviour is crucial to system design. Site measurements have the advantage of accounting for all parameters without preassumptions. They are, however, expensive and time consuming. It is therefore necessary to develop effective propagation models for wireless system design, based on a generalisation of the channel characteristics. Recent UWB propagation characterisation and modelling literature has presented rigorous investigations and analysis on the behaviour of indoor UWB communication channels and transmission [5]–[9]. Hovinen *et al.* described deterministic and statistical UWB channel modelling for accurate characterisation to be applied in system designs in [5]. Pathloss models explaining large-scale fading were presented in [7, 9] based on an extensive measurement database.

On-body channel characterisation has been presented in a few references, e.g. [10] for narrowband channels. UWB body area network channel characterisation was presented in [11] for predefined sets of nodes with multihopping networking in mind to determine energy and power requirements. However, a more realistic representation of the human body behaviour such as different body positions and postures for various antenna systems was introduced in [12]–[14].

This section presents the UWB on-body radio channel measurement campaigns using different types of antennas with emphasis on antenna characteristics influence on various channel parameters regarding the on-body environment. The measurement set-ups and channel characteristics including large-scale and time-delay analyses are also presented with introduction to suitable models to describe the radio channel for WBAN. The effects antenna parameters have on the propagation channel are also highlighted.

16.2.1 Antenna Element Design Requirements for WBAN

Wireless BANs require body-worn antennas, which, at high and microwave frequencies, can suffer from reduced efficiency due to electromagnetic absorption in tissue, radiation pattern fragmentation and variations in feedpoint impedance. The significance and nature of these effects are system specific and depend on the operating frequency, propagation environment and physical constraints on the antenna itself, as well as the characteristics of the body it is close to and the mounting arrangements. Depending on the operating frequency, the proximity to the human body can lead to high losses caused by bulk power absorption, radiation pattern fragmentation and antenna not tuning properly. This can result in increased transmission errors or, in extreme cases, loss of a marginal communication link.

For the wireless BAN to be accepted by the majority of consumers, the radio system components, including the antenna, need to be somehow hidden and compact and low weight. This requires the possible integration of these systems within clothing. Some research projects have been initiated, under the concept of smart clothing/textiles, to integrate antennas and RF systems into clothes with regards to size reduction and cost effectiveness; the wearer will hence not even notice that these subsystems exist [2].

The main objective in designing a UWB antenna is to consider an antenna with small size, omnidirectional patterns and simple structure that produces low distortion but can cover a large bandwidth. For more specific applications such as wireless BANs, the antenna design becomes more complicated than for simple free-space operating scenarios due to the proximity of a human body and additional form factor constraints. This section discusses the main parameters affecting UWB antennas and how would they apply to WBAN applications with introduction to some examples.

16.2.2 Antennas for UWB Wireless BAN Applications

Antennas may be categorised as either directional or nondirectional, and they may also be classified as electric or magnetic antennas. Electric antennas include dipoles and most horn antennas, while magnetic antennas include loops and slots. Electric antennas are more likely to couple with close objects than the magnetic ones; therefore, magnetic antennas are preferred for applications involving embedded antennas, which is the case in most WBAN systems [15]–[19].

16.2.2.1 UWB Planar Inverted Cone Antenna

The planar inverted cone antenna over a horizontal groundplane is a wideband, omnidirectional, flat antenna that can be applied to UWB systems [13, 20]. The antenna is derived from the volcano antenna and the circular disk antenna concepts. It is composed of a single flat element vertically mounted above a groundplane, as shown in Figure 16.1. The antenna geometry is very simple, with an antenna dimension of 23.5 × 65 mm, conductor thickness for cone of around 0.3 mm and for ground around 0.4 mm. The ground plane is a circular conductor plate with a diameter of 80 mm. The antenna provides outstanding impedance and radiation pattern performance. The antenna's wide bandwidth is due to the fact that the radiating element is of the biconical antenna family, which has many forms with wide bandwidth. The measured antenna return loss is shown in Figure 16.2 when placed off- and on-body; a good impedance bandwidth is apparent across the UWB band. The antenna also has an excellent radiation bandwidth within the UWB band as presented in Figure 16.3.

16.2.2.2 Self-Complementary UWB Antenna

The printed horn-shaped self-complementary antenna (HSCA [13, 21]) was fabricated on RT/Duroid board with $\varepsilon_r = 3$ and a thickness of 1.524 mm. The actual antenna dimensions are 30.6 mm by 60.7 mm and a 22 mm by 29.9 mm groundplane was added for the matching transformer, as shown in Figure 16.4.

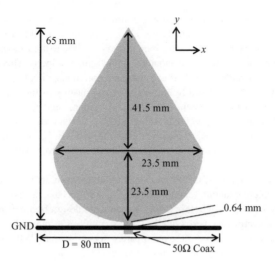

Figure 16.1 Planar inverted cone antenna above horizontal groundplane. (Source [12] reproduced by permission of © 2006 IET)

The HSCA exhibits approximately constant impedance (Figure 16.5) and absolute gain across the UWB band and its gain was found to be between 0 and 2.4 dBi. The antenna radiation patterns experienced some distortions at higher frequencies and produced peaks at directions other than the main beam (Figure 16.6). This is due to the substrate thickness and groundplane used for matching circuits causing unexpected radiation.

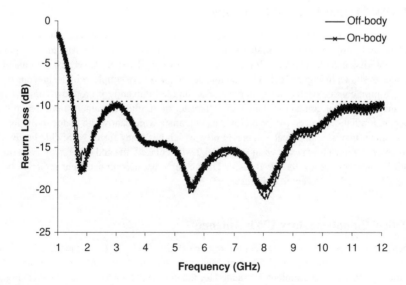

Figure 16.2 Return loss of the PICA on/off the human body (with dotted line referring to VSWR = 2)

Antenna Design and Propagation Measurements and Modelling for UWB Wireless BAN 335

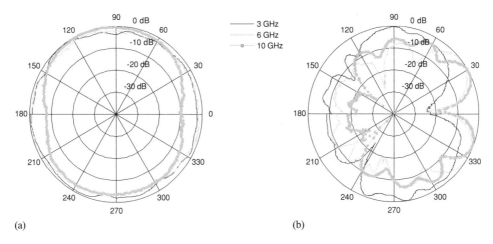

Figure 16.3 Radiation patterns of antenna over a horizontal ground plane (PICA) in free space at 3, 6 and 10 GHz for both H- and E-planes (co-planar in free space). (a) H-plane (left); (b) E-plane (right)

16.2.3 On-Body Radio Channel Measurements

To characterise the UWB radio channel, there are two possible techniques to perform the channel measurement. The channel can be measured in the frequency domain using a frequency sweep technique, or in the time domain based on impulse transmission. For frequency-domain measurement, a wide frequency band is swept using a set of narrowband signals and the channel frequency response is recorded with a network analyser. This corresponds to an S_{21}-parameter measurement set-up, where the device under

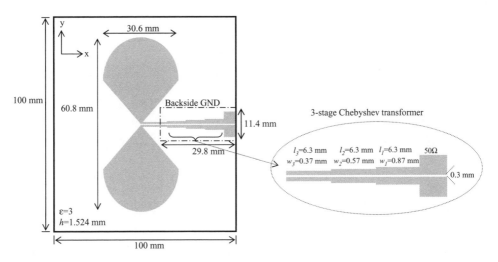

Figure 16.4 Self-complementary UWB antenna design dimensions. (Source [12] reproduced by permission of © 2006 IET)

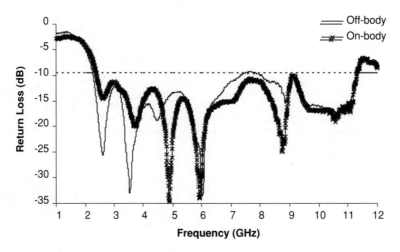

Figure 16.5 Return loss of the HSCA on/off the human body (with dotted line referring to VSWR = 2)

test is the radio channel [5]–[7]. This measurement technique was adopted for the on-body channels measurement campaign.

On the other hand, in the time-domain technique, a narrow pulse is sent through the channel and the channel impulse response is measured using a digital sampling oscilloscope. The corresponding train of impulses can also be generated using a conventional direct sequence spread spectrum based measurement system with a correlation receiver. For the frequency-domain measurements, the RF signal is generated and received by the network analyser, which simplifies the measurement set-up. It is possible to use wideband antennas instead of special impulse radiating antennas when applying frequency-domain methods.

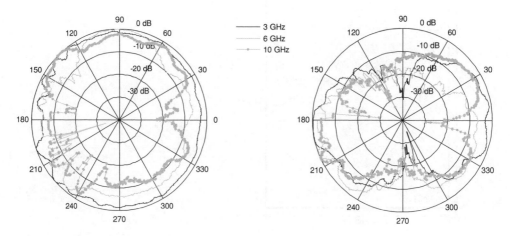

Figure 16.6 Radiation patterns of printed HSCA in free space at 3, 6 and 10 GHz for both H- and E-planes (co-planar in free space). (a) H-plane (left); (b) E-plane (right)

Table 16.1 On-body UWB measurement set-up

Frequency band	3 GHz to 9 GHz
Bandwidth	6 GHz
Sweep time	800 ms
Frequency points	1601
Transmit power	0 dBm
Average noise floor	−100 dBm

The measurement should provide sufficient information on the various on-body links for accurate propagation modelling. To this end, a vector network analyser (VNA) is set on the response mode in the range of 3 GHz to 9 GHz with intervals of 3.75 MHz, at a sweep rate of 800 ms, for UWB on-body communication channels measurement; see Table 16.1. Port-I is used as a transmit node and port-II as the receiver with two pairs of different UWB antennas to measure the channel frequency response S_{21}. Antennas are connected to the analyser by 3 and 5 m long cables. The analyser measures the magnitude and phase of each frequency component allowing the ease of obtaining time-domain response by means of the inverse discrete Fourier transform (IDFT).

Two sets of measurements were performed in the anechoic chamber to determine the channel characteristics of the human body (height 170 cm and average width of 35 cm). This resulted in 710 frequency responses for different on-body channels. Figure 16.7 shows the different antenna positions placed on the body and the measurement set-up. Different on-body scenarios are applied in the measurements, illustrating possible body movements and potential positions of body-worn devices, as listed in Table 16.2. The minimum measured distance is 15 cm, which is larger than the wavelength (10 cm) of the low frequency limit in the band (3 GHz), minimising mutual coupling effects when antennas are placed near each other, since higher frequencies result in smaller wavelength that leads to additional minimisation of mutual

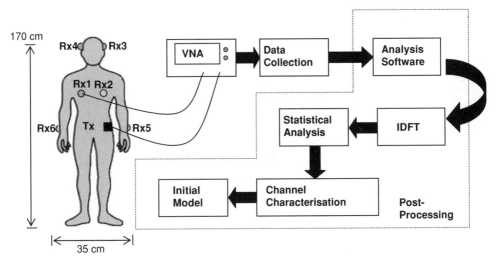

Figure 16.7 UWB on-body channel measurement set-up. (Source [12] reproduced by permission of © 2006 IET)

Table 16.2 On-body UWB propagation measurement scenarios

N	On-body scenario	N	On-body scenario
1	Rx1 – Standing still, upright	12	Rx4 – Standing upright
2	Rx1 – Standing still, body turned left	13	Rx4 – Standing upright, head turned left
3	Rx1 – Standing still, body turned right	14	Rx4 – Standing upright, head turned right
4	Rx1 – Standing still, body leaned forward	15	Rx5 – Standing upright, arm stretched
5	Rx2 – (back) Standing still, upright	16	Rx5 – Standing upright, arm above head
6	Rx2 – (back) Standing still, body turned left	17	Rx5 – Sitting still, arm along body
7	Rx2 – (back) Standing still, body turned right	18	Rx5 – Sitting still, hand on lap
8	Rx2 – (back) Standing still, body leaned forward	19	Rx6 – Standing upright, arm stretched
9	Rx3 – Standing upright	20	Rx6 – Standing upright, arm above head
10	Rx3 – Standing upright, head turned left	21	Rx6 – Sitting still, arm along body
11	Rx3 – Standing upright, head turned right	22	Rx6 – Sitting still, hand on lap

coupling. Fifteen frequency sweeps were taken for each static on-body channel; see Figure 16.8. All measurements in this study were taken using an effective isotropic radiated power (EIRP) of the order of 0 dBm.

16.2.4 Propagation Channel Characteristics

The measured channel data for the different on-body scenarios were calculated and processed in both the frequency and time domain to obtain the initial statistical parameters for both pathloss and power

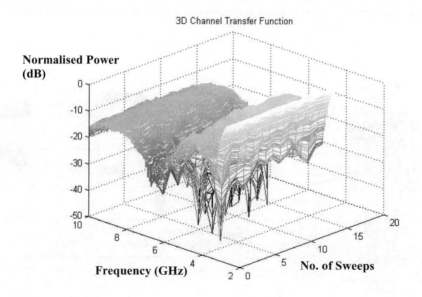

Figure 16.8 Normalised 3D channel transfer function at Tx/Rx1 channel over 15 sweeps using vertical cone antenna over a horizontal ground plane (Source: *IEEE Antenna and Wireless Propagation Letters*, **4**(1), 32, 2005 reproduced by permission of © 2005 IEEE)

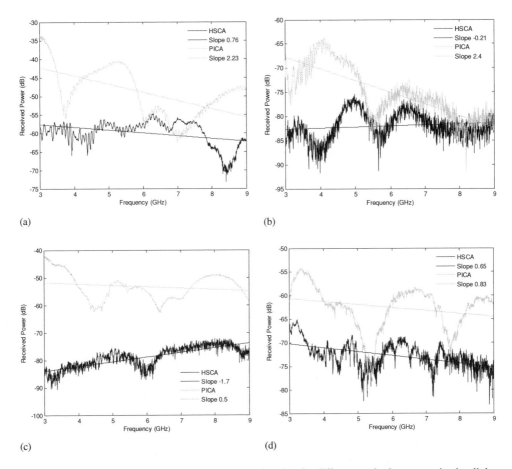

Figure 16.9 Measured channel frequency transfer function for different on-body communication links. (a) Rx1 – body turned left (top-left); (b) Rx2 – body turned left (top-right); (c) Rx4 – head turned left (bottom-left); (d) Rx5 – arm above head (bottom-right)

delay profiles, including the mean excess delay spread. Their reliability and applicability are investigated against established empirical and theoretical propagation models. The real passband technique is applied to obtain the impulse responses from the measured channel frequency transfer functions due to the carrierless nature of UWB impulse radio [6].

The average received power, $|S_{21}|^2$, is shown in Figure 16.9 for scenarios 2, 6, 13 and 16 when the receiver (Rx) is placed on the chest and body turning left, on the back with the body turning left, on the right side of the head and head turning left, and on the left wrist when arm is stretched above the head, respectively (Table 16.2). Examining the slopes, which gives a clearer picture of the behaviour of signal strength with respect to frequency, these frequency responses show the variation in slope being not only due to antenna frequency dependency and link geometry but also due to the dispersion of on-body channels regarding the placement of the receivers. An antenna over a horizontal groundplane, e.g. planar inverted cone antenna (PICA), produces more predictable slopes with received power decaying in an inverse proportional fashion with respect to frequency. However, in the HSCA case, unexpected

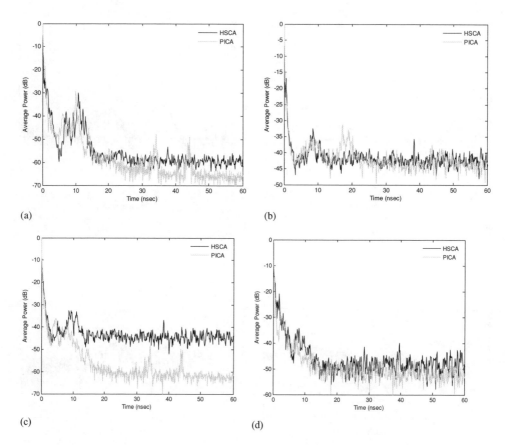

Figure 16.10 Measured power delay profiles for different on-body communication links. (a) Rx1 – body turned left (top-left); (b) Rx2 – body turned left (top-right); (c) Rx4 – head turned left (bottom-left); (d) Rx5 – arm above head (bottom-right)

behaviour is noticed and that is due to antenna gain variations at higher frequencies in addition to the antenna directivity changes with angular radiation directions.

Time-domain analysis is performed by obtaining channel impulse responses which are calculated from the measured frequency transfer functions that consist of 1601 frequency points. The real passband technique, a Hamming window and inverse fast Fourier transform (IFFT) are applied [6]. Power-delay profiles (PDPs) were produced by averaging all impulse responses and determining the noise threshold. Figure 16.10 presents delay profiles for on-body scenarios 2, 6, 13 and 16 (as defined in Table 16.2). It can be seen that most energy is received via the direct path with some multipath reflections at the late time due to the presence of human arm and clothing, in addition to parts of antenna structure. Figure 16.10(c) shows an offset in the average power values between HSCA and PICA cases, which is due to the presence of the direct LOS link between the two antennas and the highly directive behaviour of the HSCA and higher frequencies at specified angular directions.

Figure 16.11 presents impulse responses for on-body scenarios 2, 6, 13 and 16 (Table 16.2). The changes in channel characteristics due to different antenna (HSCA and PICA) properties can also be

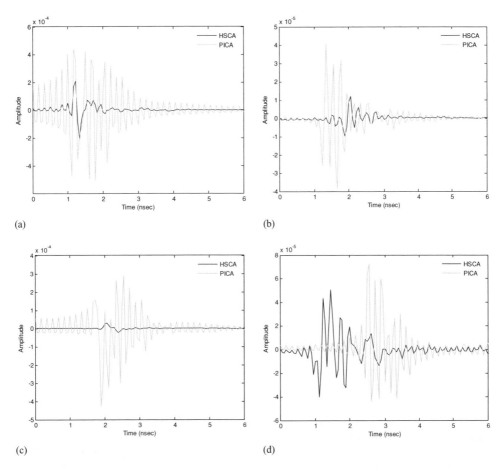

Figure 16.11 Measured impulse responses for different on-body communication links. (a) Rx1 – body turned left (top-left); (b) Rx2 – body turned left (top-right); (c) Rx4 – head turned left (bottom-left); (d) Rx5 – arm above head (bottom-right)

found in these figures. The main dissimilarity is that more strong echoes and ringing effects appear in the PICA case. This can be explained by the fact that the PICA has a narrower bandwidth and more resonance frequencies within the measured band which increases both ringing and pulse width that ultimately impacts the data rate.

16.2.4.1 Large-Scale Pathloss

Within any wireless network, as the distance between the transmitter and receiver increases, the received signal becomes weaker due to the growing propagation attenuation with the distance. Large-scale pathloss characterises the local average of the path loss. The log-distance pathloss model is a popular empirical model for narrowband systems in both indoor and outdoor radio environments (see Chapter 14).

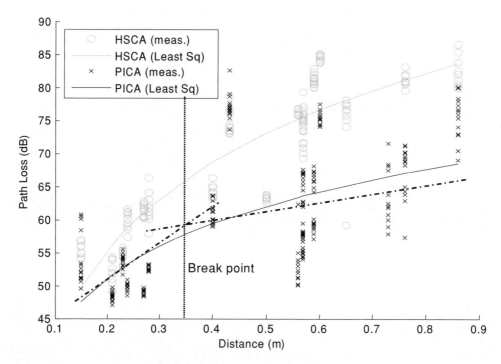

Figure 16.12 Measured and modelled pathloss for HSCA and PICA with proposed dual-slope model for PICA case. (Source [12] reproduced by permission of © 2006 IET)

Measurement results indicate that the model is also valid for UWB indoor propagation [6]–[9]. The pathloss of the channel is calculated directly from the measurement data using averaging over the measured frequency transfers at each frequency points. It is known that the mean pathloss referenced to a distance d_o (reference distance is 1 m as calibrated for VNA measurements) can be modelled as a function of distance using [7],

$$PL_{dB}(d) = PL_{dB}(d_o) + 10\alpha \log\left(d/d_o\right), \tag{16.1}$$

where α is the path loss exponent (loss is presented as a positive value) and $PL_{dB}(d_0)$ is the mean pathloss at 1 m. A least square fit computation is performed on the measured pathloss results to get the mean pathloss at d_0 and exponent α. Figure 16.12 presents the measured values and modelled pathloss for both antenna cases. The two types of antennas give different pathloss exponents. For HSCA,

$$PL_{dB}(d) = 86.5 + (4.4)10\log\left(d/d_o\right) \text{ for } 15\,\text{cm} \leq d \leq 100\,\text{cm}, \tag{16.2}$$

and for PICA as,

$$PL_{dB}(d) = 70.3 + (2.7)10\log\left(d/d_o\right) \text{ for } 15\,\text{cm} \leq d \leq 100\,\text{cm}, \tag{16.3}$$

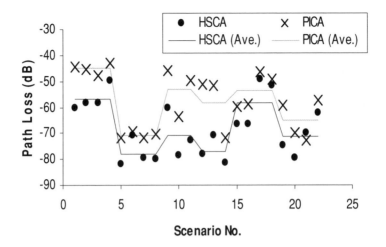

Figure 16.13 Changes in on-body channel pathloss for various body postures

where $\alpha = 4.4$ and 2.7 are the pathloss exponents for HSCA and PICA models, respectively. It can be deduced from Figure 16.12 that on-body channel pathlosses can be approximated using the conventional dual-slope model. It can be calculated that for the UWB band, the far-field will be at distances equal to and greater than 37 cm for PICA, which agrees with that from the measurements in Figure 16.12 with a breakpoint at around 37 cm.

These high exponent values are due to the non-reflecting environment in the anechoic chamber; on the other hand, lower values are expected for indoor (e.g. office) environments with more reflections and lower pathloss [11, 25]. In the PICA case, the exponent value is higher than for the HSCA case, which is due to multipath propagation from the human body and clothes caused by the PICA's excellent omnidirectional radiation and gain across the UWB range and also due to a higher antenna gain. In addition to pathloss variation as a function of distance, Figure 16.13 presents the pathloss as a function of the on-body channel measurement scenario (Table 16.2). The average pathloss for each receiver location (Rx1, Rx2, Rx3, Rx4, Rx5 and Rx6) is also presented with maximum variation in path gain values when transferring from one location to another of around 30 dB, and that is when receiver changes from chest to the back. This explains the large drop in path gain value.

Figure 16.14 shows the CDF (cumulative distribution function) of the deviation of the measured received power from the calculated average. The curves in general fit a normal distribution fairly good, however, the HSCA scenario seem to have a little more of a deviation from this distribution. This shows the unpredictability of the on-body channel and the suitability of the omnidirectional vertical antenna for this type of communication. These distributions present shadowing effects of the human body. The probability distribution of the signal strength is obtained from the measured data for both HSCA and PICA. The PDF (probability distribution function) of the pathloss is shown in Figure 16.15. The probability distributions are approximately normal for PICA case with mean $\mu = 60.24$ dB and standard deviation of $\sigma = 9.3$ dB. However, it is noticeable from the distribution of the HSCA case that two peaks exists which can be related to body posture changes effects on link geometry and antenna radiation performance; hence, a wider distribution is obtained.

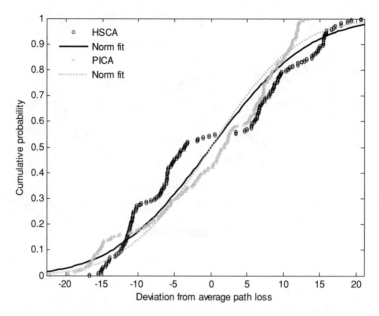

Figure 16.14 Deviation of measured received power from average power for both self-complementary and vertical antenna cases fitted to a normal distribution

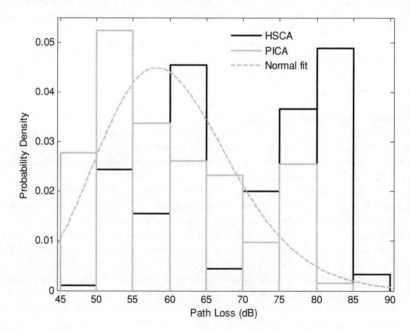

Figure 16.15 Probability distribution of measured pathloss data for both vertical antenna and self-complementary antenna

16.2.4.2 Time-Delay Analysis

The PDP is characterised by the first central moment (mean excess delay, τ_m) and the square root of the second moment of the PDP (RMS delay spread, τ_{rms}). The RMS delay spread provides a figure of merit for estimating data rates for multipath channels [9]. The mean excess delay is defined as,

$$\tau_m = \sqrt{\frac{\sum_{i=0}^{N-1} \tau_i \cdot |h(\tau_i)|^2}{\sum_{i=0}^{N-1} |h(\tau_i)|^2}}, \quad (16.4)$$

and the delay spread, τ_{rms}, as,

$$\tau_{rms} = \sqrt{\frac{\sum_{i=0}^{N-1} (\tau_i - \tau_m)^2 \cdot |h(\tau_i)|^2}{\sum_{i=0}^{N-1} |h(\tau_i)|^2}}, \quad (16.5)$$

where $h(\tau_i)$ is the time-domain impulse response obtained from the measurement data.

Delay parameters were evaluated at intervals of 80 ns, since echoes fade rapidly after this period. As expected, the mean delay due to the propagation link between the transmitter (Tx) and the receiver at the back (Rx2), Figure 16.7, link is the highest, where non-line-of-sight (NLOS) channel and propagation around the human body on the surface (creeping waves [22]) are the main propagation channel. The mean RMS delay for links with small distances and where both antennas are placed on the same part of the human body in the HSCA antenna case is smaller than the RMS spread delay for the same links with PICA; this is due to the higher directivity of HSCA at specific angular radiation directions and as well the propagation of surface waves in these cases.

Figures 16.16(a) and (b) show the cumulative distributions of the RMS delay spread and mean excess delay, respectively. The log-normal distribution is also plotted for each case with ($\sigma = 2.38$, $\mu = 3.5$) and ($\sigma = 2.12$, $\mu = 3.4$) for the RMS CDF of HSCA and PICA cases, respectively. The parameters for the log-normal distribution to fit CDF of mean excess delay are, ($\sigma = 1$, $\mu = 0.2$) and ($\sigma = 1$, $\mu = 0.05$) for HSCA and PICA cases, respectively. Other different empirical distributions were applied, however, the log-normal proved to be the best fit with a larger variance for PICA channels, which is predictable due to the random behaviour of the on-body channels.

16.3 WBAN Channel Modelling

For low-power, reliable and robust on-body communication systems, a deterministic channel model is required to provide a clear picture of the on-body radio propagation and its behaviour with regards to different scenarios and system components. There have been a number of references characterising and analysing the on-body channel and also investigating the electromagnetic wave propagation around the body [23]–[27]. However, the UWB on-body radio channels have not been well studied due to the difficulty in characterising frequency-dependent electrical properties of human tissue and other effects from antenna types and body movements etc.

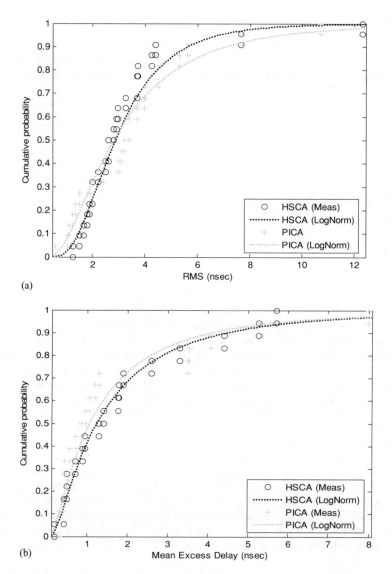

Figure 16.16 Delay parameters cumulative distribution fit to log-normal distribution (a) RMS delay spread τ_{rms} for on-body channels (top); (b) mean excess delay τ_m (bottom). (Source [12] reproduced by permission of © 2006 IET)

16.3.1 Radio Channel Modelling Considerations

Various on-body antenna positions and different body postures are applied to obtain a deterministic UWB channel model. Figure 16.17(a) shows the transmitting and receiving antenna positions as used during measurement and modelling for both a hybrid UTD/RT (uniform theory of diffraction/ray tracing) and the sub-band FDTD methods. The advantage of the sub-band FDTD method over the UTD/RT is

Antenna Design and Propagation Measurements and Modelling for UWB Wireless BAN

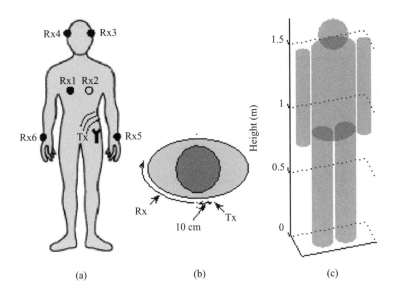

Figure 16.17 2-D and 3-D human body models: (a) Antenna positions for on-body radio channel modelling and measurement; (b) 2-D ellipse cylinder used for modelling both transmitter and receiver mounted on the trunk; (c) 3-D human body model used in both sub-band FDTD and UTD/RT models. (Source [44] reproduced by permission of © 2006 IEEE)

its accuracy when modelling complicated UWB on-body radio channels. Compared with the dispersive FDTD, the sub-band approach can be directly applied to different human tissues with any type of frequency dependence.

16.3.1.1 The UTD/Ray-Tracing Model

Basic RT techniques include two approaches: the image method [28] and the method of shooting and bouncing rays (SBR) [29]. In our UTD/RT model, the SBR method is applied. Due to the high electromagnetic absorption rate of human body tissue, the signal propagation inside the human body is neglected. The received frequency domain signal can be calculated using Equation (16.6) [28],

$$E(\omega) = \sum_{i=1}^{N} E_0(\omega) G_{ti} G_{ri} A_i \prod_l D_l(\omega) \prod_m R_m(\omega) \cdot e^{-j\frac{\omega}{c} d_i}, \quad (16.6)$$

where $E_0(\omega)$ is the transmitted frequency domain signal, G_{ti} and G_{ri} are the transmitting and receiving antenna field radiation patterns in the direction of the ith ray, A_i is a distance factor, $D(\omega)$ and $R(\omega)$ are the frequency-dependent diffraction and reflection coefficients, l and m depend on the number of diffractions and reflections before reaching the receiver respectively, $e^{-j\frac{\omega}{c} d_i}$ is the propagation phase factor due to the path length, d_i, c is the speed of wave propagation and N is the total number of received rays.

When the on-body channel is modelled using the hybrid UTD/RT approach, the RT is used to find the reflected/diffracted rays, while the UTD is applied for calculating reflected/diffracted signal strength. Depending on different receiver locations on the human body, different antenna coupling/radiation problems are considered [28]. The advantage of the UTD/RT model is that the antenna effect can be easily included in terms of radiation patterns.

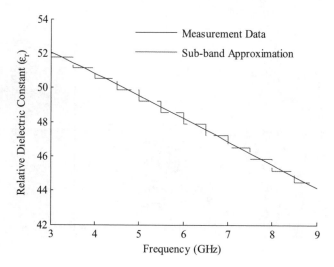

Figure 16.18 Measured permittivity of human muscle [30] and a sub-band approximation for the UTD/RT and sub-band FDTD models. (Source [44] reproduced by permission of © 2006 IEEE)

For on-body radio channels, both the transmitter and receiver are mounted on the human body; the required antenna pattern depends on the location of the transmitter. As shown in Figure 16.17(a), the transmitter is mounted on the waist and the human body is modelled in free space. Without considering the reflected rays from the environment, the on-body channel mainly contains the creeping waves outgoing from the transmitter. Therefore, only a 2-D antenna pattern in the plane tangential to the human body at the transmitter is required. The measured antenna patterns at 6 GHz for both HSCA and PICA are used in simulations, and the pattern variation at different frequencies is not considered.

16.3.1.2 Sub-band FDTD Model

The sub-band FDTD method has been proposed in [31]. To apply the method to UWB on-body radio channel modelling, the whole frequency band (3–9 GHz) is first divided into 12 sub-bands with 500 MHz bandwidth for each sub-band. The choice of number of sub-bands depends on the accuracy requirement to approximate dispersive material properties. For instance, the relative dielectric constant of human muscle ranges from 52.058 at 3 GHz to 44.126 at 9 GHz [30]. Twelve sub-bands are used to match the frequency dispersion curve by assuming the dielectric constant within each sub-band constant (obtained at the centre frequency of each sub-band). The overall error from such a curve fitting is less than 1%. Figures 16.18 and 16.19 show the frequency-dependent dielectric constant and conductivity of human muscle from measurements and their staircasing approximation used in the sub-band FDTD model.

Equation (16.7) is used to calculate the combined frequency domain signal from sub-band simulations.

$$F_r(\omega) = \sum_{i=1}^{N} F_{r,i}(\omega) \cdot A_i(\omega), \tag{16.7}$$

where $F_{r,i}(\omega)$ is the received frequency domain signal at the ith sub-band, $A_i(\omega)$ is a rectangular window function associated with the bandwidth of the ith sub-band and N is the total number of sub-bands. Finally,

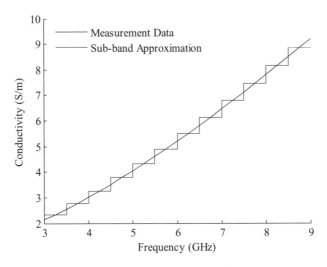

Figure 16.19 Measured conductivity of human muscle [30] and a sub-band approximation for the UTD/RT and sub-band FDTD models. (Source [44] reproduced by permission of © 2006 IEEE)

the combined frequency domain signal is inverse Fourier transformed into the time domain to obtain the time-delay profile.

Although the FDTD method has been successfully applied to the antenna design [33] for various applications, the inclusion of antenna parameters in radio propagation modelling using a global FDTD method is not feasible due to the constraints of limited computer resources. In our analysis, the antenna is approximated as a point source emitting a narrow Gaussian pulse and is deemed to be particularly viable for a vertical antenna over a horizontal groundplane, which radiates almost omnidirectionally across the frequency band. When the UTD/RT and sub-band FDTD methods are applied to the modelling of UWB on-body radio channels, a simplified human body model (Figure 16.17(c)) with only muscle is assumed.

16.3.2 Two-Dimensional On-Body Propagation Channels

The application of both sub-band FDTD and UTD/RT methods to a simple (2-D) scenario is considered. Both the transmitter and receiver are mounted on the trunk of the body. As the receiver moves along the trunk in the same horizontal plane, the scenario can be treated as a 2-D case. As shown in Figure 16.17(b), the human body (trunk) is modelled as a 2-D ellipse cylinder with semi-major axis 0.17 m and semi-minor axis 0.12 m according to the dimensions of a human candidate volunteer for the measurement. Both transmitter and receiver are placed on the 'trunk' and the transmitter is 10 cm offset from the centre. During the measurement, the receiver is always kept on the 'trunk' while moving along the route as shown in Figure 16.17(b).

Only the transversal magnetic (TM) mode is considered for both sub-band FDTD and UTD/RT models. However, the antenna pattern contribution is excluded in this analysis because, for any receiver location, the received signal only contains the contributions from two creeping waves travelling from the transmitter in opposite directions tangential to the ellipse's surface. While mutual coupling between transmitting and receiving antennas is considered [28], with the given transmitter/receiver locations and geometry dimensions (ellipse cylinder), UTD can be easily applied to calculate the diffracted signal strength. The

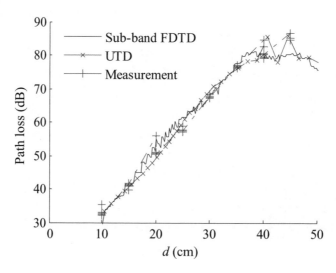

Figure 16.20 Comparison of pathloss along the trunk (see Figure 16.17(b)) from the sub-band FDTD model, UTD model and measurements. (Source [44] reproduced by permission of © 2006 IEEE)

approximate elliptic 'trunk' is also modelled using the sub-band FDTD with the cell size of 3.0×10^{-3} m and time step 5.0×10^{-12} s. The number of cells in the computational region is 140×160, which are truncated by a perfect matched layer (PML).

Figure 16.20 shows the pathloss results along the trunk (Figure 16.17(b)) from the sub-band FDTD model, UTD/RT model and measurement. Good agreement is achieved when the creeping distance of the transmitter and receiver is small. However, when the distance approaches the maximum, ripples are observed from UTD/RT and measurement, which are caused by the addition or cancellation of two creeping rays travelling along both sides of the elliptical 'trunk'. The sub-band FDTD model fails to accurately predict such phenomenon due to the staircase approximation of the curved surfaces and such a problem can be alleviated by using a conformal FDTD method [32]. Figure 16.20 indicates that for modelling simple on-body communication scenarios such as both transmitter and receiver are on the trunk; UTD proves to be very efficient and provides accurate results.

16.3.3 Three-Dimensional On-Body Propagation Channels

Both the UTD/RT and sub-band FDTD are applied to model the UWB on-body radio channel in three dimensions. As shown in Figure 16.17(a), different antenna positions are chosen due to locations of commonly used on-body communication devices, such as head-mounted display, headset and wristwatch. The human body is modelled by several different geometries: 1 sphere for the head ($r = 0.10$ m), 1 ellipse cylinder for the trunk ($a = 0.15$ m, $b = 0.12$ m, $h = 0.65$ m) and 4 cylinders for arms ($r = 0.05$ m, $h = 0.70$ m) and legs ($r = 0.07$ m, $h = 0.85$ m) according to the candidate's dimensions. The whole body is modelled as muscles and the dielectric constant and conductivity are obtained from measurement [30] (Figures 16.18 and 16.19).

For the sub-band FDTD method, the human body shown in Figure 16.17(c) is modelled in free space which is meshed by $140 \times 160 \times 630$ cells with a cell size of 3.0×10^{-3} m. The time step is chosen as

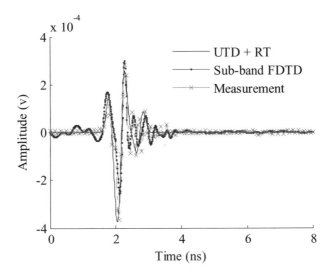

Figure 16.21 Comparison of channel impulse responses at Rx3 (see Figure 16.17(a)) using PICA from the UTD/RT model, sub-band FDTD model and measurements. (Source [44] reproduced by permission of © 2006 IEEE)

5.0×10^{-12} s according to the stability criterion. The antenna over a horizontal groundplane is modelled as a point source due to its omnidirectional radiation properties. The received signal at location Rx3 (Figure 16.17(a)) is shown in Figures 16.21 and 16.22.

With the UTD/RT method, the ray tube angle is set to be 0.5° for high accuracy [33]. RT is used to launch rays at different angles from the transmitter. When reflections/diffractions occur, UTD is applied to calculate the reflected/diffracted signal. Rays are terminated after their field strength drops 50 dB below the reference level. The threshold for on-body channel modelling is lower than that used for indoor channel modelling (30 dB) due to the non-reflecting environment (free space) and relatively low amplitude of the received signal in our analysis. The received signal at Rx3 receiver locations (Figure 16.17(a)) using two types of antennas (HSCA and PICA) and the UTD/RT model is shown in Figures 16.21 and 16.22.

At the same receiver location (Rx3 in Figure 16.17(a)), the received signal using PICA contains more multipath components compared with HSCA because of the difference between their radiation properties. Sub-band FDTD provides more accurate results than UTD/RT compared with measurements since FDTD can fully account for the effects of reflection, diffraction and radiation, while some important rays are missing in the UTD/RT model compared with measurements. The major difference between modelling results and measurements is caused by the change of antenna radiation patterns at different frequencies, which are not taken into account in numerical modelling.

16.3.4 Pathloss Modelling

In the local area of each receiver (Rx1 – Rx6 for PICA and Rx1 – Rx4 for HSCA, Figure 16.17a), two more receiver locations are considered, thus pathlosses at a total of 18 different locations are obtained. Then, the average pathloss is calculated around each receiver location and compared with measurement results. Figures 16.23 and 16.24 show the comparison of pathloss for different on-body channels using PICA and HSCA, respectively. The calculated pathloss exponent is also shown in both figures. The high

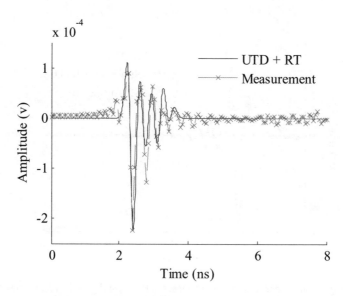

Figure 16.22 Comparison of channel impulse responses at Rx3 (see Figure 16.17(a)) using HSCA from the UTD/RT model and measurements. (Source [44] reproduced by permission of © 2006 IEEE)

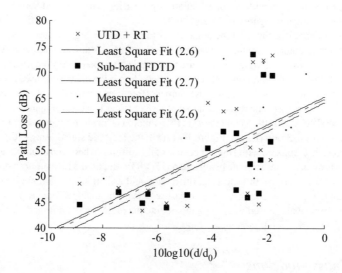

Figure 16.23 Comparison of pathloss for on-body channels using PICA from UTD/RT model, sub-band FDTD model and measurements. The least square fitted line and pathloss exponent value for each model are also shown

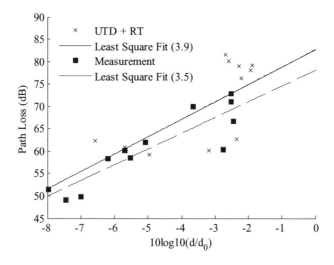

Figure 16.24 Comparison of pathloss for on-body channels using HSCA from UTD/RT model and measurements. The least square fitted line and path loss exponent value for each model are also shown

exponent values are due to the non-reflecting environment (free space) in our models. For PICA, both sub-band FDTD and UTD/RT show good agreement with measurements since the radiation characteristics of PICA are relatively stable across the UWB frequency band; thus, the approximations by using the point source in the sub-band FDTD model and the use of radiation pattern at 6 GHz for the whole ultra-wideband in the UTD/RT model are reasonable. For HSCA, the larger difference is caused by both the change of HSCA's radiation pattern at different frequencies and the inaccuracy of the UTD/RT model.

16.4 UWB System-Level Modelling of Potential Body-Centric Networks

Designing a UWB transceiver has several challenges, some of which are not shared with traditional narrowband systems. The requirement for maximum total power consumption set by the 802.15.3a specification at 110 Mb/s and 200 Mb/s is 100 mW and 250 mW, respectively [34, 35]. To meet these constraints, a transceiver must either target the lowest possible power for all data rates, or use an architecture that scales power with data rate. It is also advantageous to have an architecture that scales power consumption under optimal channel conditions (adaptive UWB system).

In this section, details of the system-level modelling applied to investigate the measured channel parameters and behaviour on currently common radio system transceivers' performance. The impulse performance of the UWB system is presented with respect to different measured channel cases and also BER evaluation introduced for specified on-body scenarios and channels.

16.4.1 System-Level Modelling

Conventional narrowband and wideband systems use radio frequency carriers to move the signal in the frequency domain from the baseband to the actual carrier frequency where the system is allowed to operate. Generally, there are two types of UWB signal designs: impulse based and pulse multi-carrier

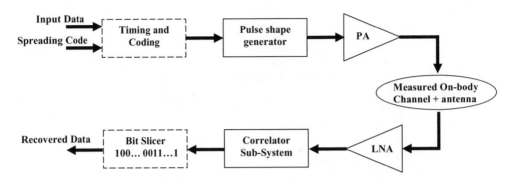

Figure 16.25 Block diagram of an impulse-based UWB radio system

(see Chapter 1). The impulse-based UWB system is of initial interest in this study to provide a clearer view of not only system performance but also propagating and received information pulses and is therefore explained in this section.

The system model presented here applies 100 Mbps data rate and operates at a low signal-to-noise ratio (SNR) of 0 dB. Figure 16.25 shows a block diagram of the radio system modelled using Agilent DSP Designer™ to investigate the effects of on-body channels on a pulse-based UWB system using different pulse modulation techniques. The modulation techniques applied in the system modelling, namely pulse position modulation (PPM) and bi-phase modulation are introduced in the following sections to highlight the main features of the modulation schemes and also performance analysis measured in terms of BER.

Impulse-based UWB systems. Impulse-based UWB has a very simple architecture. The baseband signal is modulated by the specific modulation scheme and then sent directly in impulse form. Theoretically, an impulse signal has an infinite wide spectrum, however, in realistic systems, short duration pulses such as rectangular pulse and Gaussian pulses are used [8]. There are several modulation schemes for impulse based UWB such as PAM (pulse amplitude modulation), PPM (pulse position modulation), PSM (pulse shape modulation) and bi-phase modulation.

16.4.2 Performance Analysis

The effect of the measured channel responses on the received pulses for both antenna cases was analysed and investigated. Figures 16.26 and 16.27 show the transmitted and received pulses for on-body channel scenario 1 (Scenario 1, receiver placed at Rx1 on the chest with body standing still) using PPM and bi-phase modulation, respectively. The variation of time delays between the different received pulses highlights the antenna effect on the channel behaviour of the on-body links. More echo and noise components are introduced in the case of pulses received under the PICA compared to those with the HSCA. This is due to the wider bandwidth of the HSCA antenna (as discussed earlier). However, the improved pulse shape received in the HSCA case is traded off by reduction in received energy.

The rake receiver is modelled to collect the pulse energy in multipath components. At SNR = 0 dB, the number of rake fingers was set to two (for comparison between different cases and modulation techniques) to collect the direct path component and the second finger to collect one multipath component with an average delay window of 8 ns for the investigated scenarios. With reference to Table 16.2, the applied on-body channels are scenario 1 (Rx1 on chest, standing still), scenario 8 (Rx2 on the back with body leaning forward), scenario 14 (Rx4 on the right side of the head with head turned right) and scenario

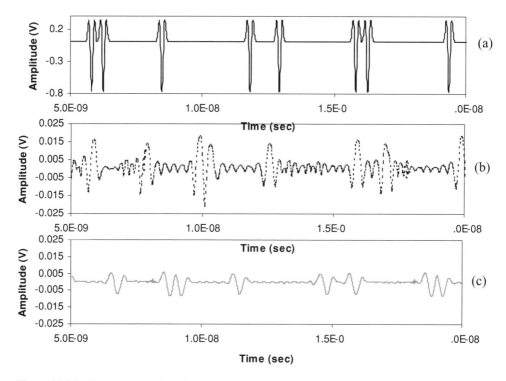

Figure 16.26 Transmitted and received pulses of the modelled PPM UWB system (a) transmitted pulse; (b) received pulse for PICA channel; (c) received pulse for HSCA channel. (Source [12] reproduced by permission of © 2006 IET)

16 (Rx5 on the left wrist with arm stretched above head). Different BER values have been obtained for the various scenarios and modulation schemes, as shown in Figure 16.28 and Table 16.3.

The BER values are given for a very low SNR environment; however, improved control of system gain and coding mechanisms can provide better performances, meeting the stability requirements of BER 0.1%. The presented values are to highlight the effects of radio channel, antennas and modulation techniques onto the system performance. Differences of around 6% in BER measures are found when PPM is used in comparison to 1.5% for bi-phase modulation systems between PICA and HSCA measured channel responses. This implies that bi-phase modulation can be adopted in designing initial potential UWB transceiver designs for wireless body-centric networks since it provides better performance of UWB systems even for higher data rates; this also agrees with results obtained in [39].

16.5 Summary

Requirements and parameters of UWB wireless BAN antennas were discussed and analysed. Radio propagation channel measurements for WBANs were also presented for different scenarios. Various antenna types and changes in body postures were also considered. Variation in both delay parameters and pathloss were observed when changes in body positions occurred for all measurement scenarios. The

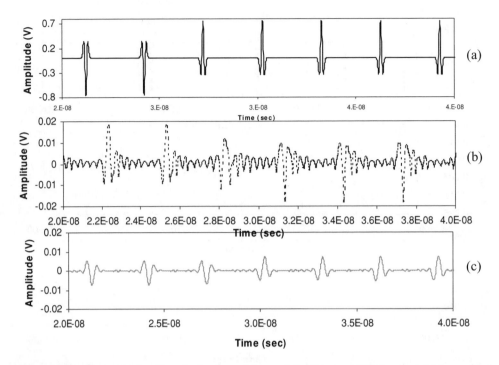

Figure 16.27 Transmitted and received pulses of the modelled bi-phase UWB system (a) transmitted pulse; (b) received pulse for PICA channel; (c) received pulse for HSCA channel. (Source [12] reproduced by permission of © 2006 IET)

received power was proven to be dependent not only on the antenna positions and distances but also on the frequency dependency of both antennas and on-body channel. Measured RMS spread delay and mean excess delay data fit to a log-normal distribution, which provides a tool for empirical on-body propagation modelling that can be combined with any other models representing different operating environments. The channel characteristics showed that the vertical cone antenna over a horizontal ground plane had the best performance.

Radio channel modelling applying various analytical and numerical techniques for wireless BAN communication links was also discussed. The on-body propagation channel modelling using ray theory and a sub-band FDTD method was presented. Both methods provide the solution to one of the main issues – material dispersion in UWB radio channels. For the sub-band FDTD, within each sub-band, conventional FDTD has been applied.

A combination technique is used at the receiver to recover all sub-band simulations. The advantage of this method is its ability of modelling materials with any type of frequency dependence. The sub-band FDTD method is applied to both 2-D and 3-D on-body scenarios and compared with the UTD/RT model and measurement results. For cases such as both transmitter and receiver mounted on the trunk, UTD/RT provides a relatively simple and reliable solution; while for more complicated scenarios, such as whole body channel modelling, sub-band FDTD is capable of providing a more general solution due to its ability of fully accounting for the effects of reflection, diffraction and radiation. A better match to measurement results for whole-body radio channel modelling has been achieved by the sub-band FDTD compared

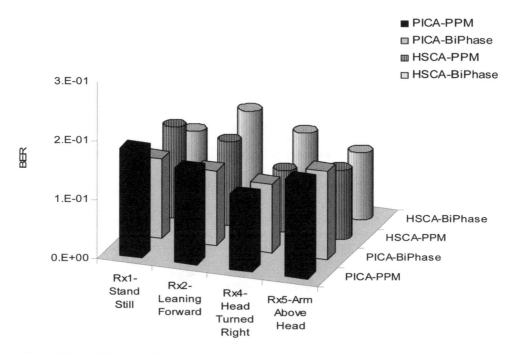

Figure 16.28 BER at specified on-body links (Rx1 with body standing still, Rx2 with body leaning forward, Rx4 with head turned right and Rx5 with arm above head) for both modulation schemes PPM and bi-phase with two Rake fingers. (Source [12] reproduced by permission of © 2006 IET)

with the UTD/RT. The effect of antenna types on on-body radio channels in terms of radiation pattern has also been investigated.

The measured and analysed data for on-body measurement applying a vertical antenna over a horizontal groundplane (PICA) and self-complementary antenna (HSCA) were used for evaluating the performance of potential UWB transceivers that could be applied to wireless body-centric networks. For very low transmitting power and low SNR, bi-phase modulation provided a better system performance in comparison to PPM for potential UWB on-body wireless communications.

Table 16.3 Comparison between BER obtained for PPM and bi-phase modulation from BER point of view using both HSCA and PICA for SNR = 0 dB

On-body scenario	HSCA		PICA	
	PPM	Bi-phase	PPM	Bi-phase
Rx1-stand still	0.154	0.114	0.182	0.135
Rx2 – Leaning forward	0.141	0.16	0.16	0.127
Rx4 – Head turned right	0.104	0.135	0.127	0.116
Rx5 – Arm above head	0.117	0.114	0.165	0.15

References

[1] J. Bernardhard, P. Nagel, J. Hupp, W. Strauss and T. von der grun, BAN – Body area network for wearable computing, presented at *9th Wireless World Research Forum Meeting*, Zurich, July, 2003.

[2] Internet resources, Smart textiles offer wearable solutions using nanotechnology, URL:http://www.fibre2fashion.com/news/NewsDetails.asp?News_id=11705.

[3] Internet resource, Ubiquitous communication through natural human actions, URL: http://www.redtacton.com/en/.

[4] J. Foerster, E. Green, S. Somayazulu and D. Leeper, Ultra-wideband for short- or medium-range wireless communications, *Intel Technology Journal*, Q2, 2001.

[5] V. Hovinen and M. Hamalainen, Ultra wideband radio channel modelling for indoors, *COST 273 Workshop*, Helsinki, 29–30 May, 2002.

[6] C-C. Chong, Y. Kim and S. Lee, UWB indoor propagation channel measurements and data analysis in various types of high-rise apartments, *Proc. IEEE Vehicular Technology Conference (VTC2004-Fall)*, Los Angeles, USA, September 2004.

[7] S.S. Ghassemzadeh, R. Jana, C.W. Rice, W. Turin and V. Tarokh, A statistical path loss model for in-home UWB channels, *IEEE Conference on Ultra Wideband Systems and Technologies, UWBST*, 59–64, 2002.

[8] Z. Chen, X.H. Wu, H.F. Li, N. Yang and M.Y. Chia, Considerations for source pulses and antennas in UWB radio systems, *IEEE Transactions on Antennas and Propagation*, **52**(7), 1739–48, 2004.

[9] J.A. Dabin, N. Ni, A.M. Haimovich, E. Niver and H. Grebel, The effects of antenna directivity on path loss and multipath propagation in UWB indoor wireless channels, *Proc. IEEE Ultra Wideband Systems and Technologies Conference*, 305–9, November 2003.

[10] Y. Nechayev, P. Hall, C.C. Constantinou, Y. Hao, A. Owadally and C.G. Parini, Path loss measurements of on-body propagation channels, *Proc. 2004 International Symposium on Antennas and Propagation*, 745–8. Sendai, Japan, August 2004.

[11] T. Zasowski, F. Althaus, M. Stager, A. Wittneben and G. Troster, UWB for noninvasive wireless body area networks: channel measurements and results, *Proc. IEEE Conference on Ultra Wideband Systems and Technologies*, 285–9, Reston, Virginia, November 2003.

[12] A. Alomainy, Y. Hao, X. Hu, C.G. Parini and P.S. Hall, UWB on-body radio propagation and system modelling for wireless body-centric networks, *IEE Proceedings Communications*, **153**(1), 107–114, 2006.

[13] A. Alomainy, Y. Hao, C.G. Parini and P.S. Hall, Comparison between two different antennas for UWB on-body propagation measurements, *IEEE Antennas and Wireless Propagation Letters*, **4**(1), 31–4, 2005.

[14] A. Alomainy, Y. Hao, C.G. Parini and P.S. Hall, On-body propagation channel characterisation for UWB wireless body-centric networks, *IEEE AP-S International Symposium on Antennas and Propagation and USNC/URSI National Radio Science Meeting*, Washington DC, USA, 3–8 July 2005.

[15] K.Y. Yazdandoost and R. Kohno, Ultra wideband qntenna, *IEEE Communications Magazine*, **42**(6), S29–S32, 2004.

[16] R.W.P. King, *The Theory of Linear Antenna*, Cambridge, MA: Harvard University Press, 1956.

[17] C.E. Baum, General properties of antennas, Sensor and Simulation Notes, note 330, Air Force Research Laboratory, 1991.

[18] M. Klemm and G. Troster, Characterisation of an aperture-stacked patch antenna for ultra-wideband wearable radio systems, *15th International Conference on Microwaves, Radar and Wireless Communications*, **2**, 395–8, May 2004.

[19] J.S. McLean, H. Foltz and R. Sutton, Pattern descriptors for UWB antennas, *IEEE Trans. on Antennas and Propagation*, **53**(1), 2005.

[20] S. Suh, W.L. Stutzman and W.A. Davis, A new ultrawideband printed monopole antenna: the planar inverted cone antenna (PICA), *IEEE Trans. on Antennas and Propagation*, **52**(5), 2004.

[21] A. Saitou, T. Iwaki, K. Honjo, K. Sato, T. Koyama and K. Watanabe, Practical realization of self-complementary broadband antenna on low-loss resin substrate for UWB applications, *2004 IEEE MTT-S International Microwave Symposium*, Fort Worth, Texas, June 2004.

[22] J. Yan, C. Xu and D. Xu, Time domain analysis of creeping wave, *Asia Pacific Microwave Conference*, Hong Kong, **3**, 1209–12. 1997.

[23] P. Hall, M. Ricci and T. Hee, Measurements of on-body propagation characteristics, *International Conference on Microwave and Millimeter Wave Technology*, 770–2, 2002.

[24] I. Kovacs, G. Pedersen, P. Eggers and K. Olesen, Ultra wideband radio propagation in body area network scenarios, *ISSSTA Proceedings*, 102–6, 2004.

[25] Y.I. Nechayev, P.S. Hall, C.C. Constantinou, Y. Hao, A. Alomainy, R. Dubrovka and C.G. Parini, On-body path gain variations with changing body posture and antenna position, *IEEE AP-S International Symposium and USNC/URSI National Radio Science Meeting*, Washington, DC, 2005.

[26] M.C. Lawton and J.P. McGeehan, The application of a deterministic ray launching algorithm for the prediction of radio channel characteristics in small-cell environments, *IEEE Transactions on Vehicular Technology*, **43**, 955–69, 1994.

[27] H. Ling. R.C. Chou and S.W. Lee, Shooting and bouncing rays: calculating the ROS of an arbitrary shaped cavity, *IEEE Transactions on Antennas and Propagation*, **37**, 194–205, 1989.

[28] D.A. McNamara, C.W.I. Pistorius and J.A.G. Malherbe, *Introduction to the Uniform Geometrical Theory of Diffraction*, Artech House, 1990.

[29] Y. Zhao, Y. Hao and C.G. Parini, Two novel FDTD based UWB indoor propagation models, 2005 *IEEE International Conference on Ultra-Wideband* (ICU 2005), Zurich, Switzerland, 5–8 September, 2005.

[30] An Internet resource for the calculation of the dielectric properties of body tissues, Institute for Applied Physics, Italian National Research Council, http://niremf.ifac.cnr.it/tissprop/.

[31] P.A. Tirkas and C.A. Balanis, Finite-difference time-domain method for antenna radiation, *IEEE Transactions on Antennas and Propagation*, **40**(3), 334–40, 1992.

[32] Y. Hao and C.J. Railton, Analyzing electromagnetic structures with curved boundaries on Cartesian FDTD meshes, *IEEE Transactions on Microwave Theory and Techniques*, **46**(1), 82–8, 1998.

[33] Y. Wang, S. Safavi-Naeini and S.K. Chaudhuri, A hybrid technique based on combining ray tracing and FDTD methods for site-specific modelling of indoor radio wave propagation, *IEEE Transactions on Antennas Propagation*, **AP-48**(5), 743–54, 2000.

[34] MBOA, MultiBand OFDM physical layer proposal for IEEE 802.15.3a, IEEE P802.15 Working Group for WPANs, September 2004.

[35] D.D. Wentzloff, R. Blázquez, F.S. Lee, B.P. Ginsburg, J. Powell and A.P. Chandrakasan, System design considerations for ultra-wideband communication, *IEEE Communication Magazine, Circuits for Communications*, **43**(8), 114–21, 2005.

[36] E.R. Bastidas-Pugo, F. Ramirez-Mireles and D. Munoz-Rodriguez, Performance of UWB PPM in residential multipath environments, *IEEE 58th Vehicular Technology Conference*, 6–9 October, 4, 2307–11, 2003.

[37] F. Ramirez-Mireles, Performance of ultrawideband SSMA using time hopping and M-ary PPM, *IEEE Journal on Selected Areas in Communications*, **19**(6), 1186–96, 2001.

[38] M. Kamoun, M. de Courville, L. Mazet and P. Duhamel, Impact of desynchronization on PPM UWB systems: a capacity based approach, *IEEE Information Theory Workshop*, 24–9 October, 198–203, 2004.

[39] W. Cheol Chung and D. Sam Ha, On the performance of bi-phase modulated UWB signals in a multipath channel, *IEEE Vehicular Technology Conference*, Jeju, Korea, 1654–8, 2003.

[40] S. Bagga, W.A. Serdijn and J.R. Long, A PPM Gaussian monocycle transmitter for ultra-wideband communications, *IEEE Joint International Workshop of UWBST and IWUWBS*, 130–4, 2004.

[41] F. Ramirez-Mireles and R. Scholtz, Performance of equicorrelated ultra-wideband pulse-position-modulated signals in the indoor wireless impulse radio channel, *IEEE PACRIM'97*, **2**, 640–4, August 1997.

[42] J. Choi and W. Stark, Performance of ultra-wideband communications with suboptimal receivers in multipath channels, *IEEE Journal on Selected Areas in Communications*, **20**(9), 1754–66, 2002.

[43] D. Cassioli, M. Win, F. Vatalaro and A.F. Molish, Performance of low-complexity Rake reception in a realistic UWB channel, *IEEE ICC'02*, **2**, 763–7, March 2002.

[44] Y. Zhao, A. Alomainy, Y. Hao, C. Parini, UWB On-Body Radio Channel Modelling Using Ray Theory and Sub-band FDTD Method, *IEEE Transactions on Microwave Theory Technology*, **54**(4), 1827–1835, 2006

17

Ultra-wideband Spatial Channel Characteristics

Wasim Q. Malik, Junsheng Liu, Ben Allen and David J. Edwards

17.1 Introduction

The Shannon capacity of a wireless channel varies logarithmically with SNR and linearly with bandwidth. The latter is especially of interest for UWB systems, which yield very high information rates as a direct consequence. As both the signal bandwidth and the transmit power, and consequently SNR, of UWB systems are controlled by regulations, increasing the capacity beyond a certain limit is not possible in the conventional manner. An alternative strategy is to exploit the spatial properties of the channel using multi-antenna arrays, which is the topic of this chapter.

Exploiting the spatial dimension, under appropriate operating conditions, offers some key advantages. Specialised UWB applications that require very high data rates can scale up the system capacity manyfold by spatial multiplexing. In addition, antenna diversity can enhance the SNR at the receiver, thereby increasing the system reliability, improving robustness to outage and extending the coverage range.

In this chapter, a detailed treatment of multiple-antenna UWB channel characteristics is presented. The achievable capacity and diversity improvement is investigated with the help of channel measurements. The relationship between the spatial system performance and the propagation environment is discussed in detail. It is demonstrated that the performance gain is very significant even with a small spatial array, underlining the value of spatial techniques for UWB systems.

17.2 Preliminaries

The use of multiple antennas at the transmitter and receiver, along with appropriate coding and decoding, has shown great promise for high data rate communications in the narrowband regime. A recent technique, referred to as multiple-input multiple-output (MIMO), makes it possible for a communications link to establish multiple, parallel end-to-end subchannels. Independent data streams can be transmitted over each of these subchannels, and the composite channel capacity increases dramatically [1–3]. As a result,

---

Ultra-wideband Antennas and Propagation for Communications, Radar and Imaging Edited by B. Allen, M. Dohler, E. E. Okon, W. Q. Malik, A. K. Brown and D. J. Edwards
© 2007 John Wiley & Sons, Ltd

a system with N_T transmitting and N_R receiving antennas can, under favourable channel conditions, achieve $N_{min} = \min\{N_T, N_R\}$ times the capacity of a traditional single-antenna system. As the radio spectrum is an increasingly scarce and expensive resource, such a dramatic increase in the spectral efficiency is of remarkable value. Closely associated with this concept is the somewhat more traditional idea of antenna diversity [4], in which multiple copies of the signal are combined to overcome fading and boost the received power, as discussed in Chapter 6. While primarily meant for fading mitigation and outage reduction, diversity systems can also be used to increase the channel capacity with appropriate signalling and combining through the increased SNR in accordance with the Shannon capacity relation. A multiple-antenna system can thus be used to obtain diversity, spatial multiplexing, or both, and the underlying tradeoffs are discussed in [5].

Multiple-antenna systems can be classified on the basis of the array configuration on either side of the link. Single-input multiple-output (SIMO) systems, with a single transmitting antenna and a receiving array, have been used for receive diversity for many decades. Multiple-input single-output (MISO) systems are constructed in the reciprocal manner, and can be used for transmit diversity. MIMO systems consist of antenna arrays at both ends, while single-input single-output (SISO) is the term used to refer to traditional single-antenna systems. Notationally, a multiple-antenna array configuration is represented as $N_T \times N_R$, where N_T and N_R denote the transmitting and receiving array size, respectively. Thus SIMO and MISO systems are represented by $1 \times N_R$ and $N_T \times 1$, respectively, while single-antenna, or single-input single-output (SISO) systems are denoted as 1×1.

Our treatment in this chapter will be based on uniform linear arrays (ULAs). We will consider the cases of the availability of channel state information at the receiver (CSIR) only, or also at transmitter (CSIT). It should be noted that the former is of greater practical significance as system constraints often render reverse channel estimation difficult. On the other hand, a system with feedback, as first studied by Shannon [6], can achieve the maximum capacity of an AWGN channel in a fading channel with appropriate encoding [7]. Significant capacity gains can be expected due to CSIT in multiple-antenna channels [8], and for this reason, we will also include CSIT in our analysis and highlight its implications for UWB systems.

17.3 UWB Spatial Channel Representation

The UWB channel is frequency selective and the channel fading coefficients are therefore frequency dependent. The multiple-antenna UWB channel can be represented as $\mathbf{H}^{(UWB)} \in \mathbf{C}^{N_R \times N_T \times N_f}$, where N_R, N_T and N_f denote the number of receiving antennas, transmitting antennas and frequency components, respectively. Here, $\mathbf{H}^{(UWB)}$ can be perceived as a frequency-domain vector each of whose elements is the flat-channel MIMO matrix, $\mathbf{H}_f$, at frequency $f \in [f_l, f_h]$, where f_l and f_h define the lower- and upper-end frequencies of the channel transfer function. For notational convenience, we will drop the subscript of $\mathbf{H}_f$ unless it is required for clarity.

Thus, for a given f, we have the familiar flat-channel MIMO relation,

$$\mathbf{y} = \sqrt{\frac{E_s}{N_T}} \mathbf{H} \mathbf{x} + \mathbf{n}, \qquad (17.1)$$

where $\mathbf{x}$ and $\mathbf{y}$ are the transmitted and received signal vectors at f, respectively, E_s is the average symbol transmit energy, $\mathbf{n}$ is the zero-mean complex Gaussian noise vector with power spectral density N_0, and $\mathbf{H}$ is the spatial channel matrix comprising the flat-fading coefficients. The channel in the above expression is normalised so that each underlying flat SISO channel has unit power. In the explicit form,

$$\begin{bmatrix} y_1 \\ \vdots \\ y_{N_R} \end{bmatrix} = \sqrt{\frac{E_s}{N_T}} \begin{bmatrix} h_{1,1} & \cdots & h_{1,N_T} \\ \vdots & \ddots & \vdots \\ h_{N_R,1} & \cdots & h_{N_R,N_T} \end{bmatrix} \begin{bmatrix} x_1 \\ \vdots \\ x_{N_T} \end{bmatrix} + \begin{bmatrix} n_1 \\ \vdots \\ n_{N_R} \end{bmatrix} \qquad (17.2)$$

where $h_{m,n}$ specifies the scalar transfer coefficient between the mth transmitter and the nth receiver. Thus, $\mathbf{H}$ is an $N_R \times N_T$ matrix, $\mathbf{x}$ is an $N_T \times 1$ vector, and $\mathbf{y}$ and $\mathbf{n}$ are $N_R \times 1$ vectors, respectively. The rank, $r_\mathbf{H}$, of $\mathbf{H}$ determines the achievable MIMO capacity and diversity gain over a SISO system. The equivalent expressions for SIMO and MISO can be obtained from the generalised MIMO channel relation by substituting the appropriate value of N_T and N_R, and using the corresponding matrix dimensions in Equation (17.2).

It is obvious that with the above representation, the UWB channel reduces to a set of flat channels centred at a single frequency component. The advantage of this methodology is that the theory of flat spatial channels, developed extensively in recent information theoretic literature, can be directly extended to UWB spatial channels. Additionally, frequency-domain analysis relates directly to the appropriate channel characterisation techniques, as UWB channel measurement is relatively easier in the frequency domain due to the complexities associated with time-domain pulse generation and detection. While time-domain treatment of frequency-selective spatial channels is also possible [9], this chapter will use the frequency-domain representation for the above reasons.

For the frequency-flat spatial channel $\mathbf{H}$, we can now define the matrix

$$\mathbf{G} = \begin{cases} \mathbf{H}^H \mathbf{H}, & \text{if } N_R > N_T \\ \mathbf{H} \mathbf{H}^H, & \text{if } N_R \leq N_T \end{cases} \qquad (17.3)$$

where $(.)^H$ denotes the Hermitian transposition. The matrix $\mathbf{G}$ has dimensions $N_{min} \times N_{min}$, where $N_{min} = \min\{N_T, N_R\}$. The properties of $\mathbf{G}$ are central to MIMO performance characterisation, as discussed in detail later. Note that $\mathbf{G}$, or $\mathbf{G}_f$ to explicitly indicate the single-frequency specification, corresponds to $\mathbf{H}_f$, and the frequency-selective equivalent of $\mathbf{G}$ is denoted by $\mathbf{G}^{(UWB)}$.

17.4 Characterisation Techniques

From previous discussion, it is clear that UWB channel characterisation and modelling must take into consideration the repercussions of the large signal bandwidth. These often amount to the peculiarities introduced by the channel's frequency selectivity, which is a consequence of rich scattering and high multipath resolution. The same holds true in the context of the spatial channel, formed by multiple antennas at each end of the link. While channel characterisation can be performed in either time or frequency domain, the latter is more popular in the wideband regime due to its experimental and mathematical convenience. Frequency-domain coherent channel sounding can be performed with the help of vector network analysers (VNAs), and the amplitude and phase of the channel response at various discrete frequency points in the band of interest is measured. The frequency-domain sample spacing must be smaller than the coherence bandwidth of the channel.

In a wide-sense stationary environment, the spatial channel can be measured using MIMO channel sounders, which transmit and receive multiple signals from multiple antennas sequentially. While this does not mimic the actual operating scenario with concurrent transmission and reception, it does capture the essential channel characteristics allowing for the reconstruction of the actual spatial channel with simultaneous signalling from all antennas. A further scaled down measurement configuration requiring only a SISO channel sounder involves synthesised transmit and receive arrays, but requires a more carefully controlled experimental environment with a longer channel coherence time.

The VNA measures the complex frequency transfer coefficients $h_{r,t,f}$ for the link between transmit antenna t and receive antenna r, where $t = \{1,\ldots,N_T\}$, $r = \{1,\ldots,N_R\}$, and $f = \{1,\ldots,N_f\}$. The measured channel matrix for the $N_R \times N_T$ antenna array is thus recorded as $\mathbf{H}_m^{(UWB)} \in \mathbf{C}^{N_R \times N_T \times N_f}$ with elements $h_{r,t,f}$. The power normalisation of the spatial channel is performed such that

$$\mathbf{H}^{(UWB)} = \left(\frac{1}{N_T N_R N_f} \sum_{r=1}^{N_R} \sum_{t=1}^{N_T} \sum_{f=1}^{N_f} |h_{r,t,f}|^2\right)^{-\frac{1}{2}} \mathbf{H}_m^{(UWB)}. \tag{17.4}$$

It is obvious that with this normalisation, we have

$$\left\|\mathbf{H}^{(UWB)}\right\|_F^2 = \sum_{f=1}^{N_f} \mathrm{Tr}\{\mathbf{G}_f\} = N_T N_R N_f, \tag{17.5}$$

where $\|.\|_F$ denotes the Frobenius norm, and $\mathrm{Tr}\{.\}$ denotes the matrix trace.

The measurement results described in this chapter are based on VNA-assisted channel sounding in the FCC-allocated UWB frequency band, i.e. 3.1–10.6 GHz, in an indoor environment. The channel, with bandwidth $B = 7.5$ GHz, is sounded at $N_f = 1601$ points at intervals of 4.6875 MHz, which is smaller than the UWB channel's coherence bandwidth. Transmitter and receiver ULAs are virtually synthesised with up to three omnidirectional discone antennas [10]. The inter-element spacing is chosen to be 6 cm, so that it is larger than the half-wavelength distance for all signal components, avoiding MIMO subchannel correlation. The arrays are oriented to each other's broadside direction. An ensemble of $N_S = 900$ spatial channel measurements is collected in this manner over a 1 m^2 area for statistical reliability using an automated positioning grid. Antenna coupling effects are not considered in our analysis, since the mutual coupling of antenna elements is insignificant for capacity and diversity performance evaluation when the inter-element spacing is larger than 0.4 of the wavelength [11].

17.5 Increase in the Communication Rate

The throughput of a communications link is determined by the achievable channel capacity with a vanishingly small error rate under appropriate signalling. Thus, the evaluation of channel capacity is central to the design of very high data-rate devices. Also referred to as the spectral efficiency, the MIMO channel capacity is the overall capacity in bps/Hz provided by a multiple-antenna system, and is higher than single-antenna capacity under suitable conditions. Similar to the Shannon capacity for SISO channels, the SIMO and MISO capacity improvement is a consequence of the improved SNR. MIMO systems, on the other hand, can use spatial multiplexing with the BLAST technique [12] to effectively create parallel wireless channels and communicate over the spatial eigenmodes, providing a sizeable capacity gain.

17.5.1 UWB Channel Capacity

The single-antenna channel capacity, C, is the mutual information, $I(X;Y)$, maximised over the probability distribution, $p(x)$, of the transmitted symbol vector, X [13]. Mathematically,

$$C = \max_{p(x)} \{I(X;Y)\}. \tag{17.6}$$

The capacity of a frequency-selective channel can be evaluated as the expectation of the narrowband capacity over the frequency band [9]. In discrete frequency terms, this can be achieved by averaging the respective single-frequency capacities evaluated at frequency intervals smaller than the channel's coherence bandwidth. Thus, if C_f represents the capacity of a flat-fading, or narrowband, deterministic channel centred at frequency f, then the capacity, $C^{(UWB)}$, of the UWB channel spanning a large set of such narrowband channels is calculated as [9]

$$C^{(UWB)} = \frac{1}{N_f} \sum_{f=1}^{N_f} C_f. \tag{17.7}$$

As the UWB channel capacity is directly related to the capacities of the constituent flat channels, we next review the concepts related to the latter. We will again drop the subscript of C_f for simplicity of presentation.

17.5.2 Capacity with CSIR Only

The capacity of a multiple-antenna flat-fading channel with side information at the receiver can be calculated as [2]

$$C_{MIMO} = \log_2 \det \left(I_{N_R} + \frac{\rho}{N_T} \mathbf{G} \right), \tag{17.8}$$

where $\rho = E_s/N_0$ is the average SNR at each of the N_R receiving antennas. $\mathbf{G}$ is obtained from the corresponding flat channel matrix $\mathbf{H}$ as described by Equation (17.3), and can be eigen-decomposed as

$$\mathbf{G} = \mathbf{U}\Lambda\mathbf{V}^H, \tag{17.9}$$

where $\Lambda = \mathrm{diag}\{\lambda_i\}, i = \{1, \ldots, N_{min}\}$, is a diagonal matrix whose elements along the principal diagonal are the eigenvalues of $\mathbf{G}$, while $\mathbf{U}$ and $\mathbf{V}$ are unitary matrices. $\mathbf{G}$ is a positive semi-definite Hermitian matrix with dimensions $N_{min} \times N_{min}$, therefore its N_{min} eigenvalues are real and non-negative, i.e., $\lambda_i \geq 0$. Using this decomposition, we can express Equation (17.8) with some algebraic manipulation as

$$C_{MIMO} = \log_2 \det \left(I_{N_R} + \frac{\rho}{N_T} \Lambda \right), \tag{17.10}$$

which further simplifies to

$$C_{MIMO} = \sum_{i=1}^{r_\mathbf{H}} \log_2 \left(1 + \frac{\rho}{N_T} \lambda_i \right). \tag{17.11}$$

It should be noted that $r_\mathbf{H} \leq N_{min}$. The equality holds under rich scattering conditions so that at large N_{min} the MIMO capacity approaches

$$C_{MIMO} = N_{min} \log_2 (1 + \rho), \tag{17.12}$$

depicting a linear increase in capacity with N_{min} at constant ρ.

SIMO is a special case of MIMO with $N_T = 1$, for which

$$C_{SIMO} = \log_2 \left(1 + \rho \|\mathbf{H}\|_F^2 \right). \tag{17.13}$$

where $\mathbf{H} = \{h_{n,1}\}$, $n = \{1, \ldots, N_R\}$ is the SIMO channel vector. Since $N_T = N_f = 1$, we have $\|\mathbf{H}\|_F^2 = N_R$, or equivalently, $\sum_{i=1}^{N_R} |h_{n,1}|^2 = N_R$, so that we can write

$$C_{SIMO} = \log_2 (1 + \rho N_R), \qquad (17.14)$$

which indicates a logarithmic increase in SIMO capacity with increase in N_R at high SNR.

In the case of MISO, $\|\mathbf{H}\|_F^2 = N_T$, and the capacity expression therefore reduces to

$$C_{MISO} = \log_2 \left(1 + \frac{\rho}{N_T} \|\mathbf{H}\|_F^2\right) = \log_2 (1 + \rho), \qquad (17.15)$$

showing that an increase in the number of transmit antennas in a system with CSIR but no CSIT provides no capacity improvement over a single-antenna system.

17.5.3 Capacity with CSIT

Given both CSIT and CSIR, the MIMO signals can be differentially weighted across the transmit and receive arrays using the $\mathbf{U}$ and $\mathbf{V}$ matrices in order to distribute power more efficiently among the available eigenmodes using the water-filling scheme [13]. The resulting capacity is given by [2]

$$C_{MIMO} = \sum_{i=1}^{r_\mathbf{H}} \left\{ \log_2 \left(\frac{\rho}{N_T} \mu \lambda_i \right) \right\}^+, \qquad (17.16)$$

where

$$\mu = \frac{N_T}{r_\mathbf{H}} \left(1 + \frac{1}{\rho} \sum_{i=1}^{r_\mathbf{H}} \frac{1}{\lambda_i} \right), \qquad (17.17)$$

and the notation $\{.\}^+$ denotes the set of positive values. This expression, valid under CSIT, yields higher system capacity than that obtained with CSIR and equal power distribution when the MIMO subchannels have unequal power gains. The capacity improvement, however, comes at the cost of increased system complexity arising from the requirement for a feedback link and an optimal power allocation unit at the transmitting antenna array.

The SIMO capacity is unaffected by the availability of CSIT as long as CSIR is present. The MISO capacity, on the other hand, registers an increase with CSIT, as we can pre-weight the transmit signal power for each antenna to maximise the received signal SNR using maximal-ratio transmission [14]. The resulting MISO capacity is then given by

$$C_{MISO} = \log_2 (1 + \rho N_T), \qquad (17.18)$$

which is the dual of Equation (17.14). Thus MISO and SIMO yield similar performance in spatially uncorrelated channels, but the former is useful when it is desired to transfer the complexity from the receiver to the transmitter.

17.5.4 Statistical Characterisation

The above expressions assume $\mathbf{H}$ to be deterministic. Wireless fading channels are, however, random due to fluctuations in the received signal and noise levels. In particular, multipath interference introduces small-scale fading which can be modelled as a random variation overlaid on the mean signal strength determined by the large-scale pathloss [15]. As a result of this random variation of the channel fading coefficients, the spatial channel matrix, **H**, is a stochastic matrix. Quantities that are derived from **H**, such as the subchannel correlation coefficients, eigenvalues and capacities are also therefore random, and can be described by probability distributions.

The channel is referred to as an ergodic channel if its random fading characteristics vary over the transmitted signal codeword, but all of its moments remain constant from codeword to codeword. The corresponding capacity, referred to as ergodic capacity, can be computed as the ensemble average of the capacity over the distribution of the elements of **H**, i.e.,

$$\bar{C} = \mathrm{E}\{C_{(k)}\}, \tag{17.19}$$

where $C_{(k)}$ represents the capacity of the kth realisation of the spatial channel denoted by $\mathbf{H}_{(k)}$.

In contrast, the non-ergodic channel is modelled with channel realisations chosen randomly before transmission and constant over the duration of the codeword transmission. There is, thus, a finite probability, referred to as the rate outage probability, that the channel will not support a given communication rate so that system outage, defined as the event $C \leq C_o$, will occur. The maximum capacity supported by the channel in that case, for a given outage probability, is often referred to as the outage capacity, C_o. Thus for a $p\%$ outage probability, the cumulative distribution, $P(C)$, of the capacity gives $P(C > C_o) = (100 - p)\%$. The outage capacity thus determines the achievable capacity for a given bit error rate, and is therefore of greater practical interest than the ergodic capacity in realistic wireless channels. Some specific values of the outage probability useful for system design and comparative analysis are 1 % and 10 %, which indicate the likelihood of the system not achieving a given capacity bound.

The ergodic capacity asymptotically approaches the outage capacity with bandwidth, i.e., $\bar{C} \to C_o$ as $B \to \infty$. This is due to the vanishing variance of the capacity distribution at large B, arising from the infinite frequency diversity in the limiting case. Thus the outage capacity of a UWB channel is higher than that of a narrowband channel while the ergodic capacity remains approximately constant. This holds for any array size, and is indeed true even for a single-antenna link, as it results from the low signal fluctuation in UWB channels. Recent results in MIMO literature involving wideband and UWB signalling confirm the tightening of the capacity bounds with bandwidth [16, 17]. It is worth emphasising that this extra improvement in the channel capacity, specified in bps/Hz, is in addition to the linear gain in the data rate, expressed in bps, that results from the use of a large bandwidth. This increase in the outage capacity is an inherent benefit of UWB channels for high data-rate communications systems.

17.5.5 Experimental Evaluation of Capacity

The UWB spatial channel capacity is derived from the channel measurements using the above formulation. Figure 17.1 shows the empirical CDFs of MIMO capacity with CSIR and uniform power allocation, evaluated at $\rho = 10$ dB. It can be seen that the UWB capacity registers only a logarithmic increase with N in the case of SIMO arrays as predicted theoretically, while a much larger capacity gain is observed with MIMO configurations. Thus 10 dB SNR, a 3×3 UWB system without CSIT provides 6.84 bps/Hz at 1 % outage. The large gradient of the CDFs in the figure shows that the variance of UWB capacity is lower than that usually observed for narrowband channels [16]. From the measurement, the difference between the ergodic and outage capacity is 7.15 % for the 1×1 system, and falls with the array size, reaching 5.39 % for the 3×3 system. Thus C_o converges to $\bar{C}$ with both bandwidth and array size. Table 17.1 lists the 1 % outage capacity for a variety of antenna configurations evaluated using UWB channel measurements.

Figure 17.2 shows the capacity CDFs when CSIT is available and optimal power allocation is used with MISO and MIMO arrays, once again assuming $\rho = 10$ dB. The MISO capacity in the figure

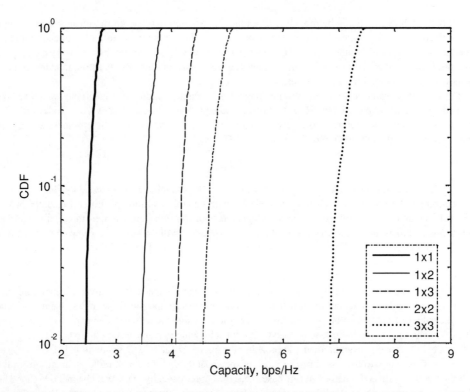

Figure 17.1 Measured capacity of the UWB spatial channel with CSIR and with uniform power allocation at 10 dB SNR

Table 17.1 The 1 % outage capacity, in bps/Hz, of multiple-antenna systems normalised at 10 dB SNR obtained from indoor UWB channel measurements

Array configuration	Array size, N		
	1	2	3
SISO (1×1)	2.47	—	—
SIMO ($1 \times N$)	—	3.48	4.09
MISO ($N \times 1$)	—	3.47	4.14
MIMO ($N \times N$), without CSIT	—	4.58	6.85
MIMO ($N \times N$), with CSIT	—	5.09	8.24

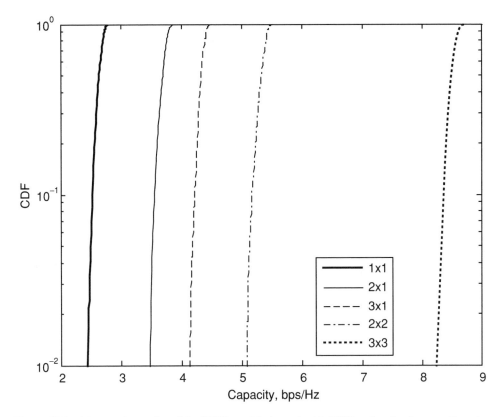

Figure 17.2 Measured capacity of the UWB spatial channel with CSIT and optimal power allocation at 10 dB SNR

is comparable to the SIMO capacity in Figure 17.1, as predicted theoretically. The MIMO capacity increases considerably under CSIT, and a 3 × 3 array provides 8.24 bps/Hz, which is a 20 % improvement over a system without CSIT. It is also observed that the capacity gain with water filling increases with the array size.

Due to low transmit power levels, UWB systems operate at low SNRs. Therefore we now study the effect of SNR on the capacity gain achieved with spatial multiplexing. Figure 17.3 shows the variation of the 1 % outage capacity with SNR. As noted above, the SIMO capacity without CSIT is equal to the MISO capacity with CSIT, so the former is not shown. It can be observed that as $\rho \to \infty$, the advantage of CSIT vanishes, and the capacities of a spatial UWB system with and without CSIT converge. It is found that the asymptotic capacity tangent approaches the array size. This behaviour is in agreement with theoretical predictions [18]. At low SNRs, however, the MIMO capacities with and without CSIT diverge considerably, as noted in [9], with the latter approaching the capacities of MISO and SISO configurations. Thus at $\rho = 0$ dB, the 2 × 2 CSIT system has higher capacity than a 3 × 3 CSIR system, which indicates that for low SNR systems, it can be more advantageous to provide a feedback link than to increase the MIMO array size. Also, for a 3 × 3 system at this SNR, the availability of CSIT increases the capacity by 3.2 bps/Hz, or 200 %.

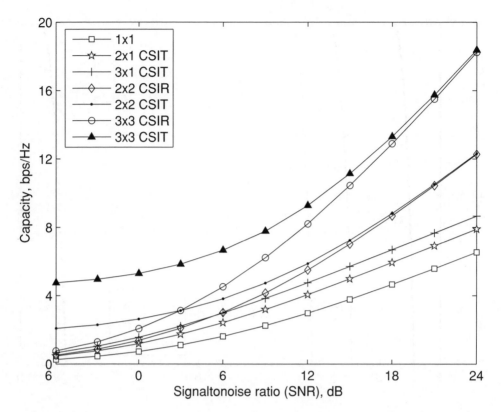

Figure 17.3 The measured 1 % outage capacity for the UWB spatial channel as a function of SNR. The SIMO capacity with CSIR is identical to the MISO capacity with CSIT shown

17.6 Signal Quality Improvement

Diversity systems aim to mitigate spatial or temporal fading, and thus reduce the outage probability and error rates. Traditional space diversity is achieved with the help of antenna arrays at the receiver or transmitter. The diversity improvement in conventional systems relates to the increase in the diversity order when the array size is increased [19]. Other parameters, such as the reduction in the level crossing rate, fade duration and fade depth, are also sometimes used to gauge diversity performance.

As the system bandwidth increases, the fading phenomena get progressively less severe. A modified definition of diversity gain is thus in order for UWB systems. We define the improvement in the SNR at the combiner output as the UWB diversity gain. With optimal linear combining, this includes the array gain achieved by the coherent combining of the independently fading incident signals [20]. It is obvious that the results can be directly related to symbol error rate reduction. In addition, the SNR gain can be translated into coverage range extension, which is an issue of greater significance than fading for power-constrained UWB systems and provides the major motivation for antenna diversity, as noted in [17, 21]. Multiple-antenna approaches to link quality improvement are especially attractive for UWB, since boosting the transmit power beyond the EIRP limits is not permissible under the radio regulations.

The SNR gain due to antenna diversity is defined as the dB difference between the SNRs of the diversity-combined and the incident signals. For a spatial channel, it can be represented by

$$\Delta\Gamma = \frac{\Gamma}{\rho_{ref}}, \qquad (17.20)$$

where ρ_{ref} is the SNR at the receive antenna designated as the reference in a SISO configuration.

17.6.1 UWB SNR Gain

The SNR of a UWB channel, $\Gamma^{(UWB)}$, can be evaluated by averaging the SNRs, Γ_f, of the constituent flat-fading channels over the signal spectrum. Mathematically,

$$\Gamma^{(UWB)} = \frac{1}{N_f} \sum_{f=1}^{N_f} \Gamma_f. \qquad (17.21)$$

The flat-channel SNR, Γ_f, is obtained from the flat spatial channel matrix $\mathbf{H}$ at frequency f, as discussed in the next two sections. Following the convention in this chapter, we will omit the subscript of Γ_f.

17.6.2 SNR Gain with CSIR Only

When only CSIR is available, SIMO and MIMO arrays can be used to boost the received signal SNR. For SIMO diversity, the optimal combining scheme is maximal-ratio combining (MRC). The receive antenna weight vector that maximises the output SNR is $\mathbf{w} = \mathbf{H}^H$. If the SNR of the signal received at the nth antenna is ρ_n, the combined signal SNR is given by

$$\Gamma_{SIMO} = \sum_{n=1}^{N_R} \rho_n. \qquad (17.22)$$

If the average received signal SNR is uniform across the array, i.e. $\rho_1 = \ldots = \rho_{N_R} = \rho$, then

$$\Gamma_{SIMO} = \rho \|\mathbf{H}\|_F^2 = \rho N_R. \qquad (17.23)$$

A MIMO array, with the Alamouti transmission scheme [22], may also be used for diversity reception in a situation with CSIR only. This scheme can extract only the receive array gain, and the SNR gain is, therefore, of the order of

$$\Gamma_{MIMO} = \frac{\rho \|\mathbf{H}\|_F^2}{N_T} = \rho N_R. \qquad (17.24)$$

As a corollary, a MISO system does not provides any improvement in the average SNR at the receiver in the absence of CSIT.

17.6.3 SNR Gain with CSIT

When CSIT is available to a MISO system, a maximal-ratio transmission strategy [14] can be used for transmit diversity. This transmit-MRC scheme is conceptually similar to the receive-MRC scheme described by Equation (17.22). Thus with appropriate signal weighting at the transmit array, represented by the vector $\mathbf{w} = \mathbf{H}^H$, the SNR of the received signal is

$$\Gamma_{MISO} = \sum_{n=1}^{N_T} \rho_n. \tag{17.25}$$

Similarly, if the average SNR, ρ, is uniform across the transmit antenna array, then

$$\Gamma_{MISO} = \rho \|\mathbf{H}\|_F^2 = \rho N_T. \tag{17.26}$$

Thus SIMO and MISO diversity performance is equivalent under perfect CSI at the appropriate end [9]. Thus MISO can also be effectively used in UWB channels, particularly in applications where it is desired to reduce receiver complexity.

A MIMO array can now use dominant eigenmode transmission to pre-weight the signals across the transmit array, in essence extending the transmit-MRC scheme across the multiple transmission modes that are now available. The SNR at the receiver is now increased by a factor equivalent to the largest eigenvalue, λ_{max}, of $\mathbf{G}$, so that it is given by [9]

$$\Gamma_{MIMO} = \rho \lambda_{max}, \tag{17.27}$$

where λ_{max} is bounded as

$$\frac{\|\mathbf{H}\|_F^2}{\mathbf{r_H}} \leq \lambda_{max} \leq \|\mathbf{H}\|_F^2. \tag{17.28}$$

If for a MIMO configuration, $N_T = N_T = N$, and the matrix is full-rank, i.e., $\mathbf{r_H} = N$, then from Equations (17.27) and (17.28), the SNR gain with CSIT is lower-bounded by the SNR gain without CSIT. Thus CSIT always improves the performance of a MIMO diversity system.

17.6.4 Statistical Characterisation of Diversity

As detailed in Section 17.5.4, signal level fluctuations due to fading necessitate a stochastic treatment of the wireless channel characteristics. The SNR, and indeed the SNR gain due to antenna diversity, for spatial radio channels is thus also described in terms of the relevant statistical values referred to as the ergodic and outage diversity gains, denoted by $\Delta\bar{\Gamma}$ and $\Delta\Gamma_o$ respectively.

17.6.5 Experimental Evaluation

The measured SNR at a single reference receive antenna, averaged across the N_S channel transfer functions, i.e.,

$$\rho_{ref} = \frac{1}{N_S} \sum_{s=1}^{N_S} \rho_s, \tag{17.29}$$

is used to normalise the SNR at a given receive antenna. The combiner output SNR is also similarly normalised by ρ_{ref} to evaluate the SNR gain due to antenna diversity. The first-order statistics of the UWB diversity gain, denoted by $\Delta\Gamma$, are thus estimated from the measurement data.

Figure 17.4 shows the measured $\Delta\Gamma$ for UWB SIMO systems with CSIR only. In the light of the discussion earlier in this section, only SIMO arrays can provide any performance gain in the absence of CSIT, and the figure therefore shows the normalised SNR gains for $N_T = 1$ and $N_R = \{1, 2, 3\}$. Because

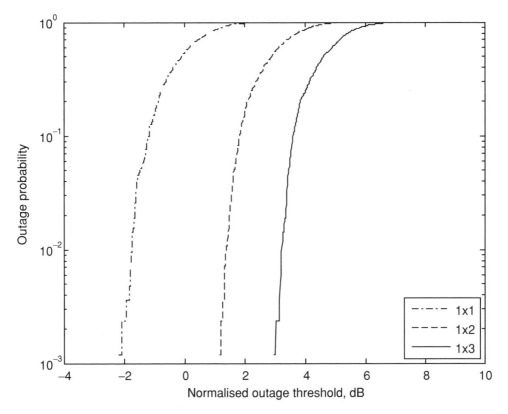

Figure 17.4 The measured output SNR of a UWB SIMO system with CSIR and maximal-ratio combining, normalised to the mean SISO SNR

of the normalisation, the outage probability for the 1×1 link is 50 % when the normalised outage SNR threshold, Γ_{th}, is 0 dB. From a comparison of the curves for the three array configurations, there is only a slight increase in the gradient, or the diversity order, with N_R. Thus, as predicted above, a large antenna array provides very little improvement in the diversity order of a UWB system, and the major benefit is in the form of an increase in the received SNR arising from the array gain. The array gain is the mean increase in the received SNR due to the antenna array. From the figure, the ergodic diversity gain, with respect to the SNR of a single-antenna system, is 3 and 4.7 dB for $N_R = 2$ and 3, respectively, which is indeed equivalent to the expected array gain. The 1 % outage diversity gain, however, is slightly larger, as listed in Table 17.2, and the difference can be attributed to the small improvement in the diversity order.

Figure 17.5 shows the diversity gain of UWB multiple-antenna systems with CSIT. The performance curves for only MISO and MIMO arrays are shown, since SIMO performance is not affected by feedback. The diversity order improvement with multiple antennas is once again almost negligible, but there is a significant increase in the outage threshold with array size. Thus with a 3×3 array, a gain of 8.3 dB is observed when the outage probability is 1 %. While the MIMO array provides better diversity performance than the MISO array, its array gain falls short of the upper bound of $N_T N_R$. Similarly, the MISO array gain is slightly lower than the SIMO array gain due to the reasons explained above. The complete set of these quantitative results is summarised in Table 17.2.

Table 17.2 The SNR gain of a multiple-antenna system normalised to the SNR of a single received signal, obtained from indoor UWB channel measurements

Array configuration	Expected array gain	Array size		SNR gain, dB	
		Transmit (N_T)	Receive (N_R)	Ergodic	1% Outage
SIMO with CSIR	N_R	1	2	3.0	3.1
SIMO with CSIR	N_R	1	3	4.7	4.9
MISO with CSIT	N_T	2	1	2.7	3.1
MISO with CSIT	N_T	3	1	4.4	4.8
MIMO with CSIT	$[\max(N_T, N_R), N_T N_R]$	2	2	5.3	5.7
MIMO with CSIT	$[\max(N_T, N_R), N_T N_R]$	3	3	8.0	8.3

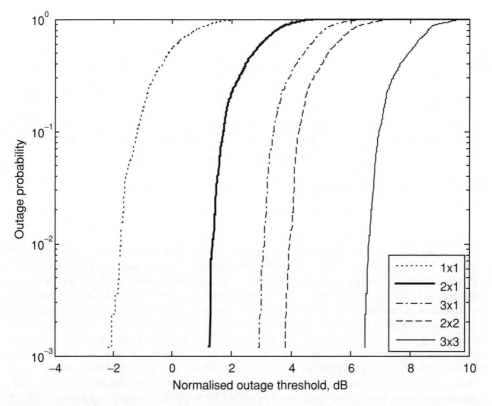

Figure 17.5 The measured output SNR of UWB MISO and MIMO systems with CSIT and optimal power transmission, normalised to the mean SISO SNR

17.6.6 Coverage Range Extension

The above discussion has investigated in detail the increase in SNR that a multiple-antenna UWB system can enjoy. In addition to improved outage performance and capacity, there is another key advantage that deserves special mention: coverage radius extension. Power-constrained UWB systems are severely range-limited, which is arguably the most major drawback of this technology that diminishes the scope of its applications. Current implementations of UWB indoor communications systems, based on single-antenna designs, are expected to offer very high data rates only up to a range of 10 m [23]. It is, however, desirable to extend the coverage range of a UWB point-to-point link beyond the current limit, so that floor-wide and inter-floor indoor wireless networks can be deployed more economically and conveniently.

Other factors being constant, the coverage range, d, depends on the pathloss experienced by the signal during propagation through the channel, since

$$PL_{dB} = 10n \log_{10} d, \tag{17.30}$$

where n is the pathloss index determined by the propagation environment, and PL_{dB} is the pathloss in dBs. An increase in the SNR, $\Delta\Gamma_{dB}$, at the receiver due to antenna diversity can be perceived as a reduction in the pathloss. Thus from Equation (17.30), we can relate the range extension, Δd, beyond the original range of operation, d_0, to the SNR gain in dBs, $\Delta\Gamma_{dB}$, in the form of the expression

$$\Delta d = d_0 \left(10^{\Delta\Gamma_{dB}/10n} - 1\right). \tag{17.31}$$

The pathloss index, n, in an indoor LOS environment is approximately 1.7 [24]. A 2×2 MIMO diversity system with CSIT provides an SNR gain of 5.7 dB, as discussed in Section 17.6.5. From (17.31), when $n = 1.7$, such a system can effectively double the coverage range. Similarly, a 3×3 system, with $\Delta\Gamma_{dB} = 8.3$ dB, can triple the range. With the propagation conditions kept constant, the latter configuration will thus increase the range of a UWB indoor system from 10 m to 30 m.

17.7 Performance Parameters

The properties of the spatial channel, and consequently the performance of a multiple-antenna system operating in that channel, are determined by several factors. Some of these are highlighted below and their impact on a multiple-antenna UWB system is quantified.

17.7.1 Spatial Fading Correlation

A high correlation between multiple-antenna signals can reduce the rank of the channel matrix and, in the limiting case, render it non-invertible. The spatial channel matrix then degenerates into a unit-rank matrix, implying a single degree of freedom or a single transmission mode as in a scalar channel. This results in annulling the spatial multiplexing gain, so that the advantage of $N \times N$ MIMO reduces to only the array gain, and the capacity is therefore upper-bounded by

$$C \leq \log_2(1 + N\rho). \tag{17.32}$$

The diversity performance is similarly affected, so that the effective diversity order is reduced to $N_{min} = 1$ in the extreme case [25]. In UWB channels, however, multiple-antenna systems do not improve the diversity order in the first place, as discussed in Section 17.6, and thus correlation does not affect the UWB diversity order. As the full array gain is realised even with correlated fading [9], the SNR gain obtained in Section 17.6 is nearly independent of the correlation.

The spatial channel correlation can be decomposed into the transmit correlation and the receive correlation under certain conditions [26]. The transmit correlation is a measure of correlation between the signals from the transmitting antenna array reaching a given receiving antenna, while the receive correlation is the dual quantity at the receiving array. Both of these components have similar impact on the channel capacity in general, and should simultaneously be sufficiently low for MIMO to provide a performance gain [9].

The loss in capacity due to correlation increases with SNR. Receive correlation is always detrimental to MIMO capacity, irrespective of the SNR, as it reduces the effective dimensionality of the channel. The same is true for transmit correlation at high SNR, but at low SNR it has a net beamforming effect [26]. The advantage of the latter phenomenon is a reduction in the required E_s/N_0, but on the downside it may now cause the signal power to exceed the EIRP limits in certain transmit directions and lead to interference to other devices. It is therefore important to carefully characterise the correlation properties of a spatial UWB system.

A correlation coefficient of up to 0.5 causes only a negligible reduction in capacity and is therefore acceptable in a multiple-antenna system [27–29]. At high SNR, if the correlation coefficient is 0.5, the capacity reduction is approximately 0.4 bps/Hz.

Signal decorrelation is introduced by multipath propagation, and is therefore higher in rich scattering environments. Sparse multipath propagation, low angular spread, closely spaced antenna elements, the keyhole effect [30] and other factors can lead to a loss in signal decorrelation. Under the Jakes' propagation model [31], with a ring of scatterers around the receiver, the relation between correlation and antenna separation is described by the Bessel function, and sufficient decorrelation is obtained at half-wavelength spacing [32]. Realistic UWB indoor channels, however, may differ considerably from this propagation scenario, depending on the specific application. In a UWB indoor wireless network, the access point may be placed close to the ceiling, so that the scatterer distribution around the transmitter is different from that at the receiver. On the other hand, in a wireless-USB application, the transmitter and receiver scattering environment is expected to be similar. While such specific scattering scenarios do affect the multipath correlation properties, the half-wavelength rule nevertheless provides a simple rule of thumb for array design.

As UWB signals operate at high frequencies, the half-wavelength antenna separation requirement translates into a smaller physical distance than required for narrowband mobile systems, making small-sized, compactly packed MIMO arrays practicable. Also, the large bandwidth of UWB signals, along with the large number of multipaths components that typically arise in indoor communications channels, result in lowering the correlation. This is verified by correlation measurements described below.

To evaluate the decorrelation properties of the channel, we analyse the correlation coefficients for all transmit and receive antenna combinations. In some measurement-based studies, where only power statistics are available, the power or envelope correlation is calculated for MIMO analysis. If vector channel data is available, such as with a complex electric field measurement, the extra phase information can be used to compute the complex correlation coefficient. The power, envelope and complex correlation coefficients are closely related [33].

The complex correlation coefficient, r_{mn}, for random variables **m** and **n** is given by

$$r_{mn} = \frac{\mathrm{E}\left[\mathbf{mn}^*\right] - \mathrm{E}\left[\mathbf{m}\right]\mathrm{E}\left[\mathbf{n}^*\right]}{\sqrt{\left(\mathrm{E}\left[|\mathbf{m}|^2\right] - |\mathrm{E}\left[\mathbf{m}\right]|^2\right)\left(\mathrm{E}\left[|\mathbf{n}|^2\right] - |\mathrm{E}\left[\mathbf{n}\right]|^2\right)}}, \qquad (17.33)$$

where (*) denotes complex conjugation. For our purpose, **m** and **n** represent the single-antenna frequency-domain channel vectors, i.e., **m** is the frequency transfer function between the tth transmitting and r^{th}

Table 17.3 The mean complex correlation coefficient magnitudes for various MIMO subchannel pairs, obtained from UWB measurements

	$T_3 R_3$	$T_3 R_2$	$T_3 R_1$	$T_2 R_3$	$T_2 R_2$	$T_2 R_1$	$T_1 R_3$	$T_1 R_2$	$T_1 R_1$
$T_1 R_1$	0.16	0.18	0.19	0.24	0.28	0.26	0.28	0.37	1
$T_1 R_2$	0.17	0.18	0.17	0.27	0.25	0.29	0.36	1	
$T_1 R_3$	0.18	0.17	0.16	0.23	0.27	0.37	1		
$T_2 R_1$	0.17	0.17	0.18	0.29	0.36	1			
$T_2 R_2$	0.16	0.18	0.19	0.35	1				
$T_2 R_3$	0.17	0.18	0.17	1					
$T_3 R_1$	0.27	0.34	1						
$T_3 R_2$	0.34	1							
$T_3 R_3$	1								

receiving antenna, or

$$\mathbf{m} = [h_{t,r,f}], \qquad f = \{1, \ldots, N_f\}, \tag{17.34}$$

and $\mathbf{n}$ is defined similarly. As r_{mn} is a random variable, we analyse its first order magnitude statistics.

Table 17.3 lists the spatially averaged $|r_{mn}|$ for all subchannel combinations. The channel $\mathbf{m} = T_t R_r$ is defined by the tth transmitting and r^{th} receiving antenna. By definition, $r_{mn} = r_{nm}^*$, and therefore only one of each pair of subchannel cross-correlation coefficients is listed in the table.

It is observed that the correlation coefficient magnitudes are smaller than 0.4 in all cases. While this non-zero correlation prevents the system from achieving the performance attainable in a perfectly uncorrelated spatial channel, it is not large enough for the impact to be significant, as discussed above. The measured correlation thus helps us predict that a large gain in capacity can be obtained through spatial multiplexing with our multiple-antenna UWB system, as has indeed been established in Section 17.5.

17.7.2 Eigen Spectrum

The number and distribution of the eigenvalues, λ_i, of $\mathbf{G}$ is a measure of the correlation between the multiple subchannels in an $N_T \times N_R$ system. High correlation between the subchannels will reduce the number of non-zero eigenvalues. In the extreme case, when all of the rows or columns of $\mathbf{H}$ are linearly dependent on each other, the number of non-zero eigenvalues reduces to unity, whereas in the case of perfect independence, there are N_{min} non-zero eigenvalues.

Let us assume that $\mathbf{H}$ is a full-rank matrix with rank $r_\mathbf{H}$. Because of the normalisation in Equation (17.4),

$$\|\mathbf{H}\|_F^2 = \sum_{i=1}^{r_\mathbf{H}} \lambda_i = N_T N_R N_f = P, \tag{17.35}$$

where P denotes the power constraint. From Equations (17.11) and (17.35), it can be shown that the optimal capacity of a MIMO system with CSIR, given by Equation (17.12), is maximised when

$$\lambda_i = \frac{P}{r_\mathbf{H}}, \quad i = \{1, \ldots, r_\mathbf{H}\}, \tag{17.36}$$

i.e., an equal power allocation strategy at the transmitter yields the optimal capacity when all eigenvalues are equal [9]. CSIT is not required in such a situation, as it does not provide any extra gain in performance.

We now investigate the properties of the eigenvalues of **G** for the measured 3×3 UWB channel. The eigenvalues of the equivalent flat channels are computed, and averaged over the frequency band to obtain the power gains of the UWB spatial channel. From Equations (17.35) and (17.36), the sum of the UWB eigenvalues should converge to $N_T N_R = 9$ for this array configuration. Also, the eigenvalues should be identically equal to $N_T N_R/r_H = 4.77$ dB for uniform power allocation to be an efficient strategy.

The mean λ_{max} from the measurements is found to be 6.87, or 8.37 dB, which is comparable to the SNR gain observable with a 3×3 MIMO system with CSIT as derived in Section 17.6.

The first-order statistics of the eigenvalues obtained from the 3×3 channel measurement ensemble are estimated. The empirical CDFs are shown in Figure 17.6, which shows that the system always has three nonvanishing eigenvalues and the channel matrix therefore has full rank. Thus the 3×3 channel is capable of supporting three orthogonal transmission modes that can be exploited for spatial multiplexing.

The three eigenvalues, however, are well separated, and therefore the power gains of the subchannels are nonuniform. This information leads us to predict that a water-filling MIMO system, that uses CSIT, will be able to exploit the channel much more efficiently than a uniform power-allocation scheme. Our earlier analysis has indeed shown this to be the case.

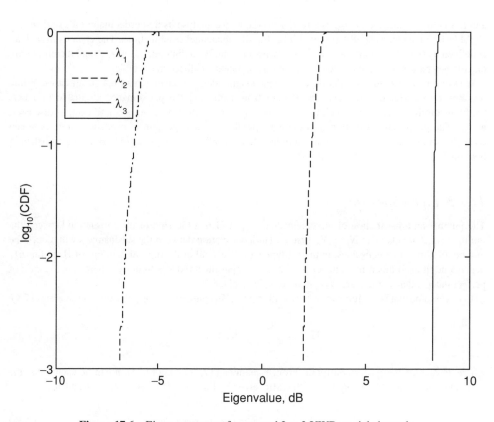

Figure 17.6 Eigen spectrum of measured 3×3 UWB spatial channels

17.7.3 Angular Spread

In a sparsely scattering environment, the multipath signals arrive at the receiver from a limited set of directions and the angular spread of the received signal is said to be low. It has been shown that the angular spread is the dominant factor contributing to the multipath fading correlation [34]. As discussed in Section 17.7.1, MIMO capacity is adversely affected by correlation. In the limiting case, when all multipath signals arrive from a single direction, the MIMO channel reduces to a unit-rank channel and thus provides no diversity order or capacity improvement. Angular clustering of the multipath components, as can be observed in some propagation environments, also impacts the achievable MIMO capacity.

If, however, the angular spread of the incident signal is larger than the beamwidth of the receiving ULA, the correlation has no impact on MIMO capacity [35]. Interpreted using the phased-array theory, this result follows from the fact that a sufficiently small beamwidth facilitates the angular resolution of the individual impinging waves.

Many MIMO channel models consider the angular spread by taking into account the direction-of-departure (DoD) and direction-of-arrival (DoA) distributions [36, 58]. The spatial channel can then be described in terms of the DoD and DoA steering matrices and the corresponding multipath intensities. If the angular multipath intensities are equal, it can be shown using a double-directional model that the asymptotic MIMO capacity is bounded by [36]

$$\lim_{\rho \to \infty} \frac{C(\rho)}{\log_{10} \rho} = \min\{s_T, s_R\} \qquad (17.37)$$

for a given number of transmit and receive antennas, where s_T and s_R denote the number of scatterers at the transmit and receive ends, respectively. Therefore the capacity is essentially limited by the amount of scattering, and is maximised when $s_T = s_R \geq N_T = N_R$.

Due to significant scattering in an indoor radio environment, s_T and s_R are often large. In addition, the diffuse multipath components arrive from a large range of directions. Thus large azimuthal angular spreads are observed in indoor UWB channels [37, 38]. A well-designed multiple-antenna system operating in such channels will not suffer significant angular correlation, and can therefore be expected to provide a good performance without any serious degradation arising from correlation.

17.7.4 Array Orientation

Among spatial antenna topologies, the linear array is the most common and well understood. It is convenient both to analyse and to deploy in practical systems. A drawback of the linear array, however, is the dependence of its performance jointly on the array orientation and the multipath angular spread. If the angular spread is low, the performance of a linear array is sensitive to the orientation, as discussed above. In such a situation, the array is referred to as broadside if the array axis is perpendicular to the principal multipath DoA, and inline if it is oriented along the DoA. The performance of a broadside spatial system is, in general, considered to be superior to an inline system. In an outdoor mobile channel, with low angular spread, an inline array requires four times as much inter-element spacing as a broadside array to achieve low correlation [32].

Several alternative topologies have been proposed in the literature to remove the dependence of MIMO performance on array orientation, such as the triangular, rectangular, hexagonal and circular antenna arrangements. Due to their symmetric forms, circular and hexagonal topologies provide significant performance improvement in low angular spread channels [39, 40].

As discussed in Section 17.7.3, indoor UWB channels have large angular spreads. For this reason, we can expect a linear UWB array to be robust against a serious degradation in performance because of its orientation.

Note that in this treatment, we have not considered the directionally asymmetric signal dispersion due to antennas [41], as a discussion of antenna distortion is beyond the scope of this chapter.

17.7.5 Channel Memory

A typical indoor UWB channel is frequency selective and has a large number of paths incident at the receiver [42]. Temporal dispersion due to multipath introduces memory into the channel, which can thus be modelled as a compound finite-state channel [43] with capacity bounds derived in [44]. The effect of memory on a single-antenna channel is to boost the channel capacity significantly [45].

The number of independent subchannels that can be supported over a MIMO system governs the asymptotic slope of the capacity as a function of SNR. The MIMO multiplexing gain in a multipath channel is bounded by [18]

$$\lim_{\rho \to \infty} \frac{C(\rho)}{\log_{10} \rho} \leq \min \{N_T, N_R, L\}, \tag{17.38}$$

where L is the number of multipath components. This relation shows that the capacity gain of a MIMO channel with SNR is virtually unbounded in a rich multipath environment where L is very large. Also, from this relation, the capacity scaling with the array size is sustained up to large arrays. The same is also true for small array configurations [46], such as those studied in this chapter. For the MIMO capacity to scale linearly with the array size, however, the number of resolved paths must undergo a quadratic increase with the array size [47]. Intuitively, this requirement demands a dense scattering environment and sufficiently high temporal resolution; both conditions are met by indoor UWB systems.

On the other hand, the diversity order achievable in a multipath MIMO channel is bounded by [19]

$$-\lim_{\rho \to \infty} \frac{\log_{10} P_e(\rho)}{\log_{10} \rho} \leq N_T N_R L. \tag{17.39}$$

Thus in a rich scattering environment, where L is much larger than N_T or N_R, an increase in the array size does not provide a significant improvement in the diversity improvement. This explanation provides the theoretical basis for the observations related to multiple-antenna diversity performance in UWB channels as undertaken in Section 17.6.

For systems with CSIT, even perfect feedback fails to increase the capacity of a channel without memory [48], whereas feedback is highly effective in channels with memory [13], which clearly demonstrates the important role played by channel memory and scattering in MIMO systems.

17.7.6 Channel Information Quality

In analysing the spatial capacity of UWB systems in this chapter, we have assumed perfect channel information. In a random fading channel, however, there is always a possibility of erroneous channel estimation. Pilot symbols are often used for channel estimation, and the correlation coefficient, r_{est}, between the channel and its estimate is a measure of the channel information fidelity [49]. If N_p pilots, each with power E_p, are used to estimate a Nakagami-m fading channel, which is the distribution followed by UWB channels [50], the correlation coefficient is given by [51]

$$r_{est} = \frac{\sqrt{N_p \varepsilon}}{\sqrt{N_p \varepsilon + \frac{1}{\rho}}}, \tag{17.40}$$

where $\varepsilon = E_p/E_s$ is the ratio of the pilot power to the data symbol power, and ρ is the average SNR. From this relation, $r_{est} \to 1$ as either $\rho \to \infty$ or $N_p \to \infty$, i.e. the estimation improves at high SNR or with a large number of pilot symbols.

The channel-side information can be regarded as perfect if the second moment of the estimation error is negligible compared to the reciprocal of the SNR [52]. Imperfect channel estimation incurs an SNR penalty with coherent diversity combining, but no diversity order penalty [51]. The achievable communication rate is also degraded by imperfect channel information, and is bounded even at infinite SNR. Erroneous CSIR is, however, still superior to the no CSIR [53]. At low SNR, the impact of imperfect receiver-side channel information is insignificant, as it does not increase the capacity which is governed by [52]

$$\lim_{\rho \to 0} C(\rho) = \rho. \tag{17.41}$$

Similar arguments apply at the transmitter; the capacity performance with imperfect CSIT is substantially higher than that without CSIT [54, 55]. This is important for low SNR situations, since, from an implementation perspective, the quality of the feedback link is unlikely to be superior to that of the forward link. Furthermore, water-filling remains the capacity-achieving power allocation strategy even when the CSIT is imperfect [56].

Practical single-antenna UWB implementations, such as MB-OFDM [57], already include pilot-assisted channel estimation. Thus channel information is available at the receiver, and, in the case of multiple-antenna systems, can be extended to the spatial dimension. Providing a feedback link will, however, increase the system complexity, but the consequent performance gain under low SNR conditions may justify the additional complexity in a given application.

17.8 Summary

This chapter has presented a comprehensive characterisation of multiple-antenna UWB channels from the perspective of information theory as well as practical implementation. An experimental investigation of the capacity and diversity performance of MIMO systems operating in UWB channels has been undertaken. Quantitative results demonstrate that the propagation characteristics of UWB channels are especially suited to multiple-antenna operation, and the enhancement in the communication rate, signal reliability and coverage range is tremendous even with small antenna arrays. It has been shown by measurements that the coverage range increases N-fold with an $N \times N$ array in a typical UWB environment. A 3×3 antenna configuration can increase the SNR by up to 8.3 dB, and the capacity by 3.3 times. The measurement results show that optimal power allocation to the antennas using channel information at the transmitter is very effective for low-SNR UWB links, boosting the capacity well beyond that achievable without this information. To conclude, spatial UWB systems, with compact elemental spacing requirements and resilience to usual performance detriments, offer the promise of delivering ultra-high performance for the wireless networks of the future.

References

[1] G.J. Foschini and M.J. Gans, On limits of wireless communications in a fading environment when using multiple antennas, *Wireless Personal Commun.*, **6**, March, 1998.
[2] I.E. Telatar and D.N.C. Tse, Capacity and mutual information of wideband multipath fading channels, *IEEE Trans. Inform. Theory*, **46**, July, 2000.
[3] A.J. Paulraj, D.A. Gore, R.U. Nabar and H. Bolcskei, An overview of MIMO communications – a key to gigabit wireless, *Proc. IEEE*, **92**, February, 2004.

[4] D.G. Brennan, Linear diversity combining techniques, *Proc. IRE*, **47**, June, 1959.
[5] L. Zheng and D.N.C. Tse, Diversity and multiplexing: a fundamental tradeoff in multiple antenna channels, *IEEE Trans. Inform. Theory*, **1**, August, 2002.
[6] C.E. Shannon, Channels with side-information at the transmitter, *IBM J. Res. Develop.*, **2**, October, 1958.
[7] M.H.M. Costa, Writing on dirty paper, *IEEE Trans. Inform. Theory*, **IT-29**, May, 1983.
[8] A. Goldsmith, S.A. Jafar, N. Jindal and S. Vishwanath, Capacity limits of MIMO channels, *IEEE J. Select. Areas Commun.*, **21**, June, 2003.
[9] A.J. Paulraj, R. Nabar and D. Gore, *Introduction to Space-Time Wireless Communications*. Cambridge, UK: Cambridge University Press, 2003.
[10] R.W.P. King, *Theory of Linear Antennas*. Cambridge, MA, USA: Harvard Press, 1956.
[11] J.W. Wallace and M.A. Jensen, Mutual coupling in MIMO wireless systems: a rigorous network theory analysis, *IEEE Trans. Wireless Commun.*, **3**, July, 2004.
[12] G.J. Foschini, Layered space-time architecture for wireless communication in a fading environment when using multiple antennas, *Bell Labs Tech. J.*, **1**, Autumn, 1996.
[13] T.M. Cover and J.A. Thomas, *Elements of Information Theory*. New York: John Wiley & Sons, Inc., 1991.
[14] T. Lo, Maximal ratio transmission, *IEEE Trans. Commun.*, **47**, 1999.
[15] T.S. Rappaport, *Wireless Communications: Principles and Practice*, 2nd ed. Prentice Hall, 2001.
[16] A.F. Molisch, M. Steinbauer, M. Toeltsch, E. Bonek and R.S. Thomä, Capacity of MIMO systems based on measured wireless channels, *IEEE J. Select. Areas Commun.*, **20**, April, 2002.
[17] W.Q. Malik and D.J. Edwards, Measured MIMO capacity and diversity gain with spatial and polar arrays in ultrawideband channels, *IEEE Trans. Commun.*, (in press).
[18] G.G. Raleigh and J.M. Cioffi, Spatio-temporal coding for wireless communication, *IEEE Trans. Commun.*, **46**, March, 1998.
[19] S.N. Diggavi, N. Al-Dhahir, A. Stamoulis and A.R. Calderbank, Great expectations: the value of spatial diversity in wireless networks, *Proc. IEEE*, **92**, February, 2004.
[20] J.B. Andersen, Array gain and capacity for known random channels with multiple element arrays at both ends, *IEEE J. Select. Areas Commun.*, **18**, November, 2000.
[21] A. Sibille and S. Bories, Spatial diversity for UWB communications, *Proc. Eur. Personal Mobile Commun. Conf.* Glasgow, UK, April, 2003.
[22] S. Alamouti, A simple transmit diversity technique for wireless communications, *IEEE J. Select. Areas Commun.*, **16**, October, 1998.
[23] S. Roy, J.R. Foerster, V.S. Somayazulu, and D.G. Leper, Ultrawideband radio design: the promise of high-speed, short-range wireless connectivity, *Proc. IEEE*, **92**, February, 2004.
[24] S.S. Ghassemzadeh, R. Jana, C.W. Rice, W. Turin and V. Tarokh, Measurement and modeling of an ultra-wide bandwidth indoor channel, *IEEE Trans. Commun.*, **52**, October, 2004.
[25] R.O. LaMaire and M. Zorzi, Effect of correlation in diversity systems with Rayleigh fading, shadowing, and power capture, *IEEE J. Select. Areas Commun.*, **14**, April, 1996.
[26] A.M. Tulino, A. Lozano and S. Verdu, Impact of antenna correlation on the capacity of multiantenna channels, *IEEE Trans. Inform. Theory*, **51**, July, 2005.
[27] M. Chiani, M.Z. Win and A. Zanella, On the capacity of spatially correlated MIMO Rayleigh-fading channels, *IEEE Trans. Inform. Theory*, **49**, October, 2003.
[28] J. Salz and J.H. Winters, Effect of fading correlation on adaptive arrays in digital mobile radio, *IEEE Trans. Veh. Technol.*, **43**, November, 1994.
[29] C. Chuah, D. Tse, J. Kahn and R. Valenzuela, Capacity scaling in MIMO wireless systems under correlated fading, *IEEE Trans. Inform. Theory*, **48**, March, 2002.
[30] D. Chizhik, G.J. Foschini, M.J. Gans, and R.A. Valenzuela, Keyholes, correlations, and capacities of multielement transmit and receive antennas, *IEEE Trans. Wireless Commun.*, **1**, April, 2002.
[31] W.C. Jakes, *Microwave Mobile Communications*. New York: John Wiley and Sons, Inc., 1974.
[32] W.C.Y. Lee, *Mobile Communications Engineering*. New York: McGraw-Hill, 1982.
[33] P. Kyritsi, D.C. Cox, R.A. Valenzuela and P.W. Wolniansky, Correlation analysis based on MIMO channel measurements in an indoor environment, *IEEE J. Select. Areas Commun.*, **21**, June, 2003.

[34] R.M. Buehrer, The impact of angular energy distribution on spatial correlation, *Proc. IEEE Veh. Technol. Conf.*, Vancouver, BC, Canada, Sepember, 2002.
[35] S. Loyka and G. Tsoulos, Estimating MIMO system performance using the correlation matrix approach, *IEEE Commun. Lett.*, **6**, January, 2002.
[36] M. Debbah and R.R. Müller, MIMO channel modeling and the principal of maximum entropy, *IEEE Trans. Inform. Theory*, **51**, May, 2005.
[37] W.Q. Malik, C.J. Stevens and D.J. Edwards, Synthetic aperture analysis of multipath propagation in the ultra-wideband communications channel, *Proc. IEEE Workshop Sig. Proc. Adv. Wireless Commun.*, New York, June, 2005.
[38] R.J.-M. Cramer, R.A. Scholtz and M.Z. Win, Evaluation of an ultra-wide-band propagation channel, *IEEE Trans. Antennas Propagat.*, **50**, May, 2002.
[39] B. Allen, R. Brito, M. Dohler and H. Aghvami, Performance comparison of spatial diversity array topologies in an OFDM based wireless LAN, *IEEE Trans. Consumer Electron.*, **50**, May, 2004.
[40] D.-S. Shiu, G.J. Foschini, M.J. Gans and J.M. Kahn, Fading correlation and its effect on the capacity of multi-element antenna systems, *IEEE Trans. Commun.*, **48**, March, 2000.
[41] W.Q. Malik, D.J. Edwards and C.J. Stevens, Angular-spectral antenna effects in ultra-wideband communications links, *IEE Proc.-Commun.*, **153**(1), 2006.
[42] W.Q. Malik, D.J. Edwards and C.J. Stevens, Optimal system design considerations for the ultra-wideband multipath channel, *Proc. IEEE Veh. Technol. Conf.*, Dallas, TX, September, 2005.
[43] A. Lapidoth and P. Narayan, Reliable communication under channel uncertainty, *IEEE Trans. Inform. Theory*, **44**, October, 1998.
[44] P. Mitran, N. Devroye and V. Tarokh, On compound channels with side-information at the transmitter, *IEEE Trans. Inform. Theory*, April, 2006.
[45] M.S. Pinsker and R.L. Dobrushin, Memory increases capacity, *Probl. Inform. Theory*, **5**, January, 1969.
[46] L. Zheng and D.N.C. Tse, Communication on the Grassman manifold: a geometric approach to the noncoherent multiple-antenna channel, *IEEE Trans. Inform. Theory*, **48**, February, 2002.
[47] K. Liu, V. Raghavan and A.M. Sayeed, Capacity scaling and spectral efficiency in wide-band correlated MIMO channels, *IEEE Trans. Inform. Theory*, **49**, October, 2003.
[48] C.E. Shannon, The zero error capacity of a noisy channel, *IRE Trans. Inform. Theory*, **IT-2**, September, 1956.
[49] M.J. Gans, The effect of Gaussian error in maximal ratio combiners, *IEEE Trans. Commun.*, **19**, August, 1971.
[50] D. Cassioli, M.Z. Win and A.F. Molisch, The ultra-wide bandwidth indoor channel: from statistical model to simulations, *IEEE J. Select. Areas Commun.*, **20**, August, 2002.
[51] W.M. Gifford, M.Z. Win and M. Chiani, Diversity with practical channel estimation, *IEEE Trans. Wireless Commun.*, **4**, July, 2005.
[52] A. Lapidoth and S. Shamai, Fading channels: how perfect need perfect side information' be?, *IEEE Trans. Inform. Theory*, **48**, May, 2002.
[53] M. Médard, The effect upon channel capacity in wireless communications of perfect and imperfect knowledge of the channel, *IEEE Trans. Inform. Theory*, **46**, May, 2000.
[54] S. Bhashyam, A. Sabharwal and B. Aazhang, Feedback gain in multiple antenna systems, *IEEE Trans. Inform. Theory*, **50**, May, 2002.
[55] E. Visotsky and U. Madhow, Space-time transmit precoding with imperfect feedback, *IEEE Trans. Inform. Theory*, **47**, September, 2001.
[56] G. Caire and S. Shamai, On the capacity of some channels with channel state information, *IEEE Trans. Inform. Theory*, **45**, September, 1999.
[57] A. Batra, *et al.*, Multiband OFDM physical layer proposal for IEEE 802.15 Task Group 3a, IEEE P802.15-03/268r0-TG3a, July, 2003.
[58] W.Q. Malik and A.F. Molisch, Ultrawideband antenna arrays and directional propagation channels, in *Proc. Eur. Conf. Antennas Propagat.* Nice, France, November 2006.

Part IV

UWB Radar, Imaging and Ranging

Introduction to Part IV

Anthony K. Brown

While there is currently considerable interest in its use in commercial communications, ultra-wideband has many other important applications in both civilian and military environments.

Many of the applications considered in this part of the book require the antenna to produce low dispersion of the pulse which, over a UWB signal, is nontrivial. One key point to achieve this is to minimise movement of the phase centre (the point on the structure from which the radiation appears to emanate) over the frequency band of interest. This immediate removes a whole class of traditional 'frequency-independent' antennas – line array log-periodic structures – which are inappropriate to impulse transmission. Low pulse dispersion can be the most demanding aspect of UWB antenna design.

So what applications are there for UWB? Of increasing importance is the whole area of localisation of radio-emitting sources. Due to the high resolution available in a UWB signal, this type of system is well suited to this problem. However, what is not obvious is how to produce accurate location information in the presence of complex multipaths such as is the case indoors. For instance, the impulse response of a typical dense multipath environment consists of a few hundred paths, where only the direct path actually carries the relevant location information. The situation is even more demanding in non-line-of-sight scenarios. This is the problem that Chapter 18 addresses where a new algorithm is proposed for dealing with the situation.

Next we turn our attention to use of UWB in imaging applications. Chapter 19 deals in some depth with the ground-penetrating radar application. Here, the unique properties of the UWB pulse are used to image objects including dielectrics which are buried beneath the surface of the earth. One obvious application is demining operations world wide where there is a real need for low-cost reliable detectors particularly for removing anti-personnel mines. After a brief review of the application, Chapter 19 deals in some depth with the antenna problem. Not only must the antenna operate over the widest possible bandwidth, but also energy must be coupled as efficiently as possible into the ground and the radiating structures must be physically compact, lightweight and low cost.

In Chapter 20 we maintain the imaging theme but now turn attention to the increasingly important area of biomedical imaging. UWB offers intriguing possibilities to distinguish noninvasively between different types of tissue with a high spatial resolution. Although in some ways similar to the ground-penetrating

---

Ultra-wideband Antennas and Propagation for Communications, Radar and Imaging   Edited by B. Allen, M. Dohler, E. E. Okon, W. Q. Malik, A. K. Brown and D. J. Edwards
© 2007 John Wiley & Sons, Ltd

radar of Chapter 19, the biomedical application is significantly different. A review is presented of both the application and the latest developments in this fast-moving field.

Chapter 21 turns to the problems faced by those systems needing high levels of UWB effective gain. Applications include radar and related high peak power systems. Following an overview of some of the areas of system performance which reflect directly on the antenna requirements, we consider the problem of high power, high fidelity impulse transmission. The most common approach to this is using reflector antennas. The fundamental properties of these structures are overviewed under impulse conditions. Suitable feed elements are highlighted, in particular spiral feeds for circular polarisation and TEM-type horns, whose basic properties are discussed. Finally, the impulse radiating antenna (IRA) is described. This unique antenna system, developed for electromagnetic compatibility testing, is capable of radiating over extremely wide frequency ranges, for example 10 MHz to 2 GHz instantaneous bandwidth, with low dispersion resulting in a high fidelity pulse.

Throughout this part of the book we retain a strong focus on the applications of UWB technology outside the pure communications arena. It is our hope that some of the ideas expounded in these chapters may result in cross-disciplinary ideas being developed and implemented in the future.

18

Localisation in NLOS Scenarios with UWB Antenna Arrays

Thomas Kaiser, Christiane Senger, Amr Eltaher
and Bamrung Tau Sieskul

18.1 Introduction

Wireless ultra-wideband (UWB) systems are characterised by an enormous bandwidth of up to several GHz and are therefore a promising candidate for high data rate indoor communications and for precise localisation of UWB signal-emitting objects. While the first products for indoor communication will soon be commercialised, the topic of UWB localisation is in its research infancy. When compared to a sparse-path radar scenario, the dense UWB multipath environment makes the reliable detection of the direct path between transmitter and receiver an ambitious challenge. For instance, the impulse response of a typical dense multipath environment consists of a few hundred paths, where only the direct path carries the relevant location information, since the indoor environment is generally unknown. Even more demanding is a non-line-of-sight (NLOS) scenario, where the receiver does not 'see' the transmitter, so that the direct path will be substantially attenuated. Within this chapter, a new infra-structure-aided approach – the so-called *BeamLoc* – is proposed for tackling the demanding NLOS scenario. The localisation task is enriched by a unidirectional communication from the transmitter to the receiver. It relies on UWB antenna arrays and solves the NLOS challenge in a straightforward way that requires neither *a priori* environmental information nor *a priori* sources of errors.

The topic of indoor localisation can be classified as shown in Figure 18.1. If the *environment is partially or completely known* such information is exploited for solving the localisation task typically as follows: in the first step, the known environment is quantised with the required accuracy – meaning that a grid is rolled out, where the grid points are the quantised location points. In the next step, a series of measurements for all points is conducted, and the results are stored in a database. The original location problem is then solved in the final step by comparing the actual measurement with the entries in the database, e.g. by selecting that grid point as the true location, which shows the least mean quadratic error.

Ultra-wideband Antennas and Propagation for Communications, Radar and Imaging Edited by B. Allen,
M. Dohler, E. E. Okon, W. Q. Malik, A. K. Brown and D. J. Edwards
© 2007 John Wiley & Sons, Ltd

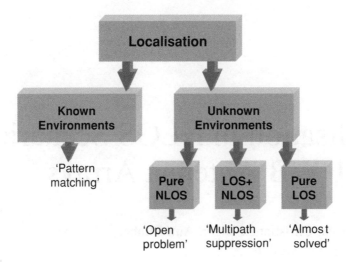

Figure 18.1 Classification of the indoor localisation problem

This basic approach is nothing more than a pattern-matching method and also has been recently proposed for UWB localisation (e.g. [1], [2], [7]).

In contrast to known environments, the techniques for solving the localisation in *unknown environments* are entirely different. If a greatly dominant line-of-sight (LOS) exists, i.e. the receiver 'sees' the transmitter, then the localisation problem is generally solved by time-delay estimation, which is widely used for radar applications. If the LOS is not dominant and other paths caused by reflections cannot be neglected, the challenge is to suppress the interfering paths because they do not carry relevant information about the transmitter location in unknown environments. The remaining and most challenging case is the pure NLOS scenario. If the straight line (SL) between the transmitter and receiver is completely blocked, either the location accuracy suffers from a systematic NLOS error (e.g. [17]), or, if the SL is substantially attenuated but not blocked, so-called *first path detection algorithms* are suggested (e.g. [13], [4]). However, the first path does not necessarily correspond to the SL (see later), so that a systematic average error can be predicted here also. In contrast, our approach tackles the NLOS scenario without such systematic error as long as the SL arrives with some minimum amplitude at the receiver.

Before we proceed with UWB localisation systems in more detail, we will briefly shed some light on other techniques for solving the localisation problem. Table 18.1 summarises the major alternatives and relates them to UWB systems.

Table 18.1 Techniques for solving the localisation problem

Ultrasonic	Laser	UWB
+ High resolution	+ High resolution	+ High resolution
+ Low cost	+ Weak multipath propagation	+ Reusability
− Susceptible mechanics	− Worse penetration	+ Beneficial polarisation
− No polarisation	− Low overall power efficiency	+ Good penetration
− Dense multipath propagation	− Eye safety	− Dense multipath propagation
		− High sampling rate

Thanks to the beneficial ratio of speed of wave propagation vs. bandwidth all approaches allow a high resolution in the order of centimetres or even less, which is generally considered to be sufficient for the targeted indoor applications. While an ultrasonic localisation system is usually of low cost, it suffers from susceptible mechanical components and is unable to exploit favourable polarisation. Polarisation is only observed for electromagnetic waves and, if it is of circular shape and is properly deployed, it allows for the suppression of half of the interfering paths because a left-hand oriented circular wave becomes a right-hand oriented wave after a single reflection (and vice versa). In contrast, laser-based systems require an LOS but additional reflections are only of marginal impact on accuracy. However, signal impenetrability of most relevant materials, restricted transmit power for guaranteeing human eye safety and low overall power efficiency severely limit laser beams' region of applicability in indoor environments. In contrast, UWB-based systems show several pros, like reusability for high data rate communications, exploitation of polarisation and good penetration properties. The remaining challenges consist of mitigating the dense multipath propagation as well as developing a reasonable hardware realisation, since a complete digital system is currently only feasible for pure research purposes and far away from any commercial deployment.

Today, research on UWB localisation is mostly devoted to sensor networks (see [8] for an overview), where each sensor is equipped with a single antenna and the localisation is achieved by cooperation among all or by a subset of sensors. The major idea behind this scenario is that those UWB devices – or *sensors* – being commercialised must be of low cost and of small size; hence, single antenna transceivers are currently under preferred investigation [10], [13].

Opposite to that we will focus here on *precise* localisation, which is – thanks to the enormous bandwidth – principally possible to the order of centimetres or less. A typical application is the positioning of a household robot inside a building. Obviously, such positioning has to be precise in order to allow the robot to carry out the right task with sufficient success, and it does not primarily suffer from low size and low cost restrictions.

In our general set-up the battery-driven robot always represents the transmitter (TX) whereas the control centre (or access point, AP) always represents the receiver (RX). This arbitrary looking assignment is made for reason of simplicity and for supposed lower transmit than receive power consumption. The latter likely holds true due to rather simple (analogue) transmit signal processing and generally low UWB transmit power, but quite sophisticated digital receive signal processing.

Since cost and size matter here only to a minor extent, both the transmitter and the receiver are always equipped with multiple antennas. These *UWB antenna arrays* facilitate beamforming on both sides of the transmission link and, in combination with a unidirectional communication, enable significant suppression of the deteriorating impact of multipath propagation. Multipath propagation is generally quite pronounced in typical UWB indoor scenarios, where even hundreds of paths are frequently encountered. Since positioning relies on capturing the single direct path among the other hundreds of interfering multipaths, any multipath cancellation technique is highly welcome. Beamforming is principally capable of reducing the delay spread of a wireless channel, but this applies here only if the transmitter and the receiver beams 'see' each other. Such a requirement can be complied by collateral communication from the transmitter to the receiver, which is the major idea behind our new approach. It will be explained in more detail in the succeeding sections.

As mentioned earlier the further ambitious challenge in positioning is the NLOS scenario, where in view of our target application, the robot is located in another room than the access point. Then, a signal travelling along the SL connecting the transmitter and the receiver suffers from attenuation, refraction and an additional delay caused by the separating wall. Figure 18.2 shows the four different scenarios that can be encountered in reality. In order to express them also in mathematical terms, consider the impulse response

$$h(t) = \sum_{l=1}^{L_p} \alpha_l \delta(t - \tau_l) \qquad (18.1)$$

from the transmitter to the receiver, where α_l is the attenuation of the lth path and τ_l the corresponding delay with $\tau_1 \leq \tau_2 \leq \ldots \leq \tau_{Lp}$. Note first that the SL does not always coincide with a physically existing path because the spatial signal propagation also suffers from Snell's law. This means that the shortest distance between the transmitter and the receiver becomes only *piecewise* linear if there is an obstacle between them. For that reason we use in the following the more appropriate term *direct path* (DP) for the physically existing path on behalf of SL.

In the first scenario (Figure 18.2(a)), an unobstructed DP does exist, which equals the LOS path; therefore, the first path represents the DP so that

$$\alpha_1 > \alpha_l \quad \forall l > 1.$$

In this rather ideal scenario known from radar applications, the DP can be easily detected. In the second scenario (Figure 18.2(b)) the DP still arrives first, but the transmit antenna does not steer its mainlobe towards the receiver. Hence, another path, here reflected by metal, might show a larger magnitude, which complicates reliable DP detection.

The third scenario (Figure 18.2(c)) reveals the major challenge in NLOS environments: the DP still arrives first, but now with a distinct smaller magnitude due to penetration loss.

The last scenario may not occur often in practice, but it demonstrates a further challenge in NLOS propagation: while penetrating through a material, an electromagnetic wave may change its speed of propagation significantly. This property is closely related to the material's dielectric constant. Although the scenario shown in Figure 18.2(d) may look somewhat artificial, it happens in the real world: first, note that the dielectric constant of concrete is $\varepsilon_y = 9$, meaning that a wall slows down the wave speed by a third. Second, taking into account hundreds of interfering paths and a door separating the rooms, a reflection passing through this door likely arrives earlier than the DP under NLOS. In conclusion, the DP is in general neither the first nor the strongest path.

Since there is no straightforward solution for localisation in NLOS scenarios, it becomes evident that reliable localisation in dense multipath environments may rely on combining different criteria, a so-called *data fusion*. The major criteria to be considered are:

- exploiting of common first arrival of the DP;
- exploiting polarisation (halves the number of interfering paths);
- deployment of beamforming (mitigates interfering paths);
- evaluation of the signal strength (allows coarse localisation);
- evaluation of the signal shape (improves performance in LOS scenarios).

Although the latter two items are intuitively clear, they are in a dense multipath environment of rather limited value. First, because of time-selective, frequency-selective and, in particular, spatial fading, the signal strength is not a reliable measure for distances. Second, even if the transmit signal shape is perfectly known at the receiver, a matched filter is mostly useful for easy-to-solve LOS but not for NLOS situations, because the signal shape is normally altered by passing an obstacle. Note that opposite to the area of wireless communications noise does not represent here a major challenge since sufficient probing time is available. Hence, the signal-to-noise ratio (SNR) can be usually improved by coding, e.g. in the simplest case by *repetition* coding, to achieve a significant coding or spreading gain. In contrast, the signal-to-interference ratio (SIR) – where an interferer is here any path except of the DP – has to be large for precise localisation and cannot be almost arbitrarily increased.

Our contribution is organised as follows. In Section 18.2 the underlying signal model is introduced in order to set up the mathematical framework. Section 18.3 highlights some relevant properties of UWB beamforming. The succeeding two sections present our approach first in an illustrative way and second

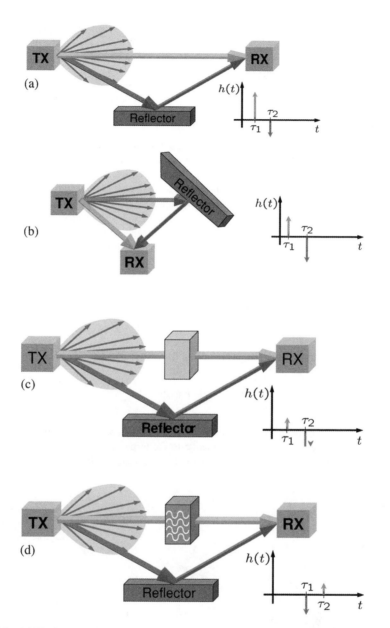

Figure 18.2 (a) Unobstructed DP in an LOS scenario; (b) attenuated DP caused by the antenna pattern; (c) obstructed attenuated DP; (d) obstructed attenuated and delayed DP

by deriving all relevant equations. By revisiting the state of the art, a further section is devoted to time-delay estimation, which is a substantial part of our approach. Before pointing out the results obtained by numerical simulations, we will briefly discuss several open questions and future research directions. A conclusion finishes this contribution.

18.2 Underlying Mathematical Framework

In this section, we introduce the transmit and receive signals by taking the typical UWB dense multipath propagation adequately into account. Since a wireless channel is usually randomly fluctuating and noise is always present in electronic circuits and systems, some signals are assumed to behave like random processes. Moreover, because an indoor environment does not change rapidly with time (large coherence time), stationarity is assumed and Doppler effects are ignored. As already mentioned in the previous section our approach relies on *UWB beamforming*, so that either an array consisting of omnidirectional antennas or, alternatively, a mechanically rotating single antenna with a lobe-shaped array pattern, is applied. We start with the latter case and continue with the antenna array.

Let us first introduce the generalised beampatterns of a beamformer in the time and frequency domains, respectively

$$b(t, \theta, \phi) = z(t, \theta, \phi), \tag{18.2}$$
$$B(f, \theta, \phi) = F\{z(t, \theta, \phi)\}, \tag{18.3}$$

where $z(t, \theta, \varphi)$ is the beamformer's output signal and the unit energy signal

$$\int_{-\infty}^{\infty} |u(t, \theta, \phi)|^2 \, dt = 1 \tag{18.4}$$

is the beamformer's input. Note that F is the Fourier transform operator, θ represents the mainlobe direction and φ the angle of the wave front relative to the array orientation (see Figure 18.3 for the transmit case). In order to define a quantity representing the average over *time* or *frequency*, we integrate

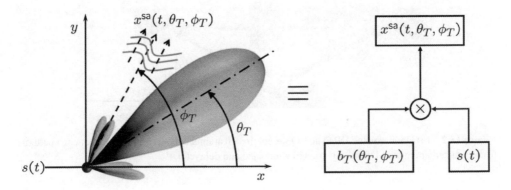

Figure 18.3 The transmitted signal $x^{sa}(t, \theta_T, \varphi_T)$ excited by a single lobe-shaped antenna (left-hand side) and a simplified mathematical model (right-hand side)

over one of these variables to end up with

$$b(\theta, \phi) = \sqrt{\int_{-\infty}^{\infty} b^2(t, \theta, \phi) dt} = \sqrt{\int_{-\infty}^{\infty} B^2(f, \theta, \phi) df}, \qquad (18.5)$$

where the latter equality holds from the Parseval's theorem.

Now, the principle of a mechanically rotating antenna, well known from classical radar applications, can be easily expressed in mathematical terms. Because the receiver requires more sophisticated processing for signal separation and detection, a mechanically rotating antenna is likely only deployed at the transmitter, whereas at the receiver an antenna array is obligatory. Hence, the transmit signal via a single rotating antenna can be written as

$$x^{sa}(t, \theta_T, \phi_T) = b_T(\theta_T, \phi_T) s(t), \qquad (18.6)$$

where $s(t)$ and $x^{sa}(t, \theta_T, \phi_T)$ are the beamformer's input signal and output signal, respectively and the lower index T represents the *transmit* beampattern. This relation is visualised in the following figure. Here that the upper index 'sa' indicates the *single antenna* case.

Note that the above model does not include noise, because the high transmit power leads to a large transmit SNR. Note further that for simplicity, Equation (18.6) relies on a frequency-independent beampattern, meaning

$$b^2(t, \theta, \phi) = b^2(\theta, \phi) \delta(t). \qquad (18.7)$$

Such an idealised beampattern could be approximately realised for a certain frequency range by adequate filtering in each antenna branch [5], which is for the huge UWB frequency range still a demanding challenge [9].

After the single antenna case let us now focus on antenna arrays. As will be pointed out later and opposite to the previous paragraph, such an antenna array will be of use mainly on the receiver side. Hence, we use the lower index 'R' and the upper index 'aa', which is illustrated in Figure 18.4.

Note that we assume throughout this contribution a *uniform circular array* (UCA), since there is no reason for any nonsymmetrical topology for localisation in unknown environments. Except for its spatial orientation, such a UCA is completely described by the radius r_R and the number of antennas N_R. At the n_Rth receive antenna element, the received deterministic signal can be written in the noiseless case and for an ideal single-path channel as

$$r_{n_R}(t) = y(t + \tau_{n_R}(\phi_R)), \quad n_R = 1(1)N_R, \qquad (18.8)$$

where $\tau_{n_R}(\phi_R)$ is the time difference for the n_Rth receiver antenna element to an arbitrarily chosen reference antenna under the incident angle φ_R. Note that for reasons of simplicity we consider here planar waves only corresponding to far-field scenarios. In contrast, spherical waves, originating from near-field sources, will be the subject of future research. To be more specific, the time difference can be expressed as

$$\tau_{n_R}(\phi_R) = \frac{1}{c} r_R \cos(\phi_R - \rho_{n_R}), \qquad (18.9)$$

where c is the speed of light, r_R is the radius of receive circular antenna array, and $\rho_{n_R} \in (0, 2\pi]$ is the position angle of the n_Rth receive antenna element with respect to the x-axis. Note further that the origin of the coordinate system is for our purpose of particular relevance. While for a fixed device this could be realised by a single calibration procedure, for a mobile device – like the robot – a compass is needed so that the TX and the RX signals are processed against a common coordinate system.

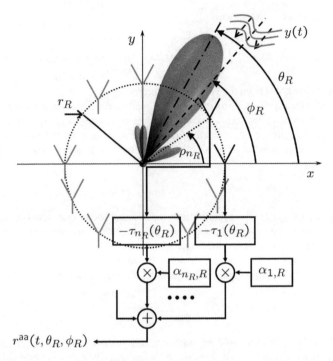

Figure 18.4 The received signal $r^{aa}(t, \theta_R, \varphi_R)$ of a receive beamformer with steering direction θ_R and a wavefront impinging from φ_R

If we apply an ideal delay-and-sum beamformer (as shown in Figure 18.4), the output can be written as

$$r^{aa}(t, \theta_R, \phi_R) = \sum_{n_R=1}^{N_R} \alpha_{n_R} r_{n_R}(t - \tau_{n_R}(\theta_R))$$
$$= \sum_{n_R=1}^{N_R} \alpha_{n_R} y(t + \tau_{n_R}(\phi_R) - \tau_{n_R}(\theta_R)), \qquad (18.10)$$

where θ_R represents the mainlobe direction and φ_R the direction of arrival (DoA) of the impinging wavefront. α_{n_R} is the *weighting coefficient*, which is independent from the angle θ_R, so the beamformer output power $E\left\{|r^{aa}(t, \theta_R, \phi_R)|^2\right\}$ equals the average of all input powers. In order to introduce here an antenna array pattern similar to the single antenna case, we consider not the individual outputs of the array antennas, but the beamformer output $r^{aa}(t, \theta_R, \varphi_R)$, so that again a single input and single output system remains. Similar to the previous definitions we introduce the maximum instantaneous beampattern of the antenna array as

$$b_{\max}(\theta_R, \phi_R) = \max_t \sqrt{\int_{t-\frac{1}{2}T_p}^{t+\frac{1}{2}T_p} |r^{aa}(t, \theta_R, \phi_R)|^2 dt}, \qquad (18.11)$$

which shows several beneficial properties discussed later. One can summarise that given a mainlobe direction θ_R, $b_{\max}(\theta_R, \varphi_R)$ is the beampattern or the sensitivity according to any arbitrary DoA's φ_R.

After describing the transmitter and the receiver signal model, the channel must now be determined. It is known and was observed in extensive measurement campaigns [14], [22], that in UWB indoor environments the numerous paths occur in *clusters*, i.e. the overall L_p paths can be divided in a few L_C subsets denoting the number of clusters, where each cluster contains L_R rays. Accordingly, we augment the impulse response given in Equation (18.1) with a spatial component

$$h(t, \theta) = \sum_{l=1}^{L_C} \sum_{k=1}^{L_R} \alpha_{k,l} \delta(t - T_l - \tau_{k,l}, \theta - \Phi_l - \varphi_{k,l}), \qquad (18.12)$$

so that each path is assigned with an angle $\Phi_l + \varphi_{k,l}$ under which the particular wavefront impinges, where Φ_l is the mean cluster angle and $\varphi_{k,l}$ gives the angle of the kth ray in cluster l relative to the mean cluster angle. The arriving time of cluster l is given by T_l and the arriving time of all other rays within this cluster relative to the cluster arriving time is given by $\tau_{k,l}$. The paths within one cluster impinge from roughly the same direction, which is modelled by defining a standard deviation for $\varphi_{k,l}$, namely σ_l. In communication channel modelling, several of these parameters are considered as random variables with a certain probability density function (PDF), see [14] for more details. Assuming a pulse width of about 1 ns, paths with a path-length difference of 30 cm are separated at the receiver, which holds generally true for different clusters but often also for rays within one cluster coming from nearly the same direction. The clustering behaviour is illustrated in Figure 18.5, where a transmitter (TX) emits a signal in all directions (omnidirectional). Two clusters are shown being reflected at the walls. The black lines mark the averaged

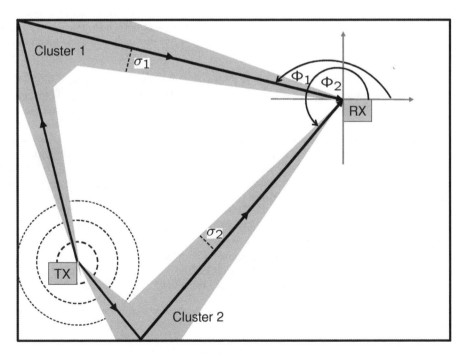

Figure 18.5 Clustering phenomenon in UWB indoor environments

cluster path and the grey areas are the regions for potential paths within the clusters, so that the width of the grey areas is determined by σ_l. Note that due to scattering at the walls, not only the averaged cluster path is captured by the receiver but also several reflections being assigned to the clusters.

After the introduction of the mathematical framework, we will point out the most relevant aspects of UWB beamforming with relation to the localisation task.

18.3 Properties of UWB Beamforming

UWB beamforming is different from narrow and broadband beamforming in several respects [18], [26], simply because the travel time of a UWB pulse across an array is longer than its own duration. For that reason, only antenna arrays are considered in the following – a lobe-shaped single UWB antenna does not show all beneficial properties. The major properties are highlighted in this section, where special emphasis is given to the localisation task.

Suppose again there is a delay-and-sum beamformer with a single delay $\tau_n(\theta)$ in the nth antenna branch, and a UCA with radius r. Note that we omit further indices since – thanks to antenna reciprocity – the following applies to the transmitter and the receiver as well. The delays are chosen, so that the resulting beampattern shows a mainlobe in the θ direction; see Equation (18.7).

Property 1: The *mainlobe width* is mainly controlled by the ratio of centre wavelength to array size and marginally by the pulse shape. The -3 dB *worst-case mainlobe width* is given by

$$\theta_{-3dB} = 2\arcsin\left(\frac{\lambda_c}{\sqrt{2}L}\right), \quad (18.13)$$

where λ_c is the centre wavelength and L is the array size [18]. Note that the *exact* mainlobe width depends on the pulse shape as well, so Equation (18.13) should be seen only in an approximate way. In Figure 18.6, the -3 dB pulse width is shown by the double arrow. See later on how to exploit this property in combination with the following Property 2.

Property 2: Unfavourable ambiguities in the beampattern caused by the well-known *grating lobes* limit the spatial antenna separation in narrowband beamforming to $\lambda_c/2$, where λ_c is the wavelength of the carrier frequency. This restriction is no longer required in UWB beamforming, because the UWB beampattern does not show ambiguities due to a broad averaging of frequency (see Figure 18.6). In other words, assume an antenna spacing of $2\lambda_c$ and a signal arriving from the end-fire direction (e.g. $\Phi = -90°$). Suppose that this signal is of narrowband nature, then the output of the beamformer does not change if the same signal alternatively impinges from $\Phi = 0°$. This is due to perfectly overlapping sinusoidals shifted by one full wavelength. Hence, the beamformer cannot distinguish between these two completely different directions. For pulse-shaped UWB signals it becomes apparent that this undesirable property no longer holds true.

The combination of Properties 1 and 2 opens a new way to sharpen the mainlobe, simply by increasing the antenna spacing, where the number of antennas remains constant and the beampattern is nevertheless grating-lobe free. In contrast, in narrowband scenarios more antenna elements have to be provided to sharpen the mainlobe if an unambiguous beampattern is desired. When applied to localisation, a better ranging accuracy is expected for increased array size as a result of a reduced mainlobe width.

Property 3: As shown in [18], the main to *fixed* sidelobe ratio – or simply *the beamforming gain*, is given by

$$BG = \frac{b_{max}^2(\phi,\phi)}{SLL^2}, \quad (18.14)$$

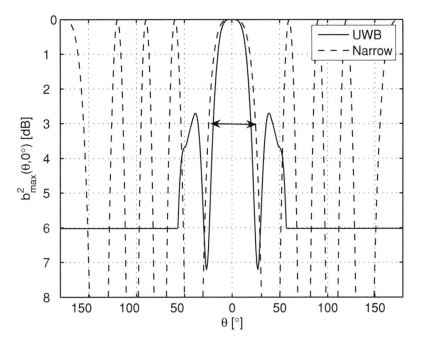

Figure 18.6 Narrowband beampattern (solid line), UWB beampattern (dashed line), $L = 2\lambda_c$, $f_c = 6.85$ GHz, $N = 2$

with the fixed sidelobe-level (*SLL*) defined as $SLL^2 = b_{max}^2(\theta, \phi)$ in the case of disjoint intervals $[\tau_n(\theta, \phi) - T_p/2, \tau_n(\theta, \phi) + T_p/2,]$ with $\tau_n(\theta, \phi) = \tau_n(\theta) - \tau_n(\phi)$, i.e. the pulses impinging at the antennas do not overlap.

Interestingly, for equal prefilters the *BG* equals N^2, which means on a decibel scale a 'double-dB gain' compared to the narrowband scenario. This rather relevant property is also shown in Figure 18.6 and can be explained in an illustrative way: suppose a delay-and-sum beamformer with two antennas, large antenna spacing and zero delays. Under this set-up a pulse-shaped wavefront arriving from the broadside is added constructively by the beamformer leading to an output signal with doubled magnitude and therefore four times the peak (or instantaneous) power of the signal itself. In contrast, for another pulse-shaped wavefront arriving from end-fire, the two pulses travelling along the two antenna branches do not overlap at the beamformer output (because of the large antenna spacing), so that the peak power equals the single signal power only. Hence, even with only two antennas, the *peak power ratio* is four, which results in the *double-dB gain*. Note that the double-dB gain occurs for two antennas if

$$\frac{L}{T_p} = c \quad \Rightarrow \quad LB = c \quad for \quad B \approx \frac{1}{T_p}. \tag{18.15}$$

Equation (18.15) gives only the worst-case situation, since the ratio also depends on the mainlobe direction, i.e. for $\Phi = 0°$ the left-hand side can be multiplied by two and still hold true. A more stringent mathematical derivation can be found in [18]. The conclusion is that for UWB signals the undesired multiple paths can be cancelled to a much larger amount than for narrowband signals, simply because the travel time of the pulse across the array is larger than its own duration.

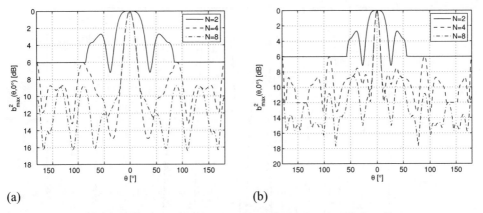

Figure 18.7 Normalised and squared BP: (a) $L = \lambda_c$; (b) $L = 2\lambda_c$

Observe that opposite to the mainlobe width, the *BG* is enhanced *only* with an increasing number of antennas N, which is irrespective of the array size. For this reason, with larger N and fixed L, the ranging accuracy will increase in low SINR (signal to inference plus noise ratio) scenarios as interference and noise are better suppressed. An SINR threshold will exist, the point at which increasing the number of antennas will not result in further accuracy enhancement. Below the threshold 'out of mainlobe' noise and interference are completely suppressed or even not present, but being inherent in the mainlobe they are still a factor. The only chance to further increase ranging accuracy under such high SINR is to increase the array size L, mentioned in Property 1.

Figures 18.7 and 18.8 show the behaviour of the normalised squared beam power pattern for a UCA. The wavelength $\lambda_c = 4.4$ cm corresponds to the UWB centre frequency $f_c = 6.85$ GHz and the pulse shape is the second derivative of the Gaussian pulse

$$p(t) = \left(1 - 2\left(t/\tau_p\right)^2\right) \exp\left(-\left(t/\tau_p\right)^2\right), \tag{18.16}$$

where $\tau_p = 0.0233$ ns determines the pulse width. It is assumed that a pulse arrives from $\Phi = 0°$ and, for the moment, multipath components and noise are not present. In order to highlight the above-mentioned

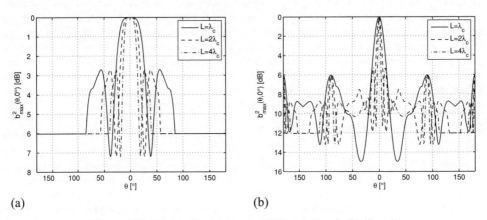

Figure 18.8 Normalised and squared BP: (a) $N = 2$; (b) $N = 4$

properties, either the number of antennas or the array size is varied while keeping the other factor fixed. The major observations can be summarised as follows:

- sharper mainlobe with increasing array size;
- no grating lobes;
- double-dB gain for the fixed sidelobe level;

This thereby confirms all the Properties 1–3. The influence of the beampattern properties on the ranging accuracy is further discussed in Section 18.7.

18.4 Beamloc Approach

The major aim of this contribution is to present a new approach called 'BeamLoc'. BeamLoc means being principally able to locate a transmitter even in NLOS environments as long as the DP arrives with a minimum so-called 'threshold' amplitude. BeamLoc is mainly based on the above-mentioned beneficial properties of UWB beamforming, like the double-dB gain. For instance, a realisable UWB antenna array with eight antennas could achieve a gain of up to 18 dB. Applying a beamformer on both sides of the link allows a gain of up to 36 dB, which means that a wall width of approximately less than 36 cm could be principally overcome (a rule of thumb is 1 dB loss per centimetre of wall width). Hence, even if the DP is severely attenuated, it might be still detectable as long as the *transmitter mainlobe* and the *receiver mainlobe* 'see' each other. This latter target situation, which we call the 'lock mode', could be effectuated as illustrated in Figure 18.9.

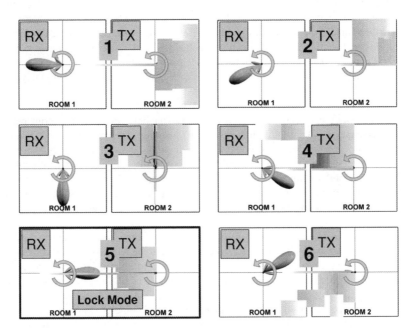

Figure 18.9 Effectuating the lock mode

All six subplots show two 2D-rooms separated by a wall, where the receiver is firmly installed in the left room (Room 1) and the transmitter is somewhere moving in the right room (Room 2). Note first that without any restriction of generality, both TX and RX are located in the middle of the room only for simplicity. Note also that both the receiver as well as the transmitter must know the orientation of an underlying coordinate system. This can be simply complied by enriching the moving robot with a compass and by a nonrecurring calibration during the access point's installation procedure. For the moment, we only use the cardinal points 'North' (N), 'West' (W), 'South' (S), 'East' (E), and any combination of them as representative for the required coordinate system. Such coarse granularity should obviously become finer in succeeding studies, i.e. with a resolution in the order of a degree or even less. Observe that all the subplots show a simplified sketch of the beampattern (in general, it depends on angle *and* frequency) to more easily illustrate the impact of the mainlobe and the sidelobe.

Let us start with subplot no.1. In room 2, the transmitter continuously emits a signal, which somehow includes the information about the current mainlobe direction (here 'E'). Of course, this signal is not only transmitted via the mainlobe, but also spuriously via the sidelobes. This property is indicated in the first subplot by three signals leaving the transmitter in three different directions with different magnitudes corresponding to the individual lobe magnitudes. Note that the shading along the signal paths indicates their attenuation according to the specific path loss. The RX located in room 1 also receives continuously, where its mainlobe direction (here 'W') is calculated according to the information carried by the transmitted signal. Note further that during the initialisation phase of our localisation approach only a single path among all the multiple paths has to reach the receiver in order to steer the mainlobe to the 'right direction'. where this 'right direction' is simply the direction of the transmitter's mainlobe plus (or minus) 180 degrees. In other words, if the transmitter steers towards 'N', the receiver steers towards 'S' etc. Now the transmitter's mainlobe starts to rotate counterclockwise from 'E' to 'NE' etc. (see subplot 2), so that the receiver's mainlobe rotates accordingly. This is demonstrated in the succeeding subplots. After one complete rotation of 360 degrees (not completely shown here), the orientation-dependent instantaneous power (after matched filtering) at the receive beamformer's output is evaluated with respect to the mainlobe direction. The direction corresponding to the maximum instantaneous power is selected as the wanted direction. This situation is called the 'lock mode' (LM) shown in subplot no. 5. Once the LM is determined, the conventional time-delay estimation begins in order to determine the distance between the robot and the AP. Knowledge of the distance and the direction uniquely determines the robot's position.

One can argue that in a typical UWB dense multipath environment, hundreds of paths arrive at the receiver, so that even for UWB beamforming on both sides of the link, numerous interfering paths remain for reliable DP separation and detection. It is the aim of our future research to investigate this situation in detail and to gain quantitative results. For simplicity, however, we will here focus on a sparse path model, so that we will demonstrate the principal functioning of our approach by preliminary results. A similar criticism regarding multiple antennas for UWB localisation is given along the lines of [8, p. 2, 'AoA section'] with respect to DoA estimation. However, note that here – due to direct transmission of the current receive mainlobe direction – there is no need for sensitive DoA estimation algorithms at the receiver.

One can also argue that the installation of an access point in any room could circumvent the demanding NLOS situation in an easy way. However, in a typical household the NLOS situation will further persist because of other obstacles, like a cupboard etc., inside a room. Moreover, even if the robot is in LOS with one access point, another access point being in NLOS could add further localisation information for an even more precise estimate. This holds especially true because several available APs open up the additional option of a conventional triangularisation approach yielding an improved ranging accuracy by evaluating intersections. The above chosen scenario with one robot in room 2 and one AP in room 1 should be understood against this background.

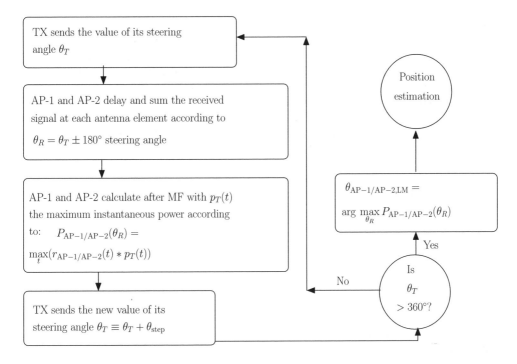

Figure 18.10 Flowchart of the BeamLoc approach

18.5 Algorithmic Framework

After this more descriptive explanation of the BeamLoc approach the algorithmic framework follows now. Figure 18.10 shows a coarse flowchart of our BeamLoc approach. BeamLoc is mainly based on the above-mentioned beneficial properties of UWB beamforming, like the double-dB gain, and can be summarised in the following two steps.

Searching for the *lock mode* (SLM)

- TX emits a signal carrying the current mainlobe direction θ_T. The mainlobe rotates with an angular step size of θ_{step}. Note that TX is equipped with a compass in order to achieve a common reference, while the APs are only needed to be calibrated once during its installation procedure.
- After receiving the value of θ_T, AP-1 and AP-2 rotate their beams toward $\theta_R = \theta_T \pm 180°$ and calculate the received maximum instantaneous power $P_{\text{AP-1/AP-2}}(\theta_R)$ after matched filtering (MF) between the received signal $r_{\text{AP-1/AP-2}}(t)$ and the pure symmetric transmitted pulse $p_T(t)$. Note that the sign '*' denotes a convolution. Note also that θ_R is measured from the positive x-axis as a common reference.
- TX increases its steering angles by θ_{step} and transmits the update, $\theta_T + \theta_{\text{step}}$. Again AP-1 and AP-2 calculate the received maximum instantaneous power at $\theta_R = (\theta_T + \theta_{\text{step}}) \pm 180°$, after one complete rotation the individual maximum of $P_{\text{AP-1/AP-2}}(\theta_R)$ with respect to θ_R determines the two *lock mode* (LM) angles $\theta_{\text{AP-1,LM}}$, $\theta_{\text{AP-2,LM}}$. Observe that within the LM both APs should steer their mainlobes toward the transmitter, so that multipath propagation is now mitigated and the position estimation can start now.

Exploiting the Lock Mode (ELM)

- Once both locked modes are determined, AP-1 and AP-2 are equipped with each DoA, so that the TX position can be calculated principally by intersection. Note that this approach is vulnerable to any model mismatch since a small angular variation may result in a distinct positioning error; this could be denoted as a near-far problem. For that reason, the merge of two APs with DoA and especially with ToA estimation is worth being further investigated.

At the end of the algorithm description we derive a simplified formula for the RX beamformer output signals $r(t)$ in the lock mode. Note that we do not distinguish between AP-1 and AP-2, since the equation holds true for both. Let θ_{LM} be the lock (fixed) angle measured with respect to the positive x-axis at the transmitter. At the locked situation, the transmit direction of the mainlobe becomes

$$\theta_T = \theta_{LM}. \tag{18.17}$$

Assume that the direct path is straight during the lock period. The receive direction of the mainlobe can then be written, according to the transceiver geometry (see subplot no. 5 in Figure 18.9), as

$$\theta_R = \theta_{LM} - \pi. \tag{18.18}$$

For simplicity, we shall consider here the single antenna scenario only. Furthermore, we assume that in the receiver noisy signal model $r^{sa}(t) = b_R(\theta_R, \varphi_R) y(t)$, there are beampatterns $b_T(\theta, \varphi)$ and $b_R(\theta, \varphi)$ for each single antenna at the transmitter and the receiver, respectively. Towards this end, it can mathematically be expressed as

$$\begin{aligned} r^{sa}(t) = &\, b_T(\theta_{LM}, \theta_{LM}) s(t - \tau_{DP}) \alpha_w b_R(\theta_{LM} - \pi, \theta_{LM} - \pi) \\ &+ \sum_{l=1, l \neq DP}^{L_R} b_T(\theta_{LM}, \theta_l) s(t - \tau_l) \alpha_w \alpha_l b_R(\theta_{LM} - \pi, \theta_l) \\ &+ n(t), \end{aligned} \tag{18.19}$$

where τ_{DP} is the time delay of the direct path, α_w is the attenuation caused by the wall, α_l is the reflection coefficient due to the lth path, and $s(t)$ is the transmitted signal.

18.6 Time-delay Estimation

Other than signal strength and the DoA, the time of arrival (ToA) is a key feature typically employed in geolocation systems (e.g. [19], [20], [3], [15], [16]). The way to estimate time delay can be, in fact, divided into two consecutive portions. The first step is the effort to identify the direct path at the receiver. In the presence of dense multipath, a search algorithm is proposed to detect the signal due to the direct path [Chapter 3, 11]. This issue is addressed above and is therefore outside the scope of this section. For the second step, we certainly need to estimate the *time delay* of this direct path. In general, there are several different algorithmical classes available for time delay estimation (TDE), for example maximum likelihood [6], MUSIC [23] and ESPRIT [21].

Assume that the signal waveform $s(t)$ is known *a priori* at the receiver. This assumption seems to be justifiable, because it can be preassigned between the transmitter and the receiver. Note that as earlier mentioned a propagation of $s(t)$ through a wall may change its waveform significantly, thus leading to a potential source of further error. However, once we enter the LM, we may assume that the transmit and receive beamformer suppress the interfering paths to a large extent so that we approximately obtain

$$r^{sa}(t) = b_T(\theta_{LM}, \theta_{LM})s(t - \tau_{DP})\alpha_w b_R(\theta_{LM} - \pi, \theta_{LM} - \pi) + n(t). \quad (18.20)$$

In this rather ideal situation the only missing task is to determine the delay τ_{DP} of the DP. This can be achieved by simply cross correlating the received signal $r^{sa}(t)$ with the known template signal $s(t)$, i.e.

$$c_{rs}(\tau) = \int_{-\infty}^{\infty} r^{sa}(t)s(t + \tau)dt, \quad (18.21)$$

so that the crucial time delay is estimated by

$$\hat{\tau}_{DP} = \arg\max_{\tau} c_{rs}(\tau). \quad (18.22)$$

Obviously, while the approach seems to be feasible for a single path channel, it is quite vulnerable to dense multipath propagation. Even if both beamformers suppress the interfering paths significantly, the large number of paths in a typical indoor environment may severely impact the accuracy of this method. There are several ways to further improve our approach. A simple but costly one is given by more antennas, meaning higher BG, or a larger aperture, meaning a sharper mainlobe. Another approach is to take the statistical properties of the interfering paths in the optimisation into account by abbreviating the interfering term as

$$w(t) = \sum_{l=1, l \neq DP}^{L_R} b_T(\theta_{LM}, \theta_l)s(t - \tau_l)\alpha_w \alpha_l b_R(\theta_{LM} - \pi, \theta_l) + n(t). \quad (18.23)$$

While Equation (18.21) relies on a matched filter to the transmitter pulse and is therefore optimum only for an additive white Gaussian noise channel in terms of maximum SNR, different local optimum detectors are available for other types of noise (for more details, see Chapter 7 in [12]). From another perspective, keeping in mind that UWB signals are typically modulated by pulse position modulation, the interfering paths of a single user can be also interpreted in a first order as further single path channel distortions caused by other users. For the latter scenario, a closed form approximation has been recently derived for the PDF of the multi-user interferences as follows (see [24] and [25] for more details)

$$f_X^{MUI}(x) = \sum_{\mu=0}^{n} \frac{A^{\mu} e^{-A}}{\mu! \sqrt{2\pi \sigma_{\mu}^2}} e^{-\frac{x^2}{2\sigma_{\mu}^2}}, \quad (18.24)$$

with

$$\sigma_{\mu} = \sqrt{\frac{\mu}{3}} \int_{-\infty}^{\infty} p_T(t)(p_T(t) - p_T(t - \delta))dt, \quad (18.25)$$

and

$$A = N_s(U - 1)\frac{\frac{1}{T_f}\int_{-T_p}^{T_p}(R_p(\tau) - R(\tau + \delta))^2 d\tau}{(R_p(\tau) - R(\tau + \delta))^2}, \quad (18.26)$$

where $p_T(t)$ is the transmitter pulse shape, U is the number of users, N_s is the number of pulses used to transmit a single information bit, T_f is the duration of one frame, δ is the PPM parameter and T_p is the pulse width. Hence, the statistical properties of $w(t)$ are approximately known, so that an improved time delay estimator becomes feasible; this is one vital topic of future research.

18.7 Simulation Results

In this section, we illustrate the principal behaviour of the proposed ranging algorithm by numerical simulation results. Before we proceed further we briefly revisit general performance bounds in order to understand the principal limits as a function of the relevant system parameter for any localisation method. Assume a single path channel and additive white Gaussian noise at the receiver. Then the standard deviation of an estimator $\hat{d}$ for the true value d can be written as [8]

$$\sqrt{\text{Var}(\hat{d})} \geq \frac{c}{2\sqrt{2\pi}\sqrt{SNR}\beta}, \qquad (18.27)$$

where c is the speed of light, and β is the signal's effective bandwidth given by

$$\beta = \sqrt{\frac{\int_{-\infty}^{\infty} f^2 |S(f)|^2 df}{\int_{-\infty}^{\infty} |S(f)|^2 df}}, \qquad (18.28)$$

with the Fourier transform $S(f)$ of the signal $s(t)$. Observe that this bound is inversely proportional to bandwidth and SNR, which confirms the intuitive claim of a raised accuracy for higher SNR and higher bandwidth. A quantitative evaluation of Equation (18.27) results in a possible sub-centimetre accuracy for UWB signals. However, the bound has been derived under several ideal conditions, so it has to be proven further under harsh UWB environments.

In the following simulation scenarios AP-1 and AP-2 are considered as two UWB receivers being located in the centre of the first and the second room, respectively. For reference purposes we set the lower left corner of the first room in the origin of the coordinate system. Wall thickness is set to 0.3 metres. TX, by definition, is assumed to be in the first room at an unknown position (x_T, y_T). We assume that the size of each room is small enough to be covered by one access point but large enough so that the plane wave assumption is reasonable. Furthermore, both the transmitter and the APs are able to perform beamforming. For all following simulation results, assume $N_T = N_R = N$ and $\lambda_c = 4.4$ cm. The impulse response consists of the DP and the reflections at the walls, where the channel is distorted by additive white Gaussian noise. The simulations are based on the scenario shown in Figure 18.11, where the circles represent the AP positions and the crosses represent 50 randomly chosen TX positions. The error is averaged over these TX positions. 100 runs per TX position and per SNR are carried out.

Figure 18.12 shows the behaviour for different array sizes. It can be observed that with increasing array size the estimation error decreases, which results from the sharpened mainlobe width for increased array size. Observe that the results for AP-2 are similar to the one for AP-1 except for a shift along the SNR axis. This behaviour is quite reasonable since the wall causes additional attenuation. However, it becomes clear that even under NLOS a localisation might be successfully performed. Note further that even with rather simplistic signal processing the accuracy already tends to be 10 centimetres and below for high SNRs.

As mentioned earlier, the bandwidth – or in the time domain the pulse width – also plays a dominant role in highly accurate indoor positioning. A decreasing pulse width behaves similarly to an increased array size (compare with Equation (18.13)), which is demonstrated in Figure 18.13.

For an approximate bandwidth of 2 GHz, the optimum accuracy tends toward 3 cm for large SNR. This means that for a full UWB bandwidth of 7.5 GHz even subcentimetre resolution might be feasible.

A further experiment illustrates the estimation error for a different number of antennas. Note that while the double-dB gain applies for the SIR, the SNR is increased by conventional 3 dB per doubling of the number of antennas; this can be clearly seen in Figure 18.14.

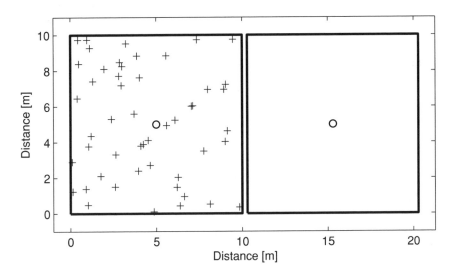

Figure 18.11 50 random TX positions in room 1

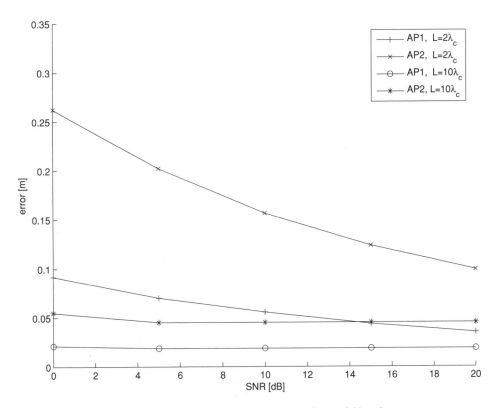

Figure 18.12 Position error for $T_p = 0.5$ ns and $N = 4$

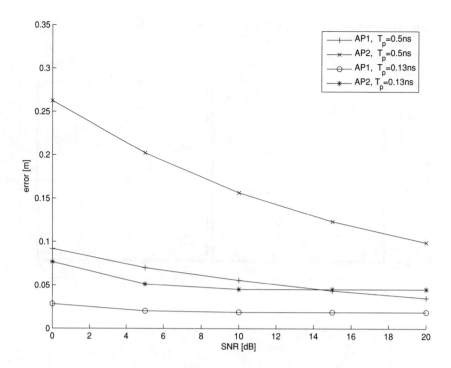

Figure 18.13 Position error for $N = 4$ and $L = 2\lambda_c$

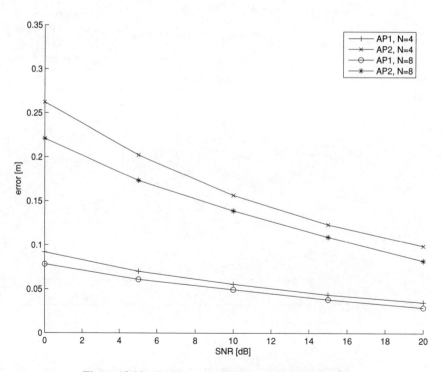

Figure 18.14 Position error for $T_p = 0.5$ ns and $L = 2\lambda_c$

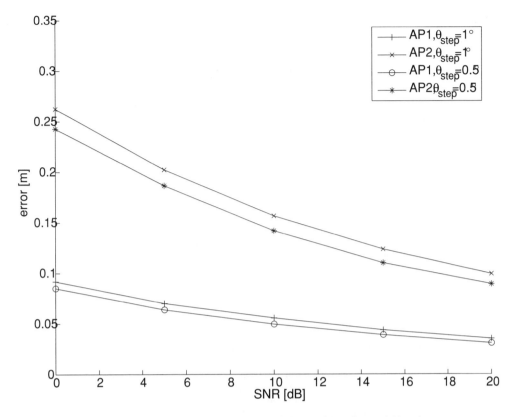

Figure 18.15 Position error for $T_p = 0.5$ ns and $L = 2\lambda_c$ and $N = 4$

In all above considered simulation results the rotation step size θ_{step} equals $1°$. Since the quality of the estimation depends on the step size as well, Figure 18.15 shows the behaviour of the estimation error for different step sizes. It can be seen that with decreasing step size the error decreases.

In general, the estimation error depends on the distance between TX and AP, d, the DoA estimation error ε_{DoA} and the ToA estimation error ε_{ToA} as follows

$$\varepsilon(d, \varepsilon_{DoA}, \varepsilon_{ToA}) = \sqrt{(\cos(\varepsilon_{DoA}/2)\varepsilon_{ToA}c)^2 + \sin(\varepsilon_{DoA}/2)^2(\varepsilon_{ToA}c + 2d)^2}. \tag{18.29}$$

This equation can be simply derived from geometric properties. Since the beamformer rotates in all our experiments with a stepsize of $1°$, the maximal DoA estimation error in noise free environment is $\varepsilon_{DoA} = 0.5°$. In combination with exact ToA estimation ($\varepsilon_{ToA} = 0$ s) the previous equation simplifies to

$$\varepsilon(d, 0.5°, 0s) = 0.0087d, \tag{18.30}$$

which gives us an upper bound for the estimation error in our simulation results in noise-free environment.

18.8 Conclusions

The aim of this contribution was to introduce a novel approach for localisation in NLOS environments. We call this approach *BeamLoc* in order to reflect the inherent beamforming and the vital lock mode. The device to be located emits its current mainlobe direction within the transmit signal, so that despite multipath propagation, DoA estimation reduces to a receiver power maximisation. Preliminary simulations show the principal feasibility of BeamLoc, but much more work is needed for a truly meaningful evaluation. Several issues, like the inclusion of more realistic beampatterns, adequate handling of near-field propagation, derivation of performance bounds being valid for UWB environments, more precise channel modelling and improved time-delay estimation are obvious topics for future research. Many of them are being currently tackled in our research group, and the results will be presented in succeeding publications.

References

[1] F. Althaus, F. Trösch and A. Wittneben, Geo-regioning in UWB networks, *14th ISE Mobile & Wireless Communications Summit*, Dresden, Germany, June 2005.
[2] P. Bahl and V.N. Padmanabhan, RADAR: an in-building RF-based user location and tracking system, *Proc. IEEE Infocom 2000*, Tel-Aviv, Israel, March 2000.
[3] J.J. Caffery and G.L. Stuber, Overview of radiolocation in CDMA cellular systems, *IEEE Commun. Mag.*, 38–45, 1998.
[4] B. Denis, J. Keignart and N. Daniele, Impact of NLOS propagation upon ranging precision in UWB systems, *Proc. IEEE Conf. Ultra Wideband Systems and Technologies* (UWBST'03), Reston, VA, November 2003.
[5] T. Do-Hong and P. Russer, Signal processing for wideband smart antenna array applications, *IEEE Microwave Mag.*, 2004.
[6] M. Feder and E. Weinstein, Parameter estimation of superimposed signals using the EM algorithm, *IEEE Trans. Acoustics, Speech, and Signal Processing*, **36**(4), 477–89, 1988.
[7] S. Gezici, H. Kobayashi and H.V. Poor, A new approach to mobile position tracking, *Proc. IEEE Sarnoff Symp. Advances in Wired and Wireless Communications*, 204–7, Ewing, NJ, March 2003.
[8] S. Gezici, Z. Tian, G.B. Gainnakis, H. Kobayashi, A.F. Molisch, H.V. Poor and Z. Sahinoglu, Localisation via ultra-wideband radios, *IEEE Signal Processing Magazine*, July 2005.
[9] M.G.M. Hussain, Principles of space–time array processing for ultra-wideband impulse radar and radio communications, *IEEE Trans. Vehicular Technology*, 2002.
[10] D.B. Jourdan, J.J. Deyst, M.Z. Win and N. Roy, Monte Carlo localisation in dense multipath environments using UWB ranging, *Proc. IEEE ICU 2005*, Zürich, Switzerland, September 2005.
[11] J.-Y. Lee, Ultrawideband ranging in dense multipath environments, Ph.D. Dissertation, University of Southern California.
[12] S.A. Kassam, *Signal Detection in Non-Gaussian Noise*, Springer Verlag, Dowdwn & Culver Inc., 1988.
[13] J.Y. Lee and R. Scholtz, Ranging in a dense multipath enviornment using a UWB radio link, *IEEE Trans. Select. Areas Com.*, **20**(9), 1677–83, 2002.
[14] A.F. Molisch, J.R. Foerster and M. Pendergrass, Channel models for ultrawideband personal area networks, *IEEE Wireless Commun.*, 2003.
[15] K. Pahlavan, P. Krishnamurthy and J. Beneat, Wideband radio propagation modeling for indoor geolocation applications, *IEEE Commun. Mag.*, 60–45, 1998.
[16] K. Pahlavan, X. Li and J.-P. Makela, Indoor geolocation science and technology, *IEEE Commun. Mag.*, 112–18, 2002.
[17] Y. Qi and H. Kobayashi, On time of arrival positioning in a multipath environment, submitted to *IEEE Trans. Veh. Technol.* (downloadable from www.princeton.edu/~sgezici/Qi%20et%20al%20TOA%20Positioning%20in%20MP.pdf).
[18] S. Ries and T. Kaiser, Ultra wideband impulse beamforming: it's a different world, *Special Issue on 'Signal Processing in UWB Communications'*, invited paper, Elsevier Science, 2005.

[19] T.S. Rappaport, J.H. Reed and B.D. Woerner, Position location using wireless communications on highways of the future, *IEEE Commun. Mag.*, 33–41, 1996.
[20] J.H. Reed, K.J. Krizman, B.D. Woerner and T.S. Rappaport, An overview of the challenges and progress in meeting the E-911 requirement for location service, *IEEE Commun. Mag.*, 30–7, 1998.
[21] H. Saarnisaari, TLS-ESPRIT in a time delay estimation, *Proc. VTC '97*, **3**, 1619–23, 1997.
[22] Q.H. Spencer, B.D. Jeffs, M.A. Jensen and A.L. Swindelhurst, Modeling the statistical time and angle of arrival characteristics of an indoor multipath channel, *IEEE Journal on Selected Areas in Communications*, 2000.
[23] C.W. Therrien, S.D. Kouteas and K.B. Smith, Time delay estimation using a signal subspace model, *Proc. 34th Asilomar Conference on Signals, Systems, and Computers*, **2**, 837–41, Monterey, CA, October 2000.
[24] Y. Dhibi, *Impulsive Kanalstörungen und deren Einfluss in der ultrabreitbandigen Übertragung* (Impulsive Noise and its Impact in the Ultra WideBand Communications), PhD thesis, University Duisburg-Essen, July 2005.
[25] Y. Dhibi and T. Kaiser, On the impulsiveness of multi-user interferences in TH-PPM-UWB-systems, to appear in *IEEE Trans. Signal Processing*.
[26] M.G. di Benedetto, T. Kaiser, A. Molisch, I. Oppermann, C. Politano and D. Porcino, *UWB Communication Systems – A Comprehensive Overview*, EURASIP Book Series, Hindawi, 2005.

19

Antennas for Ground-penetrating Radar

Ian Craddock

19.1 Introduction

Ground-penetrating radar (GPR) is a catch-all term that encompasses many applications, not all of which actually involve the ground – through-wall imaging being an example. This chapter considers some of these applications; however, the subject of wideband biomedical imaging (which may also be regarded as a form of GPR) is reserved for Chapter 20.

GPR antennas are generally designed to operate over the widest possible bandwidth, and frequently designed to radiate pulse waveforms. In these respects they have much in common with current designs for other UWB systems.

The emphasis of this chapter is on antennas for GPR, however, in order to put this discussion into context the chapter commences with a survey of some of the situations in which GPR has been applied. Although the applications are diverse, most share a number of features that result in similar constraints for the antenna designer. A number of wideband antenna designs that attempt to meet these constraints are then presented later in this chapter, along with some notes on antenna measurement and analysis methods. The reader interested in a more detailed description of GPR systems and their applications is referred to [1].

19.2 GPR Example Applications

19.2.1 GPR for Demining

An especially well-known GPR application is the location of buried landmines (and other unexploded ordinance). The devastating human, social, environmental and financial impact of mines in many countries around the world [2] has understandably resulted in a strong focus within the GPR community on demining activities.

Figure 19.1 Anti-tank and anti-personnel mines, and test sphere

The technical challenge is, however, considerable, and after decades of research, there is still much work to be done. The difficulty involved may be appreciated when it is noted (see Figure 19.1) that mines may vary from substantial, largely metal anti-tank devices, buried often at some depth, through to hand-sized, mostly plastic anti-personnel mines, generally quickly and shallowly buried among the inhomogeneity and vegetation of the soil surface.

Further complicating the detection process is the fact that the ground itself is inhomogeneous and may vary from dry sand (which gives very poor dielectric contrast between the mine and its surroundings) to water-logged clay (in which the radar signal suffers high attenuation); see an example depicted in Figure 19.2. The resulting variation in the impedance, attenuation and propagation velocity of the medium makes the detection process more difficult.

One specific challenge is unique to this application: mines are explosive devices that are readily triggered by pressure or movement on the ground. This precludes in-contact operation, and so embedding the antenna in the ground, or in an impedance-matching medium in contact with the ground, is impractical. The antenna system must therefore be suspended above the ground and, on transmitting, will inevitably experience a dominant reflection from the air/ground interface; see Figure 19.3. While the other GPR applications do not face this particular hazard, in any mobile scenario, preventing damage to the antenna requires that a separation is maintained between the antenna and the uneven ground beneath.

19.2.2 Utility Location and Road Inspection

Locating buried utilities (pipes and cables for gas, water, sewage, electricity and communications) is an extremely important exercise, and one which costs billions of pounds each year. This fact may be readily appreciated by a report on this issue in and around Denver, Colorado, which noted that in 1998 approximately 2.5 million buried utility lines had needed to be located and that in the same year one local cable company had spent $3 million locating utilities (http://www.nal.usda.gov/ttic/utilfnl.htm).

The location of many such utilities, especially in western cities where the underlying infrastructure may be hundreds of years old, is often not known precisely and more modern utilities are more difficult

Figure 19.2 Minefield after rain (reproduced by permission of ERA Technology)

Figure 19.3 Training with the *Minehound* combined GPR/metal detector (reproduced by permission of ERA Technology)

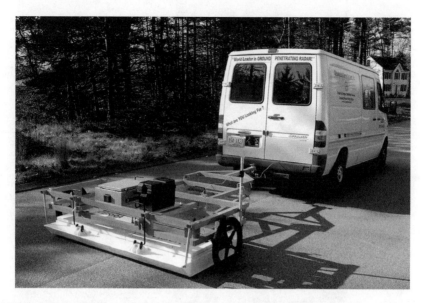

Figure 19.4 Road and utility inspection GPR in operation (reproduced by permission of Geophysical Survey Systems, Inc. (GSSI))

to detect as a result of being non-metallic. Accidental damage resulting from excavation or building work may be expensive or even disastrous.

The targets of interest are generally somewhat larger than in the landmine application and, in any event, consistent along considerable lengths. The properties of the ground are basically the same. Although there is of course no danger of explosion, non-contact operation is still desirable since time is of the essence in many applications and hence operation from a moving platform is the objective.

An addition to detecting the presence of pipes, the possibility of detecting *leaks* from the pipes is especially interesting, for a number of environmental reasons. Related to this topic is the use of GPR for detecting hazardous waste – such as buried drums, seeping contaminants [3] [4] and buried landfill.

A related application is that of road and pavement condition monitoring – here simply the thickness of the various layers is of concern, rather than imaging buried objects [5, 6]. Here, as shown in Figure 19.4, GPR provides a measurement capability that is sufficiently rapid to avoid interfering with traffic flow.

19.2.3 Archaeology and Forensics

The use of GPR in archaeology provides an invaluable and unique technique for non-invasively mapping the ground in 3D [7]. GPR systems have been used at sites as diverse as the Great Pyramids in Cairo and the Siberian tundra.

Figure 19.5 shows a GPR survey in progress at Petra in Jordan. The image produced from this survey is shown in Figure 19.6, revealing the walls of a buried temple.

In police forensic work, GPR has proven useful in detecting buried bodies, as documented [1] in the infamous West murder case in the UK. Often the detection of the body is in fact indirect, since decaying biological tissues often do not have strong contrast to the surrounding soil, and the GPR actually relies on detecting the disturbance to the ground associated with the burial [8].

Figure 19.5 GPR survey at Petra (reproduced by permission of © Alta Mira Press 2004)

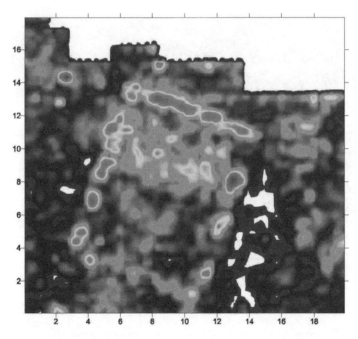

Figure 19.6 GPR image at 1 m depth from the Petra site, scales in metres (reproduced by permission of Lawrence Conyers, University of Denver)

19.2.4 Built-structure Imaging

GPR systems for inspecting the condition of concrete structures, and identifying the location of reinforcements, have been devised, and employed for bridge and runway inspection, as shown in Figure 19.7.

Penetration through structural and nonstructural walls is also possible, and the possibility of through-wall imaging has been investigated. The low water content of these structures is more favourable for radio-wave propagation and there has been interest in the law-enforcement and military applications of such systems [9, 10].

It is hoped that the above discussion, while not in any sense a definitive list of applications, gives some idea of the very wide range of scenarios where GPR finds useful application. In general, it is clear that there are significant similarities between most of the applications: a radio wave must be transmitted into and through a dense medium which, for practical reasons, is generally distanced from the transmitting and receiving antennas; the material constituents of this medium are not usually homogeneous, or indeed known *a priori*.

A further similarity between all the applications (with the possible exception of through-wall imaging of human subjects) is that the target of interest is stationary with respect to its surroundings. These surroundings (the ground surface, vegetation, surface debris, soil inhomogeneity – anything within range that presents a different refractive index to the transmitted wave) generate reflections, called 'clutter', that compete with the wanted signal from the target. Unlike many other radar applications, there is no Doppler shift that can be exploited to separate the target waveform from the clutter, and this goes some way to explaining the difficulty inherent in most GPR imaging.

Perhaps the most notable similarity, however, is that in the pursuit of high-resolution images of the medium, all the GPR applications require operation over a wide bandwidth. Unlike operation -in-air-,

Figure 19.7 GPR inspection of concrete bridge deck (reproduced by permission of Geophysical Survey Systems, Inc. (GSSI))

however, wide-bandwidth operation is strongly limited by the RF propagation characteristics of the ground, and most GPR systems are designed as a compromise between resolution and penetration depth. This is considered in the next section.

19.3 Analysis and GPR Design

19.3.1 Typical GPR Configuration

Figure 19.8 shows, in two dimensions, a typical scenario: GPR systems are generally interested in targets at very short ranges from the antenna (for example, a buried mine at 10 cm depth), and for this reason usually employ separate transmit and receive antennas in order to avoid the very fast RF switching that would be required when using a single antenna. This type of radar system is commonly known as 'bistatic' – although in practice both antennas are likely to be in motion (usually the same motion) and, in fact, it is perfectly possibly to consider more than one pair of transmit/receive elements [11].

The transmit and receive antennas are usually in air, and the ground is therefore a higher dielectric medium, so the free-space radiation patterns are narrowed by refraction at the surface as shown. RF energy radiated from the transmit antenna will be reflected from any discontinuity in the wave impedance of the ground back to the receive antenna, provided that the two antennas' patterns overlap at that point. This permits an image to be formed.

In Figure 19.8 the respective patterns are shown for clarity as solid shaded areas, whereas the overlap between the patterns is of course more complex, especially if the near-field coupling is included. Nor, of course, does the pattern truncate completely at a fixed depth. The figure does, however, illustrate the interdependency between the 'stand off' from the surface, the antenna beam pattern ('footprint', or directionality) and the ability to achieve coverage in the ground at a range of depths.

High lateral resolution images may to some extent be produced using narrow antenna beams or synthetic aperture techniques [12], however, range/depth resolution is largely dependent on the bandwidth of the received waveform – or, put simply in the case of an impulse radar, the narrower the received pulse, the more accurately the origin of the reflection can be determined. Large operational bandwidths are also helpful in reducing the effect of clutter, since time-gating (or similar) can be used to exclude reflected signals from different ranges.

The 'ground' is, however, usually an attenuating medium, and this limits the range of frequencies that can be used, as will be shown in the subsequent section.

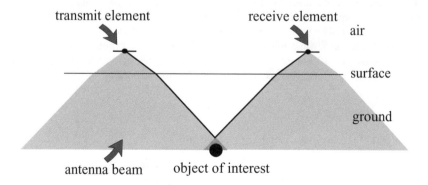

Figure 19.8 Typical GPR scenario

19.3.2 RF Propagation in Lossy Media

Unlike most antenna applications, the GPR antenna must radiate into a lossy half-space. The propagation of radio waves in lossy media is not usually covered in antenna textbooks and so for completeness it is presented here, derived from Maxwell's equations. Readers more interested in results than derivations may wish to merely note the expressions for 'skin depth', Equations (19.7) and (19.10).

The electric flux density **D**, the electric field intensity **E**, magnetic flux density **B**, magnetic field intensity **H** and the conduction current density **J** are related at each point in the ground by the time-harmonic version of Maxwell's curl equations (see also Chapter 2):

$$\nabla \times \mathbf{E} = -j\omega \mathbf{B} \tag{19.1a}$$
$$\nabla \times \mathbf{H} = \mathbf{J} + j\omega \mathbf{D} \tag{19.1b}$$

The medium is usually linear, hence $\mathbf{D} = \varepsilon_0 \varepsilon_r \mathbf{E}$ and $\mathbf{B} = \mu_0 \mu_r \mathbf{H}$. In general, μ_r and ε_r may be tensors, however, in most GPR scenarios the medium may be considered isotropic, in which case they are scalars. The medium is usually also nonmagnetic and so $\mu_r = 1$. The relative permittivity, ε_r, may, however, be complex and a function of frequency, hence:

$$\nabla \times \mathbf{E} = -j\omega \mu_0 \mathbf{H} \tag{19.2a}$$
$$\nabla \times \mathbf{H} = \mathbf{J} + j\omega \varepsilon_0 \varepsilon_r(\omega) \mathbf{E} \tag{19.2b}$$

The ground is usually not a perfect insulator (although where the 'ground' is dry it may be quite insulating) and, assuming an applied static electric field would cause a conduction current to flow in accordance with Ohm's law, this conduction current density is related to the electric field by the conductivity σ. Therefore:

$$\nabla \times \mathbf{H} = (\sigma + j\omega \varepsilon_0 \varepsilon_r(\omega)) \mathbf{E} \tag{19.3}$$

Typically the conductivity and permittivity are grouped together:

$$\nabla \times \mathbf{H} = j\omega \left(\varepsilon_0 \varepsilon_r(\omega) - j\frac{\sigma}{\omega} \right) \mathbf{E} = j\omega(\varepsilon' - j\varepsilon'') \mathbf{E}$$
$$= j\omega \varepsilon_c \mathbf{E}, \tag{19.4}$$

where ε_c is the complex permittivity, with real part ε' and imaginary part ε''.

Energy is dissipated in the medium as the radio wave travels through it, and this, as will be seen, arises from the above equation purely because of the imaginary part of ε_c. In the case where the relative permittivity is real, as is often assumed, the only loss mechanism will be the heat dissipated as a result of the conduction current flowing. Regardless of the origin of these losses, however, it is customary to describe the 'lossiness' using a parameter called the loss tangent, $\tan \delta_c$, which simply describes the relative size of the real and imaginary parts of ε_c and is therefore defined as:

$$\tan \delta_c = \frac{\varepsilon''}{\varepsilon'}$$

Taking the curl of (19.2a) and substituting (19.4) yields:

$$\nabla \times (\nabla \times \mathbf{E}) = -j\omega \mu_0 (\nabla \times \mathbf{H})$$
$$= -j\omega \mu_0 (j\omega \varepsilon_c \mathbf{E})$$
$$= \omega^2 \mu_0 \varepsilon_c \mathbf{E} \tag{19.5}$$

The left-hand side of this relation can be rewritten using the vector identity $\nabla(\nabla \cdot \mathbf{E}) - \nabla^2 \mathbf{E}$, and this may be simplified to $-\nabla^2 \mathbf{E}$ if $\nabla \cdot \mathbf{E} = 0$, in other words if there is negligible charge present in the ground. Hence:

$$\nabla^2 \mathbf{E} + \omega^2 \mu_0 \varepsilon_c \mathbf{E} = 0.$$

Solutions to this equation (the Helmholtz equation) describe wave propagation in the medium, in all possible directions, and with all possible polarisations. To determine the general effect of losses in the medium it is, however, sufficient to consider only one polarisation (arbitrarily here the x-direction) and one direction of propagation (z). The partial differential equation then reduces to the scalar equation:

$$\frac{d^2 E_x(x)}{dz^2} + k^2 E_x(x) = 0, \tag{19.6}$$

where, for brevity, $k = \omega\sqrt{\mu_0 \varepsilon_c}$ (a quantity known as the 'wavenumber').

Any function $E_x(z) = A \exp(-jkz)$ can be seen to be a solution to Equation (19.6), where A is an arbitrary constant. In a lossy medium, since k arises from the square root of the complex permittivity, it too will be complex and so the exponential solution will comprise both a real and an imaginary term. Hence:

$$E_x(z) = A \exp(-j\mathrm{Re}(k)z) \exp(\mathrm{Im}(k)z)$$

On inspecting this solution it is clear that if the imaginary part of the wavenumber k is non-zero (and negative, as is required by energy conservation in this case) the propagating wave will attenuate exponentially as it travels in the z direction. The size of the imaginary part of the wavenumber k will determine the rate of decay (the real part determines the rate of change with distance and so is related to the wavelength).

The distance travelled by the wave before it reduces in amplitude by a factor of e^{-1} is a convenient and common measure of the rate of attenuation and is known as the skin depth, D, which, from the above wave solution, may be seen to be simply:

$$D = \frac{1}{\mathrm{Im}(k)} = \frac{1}{\mathrm{Im}(\omega\sqrt{\mu_0 \varepsilon_c})} \tag{19.7}$$

In order to determine the actual value of D for a given ground and frequency, and perhaps more importantly to appreciate how it depends on the chosen frequency, it is therefore necessary to consider the complex permittivity:

$$\varepsilon_c = (\varepsilon' - j\varepsilon'') = \left(\varepsilon_0 \varepsilon_r(\omega) - j\frac{\sigma}{\omega}\right) \tag{19.8}$$

The frequency-dependent, and possibly complex, relative permittivity, ε_r, embodies the response of the molecules in the medium to the electric field. Various different models exist for describing this interaction, which is very involved [13]. In the wet materials which comprise the ground in typical GPR applications, for example, the polar water molecules in the soil attempt to align with the applied electric field, in general this takes a finite amount of time and its effect on the permittivity is modelled by expressions such as the Debye formula [13]:

$$\varepsilon_{r(\text{water})}(\omega) = \varepsilon_{\infty(\text{water})} + \frac{\varepsilon_{s(\text{water})} - \varepsilon_{\infty(\text{water})}}{1 + j\omega\tau}, \tag{19.9}$$

where $\varepsilon_{s(\text{water})}$ is the value of permittivity that would be observed at low-frequencies. $\varepsilon_{\infty(\text{water})}$ the value for 'infinite' frequency. τ is the time constant of the response of the water molecules, called the 'relaxation time'.

Water is just one ingredient of the ground and although it can dominate the dielectric behaviour in wet media, in order to determine the actual relative permittivity the values for the dry components must also be considered.

The relaxation time of the water molecules (i.e. the time constant of their reorientation to the applied field) is strongly temperature-dependent but of the order of 10 ps, and therefore at angular frequencies much below 100×10^9 rad/s (i.e. below 15 GHz), $\omega\tau \ll 1$ in Equation (19.8). As a result the imaginary, frequency-dependent part of ε_r will be negligible. Most GPR systems operate well below this frequency and so, for a rough appreciation of the skin depth effect it is reasonable to neglect the frequency-dependent variation of the permittivity. Henceforth, therefore, the relative permittivity is assumed to be the real, low-frequency value, so:

$$\varepsilon_c = (\varepsilon' - j\varepsilon'') = \left(\varepsilon_0 \varepsilon_s - j\frac{\sigma}{\omega}\right),$$

where ε_s is the low-frequency (or 'static') relative permittivity. Then, the skin depth (19.7) is:

$$D = \frac{1}{\operatorname{Im}(\omega\sqrt{\mu_0 \varepsilon_c})} = \frac{1}{\omega\sqrt{\mu_0 \varepsilon_s \varepsilon_0} \operatorname{Im}\sqrt{1 + j\sigma/\omega\varepsilon_s\varepsilon_0}}$$

$$= \frac{1}{\omega\sqrt{\mu_0 \varepsilon_s \varepsilon_0}\left(1+(\sigma/\omega\varepsilon_s\varepsilon_0)^2\right)^{1/4} \operatorname{Im}(\exp(j\theta))} = \frac{1}{\omega\sqrt{\mu_0 \varepsilon_s \varepsilon_0}\left(1+(\sigma/\omega\varepsilon_s\varepsilon_0)^2\right)^{1/4} \sin\theta}$$

where, from application of De Moivre's theorem, $\theta = \frac{1}{2}\arctan(\sigma/\omega\varepsilon_s\varepsilon_0)$.

To find $\sin\theta$, we note:

$$\tan 2\theta = \frac{\sigma}{\omega\varepsilon_s\varepsilon_0} \quad \text{and therefore} \quad \cos 2\theta = \frac{\omega\varepsilon_s\varepsilon_0}{\sqrt{\sigma^2 + (\omega\varepsilon_s\varepsilon_0)^2}} = \frac{1}{\sqrt{1+(\sigma/\omega\varepsilon_s\varepsilon_0)^2}};$$

then, using the trigonometric identity $\sin^2\theta = (\frac{1}{2}(1-\cos 2\theta))$:

$$D = \frac{1}{\omega\sqrt{\mu\varepsilon_s\varepsilon_0}\left(1+(\sigma/\omega\varepsilon_s\varepsilon_0)^2\right)^{1/4}\sqrt{\frac{1}{2}\left(1-\frac{1}{\sqrt{1+(\sigma/\omega\varepsilon_s\varepsilon_0)^2}}\right)}}$$

$$= \frac{1}{\omega\sqrt{\mu\varepsilon_s\varepsilon_0/2}\sqrt{\left(\sqrt{1+(\sigma/\omega\varepsilon_s\varepsilon_0)^2}-1\right)}}$$

This equation permits the skin depth to be calculated from the frequency and the parameters of the ground. It is clear from the expression that increases in conductivity reduce the skin depth (depth of penetration). Increases in frequency also reduce the skin depth, although the relationship is slightly more complex since frequency appears twice in the numerator of the expression.

The skin-depth expression is often simplified for 'good conductors', where $\sigma \gg \omega\varepsilon$. Then:

$$D = \frac{1}{\omega\sqrt{\mu\varepsilon_s\varepsilon_0/2}\sqrt{\left(\sqrt{(\sigma/\omega\varepsilon_s\varepsilon_0)^2}-1\right)}} = \frac{1}{\omega\sqrt{\mu\varepsilon_s\varepsilon_0/2}\sqrt{(\sigma/\omega\varepsilon_s\varepsilon_0 - 1)}}$$

$$D = \frac{1}{\omega\sqrt{\mu\varepsilon_s\varepsilon_0/2}\sqrt{(\sigma/\omega\varepsilon_s\varepsilon_0)}} = \frac{1}{\sqrt{\mu\sigma\omega/2}} \qquad (19.10)$$

This rather simpler expression makes the dependence on conductivity and frequency more clear, although the assumptions made in its derivation should be noted.

Dielectric properties of the ground have been measured [1] [14] [15] and the increasing attenuation with frequency, predicted by the above skin-depth expression, have been observed. Moisture content and soil type considerably affect these properties: ε_r values between 10 and 25 are typical, with attenuations varying from 10 dB/m at 500 MHz to as much as 100 dB/m at 2 GHz in wet soil [1].

While bandwidth has the desirable property of improving the radar system's resolution, it is apparent from the preceding discussion that attenuation in the medium poses a practical limit to the upper frequency that can be used, depending on the depth of penetration required. The penetration depth required for demining is limited (perhaps 30 cm for anti-personnel mines) and the targets are small, so high frequencies up to 4 GHz are used, however, for archaeological surveys many metres of penetration may be required and for this, frequencies in the MHz range are required.

Similar analysis of the role of the *real* part of the wavenumber would show that this is not a linear function of frequency, even for the simplified lossy ground considered here. This results in different frequencies travelling at different phase velocities, a phenomenon known as 'dispersion'. Attenuation is, however, usually the main concern of the GPR designer.

19.3.3 Radar Waveform Choice

As with a communications system, the available bandwidth (limited here by the rising attenuation with frequency in the lossy medium) may be utilised in a number of ways. Broadly speaking the transmitted broadband waveform in a GPR system will either be swept/stepped frequency or an impulse. Other techniques for generating broadband waveforms, such as M-sequences, are also possible of course, though not as common in GPR.

The impulse scheme is the traditional choice for GPR systems since it is conceptually simple, inexpensive to implement at low frequencies and the different origins of reflected signals may in the time domain be readily determined from their time delays. A frequency sweep (in which category we include frequency modulated continuous wave (FMCW) and stepped-frequency systems) may be employed instead, and the time-domain signal synthesised if desired, from, essentially, an inverse Fourier transform of the received broadband signal [1].

Consider a hypothetical impulse radiated from an antenna shown in Figure 19.9. Ideally the antenna radiates a well-defined pulse (top waveform) and reflected signals, consisting of, roughly speaking, attenuated and time-delayed versions of the original pulse, will be simple to detect and separate one from another. If, however, the radiated pulse looks more like the bottom waveform, then it will be considerably more difficult to resolve reflected signals, since the longer 'tail' of the impulse is more likely to result in signals overlapping each other.

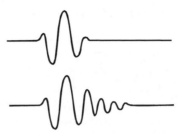

Figure 19.9 Radiated impulses

Radiating a well-defined pulse without excessive ringing (or 'time-sidelobes' as they are sometimes known) is often an important issue for the GPR antenna designer. It is well known that this can be difficult to achieve, however, because:

1. Many electrically large antenna designs effectively radiate different frequencies from different parts of their structure (and, indeed, the different modes present will result in each frequency being radiated with a different pattern). The 'point' from which the radiation appears to originate is called the 'phase centre'. If this position is frequency dependent, the impulse waveform will be dispersed, since each frequency component will travel a different distance in its journey from the antenna.
2. The pulse, applied at the feedpoint, will usually experience reflections along the length of the antenna and most notably at the ends of the antenna. The resultant reverberation will again cause a trailing 'tail' in the time-domain response of the antenna.

Effect (2) is in essence the result of the antenna being a resonant structure with a Q that is too high (i.e. a bandwidth that is too narrow) to accommodate the bandwidth of the input waveform. It may be ameliorated by resistive loading, for example (see later in this chapter). Effect (1) is different, however, since it persists regardless of the Q of the antenna – although if the phenomena is measured or otherwise characterised, the frequency dependence of the phase centre may be countered by matched filtering or by signal processing in the frequency domain.

19.3.4 Other Antenna Design Criteria

As may be seen from the preceding sections, the GPR antenna designer's overriding concern is operating over the widest possible bandwidth permitted by the attenuation in the medium. As in UWB systems, operational bandwidth must be defined not only in terms of input match but also in terms of pattern performance, since it will be of no use if the antenna is impedance matched over the required band, but radiates backwards, or with significant nulls in the pattern.

If the lowest operational frequency is f_{min} and the highest f_{max}, bandwidth may be defined as:

- fractional bandwidth $\dfrac{f_{max} - f_{min}}{(f_{max} + f_{min})/2}$, or
- the ratio $f_{max}:f_{min}$.

The diverse GPR applications generally arise outside of a laboratory environment and the GPR system will need to be human portable or at least vehicle portable. The antenna design will have a significant influence on the size and weight of the final product, and the requirement for operating bandwidth is particularly difficult to achieve within limited size and weight parameters.

Antenna efficiency is another criterion, but often neglected and in any event extremely difficult to determine in a realistic measurement environment such as that shown in Figure 19.7. The ground medium is likely to be heavily attenuating in any case, so the loss in the antenna might not be thought to be significant. Nonetheless, if depth of penetration, soil characteristics and time of scan are maintained as constant, an inefficient antenna will inevitably require extra RF transmit power, hence higher power RF components, extra heatsinking, larger power supplies and, ultimately therefore, cost and weight. If deep penetration is not required, however, the GPR system is far more likely to be limited by clutter than by attenuation, in which case trading efficiency for bandwidth is often an acceptable compromise.

GPR antenna design is certainly a challenging area. The subsequent sections consider some of the antenna designs employed and newly proposed within the field.

19.4 Antenna Elements

GPR is a very mature subject area and not all designs can be included here. Nonetheless there are a number of popular categories of antenna design, and these are represented herein as fully as possible.

GPR systems and their associated antennas do not all use the same frequency ranges, for reasons detailed in the applications section of this chapter, therefore quoting absolute antenna sizes would not be particularly enlightening. Size is broadly dominated by the in-air wavelength at the lowest operational frequency, and so the electrical dimensions at this frequency are the ones quoted. This also permits a simple estimation of the size of the antenna design if scaled to work at a different frequency. For similar reasons, fractional bandwidth and bandwidth ratios are quoted rather than absolute values.

19.4.1 Dipole, Resistively Loaded Dipole and Monopoles

The basic antenna element is the dipole, and dipole elements have certainly been used for GPR systems. The dipole antenna is reasonably compact and lightweight, however, its bandwidth is not very large and, when excited by an impulse, the reflections of the impulse from the ends of the dipole are evident as a long, ringing, impulse response which, as discussed above, is undesirable.

The end-reflection problem can be eased by placing lossy material at the dipole ends in order to reduce the reflected wave (and hence the Q of the antenna), or by placing resistors at a quarter-wavelength distance from the ends of the antenna. The latter technique was used as far back as the early 1960s to create a travelling-wave antenna, although the varying electrical distance from the resistors to the ends of the dipole arms makes this a rather bandwidth-limited approach [16] and it also requires the length of the dipole to be greater than a half-wavelength.

A more common, though more complex, approach is to introduce a varying resistive-loading profile along the antenna, for example by constructing the dipole itself out of a lossy coating deposited on a nonconducting rod. The classic paper on this technique was published in 1965 [17] and the resulting resistivity distribution is known after the authors as the Wu–King profile. The optimum loading profile is in fact complex (i.e. it is an impedance profile), but for practical reasons this is normally approximated as a purely resistive quantity:

$$r(z) \propto \frac{1}{(1 - z/h)},$$

where r is the resistance per unit length of a dipole with arm length h, as a function of distance z from the feedpoint. Careful scaling of this profile is important in order to avoid a strong impedance discontinuity at the feedpoint, and it can be difficult to obtain a 50 ohm impedance – in which case the use of a higher impedance feedline may be required.

One variant is the resistively loaded vee-dipole [18], where the arms of the dipole flare gradually out from the central feedpoint. Recently published work [19] on an exponentially tapered version loaded using surface-mount resistors mounted on a thin substrate, demonstrates a bandwidth of the order of 6:1 along with light weight and a length of around one wavelength. Efficiency figures are not quoted, however, a point of interest is that the performance is significantly constrained by the balun which is required to transform from an unbalanced 50 ohm feedline to the (balanced) 200 ohm at the dipole feedpoint.

A resistively loaded monopole constructed by soldering together chains of 32 resistors has also been presented [20], with a quoted efficiency of 25% and very good impulse-radiation characteristics. In general, as the electrical size of the antenna reduces, the amount of resistive loading must be increased to maintain bandwidth, and efficiency reduces.

Resistive loading is also a popular technique in wideband medical imaging, and further discussion can be found in Chapter 20.

19.4.2 Bicone and Bowtie

A conventional approach for widening the bandwidth of a dipole antenna is to extend the two arms of the dipole into cones. The resulting biconical dipole [21], see Figure 19.10 (left), is well known. The disadvantage of the bicone is that it is bulky and heavy, and within the GPR community its planar equivalent, the 'bowtie' is more commonly employed, where printed circuit techniques are utilised to construct a planar geometry on a suitable substrate, see Figures 19.10 (right) and 19.11. In the presence of a ground plane, conical monopoles and bowtie monopoles can be formed instead of their dipole equivalents, although if the ground plane is parallel to the ground surface then the resulting monopole's radiation pattern is not usually suitable for a GPR system.

A typical bowtie dipole design is shown in Figure 19.11. This bowtie is fabricated on a 2.2 dielectric constant RT/Duroid substrate and employs a balun to transform from the coaxial feed [22]. The antenna is matched to the coaxial feed (Figure 19.12) from 770 MHz to 1.6 GHz (a fractional bandwidth of 70%), and is relatively compact, being half-a-wavelength long at 770 MHz.

In the GPR application this bowtie was mounted through a ground plane, however, the electrical distance between the dipole and the ground plane was found to result in distortions to the radiation pattern at higher frequencies (due to destructive interference between the dipole and its image). It was decided for this reason to line the ground plane with conventional radar-absorbing material – a form

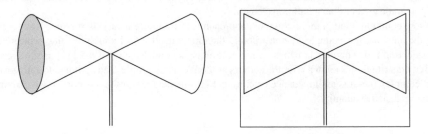

Figure 19.10 Bicone (left) and planar version, the bowtie (right)

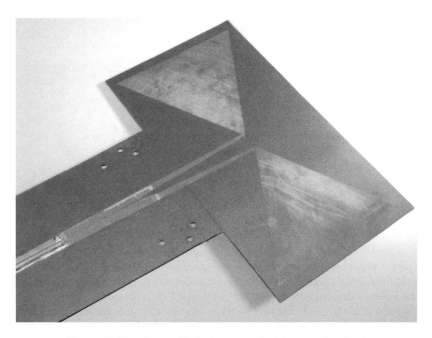

Figure 19.11 Bowtie dipole (courtesy of University of Bristol)

of resistive loading – and, at the expense of the loss of the gain associated with the ground plane, this resulted in a very stable radiation pattern over frequency. Another advantage of this approach is that it damps reverberations (multiple bounces) between the ground plane and the surface of the ground.

Resistive loading of the bowtie may also be accomplished in a similar fashion to the resistive loading of dipoles and monopoles. This is slightly more complex for a bowtie since the conductor area increases

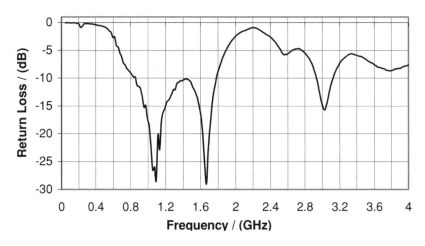

Figure 19.12 Bowtie S11 (courtesy of University of Bristol)

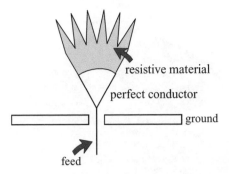

Figure 19.13 Practical arrangement for producing resistivity distribution in a bowtie monopole (after [23])

with distance from the feedpoint, however, broadly the same Wu–King profile can be employed, although the practicalities of construction may require some innovation. In [23] for example, a serrated resistive sheet is used, instead of a graduated resistance profile, in order to produce the required distribution in a bowtie monopole; see Figure 19.13. An efficiency of around 50% is quoted.

A combination of resistive and capacitive loading of the bowtie antenna has also been described [24], using volumetric absorbing material along one side of the antenna, and annular slots in the metallisation to create a tapered series capacitive loading – 160% fractional bandwidth (bandwidth ratio 10:1) and excellent pulse-radiation properties are reported, however, at low frequencies the antenna design approaches a wavelength in overall length (and almost the same in width).

19.4.3 Horn Antennas

The TEM horn is a popular antenna design in GPR applications, consisting simply of a flared pair of conductors that support a TEM wave and gradually accomplish the transition from the feedline to a travelling wave in space. The antenna is capable of radiating an impulse with little distortion, and their directionality results in useful gain. This type of antenna is commonly used in commercial systems, including those produced by Geophysical Survey Systems, Inc. (GSSI); see Figure 19.4.

Dielectric loading may be applied to reduce the size of the antenna, however, this carries a penalty in terms of weight. For example the University of Liverpool have developed a dielectric-loaded TEM horn with 180% fractional bandwidth (bandwidth ratio of 30:1) and length of half-a-wavelength, but which weighs 8 kg (Figure 19.14). Other related developments of the TEM horn design may be found in the literature [25] [26].

19.4.4 Vivaldi Antenna

The Vivaldi antenna has been widely used in GPR systems, and, somewhat like a TEM horn antenna, employs a tapered slot to assist the transition from a guided wave in the feed to a radiating wave in free space. Unlike the TEM horn antenna the design is planar, resulting in considerable reductions in volume and weight, however, its phase centre is a function of frequency.

Reported designs for GPR include an antipodal Vivaldi matched from 3 to 20 GHz, measuring approximately one wavelength in each of its two dimensions [27].

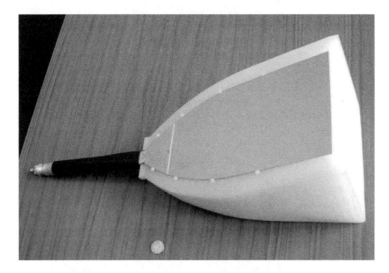

Figure 19.14 Dielectric-loaded TEM horn (reproduced by permission of Yi Huang, University of Liverpool)

The exponential tapered slot antenna (or ETSA, shown in exploded form by Figure 19.15) is a related design to the antipodal Vivaldi, and is a light-weight, multilayer design, half-a-wavelength in length and with a 170% fractional bandwidth (16:1 ratio) [28]. The input response and a measured, in-soil, radiation pattern for this antenna are provided in the section on antenna measurement (Figures 19.18 and 19.19).

19.4.5 CPW-fed Slot Antenna

Slot antennas also offer a compact planar geometry and the design shown in Figure 19.16 is matched from 1 GHz upwards, at which frequency it is little more than a quarter of a wavelength in diameter. Radiation patterns were not yet measured at the time of writing.

19.4.6 Spiral Antennas

Antennas defined solely by angles are frequency independent [29], and, though in practice this would lead to infinite-sized antennas, even when truncated to finite sizes these antennas are generally wideband. Spiral antennas, such as the logarithmic and conical spiral (as well as the aforementioned bicones) are examples of this design approach and have been used in GPR systems [30]. They can achieve considerable bandwidth within a relatively compact volume (typically a half-wavelength in diameter), although planar spirals usually radiate significantly in the 'backwards' direction, and this may need to be suppressed by enclosing the antenna in a cavity.

The conical spiral antenna, which is circularly polarised, has some interesting properties for GPR in particular [31]. It is directional and maintains to a large degree its frequency independence in the presence of the ground. An approximately 6:1 bandwidth is reported and it is insensitive to the ground reflection (which, over most angles of illumination, returns with the opposite direction of polarisation), however, for the same reason it is also insensitive to the radar returns from some types of buried objects. Like many spiral antennas it has the additional disadvantage that its phase centre depends on frequency and is therefore not well suited to impulse radiation.

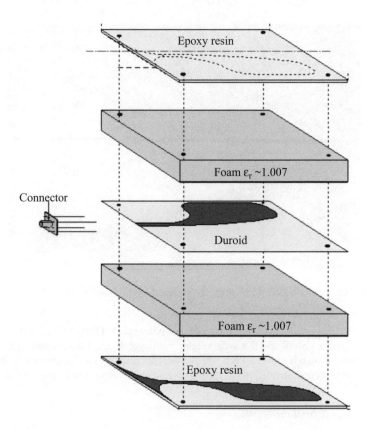

Figure 19.15 ETSA (Vivaldi) antenna (reproduced by permission of Jean-Yves Dauvignac, University of Nice-Sophia Antipolis-CNRS)

19.5 Antenna Measurements, Analysis and Simulation

19.5.1 Antenna Measurement

Any candidate GPR antenna design should be measured in order to verify its performance. The antenna may be measured 'in air' in a conventional anechoic chamber, and this allows useful determination of general pattern performance and, by careful 3D pattern measurement, an indication of gain and efficiency.

Determining the actual performance in the intended application is a good deal more complex and relatively few suitable facilities exist for this type of measurement [32]. One such facility, at the Technical University of Denmark is shown in Figure 19.17. This consists of an approximately 4 m × 3 m (cross-section) and 1 m deep, lined timber box, filled with soil. The choice of soil type rather depends on the intended application, and the soil type and especially moisture content greatly affect the propagation environment. In this case the moisture content is monitored and a moisture meter used in an attempt to maintain consistent – and reasonably uniform – moisture levels, despite the continual loss of water through evaporation from the upper surface.

A probe antenna is buried at the mid-point of the soil box, 20 cm deep, and, once buried, can obviously not be moved without disturbing the measurement set-up. The antenna under test is moved using a

Antennas for Ground-penetrating Radar 431

Figure 19.16 CPW-fed slot antenna (courtesy of National Technical University of Athens)

Figure 19.17 GPR antenna measurement facility at Technical University of Denmark (reproduced by permission of Peter Meincke, Technical University of Denmark)

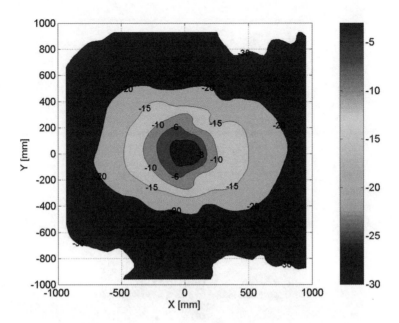

Figure 19.18 ETSA antenna copolar footprint at 910 MHz, contours in dB (reproduced by permission of Jean-Yves Dauvignac, University of Nice-Sophia Antipolis-CNRS and P. Meincke, Technical University of Denmark)

motorised two-axis planar scanning arrangement. Proximity to the soil box edges limits the angular range of the 2D scan, depending on the beamwidth of the antenna under test, and its height above the soil. The measurement is then performed using a network analyser just as in a standard anechoic chamber.

Recently, this facility was used for a series of tests on eight GPR antennas, supported by the European Network of Excellence on Antennas (ACE) [33]. Typical measured results for an ETSA (Vivaldi) antenna are shown in Figure 19.18, with the associated input response in Figure 19.19.

19.5.2 Antenna Analysis and Simulation

GPR antenna design is a difficult problem, since the antenna designs themselves are complex and not generally amenable to closed-form mathematical analysis, and additionally because the behaviour of the antenna must at some point be considered in the presence of the ground. Numerical analysis techniques are therefore of considerable value.

Traditional numerical analysis of antennas is accomplished by integral techniques which expand the current distribution on a wire antenna as a sum of known functions, each weighted by an unknown constant [34]. This approach can be very rapid, but becomes significantly more difficult when the antenna contains dielectric materials in addition to conductors, and when the presence of the ground must be considered. It is also almost invariably applied in the frequency domain, i.e. a wideband antenna must be analysed one frequency at a time across the band of interest.

In recent years, therefore, the analysis of wideband antennas for GPR has become dominated by FDTD [35] and related techniques (TLM, finite integration technique, etc). Although there are differences in the

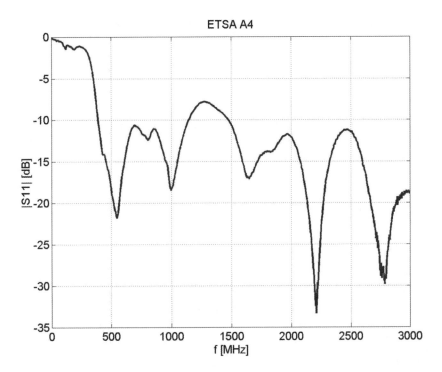

Figure 19.19 ETSA antenna input response (reproduced by permission of Jean-Yves Dauvignac, University of Nice-Sophia Antipolis-CNRS and P. Meincke, Technical University of Denmark)

various formulations, all are able to provide a time-domain solution (i.e. wideband) for antenna structures of almost arbitrary complexity in the presence of a lossy and, if desired, inhomogeneous ground.

Computational demands can become considerable for electrically large structures, however, continued advances in affordable computing have greatly eased this constraint, and even standard PCs are easily able to model single GPR antennas – if perhaps not entire arrays.

Examples of the FDTD analysis of GPR-type antennas include loaded monopoles [20], TEM horns [36, 37] and horn-fed bowties [38]. More recent papers have included a dispersive, lossy ground in the simulation and sometimes also a buried object in order that the effect of the antenna on the imaging process itself may be determined [31, 39, 40, 41] – the complexity of these scenarios is a testament to the generality of the modelling techniques employed.

19.6 Conclusions

This chapter has briefly surveyed the myriad applications of GPR. In most applications the antenna designer has similar problems to overcome. He or she is seeking to produce a portable (hence small and lightweight) antenna of reasonable gain/efficiency which radiates satisfactorily over the widest possible bandwidth. In GPR applications the bandwidth requirement is often expressed in terms of pulse-radiating performance.

Work in this field is ongoing. Aided by measurement and by powerful computer modelling techniques, GPR antenna designers are increasingly able to predict and understand the performance of proposed designs in realistic electromagnetic environments.

Acknowledgements

In writing this chapter the author would like to acknowledge the assistance provided by David Daniels (ERA Technology, UK), Andreas Kathage (GSSI, USA), Jean-Yves Dauvignac and Christian Pichot (Electronics, Antennas and Telecommunications Laboratory, University of Nice-Sophia Antipolis-CNRS, France), Rajagopal Nilavalan (Brunel University, UK), Peter Meincke (Technical University of Denmark, Denmark), Lawrence Conyers (University of Denver, USA) and Evangelos Angelopoulos (Institute of Communication and Computer Systems, National Technical University of Athens, Greece).

References

[1] D. Daniels, Ground-Penetrating Radar, *Institution of Electrical Engineers*, 2004.
[2] B. Boutros-Ghali, The land-mine crisis, *Foreign Affairs*, **73**(5), 8–13, 1994.
[3] S.A. al Hagrey, GPR application for mapping toluene infiltration in a heterogeneous sand model, *Journal of Environmental and Engineering Geophysics*, **9**, 79–85, 2004.
[4] D.L. de Castro and R.M.G.C. Branco, 4-D ground penetrating radar monitoring of a hydrocarbon leakage site in Fortaleza (Brazil) during its remediation process: a case history, *Journal of Applied Geophysics*, **54**, 127–44, 2003.
[5] J.S. Lee, C. Nguyen and T. Scullion, A novel, compact, low-cost, impulse ground-penetrating radar for non-destructive evaluation of pavements, *IEEE Transactions on Instrumentation and Measurement*, **53**, 1502–9, 2004.
[6] X. Dérobert, C. Fauchard, E. Côte, E. Le Brusq, J. Guillanton, Y. Dauvignac and C. Pichot, Step-frequency radar applied on thin road layers, *Journal of Applied Geophysics*, **47**, 317–25, 2001.
[7] Lawrence B. Conyers, *Ground-Penetrating Radar for Archaeology*, Altamira Press, 2004.
[8] W.S. Hammon, G.A. McMechan and X. Zeng, Forensic GPR: finite-difference simulations of responses from buried human remains, *Journal of Applied Geophysics*, **45**, 171–86, October 2000.
[9] E.F. Greneker, Radar sensing of heartbeat and respiration at a distance with security applications, *Proceedings of SPIE, Radar Sensor Technology II*, Orlando, Florida, 22–7, April, 1997.
[10] J. Tatoian, G. Franceschetti, H. Lackner and G. Gibbs, Through-the-wall impulse SAR experiments, *Proceedings IEEE Antennas and Propagation Symposium*, Washington DC, 2005.
[11] R. Benjamin, I.J. Craddock, G.S. Hilton, S. Litobarski, E. McCutcheon, R. Nilavalan and G.N. Crisp, Microwave detection of buried mines using non-contact, synthetic near-field focussing, *IEE Proceedings on Radar, Sonar and Navigation*, **148**, 233–40, 2001.
[12] S. Kingsley and S. Quegan, *Understanding Radar Systems*, McGraw-Hill, 1992.
[13] R.S. Elliot, *Electromagnetics: History, Theory and Applications*, IEEE Press, 1993.
[14] M.T. Hallikainen, F.T. Ulaby, M.C. Dobson, M.A. Elrayes and L.K. Wu, Microwave dielectric behaviour of wet soil (parts 1 and 2), *IEEE Transactions Geoscience and Remote Sensing*, **23**, 25–34, 1985.
[15] W.R. Scott and G.S. Smith, Measured electrical parameters of soil as functions of frequency and moisture content, *IEEE Transactions on Geoscience and Remote Sensing*, **30**, 621–3, 1992.
[16] E.E. Altshuler, The travelling wave linear antenna, *IRE Transactions on Antennas and Propagation*, **9**, 324–9, 1961.
[17] T.T. Wu and R.W.P. King, The cylindrical antenna with non-reflecting resistive loading, *IEEE Transactions on Antennas and Propagation*, **AP-13**, 369–73, 1965.
[18] T.P. Montoya and G.S. Smith, Resistively-loaded vee antennas for short-pulse ground-penetrating radar, *Proceedings IEEE Antennas and Propagation Symposium*, Baltimore, 2068–71, July 1996.

[19] K. Kim and W.R. Scott, Design of a resistively-loaded vee dipole for ultrawideband ground-penetrating radar applications, *IEEE Transactions on Antennas and Propagation*, **53**, 2525–32, 2005.

[20] J.G. Maloney and G.S. Smith, A study of transient radiation from the Wu–King resistive monopole – FDTD analysis and experimental measurements, *IEEE Transactions on Antennas and Propagation*, **41**, 668–76, 1993.

[21] C. Balanis, *Antenna Theory, Analysis and Design*, John Wiley & Sons, Ltd, 1982.

[22] R. Nilavalan, G.S. Hilton and R. Benjamin, Wideband printed bowtie antenna element development for post reception synthetic focussing surface penetrating radar, *Electronics Letters*, **35**(20), 1771–2, 1999.

[23] K.L. Shlager, G.S. Smith and J.G. Maloney, Optimization of bow-tie antennas for pulse radiation, *IEEE Transactions on Antennas and Propagation*, **42**, 975–82, 1994.

[24] A.A. Lestari, A.G. Yarovoy and L.P. Ligthart, RC-loaded bow-tie antenna for improved pulse radiation, *IEEE Transactions on Antennas and Propagation*, **52**, 2555–63, 2004.

[25] K.L. Schlager, G.S. Smith and J.G. Maloney, TEM horn antenna for pulse radiation: an improved design, *Microwave and Optical Technology Letters*, **12**, 86–90, 1996.

[26] A.G. Yarovoy, A.D. Schukin, I.V. Kaploun and L.P. Ligthart, The dielectric wedge antenna, *IEEE Transactions on Antennas and Propagation*, **50**, 1460–72, 2002.

[27] F. Guangyou, New design of the antipodal Vivaldi antenna for a GPR system, *Microwave and Optical Technology Letters*, **44**, 126–39, 2005.

[28] E. Guillanton, J.-Y. Dauvignac, C. Pichot and J. Cashman. A new design tapered slot antenna for ultra-wideband applications, *Microwave and Optical Technology Letters*, **1**, 286–9, 1998.

[29] V.H. Rumsey, *Frequency Independent Antennas*, Academic, 1966.

[30] J. Thaysen, K.B. Jakobsen and J. Appel-Hansen, Circular polarised stepped frequency ground penetrating radar for humanitarian demining, *Detection and Remediation Technologies for Mines and Minelike Targets*, **VI**, 671–9, Proceedings of the Society of Photo-Optical Instrumentation Engineers (SPIE).

[31] Thorsten W. Hertel and Glenn S. Smith, The conical spiral antenna over the ground, *IEEE Transactions on Antennas and Propagation*, **50**, 1668–75, 2002.

[32] R.V. de Jongh, A.G. Yarovoy and L.P. Ligthart, Experimental set-up for measurement of GPR antenna radiation patterns, *Conference Proceedings, 28th European Microwave Conference*, RAI Centre, Amsterdam, **2**, 539–43, 6–8 October 1998.

[33] H.-R. Lenler-Eriksen, P. Meincke, A. Sarri, V. Chatelee, B. Nair, I.J. Craddock, G. Alli, J.-Y. Dauvignac, Y. Huang, D. Lymperopoulos and R. Nilavalan, Joint ACE ground penetrating radar antenna test facility at the Technical University of Denmark, *Proceedings IEEE Antennas and Propagation Symposium*, Washington DC, 2005.

[34] J.J.H. Wang, *Generalised Moment Methods in Electromagnetics*, John Wiley & Sons, Ltd, 1991.

[35] A. Taflove, *Computational Electrodynamics: The Finite-Difference Time-Domain Method*, Artech House, 1995.

[36] K.L. Shlager, G.S. Smith and G.J. Maloney, Accurate analysis of TEM horn antennas for pulse radiation, *IEEE Transactions on Electromagnetic Compatibility*, **38**, 414–23, 1996.

[37] John B. Schneider and Kurt L. Shlager, FDTD simulations of TEM horns and the implications for staircased representations, *IEEE Transactions on Antennas and Propagation*, **45**, 1830–8, 1997.

[38] K.H. Lee, C.C. Chen, F.L. Teixeira and R. Lee, Modeling and investigation of a geometrically complex UWB GPR antenna using FDTD, *IEEE Transactions on Antennas and Propagation*, **52**, 1983–91, 2004.

[39] J.M. Bourgeois and G.S. Smith, A fully three-dimensional simulation of a ground-penetrating radar: FDTD theory compared with experiment, *IEEE Transactions on Geoscience and Remote Sensing*, **34**, 36–44, 1996.

[40] B. Lampe, K. Holliger and A.G. Green, A finite-difference time-domain simulation tool for ground-penetrating radar antennas, *Geophysics*, **68**, 971–87, 2003.

[41] U. Oguz and L. Gurel, Frequency responses of ground-penetrating radars operating over highly lossy grounds, *IEEE Transactions on Geoscience and Remote Sensing*, **40**, 1385–94, 2002.

20

Wideband Antennas for Biomedical Imaging

Ian Craddock

20.1 Introduction

This chapter considers wideband radio-frequency (including microwave-frequency) biomedical imaging. This is somewhat similar to ground-penetrating radar (GPR), which is discussed in the previous chapter; however, there are some significant differences.

Radio waves may be used in many ways to image the human body, for example microwave radiometry may be used to map tissue temperature and hence detect changes associated with tumours [1]; however, this technique does not require wideband antennas and so will not be considered here. Neither does this chapter consider therapeutic applications of microwaves such as hyperthermia (the deliberate heating of tissue), or the general subject of radio-wave interactions with the body. Readers interested in these wider medical applications are referred in the first instance to [2].

In principle radio waves can be used to image any part of the human body where the structure of interest has a different refractive index from its surroundings. In practice, however, imaging the human body is an extremely difficult task, owing to the attenuation suffered by the radio signals in human tissue, the very small size of features of interest, and the cluttered, inhomogeneous, radio propagation environment.

20.2 Detection and Imaging

20.2.1 Breast Cancer Detection Using Radio Waves

Breast cancer is the most common cancer in woman (excluding skin cancers) – for example, in the USA in 2005, over 40 000 women have died from the disease. Early detection significantly increases the likelihood of successful treatment and long-term survival [3]. X-ray mammography is currently the most effective method for detecting breast tumours [4], however, this technique suffers from relatively

---

Ultra-wideband Antennas and Propagation for Communications, Radar and Imaging Edited by B. Allen, M. Dohler, E. E. Okon, W. Q. Malik, A. K. Brown and D. J. Edwards
© 2007 John Wiley & Sons, Ltd

high false-negative and false-positive detection rates, and it involves uncomfortable compression of the breast – frequent ionising X-ray screenings are also a matter of concern for many women.

Fortunately, the breast is one of the easiest parts of the body to image with radio waves, since it is accessible and relatively homogeneous (compared, at least, to other organs). As a consequence, by far the most common application of wideband radio-frequency imaging to the human body is the detection of breast tumours. This is reflected by the considerable amount of literature on the subject, and the existence around the world of a number of established research groups. The exclusive focus of this chapter is therefore the breast cancer detection application, and the antenna designs proposed for this purpose.

20.2.2 Radio-wave Imaging of the Breast

Traditional microwave breast imaging is posed as an inverse scattering problem [5], where radio-frequency energy illuminates the breast from one or more transmitters and the scattered field is measured at a number of external locations. From this information the permittivity distribution inside the breast may be obtained by essentially inverting the observations made at particular frequency [6, 7]. Unfortunately, solving this inverse scattering problem is extremely challenging. Furthermore, since the inverse problem is formulated in the frequency domain, it is not immediately obvious how to include information from a wide band of frequencies.

Inverse scattering research for medical imaging is ongoing [8, 9]. An alternative approach is being pursued by other groups, in which, rather than solve an inverse scattering problem, the problem is considered in a similar fashion to a traditional wideband GPR scenario. In other words, narrow pulses (or, equivalently a narrow pulse synthesised from a frequency sweep) are transmitted into the medium and the reflections analysed, basically in the time domain.

The radar approach is conceptually rather simpler, avoids a great deal of mathematics and straightforwardly incorporates information from a wide band of received signals. Recent notable work in this field has been undertaken by groups such as those at the Universities of Wisconsin-Madison (USA), Calgary (Canada) and Bristol (UK).

The general scenario may be considered as in Figure 20.1, showing a cross-section though a patient lying face upwards (a face-down posture is an alternative, but the principles remain the same). Unlike the GPR applications considered in the previous chapter, there is no objection to operating the antenna

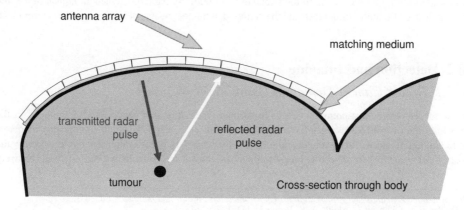

Figure 20.1 Breast-imaging scenario

system in direct contact with the human body, or to embedding it in a medium which itself is in direct contact with the body. Most groups in fact propose such an arrangement.

The transmitting and receiving antennas may be one and the same (monostatic) or they may be separate (bistatic) and the antennas may be part of an array (as shown in Figure 20.1) or indeed they may be mechanically scanned over the breast. The antennas may be placed in direct contact with the skin (if not mechanically scanned) or alternatively placed a short distance from the skin, embedded in a soft (or liquid) immersion medium.

Use of an immersion medium is a generally popular approach, since operating the antenna in air, outside the body results in a very large reflection as the transmitted radio wave encounters the body. The immersion medium is closer to the dielectric properties of the breast and hence removes some of this reflection. Use of the immersion medium also helps to reduce the size of the antenna, since the antenna size is dictated largely by the wavelength and this is reduced by the dielectric constant of the medium. A smaller antenna then permits more antennas to be placed near to the breast.

Soft body tissues generally have a substantial water content, and the dielectric constant is therefore significantly influenced by dielectric properties of water. As in the GPR scenario described in the previous chapter, the complex permittivity of the various tissues may therefore be considered as following a Debye model:

$$\varepsilon_c = \left(\varepsilon' - j\varepsilon''\right) = \left(\varepsilon_0 \varepsilon_r(\omega) - j\frac{\sigma}{\omega}\right),$$

where

$$\varepsilon_r(\omega) = \varepsilon_\infty + \frac{\varepsilon_s - \varepsilon_\infty}{1 + j\omega\tau},$$

where ε_s is the value of permittivity that would be observed at low frequencies, ε_∞ the value for 'infinite' frequency, τ is the relaxation time and σ is the conductivity.

The tissue properties of the Debye model have been determined by fitting against measured values [10, 11] up to 3 GHz and found to be for normal tissue: $\varepsilon_s = 10$, $\varepsilon_\infty = 7$, $\tau = 7$ ps and $\sigma = 1.15$ S/m. Whereas for breast tumours: $\varepsilon_s = 50$, $\varepsilon_\infty = 4$, $\tau = 7$ ps and $\sigma = 0.7$ S/m.

To determine the permittivity values for higher frequencies, where there is little available data, extrapolations using more complex material models (e.g. a four-term Cole–Cole expression) have been performed [11, 12, 13].

As might be imagined, measurement of the dielectric properties of these tissues is not an easy matter [14] and there is understandably not complete agreement on the values, especially once extrapolated beyond measured frequency ranges. Furthermore, the breast is not homogeneous (so, regardless of the 'average' value, there will be variations across the breast) and there will be variation between different women.

What may be appreciated from the above values, however, is that there seems to be a substantial difference in permittivity between healthy and malignant tissue – a contrast of perhaps 5:1 in both permittivity and conductivity. This difference results in the possibility of detecting tumours by sensing reflections of the transmitted radio wave, although with younger women (who have denser breast tissue) this difference will be diminished.

Penetration into the breast is limited, as in the GPR scenario, by attenuation, which rises with frequency in much the same way as was described in the preceding chapter. Fortunately the required penetration depth is only a few centimetres and although attenuation remains an important consideration, frequencies up to 10 GHz can be used without difficulty. Above this frequency, attenuation rises steeply and it also becomes difficult to achieve the required bandwidth from the antenna. In practice, therefore, most work has been carried out in the 1 GHz to 10 GHz range.

A further obstacle to penetration is the skin. This has a high dielectric constant (approximately $\varepsilon_r = 36.0$) and consequently a large reflection from the skin is inevitable, although somewhat lessened by operation in an immersion medium. This must be overcome by whatever signal processing is applied to the received radar signals, however, this falls outside the scope of this chapter [15].

20.3 Waveform Choice and Antenna Design Criteria

As with GPR systems the waveform transmitted from the antenna can be a pulse or an equivalent broadband waveform, such as a swept- or stepped-frequency sinusoid or an M-sequence. In practice, however, generating pulses with frequency content at 10 GHz and, more importantly, sampling the reflected signals, is a difficult exercise and invariably the radars in this application utilise the stepped-frequency arrangement, often using a vector network analyser as a convenient, high-performance RF source and receiver.

While in the frequency domain, the recorded data may be processed to compensate for frequency-dependent antenna effects, and frequency-dependent attenuation. Generally, however, the data is transformed back to the time domain for some sort of time-shifting process that, based on the time-of-flight on the path via any point of interest in the medium, aligns the reflected waveforms from that point and computes the associated energy. Various refinements are possible, such as subsequent FIR filtering to compensate for frequency-dependent propagation effects [12].

The demands on the antenna itself include an acceptable input match and a stable radiation pattern which is largely directed into the breast. Both these characteristics are required over a wide band of operation, and, as in the GPR application, there is also emphasis on the time-domain characteristics of the antenna – in particular its ability to radiate a pulse with a minimum of late-time ringing.

As discussed in the GPR chapter, if the radiated pulse looks more like the bottom waveform than the top one in Figure 20.2, then it will be considerably more difficult to resolve reflected signals from the breast, since the longer 'tail' of the impulse is more likely to result in signals overlapping each other. This applies especially to the large skin reflection which may completely obscure the tumour signal unless the transmitted pulse is well defined.

A measure frequently quoted is the 'fidelity' F of the transmitted pulse [16] and this is simply defined as the maximum cross-correlation between the (normalised) desired transmitted signal, d, and the normalised actual signal a:

$$F = \max_\tau \int_{-\infty}^{\infty} d(t - \tau)a(t)dt$$

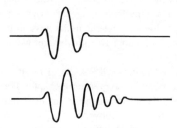

Figure 20.2 Hypothetical transmitted radio pulses

It may be noted that d, the desired signal, is not usually the same as the voltage applied to the antenna feed, since the antenna is usually a high-pass structure that radiates approximately the derivative of the excitation waveform.

As with the GPR application, system performance is usually limited by clutter rather than attenuation, and in any event the attenuation in the breast is unavoidable. Antenna efficiency is therefore, not a pressing concern, and trading off efficiency in favour of bandwidth/fidelity is usually acceptable, e.g. by introducing resistive loading.

Unlike a GPR system, the final system need not be lightweight or easily mobile, however, there is generally a need for more than one antenna to be in close contact with the breast and this does require a reasonably compact element. As described above, the antenna will usually need to be embedded in an immersion medium that more closely matches the impedance of the breast, and this must be taken into account not only in the electrical design but also in the mechanical specification.

20.4 Antenna Elements

Most of the wideband systems proposed for breast cancer detection use similar antenna designs to those encountered in GPR, which is unsurprising given their broadly similar requirements. Owing to the higher frequencies used, and the reduction in size that arises from operation in a high-dielectric medium, the antennas are of course reduced substantially in size from GPR designs.

Wideband biomedical imaging antennas fall into a small number of popular categories and these may be described quite succinctly as follows.

20.4.1 Dipoles, Resistively Loaded Dipoles and Monopoles

Monopoles and dipoles are simple antennas and attractive for this application, not least because most groups have employed numerical modelling to evaluate their approaches and these antenna types are straightforward to include in these models. They are also the easiest antennas to construct, which facilitates the difficult step from computer simulation to practical experiment.

As might be expected from the preceding chapter's discussion of UWB monopoles and dipoles for GPR, resistive loading is a popular method to improve the performance of the basic design.

One such dipole antenna [17] has been designed at the University of Calgary for operation from 2 to 8 GHz, employing a Wu–King [18] resistive-loading distribution. The resistive loading resulted in an efficiency of 2 % to 16 % over this frequency range. Resistively loaded dipoles are also used at Dartmouth College, New Hampshire, at frequencies from 300–500 MHz [19], and may be seen in Figure 20.10.

A loaded monopole, similar to the design in [20], but designed for operation in an immersion medium consisting of oil ($\varepsilon_r = 3.0$), was also investigated at Calgary. In this design, chip resistors were soldered to a high-frequency substrate in order to form a monopole with the required resistivity distribution (Figure 20.3).

This monopole was measured immersed in oil, but was found, like the dipole, to have low efficiency. Nonetheless it is compact (11 mm, approximately a quarter-wavelength in the oil medium at the centre frequency of 4 GHz) and radiates a pulse with high fidelity to the desired waveform. Tumour detection in a simple breast phantom[1] was possible [21] using this antenna.

---

[1] A 'phantom' is an object created to mimic the electrical properties of part of the body (in this case the breast). Materials are chosen for the phantom that have approximately the same relative permittivity and conductivity as

Figure 20.3 Loaded monopole (courtesy of Jeff Sill, University of Calgary)

Other types of dipole and monopole may be considered, for example the vee-dipole [22]. For these geometries the Wu–King resistivity distribution is no longer ideal. In [22] a genetic algorithm (GA) is used to optimise the resistivity profile for a modified vee-dipole, with a time-domain method of moments code used to determine the performance of each individual during its evolution. The performance criterion was the fidelity of the pulse and the GA was able to slightly improve the fidelity and produce a more constant input impedance.

20.4.2 Bowtie

The bowtie antenna is also suitable for biomedical imaging and published work has included [23] in which an FDTD model of Wu–King loaded bowtie yielded a 20 mm long antenna matched from 1.6 GHz.

In [24] a slightly modified Wu–King resistive loading was employed to a bowtie antenna of 80 mm length, operating in a medium with $\varepsilon_r = 9.0$ (similar to normal breast tissue) over a frequency range from approximately 2–8 GHz. Unlike the loaded monopoles and dipoles in the previous section, this antenna is relatively large compared to the wavelength in the medium and simulations correspondingly show that it radiates much more efficiently.

Crossed bowties are of use where sensing cross-polar elements of the scattered signal allow a reduction in the size of the reflection from planar surfaces, such as the chest wall. Crossed versions of the bowtie design are considered in [25], along with [24].

A recent variant on the bowtie design has been published [26, 27]. Called a 'slotline bowtie hybrid' it consists of a shaped slot with Vivaldi, linear and elliptical sections along its 3 cm overall length. The careful shaping of the slot permits control over internal reflections, and the bowtie plates attached to the slot provide control over antenna beamwidth. Figure 20.4 shows the antenna along with its associated microstrip

---

normal breast tissue, in this chapter the main use for the phantom is to load the antenna, rather than to precisely duplicate the properties of the heterogeneous tissues. Phantoms for this purpose may be relatively simple.

Figure 20.4 Slotline bowtie hybrid (courtesy of Chris Shannon, University of Calgary)

feed and micostrip-slot balun. The antenna was designed to operate in an immersion medium similar to breast fat ($\varepsilon_r = 9.0$) and was constructed on a dielectric substrate with relative permittivity of 10.2.

FDTD simulations of the antenna structure showed return loss better than 10 dB over the 2.5 to 10 GHz band. Practical measurements, with the antenna immersed in a medium, confirmed the simulated results.

20.4.3 Horn Antennas

An adaptation of the horn antenna for biomedical imaging has been presented in [28]. This pyramidal horn, shown in Figure 20.5, is coaxially fed via a subminiature version A (SMA) connector and hence needs no balun. The transition from the central conductor of the coaxial feed is made using a curved and tapered launching plane, ultimately terminated in two 100 ohm chip resistors connected in parallel between the launching plane and the horn itself – these provide resistive loading to suppress reflections from the end of launching plate.

Reported performance in air includes a directive radiation pattern with low sidelobes, and good input matching from 1 to 11 GHz. The dimensions of the horn are 13 mm deep with an aperture of 25 mm by 20 mm, which renders it very suitable for the intended application, and this design is employed in the imaging experiments at the University of Wisconsin-Madison.

20.4.4 Spiral Antennas

Spiral antennas have also been demonstrated [29], again featuring resistive loading to prevent reflected waves where the spiral is truncated, and, as with all such antennas, the resulting single-arm spiral antenna is circularly polarised. FDTD modelling enables a careful examination of antenna performance, losses and penetration depth, and the resulting simulated design has a diameter of 3 cm and operates well from

Figure 20.5 Pyramidal horn (Source: [34] reproduced by permission of © 2003 IEEE)

approximately 2 GHz to 8 GHz. Its construction, consisting of a single arm over a ground plane, needs no balun and is inherently directive.

20.4.5 Stacked-patch Antennas

While stacked-patch antennas are well known to have good operating bandwidths, the bandwidths achieved are usually of the order of 30 % [30]. The stacked patch antenna developed at the University of Bristol was designed from the outset to radiate directly into breast tissues, and furthermore achieves a bandwidth of approximately 77 %. It achieves this without resistive loading, and, in fact, FDTD models demonstrate that even if the losses in the surrounding tissues are removed, the bandwidth is practically unchanged.

The antenna consists of two stacked patches printed on a dielectric substrate of $\varepsilon_r = 2.2$ and separated from the ground plane by a second substrate of $\varepsilon_r = 10.2$. The antenna is fed using a microstrip line on the reverse face of the antenna via a slot. The top face of the antenna is protected from the chemical and electrical effects of immersion by a radome made of 10.2 permittivity substrate. Antenna dimensions were carefully optimised using an FDTD code, in order to obtain optimum matching and radiation properties. The resulting antenna is shown in various stages of assembly in Figure 20.6.

The dimensions of the actual patches are less than 7 mm and, even allowing for some substrate around the periphery, the cross-section of the overall antenna can be easily made 15 mm × 15 mm. The total thickness of the antenna is 4.5 mm, which includes the feed substrate behind the ground plane and the radome on top.

Antenna input responses from FDTD simulations and practical measurements are shown in Figure 20.7. The practical input measurements were carried out with the antenna radiating into a dielectric phantom and the FDTD model also included appropriate dielectric values to represent the body.

Front-to-back ratios and cross-polar levels, found using FDTD and by measurements where practical (see the subsequent section) are 15 dB below the main copolar pattern, which itself is steady over frequency, apart from a slight boresight null at 9.5 GHz.

Figure 20.6 Stacked patch antenna (left to right: assembled, reverse view, intermediate substrate, bottom substrate)

20.5 Measurements, Analysis and Simulation

20.5.1 Antenna Measurement

As with the GPR antennas, characterising biomedical imaging antennas is not straightforward. The antenna may be measured 'in air' in a conventional anechoic chamber, and this gives some insight into the operation of the antenna, its efficiency, and so on.

The body is a high-dielectric medium, however, and the influence of the tissue on the antenna must be included. An immersion medium is also likely to be required in practice, and so this too must be incorporated. In order to do this, some sort of phantom must be created with similar dielectric properties to the human breast.

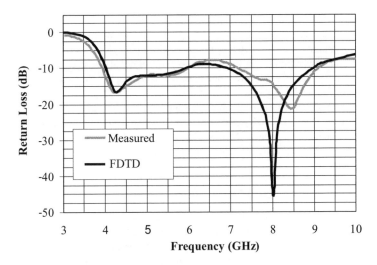

Figure 20.7 Stacked-patch antenna input response

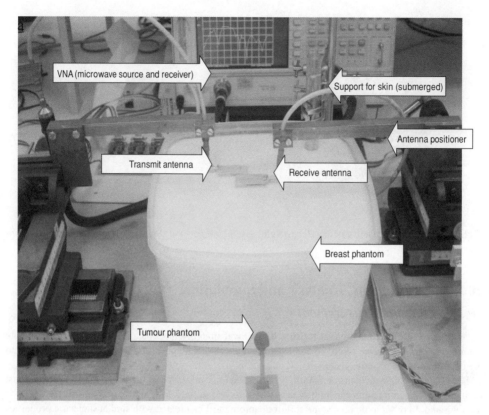

Figure 20.8 University of Bristol phantom and antenna measurement set-up (Source [33] reproduced by permission of © 2005 IEEE)

Figure 20.8 shows the initial arrangement developed at the University of Bristol for validation of the previously discussed stacked-patch design. Stepper motors permit movement of two antennas in a plane. The breast phantom [31] is an oil/water emulsion which very closely mimics the average electrical properties of the breast over frequency. In addition to permitting various imaging experiments, the two antennas may be arranged to face each other and one scanned in a plane, thereby measuring the radiation pattern in medium.

This experimental procedure is valuable, but the limited size of the phantom, not to mention the difficulties in operating antennas essentially underwater, makes characterising the antenna by measurement alone a difficult and imprecise exercise. The Bristol team, and most of the other groups engaged in similar work, have therefore found computer modelling an essential tool for refining, and indeed understanding, their antenna designs.

20.5.2 Antenna Analysis and Simulation

The analysis of wideband antennas for breast cancer imaging has become dominated by FDTD-based techniques [10], due to its simplicity, generality and inherently wideband nature. The well-known strength of the FDTD model to include inhomogeneity has made it possible for very complex models to be developed – for example, researchers at the University of Wisconsin-Madison have developed FDTD models that include tissue heterogeneity derived from MRI scans of real breasts [12]. The frequency dependence of the electrical properties of the breast may also be included [10] and this has been found to be an important effect to consider [32].

The real difficulties encountered in phantom construction and practical antenna measurement will continue to require the use and development of wideband, full-wave, computer analysis techniques in the development of novel antenna designs for this application.

20.6 Conclusions

Devising suitable wideband antenna elements for breast cancer detection is a significant challenge for engineers and scientists involved in biomedical imaging. Good designs are, however, available, with compact size, good bandwidth and optimised to radiate useful pulses. Many (though not all) designs are resistively loaded and hence somewhat inefficient – this is generally acceptable if clutter, rather than signal-to-noise, is the limiting factor.

Although much work remains to be done in the field of antenna design, the various research groups are increasingly producing imaging prototypes that are either able to be applied to patients (e.g. Figure 20.10) or just a short step away from that objective (Figure 20.9). Given the sobering mortality statistics associated with breast cancer, it is profoundly to be hoped that the developments described herein are paving the way for the introduction of radio-wave imaging to breast cancer screening.

Figure 20.9 Automated stacked-patch antenna array at University of Bristol

Figure 20.10 Dipole array at Dartmouth College (Reproduced by permission of © Paul Meaney)

Acknowledgements

In writing this chapter, the author gratefully acknowledges the assistance provided by Elise Fear and Jeff Sill (University of Calgary, Canada), Susan Hagness (University of Wisconsin-Madison, USA), Paul Meaney (Dartmouth College New Hampshire, USA) and Rajagopal Nilavalan (Brunel University, UK).

References

[1] Y. Leroy, A. Mamouni, J.C. can de Velde, B. Bocquet and B. Dujardin, Microwave radiometry for noninvasive thermometry, *Automedica*, **8**, 181–202, 1987.
[2] A. Rosen, M.A. Stuchly and A. Vander Vorst, Applications of RF/microwaves in medicine, *IEEE Transactions on Microwave Theory and Techniques*, **50**, 963–74 2002.
[3] American Cancer Society, Breast cancer facts and figures 2005–2006, www.cancer.org.
[4] M. Brown, F. Houn, E. Sickles and L. Kessler, Screening mammography in community practice, *Amer. J. Roentgen*, **165**, 1373–7, 1995.
[5] P.M. Meaney, M.W. Fanning, D. Li, S.P. Poplack and K.D. Paulsen, A clinical prototype for active microwave imaging of the breast, *IEEE Transactions on Microwave Theory and Techniques*, **48**, 1841–53, 2000.
[6] A.E. Souvorov, A.E. Bulyshev, S.Y. Semenov, R.H. Svenson and G.P. Tatsis, Two-dimensional computer analysis of a microwave flat antenna array for breast cancer tomography, *IEEE Transactions on Microwave Theory and Techniques*, **48**, 1413–15, 2000.
[7] P.M. Meaney and K.D. Paulsen, Nonactive antenna compensation for fixed-array microwave imaging – Part II: imaging results, *IEEE Transactions on Medical Imaging*, **18**, 508–18, 1999.
[8] Q.Q. Fang, P.M. Meaney, S.D. Geimer, A.V. Streltsov and K.D. Paulsen, Microwave image reconstruction from 3-D fields coupled to 2-D parameter estimation, *IEEE Transactions on Medical Imaging*, **23**, 475–84, 2004.
[9] M. El-Shenawee, Resonant spectra of malignant breast cancer tumors using the three-dimensional electromagnetic fast multipole model, *IEEE Transactions on Biomedical Engineering*, **51**, 35–44, 2004.

[10] A. Taflove and S.C. Hagness, *Computational Electrodynamics: The Finite-Difference Time-Domain Method*, 2nd ed., Artech House, 2000.
[11] C. Gabriel, S. Gabriel, R.W. Lau and E. Corthout, The dielectric properties of biological tissues: Parts I, II, and III, *Phys. Med. Biol.*, **41**, 2231–49, 1996.
[12] E.J. Bond, X. Li, S.C. Hagness and B.D. Van Veen. Microwave imaging via space-time beamforming for early detection of breast cancer, *IEEE Transactions on Antennas and Propagation*, **51**, 1690–705, 2003.
[13] An Internet resource for the calculation of the dielectric properties of body tissues in the frequency range 10 Hz–100 GHz, http://niremf.ifac.cnr.it/tissprop.
[14] D. Popovic, L. McCartney, C. Beasley, M. Lazebnik, M. Okoniewski, S.C. Hagness and J.H. Booske, Precision open-ended coaxial probes for in vivo and ex vivo dielectric spectroscopy of biological tissues at microwave frequencies, *IEEE Transactions on Microwave Theory and Techniques*, **53**, 1713–22, 2005.
[15] T.C. Williams and E.C. Fear, Tissue sensing adaptive radar for breast tumor detection: using a deconvolution method for enhanced skin sensing, *Proceedings IEEE Antennas and Propagation Symposium*, Washington DC, 2005.
[16] D. Lamensdorf and L. Susman, Broadband-pulse-antenna techniques, *IEEE Antennas Propagation Magazine*, **36**, 20–30, 1994.
[17] E.C. Fear and M.A. Stuchly, Microwave detection of breast cancer, *IEEE Transactions on Microwave Theory and Techniques*, **48**, 1854–63, 2000.
[18] T.T. Wu and R.W.P. King, The cylindrical antenna with non-reflecting resistive loading, *IEEE Transactions on Antennas and Propagation*, **3**, 369–73, 1965.
[19] P.M. Meaney, M.W. Fanning, D. Li, S.P. Poplack and K.D. Paulsen, A clinical prototype for active microwave imaging of the breast, *IEEE Transactions on Microwave Theory and Techniques*, **48**, 1841–53, 2000.
[20] J.G. Maloney and G.S. Smith, A study of transient radiation from the Wu–King resistive monopole, *IEEE Transactions on Antennas and Propagation*, **41**, 668–76, 1993.
[21] J.M. Sill and E.C. Fear, Tissue sensing adaptive radar for breast cancer detection: experimental investigation of simple tumor model, to appear in *IEEE Transactions on Microwave Theory and Techniques*.
[22] M. Fernandez-Pantoja, S. Gonzalez-Garcia, M.A. Hernandez-Lopez, A. Rubio Bretones and R. Gomez-Martin, Design of an ultra-broadband V antenna for microwave detection of breast tumors, *Microwave and Optical Technology Letters*, **34**, 164–6, 2002.
[23] M.A. Hernandez-Lopez, M. Quintillan-Gonzalez, S. Gonzalez-Garcia, A. Rubio Bretones and R. Gomez-Martin, A rotating array of antennas for confocal microwave breast imaging, *Microwave and Optical Technology Letters*, **39**, 307–11, 2003.
[24] S.C. Hagness, A. Taflove and J.E. Bridges, Three-dimensional FDTD analysis of a pulsed microwave confocal system for breast cancer detection: design of an antenna-array element, *IEEE Transactions on Antennas and Propagation*, **47**, 783–91, 1999.
[25] X. Yun, E.C. Fear and R.H. Johnston, Compact antenna for radar-based breast cancer detection, *IEEE Transactions on Antennas and Propagation*, **53**, 2374–80, 2005.
[26] C.J. Shannon, E.C. Fear and M. Okoniewski, Dielectric-filled slotline bowtie antenna for breast cancer detection, *Electronics Letters*, **41**, 388–90, 2005.
[27] A.K.Y. Lai, A.L. Sinopoli and W.D. Burnside, A novel antenna for ultra-wideband applications, *IEEE Transactions on Antennas and Propagation*, **40**, 755–60, 1992.
[28] X. Li, S.K. Davis, S.C. Hagness, D.W. van der Weide and B.D. Van Veen, Microwave imaging via space-time beamforming: experimental investigation of tumor detection in multilayer breast phantoms, *IEEE Transactions on Microwave Theory and Techniques*, **5**, 1856–65, 2004.
[29] S. Jacobsen, H.O. Rolfsnes and R.R. Stauffer, Characteristics of microstrip muscle-loaded single-arm Archimedean spiral antennas as investigated by FDTD numerical computations, *IEEE Transactions on Biomedical Engineering*, **52**, 321–30, 2005.
[30] F. Croq and D.M. Pozar, Millimeter wave design of wideband aperture-coupled stacked microstrip antennas, *IEEE Transactions on Antennas and Propagation*, **39**, 1770–6, 1991.
[31] J. Leendertz, A. Preece, R. Nilavalan, I.J. Craddock and R. Benjamin, A liquid phantom medium for microwave breast imaging, *Proceedings 6th Congress of European Bioelectromagnetics Association (EBEA)*, Budapest, Hungary, 2003.

[32] P. Kosmas, C.M. Rappaport and E. Bishop, Modeling with the FDTD method for microwave breast cancer detection, *IEEE Transactions on Microwave Theory and Techniques*, **52**, 1890–7, 2004.

[33] I.J. Craddock, R. Nilavalan, J. Leendertz, A. Preece and R. Benjamin, Experimental Investigation of Real Aperture Synthetically Organised Radar for Breast Cancer Detection, *IEEE Antennas and Propagation Society International Symposium,* Volume **1B**, 179-182, 3–8 July 2005

[34] X. Li, S. Hagness, M. Choi, D. van der Weide, Numerical and Experimental Investigation of an Ultrawideband Ridged Pyramidal Horn Antenna With Curved Launching Plane for Pulse Radiation, *IEEE Antennas and Wireless Propagation Letters,* **2**, 2003

21

UWB Antennas for Radar and Related Applications

Anthony K. Brown

21.1 Introduction

The range of applications that can be considered as UWB is extensive[1] including, for example, ground-probing radar, electronic surveillance monitoring (ESM), long-range UWB radar, electronic countermeasures (ECM) and directed energy weapons (DEW). Each of these applications has different requirements from system considerations and hence places different constraints on the antenna. Some of the important system questions that impact the antenna designer include:

- Is the system concerned with the impulse response of the antenna and environment? Some impulse radar and other applications require minimum distortion to the transmitted pulse. Many monitoring systems, on the other hand, use a UWB antenna backed by a suitable low-noise amplifier but immediately 'channelise' the receive chain Figure 21.1(a) – that is split the received signals into a number of relatively narrow bands before receiving/processing. This can help the antenna designer in some respects – for example phase-centre movement in the antenna may be corrected – but equally often requires the antenna pattern characteristics to be maintained across the entire UWB band rather than averaged over the band as may be the case with impulse systems.
- Is the off-axis performance important? Some systems only concern themselves with performance close or on the boresight of the antenna, with only relatively loose requirements at other angles. A surveillance system, for example, may require a reasonably low dispersion in the antenna (that is to preserve the impulse shape) on axis, but only requires off-axis performance to be below a specified mask. Alternatively an area-monitoring application such as some ESM systems may well require the antenna to provide a broad beam which has an approximately constant beamwidth over the entire frequency range of interest.

---

[1] UWB is normally taken in military circles as systems covering at least one octave of frequency range.

Ultra-wideband Antennas and Propagation for Communications, Radar and Imaging   Edited by B. Allen, M. Dohler, E. E. Okon, W. Q. Malik, A. K. Brown and D. J. Edwards
© 2007 John Wiley & Sons, Ltd

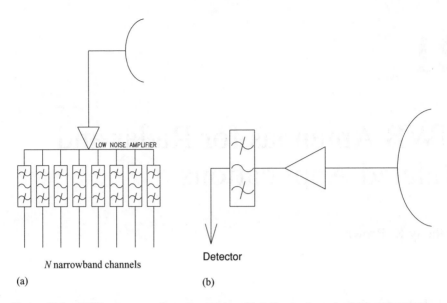

Figure 21.1 UWB system configurations: (a) multiple linear channels; (b) impulse system

- Input impedance variation. Variations in input impedance across the band will normally cause the antenna to be mismatched to the transmit/receiver by differing amounts over the band. This obviously affects the efficiency of the radiation, that is how much power is coupled from the source (on transmit) or delivered to the load (on receive). However, impedance effects can have more subtle problems: as examples we note for transmitters working with high power the reflected power can cause difficulties within the source, causing the source to oscillate unexpectedly. Furthermore the antenna/source mismatch effect leads to 'ringing' in the antenna, that is transmitted power being reflected off the antenna to the source, and back again off the antenna, causing the pulse to be effectively elongated in time.
- Power handling. This varies enormously with application. Often it is assumed that UWB is synonymous with low-power applications. However, long to medium range applications very often require tens or hundreds of kilowatts of peak power. This has serious implications principally in the feed mechanisms for the antenna. For these applications it is commonplace to adopt a ridged waveguide as the basic feeder mechanism.
- Antenna gain – a fundamental requirement. The allowable variation in antenna gain over the frequency band, which affects the transmitted impulse, is one criterion whereas the average gain (normally quoted at the centre band of the system) is often used for basic system calculations.

This chapter concentrates on longer range applications where the problems facing the designer are somewhat different from many of the applications discussed elsewhere in this text. One such important class is medium- and long-range radar.

21.2 Medium- and Long-Range Radar

In a pulsed radar, a stream of radio frequency pulses is sent from the radar. Targets reflect a proportion of this energy back. By timing the difference between transmit and received pulses the range can be derived. Principally, the pulse width defines the range resolution (that is the ability to distinguish between two targets in range) and is highly important in limiting the amount of unwanted power received back from sources such as rain (clutter). Use of the narrow duration pulses of a UWB system allows high-range resolution, increased resilience to multipath and low susceptibility to clutter and chaff [1] In turn the high-range resolution allows, in principle, improved target recognition. Some, but not all, of these advantages can also be gained using pulse-compression approaches, however, extremely high compression ratios to achieve very narrow pulses bring their own difficulties resulting in cost and performance limitations [2].

The difficulties of producing extremely narrow (<<1 ns) pulses at high power are significant. Referring first to a more traditionally narrowband system, the well-known radar range equation gives (in free space and clear weather):

$$P_r = \frac{P_t G^2 \sigma \tau \lambda^2}{(4\pi)^3 R^4},$$

where P_r is the power received back at the radar from a target at range R, P_t is the peak transmitted power, G the antenna gain, σ the target size (RCS), τ the pulse width and λ the operational wavelength.

In UWB radar this equation can usefully be recast as:

$$E_r = \frac{E_t A_{eff} G_e \sigma}{(4\pi)^2 R^4}.$$

Here, E_r is the total energy received from the target, E_t the transmitted energy ($= P_t \tau$ for a square pulse), A_{eff} the antenna effective area, G_e the total energy gain defined as the energy incident on the target compared to the energy which would have occurred if the same energy had been applied to a ideal, zero dispersion antenna. Writing the equation in this fashion assumes the pulse shape is unimportant and that all available energy is captured by the receiver. However, in practice this is not true, at least the receiver front end will encompass a bandpass filter to limit the noise bandwidth, but some signal processing/receiver architectures correlate the received pulse waveform with a known reference pulse shape, often taken as the transmitted pulse. For these circumstances pulse distortion caused by the antenna and propagation environment becomes a major issue. These considerations lead to the concept of 'useful energy', that is the energy received which will be used by the system. While this is a topic of significance in some systems, for this discussion we merely note the point and comment that the receiver architecture interacts with the antenna to produce the effective system G_e and, radiation pattern performance.

Even under the somewhat simplistic assumption of whole energy capture, UWB radar (except for short-range specialist applications) often requires the antenna to produce substantive energy gain. Even then without significant transmit power only medium ranges are typically sited.

Some work in the open literature [3] considers extremely high-power UWB radar sources. In many UWB radar cases an electrically large antenna, even with the lowest frequency-significant spectral components, is needed.

21.3 UWB Reflector Antennas

Perhaps the easiest and certainly the most widely used method of implementing electrically large antennas is to use a focusing parabolic reflector. For simplicity we initially separate the discussion of the reflector itself from common feed types used in UWB systems.

21.3.1 Definitions

The following terminology is useful to many UWB antennas and reflectors in particular:

21.3.1.1 Prompt Aperture Efficiency [4]

The ratio between the peak radiated power on boresight of an antenna and the peak radiated power on boresight from an ideal reference antenna of the same aperture with uniform electric field distribution of magnitude equal to the peak aperture field of the antenna under consideration.

21.3.1.2 Blockage

In a axi-symmetric reflector antenna the feed and any support structure 'shadows' the main reflector hence reducing performance. This is known as blockage and may be removed by using an offset parabola configuration [5].

21.3.1.3 Transfer Function

The transfer function is defined either on transmit or receive and may be defined in either the frequency or time domain. In the frequency domain, a transmitting antenna transfer function, $H_{TA}(\theta, \varphi, \omega)$ is defined as the ratio of the radiation intensity at the spatial test point (θ, φ) to the signal excitation at the transmitting terminal. Similarly, a receiving transfer function $H_{RA}(\theta, \varphi, \omega)$ is defined as the ratio of received signal to the incident-field intensity (here ω is $2\pi f$, with f the frequency component being considered). If two antennas are separated by a distance R in free space, the overall signal-transfer function between two antennas is the product of $H_{TA}(\theta, \varphi, \omega) H_{RA}(\theta, \varphi, \omega) e^{-j\beta R}$.

A time-domain definition of the transfer function for a transmitting antenna is the time-varying signal at a spatial test point, in response to an impulse applied to the transmitting antenna terminals. For a receiving antenna, the transfer function is the complex response at the antenna output port when the receiving antenna is excited by the electromagnetic waves resulting from a delta function emitted from a spatial test point O.

21.3.1.4 Pulse Dispersion

A change in pulse shape brought about by frequency dependence in the antenna.

21.3.1.5 Integrated Cancellation Ratio

The ratio of the total power received by the radar in the unwanted hand of circular polarisation to the total power received. This is normally measured by integration of the radiation pattern over 4π sterradians.

21.3.1.6 Phase Centre

The point on the antenna from which the radiation appears to originate. This is normally a frequency-sensitive parameter so that in the time domain the phase centre is ill defined.

21.3.2 Equivalent Aperture Model for Impulse Radiation

Consider an infinite screen with a circular aperture in it as shown in Figure 21.2.

We assume the aperture is illuminated by a impulsive electric field in the $z < 0$ region of time variation $f(t)$ and polarised solely parallel to the y axis. This is assumed to be provided by a distant source excited

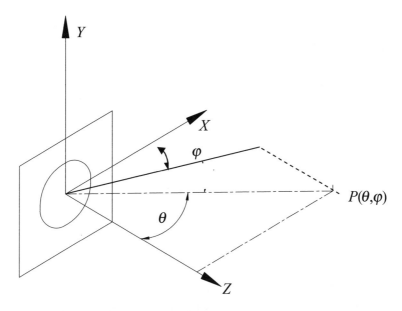

Figure 21.2 Equivalent aperture and coordinate system

by time-dependent voltage $V(t)$ arriving with equal path length over the aperture. The source results in a spatial distribution (normalised to the electric field at the origin $x = y = z = 0$) of $g(x_1,y_1)$ where x_1,y_1 is any point within the aperture. The magnitude of the electric field at the origin is E_{ap0}. The time variation, $f(t)$, is assumed constant across the aperture, so we may write the electric field in the aperture in the time domain as

$$\boldsymbol{E_{ap}}(x_1, y_1, t) = f(t)\, E_{ap0}\, g(x_1, y_1)\, \hat{\boldsymbol{y}}$$

where $\hat{\boldsymbol{y}}$ is the unit vector in the y-direction

We will consider D, the aperture diameter to be large with respect to the highest transmitted spectral component of $f(t)$. Consider P to be a distant point on the z-axis at range R and initially consider the frequency-domain performance.

We can show from straightforward aperture theory [6] that the magnitude of the electric field at point P is proportional to

$$\frac{jwe^{jkR}\, V(w)}{R}$$

Here $V(w)$ represents the frequency spectrum of the source voltage. Taking the inverse Fourier transform of this allows us to express the time-domain representation of the electric field at the on axis point P as proportional to

$$\frac{dV(t)}{R\, dt}$$

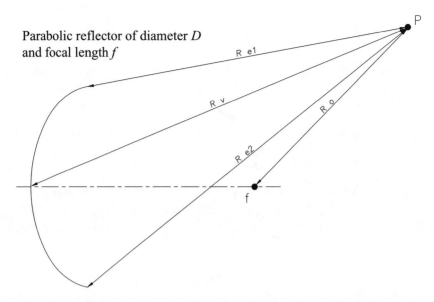

Figure 21.3 Reflector geometry. R_{e1}, R_{e2} are rays from the edge of the reflector to the point P, R_v is the ray from the vertex

Hence we see the temporal variation of the field at a distant point P is proportional to the derivative of the source pulse $V(t)$.

21.3.3 Parabolic Antenna

To generalise the simple aperture analysis to a parabolic reflector, we note the assumption in the equivalent aperture model that the path length to all points on the aperture is constant . This is true on axis for a parabola but is untrue for off-axis points. To give insight to this we use a ray-tracing approach.

Consider a parabola feed with a point source as shown in Figure 21.3. A well-known property of the parabola is that the distance from any point in the aperture plane to the focal point is a constant – indeed it is this property of the parabola which causes it to have a single point focus. We are interested in the radiating properties of such an aperture under impulse conditions. References [7]–[9] give a detailed account of the electromagnetics. To understand the basic operation we use a simple ray optic explanation [7]. Initially ignore edge diffraction and consider only the forward hemisphere of radiation. An ideal point source feed is assumed and radiation direct from the feed itself into the forward sector is ignored.

Consider a short impulse from the transmitter. The leading edge of the pulse first hits the central portion of the reflector. This excites a current flow on the reflector which in turn starts to radiate. As the reflector is symmetrical this initially excited current forms a disk whose radius grows in time until the trailing edge of the pulse hits the reflector. For a short pulse this can occur before the leading edge has reached the edge of the reflector. Under these conditions, the induced current never fully fills the reflector and forms an annulus which continues to expand in time until the edge of the reflector is met.

Now let's consider what this means to the far-field pattern. Following Hansen [7] and other authors we next consider the path length from the feed via different points on the reflector to a distant point,

$P(\theta, \varphi)$, where θ, φ are the conventional spherical coordinates as given in Figure 21.3. For this discussion we will consider $\varphi = 0$ and take the rays from the edges and the vertex to the point P. We assume the range, $R_0 \gg D$, the dish diameter, and we note a practical minimum value of f/D (focal length to diameter ratio for the parabola) is 0.25, which is where the feedpoint lies in the aperture plane of the reflector.

The distance to an off-axis point P from the vertex of the reflector can then be approximated as

$$ct_v = R_0 + f\cos(\theta)$$

Reference [8] derives the time delay for a ray to reach the outer edge of the reflector from the feedpoint to be $D^2/16fc$ where D is the dish diameter, f the focal length and c the speed of light. The total time for a ray to leave the focus, hit the edge of the reflector closest to the observation point P (the upper edge in Figure 21.3) and then arrive at P is

$$ct_{e1} = R_0 + f\cos(\theta) + D^2/16f(1-\cos(\theta)) - D/2\sin(\theta)$$

The ray hitting the lowest edge has a total time to P of:

$$ct_{e2} = R_0 + f\cos(\theta) + D^2/16f(1-\cos(\theta)) + D/2\sin(\theta)$$

We note that $t_v = t_{e1} = t_{e2} = R_0 + f$ when $\theta = 0$, that is on the axis of the parabola as expected.

For $\theta > 0$ it can be seen $t_{e1} < t_v < t_{e2}$ with a minimum value of $t_{e1} = R_0 + D^2/16f - D/2$.

From this ray optic argument, the earliest part of the pattern to form is hence at $\theta = \pi/2$ and is formed from the edge closest to point P. As time progresses the edge current contributes to angles $\theta < \pi/2$ and the currents across the reflector progressively contribute to the pattern at $\theta = \pi/2$, with the last contributions to arrive from the most distant edge of the reflector.

The quantity $D^2/16fc$ is known as the transit time of the reflector. Figure 21.4 shows transit time as a function of f/D ratio and dish diameter.

Now consider the pulse width, τ, is longer than the transit time for the reflector. Under these circumstances the current will fully fill the reflector before the trailing edge of the pulse reaches the vertex. For a period of time the entire reflector is therefore contributing to the pattern. When the trailing edge hits the reflector the radiating current forms an annulus, growing from the centre of the aperture until the pulse eventually reaches the reflector edge. However, due to the difference in path lengths noted above as far as off-axis point P is concerned the edge ray contribution disappears first, followed by the disappearance of contributions from the rest of the reflector.

The situation is even more significant when the pulse width is less than the transit time. Under these circumstances the current never fully fills the reflector so that full aperture directivity is never achieved on axis. We therefore see a fundamental limit on the minimum pulse width that a reflector will develop the peak directivity is the fill time. In practice, this is considered an absolute minimum – a more practical limit is the pulse width should be at least twice the transit time for the reflector.

An important point here is that the behaviour of the reflector during the build up of the pulse is different to the behaviour during the decay period. To understand this, consider the conditions existing just after the leading edge reaches the edge of the reflector. Only the portion of the reflector close to the edge of the reflector contributes to point P, as noted above. However, when the pulse decays, all the reflector contributes (assuming transit time conditions are met) except for the region close to the edge of the reflector. Clearly there is far more area of reflector contributing to point P during the decay then during pulse build up.

Figure 21.5 illustrates the pulse build up graphically. Here we show the pulse shape at particular angles off boresight for a particular reflector of f/D ratio 0.33. A pulse width of half the transit time has been

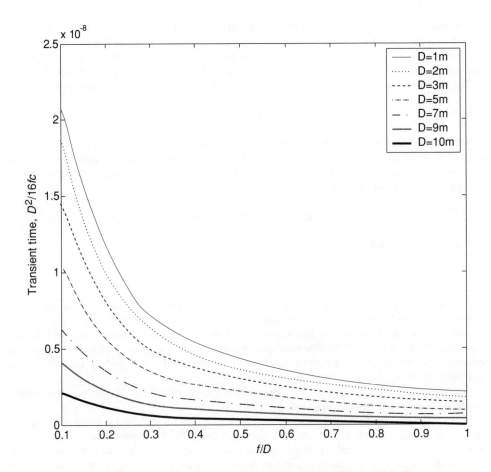

Figure 21.4 Transit time (seconds) as a function of f/D for different dish diameters

used diameter with a perfect square pulse excitation. The different rise and fall times of the pulse with angle is clearly seen.

So far, this discussion has been based on a simple ray optic approach. However, in practice this is a simplistic view. For accurate predictions it is important to consider both the actual current flow on the curved surface of the reflector (which can, for example, give rise to cross-polarisation effects) and, of course, edge diffraction.

De Olivera and Helier [10], based on earlier work by Sergey and Turchin [9], present a time-domain physical optics (TDPO) analysis of a parabolic reflector fed by a Gaussian pulse. The authors ignore the diffraction at the edge of the dish and any direct spillover by the feed which results in relatively simple to compute formulas.

To include for diffraction, Rego, Hasslemann and Moriera [11] present a treatment again based on TDPO and a time-domain formulation of the uniform theory of diffraction (TD-UTD). The authors apply this to a hyperbodial reflector although the approach is easy to generalise to the paraboloidal case. The interested reader is referred to those papers for the detailed analysis.

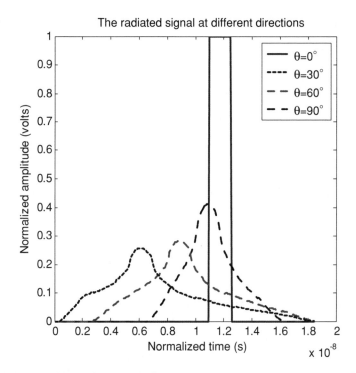

Figure 21.5 Off-axis pulse characteristics for a typical large reflector under impulse conditions. In this example the pulse width is half the transit time of the reflector

21.4 UWB Feed Designs

The results of the above section assume an ideal point source with all the radiation in the feed forward hemisphere. Obviously, this is an impractical assumption; the feed will in practice have its own radiation pattern. As we illustrate in this section, the feed must:

- to produce a pattern optimised to the optics of the reflector with little radiation direct from the rear of the feed;
- induce little pulse distortion in its own right;
- have a well-located phase centre;
- polarisation characteristics consistent with the system need;
- impedance matched to transmitter/receiver;
- power handling consistent with the application (both peak and mean power need to be considered);
- introduce minimum possible loss;[2]
- introduce minimal blockage of the main reflector.[3]

---

[2] We note here that in certain applications such as radio astronomy loss is a critical consideration due to the noise temperature requirements.
[3] One way to remove blockage is by use of a offset reflector configuration.

21.4.1 Feed Pattern Effects

In general, small antennas which may be used as feeds have radiation patterns which exhibit angularly dependent pulse dispersion. Hence, in general, different parts of the reflector will be illuminated with different pulse shapes leading to overall pulse dispersion at all angles.

Fundamentally, the energy spilling past the reflector will directly affect the efficiency of the antenna and depends on the feed illumination. Reference [5] shows the basic spillover efficiency as a function of the edge illumination for a parabolic antenna. This assumes the feed pattern is approximately parabolic in shape and has no angular dispersion, so provides a basic guide to the type of feed pattern required.

In UWB there are additional issues to be considered. It is illustrative to initially consider a frequency-domain analysis. As the frequency is altered over a typical UWB bandwidth the feed pattern itself will, in general, alter, narrowing as the frequency increases. Hence the edge illumination of the reflector alters with frequency with the result that different parts of the frequency band have considerably different efficiency. Turning this to the time domain, the result is the pulse shape will be distorted due to this effect.

We see, therefore, that a feed which maintains a substantially constant beamwidth over the frequency range of interest will produce less pulse distortion.

Unfortunately, most feed antennas, like the simple aperture assumed above, have frequency-dependent patterns, hence leading to pulse distortion.

21.4.2 Phase Centre Location

In addition to the aperture amplitude effects, to transmit a impulse with low distortion requires the effective phase centre of the feed to be well defined. The phase centre is the point on the feed that the radiation appears to be emanating, and sets where the feed should be placed with respect to the reflector focal point. Even in narrowband feeds there is often phase centre movement which leads to some defocusing of the antenna. Some wideband feeds have extreme phase centre movement which results directly is dispersion in the antenna – a point we will return to later.

21.4.3 Input Impedance

The input impedance of the feed (or strictly the reflector/feed combination) will in general change over the frequency band encompassed by a UWB pulse. Impedance mismatch has a number of effects: first it reduces efficiency and increases noise temperature on receive as all or some part of the reflected power due to the mismatch is lost to the system. Secondly, some of the power will be reflected back off the transmitter/receiver and be reradiated (the so-called 'ringing' effect), which results in the effective transmitted pulse being replicated sequentially in time. This ringing may or may not be a significant system problem, depending on the particular system being considered [12]. As a further point the impedance of practical UWB antennas is often not optimally matched to 50 ohm which has implications for optimum transmitter/receiver design and for component testing. [4]

21.4.4 Polarisation

The feed sets the polarisation of the antenna and different solutions are available for linear and circular polarisation. A key issue is the level of cross-polarisation that can be tolerated by the system. This is

---

[4] Most RF test equipment is matched to 50 ohms necessitating the use of a transformer if other impedances are required which in turn induces inaccuracies in measurement.

particularly true in a circularly polarised system. This leads to consideration of the integrated cancellation ratio (ICR), as defined above. We note that in general this is dominated by the feed performance in practical antennas. For a radar just how low the ICR needs to be is arguable. Figures of the order -20 dB are typically required.

21.4.5 Blockage Effects

Blockage occurs in axi-symmetric reflectors. It arises simply from the shadowing effect of the main reflector by the feed and its supporting structures. The result is a decrease in aperture efficiency and increase in off-axis radiation due to scattering effects. Reference [5] gives more detail on the effect of blockage. As an approximation, if the diameter of the central blockage region represents 10 % of the diameter of the main reflector then approximately 6 % decrease in aperture efficiency occurs.

21.5 Feeds with Low Dispersion

In this section we review some of the more popular feed types for UWB reflectors.

Classically, as Rumsey [13] points out, there are two basic principles that enable an antenna to exhibit frequency-independent behaviour. First, if the structure can be defined such that the geometry periodically scales then the frequency performance will also scale resulting in essentially frequency-independent radiation patterns. In effect such antennas have an active region at any one frequency which moves over the structure as the frequency is altered. This is the basis of the well-known log-periodic antenna array [13]. Unfortunately as the frequency changes the effective phase centre of the radiation moves. This is a classic example of a UWB antenna suitable for frequency-hopping systems but unsuitable for impulse transmission as the phase centre movement causes pulse dispersion. Reference [13] gives more details for the design of this type of antenna.

Of more importance to impulse transmission is the second technique for essentially frequency-independent performance, which is to define a structure without recourse to linear dimensions.

21.5.1 Planar Spiral Antennas

Planar spiral antennas are popular for the radiation of broadband circular polarisation. Even over several octaves of bandwidth the radiation pattern retains essentially constant beamwidth making it an obvious candidate as a reflector feed.

Figure 21.6 illustrates two popular variants of the spiral antenna, the Archimedean and equi-angular geometries.

The equi-angular spiral (Figure 21.6(b)) can be defined in terms of two angles [12] and hence in principle the frequency performance is only limited by the truncation effects at the feedpoint (which affects the high frequency of operation) and the diameter of the spiral (which limits the low frequency performance). The Archemedian spiral (Figure 21.6(a)) is popular as it is somewhat more compact for the same frequency band. We can define the planar spiral geometry by

$$\rho = e^{a\varphi}$$

with a being a constant. The far-field radiation pattern for the fundamental harmonic approximates to [13]:

$$|E| = \frac{\cos(\theta)\tan(\theta)\exp[a^{-1}\tan^{-1}[a\cos(\theta)]]}{(1+a^2\cos^2(\theta))^{1/2}}$$

where θ is the pattern angle defined in Figure 21.2. We note this is independent of both φ and frequency.

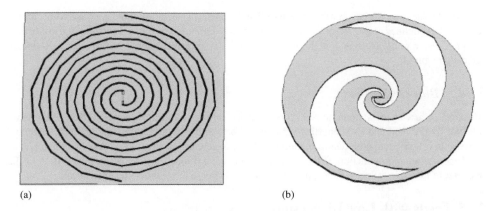

Figure 21.6 Spiral antennas: (a) Archemedian spiral; (b) equi-angular spiral

Unfortunately, the effective phase centre of the spiral lies somewhat behind the plane containing the spiral, being approximated by $\lambda a \pi^{-1}$ (for $a < 1$). Hence some movement of the phase centre occurs across the frequency band out of the plane of the spiral which therefore translates to pulse dispersion under impulse conditions.

Figure 21.7 illustrates the impulse response of an Archimedean spiral antenna. In general the equi-angular spiral shows somewhat improved dispersion characteristics.

One problem for both types of design is that the spiral naturally radiates from both sides of the sheet. Clearly this is not appropriate for a feed. To inhibit the radiation in one direction the spiral is normally backed by a cavity which limits the VSWR achievable over the frequency band. Alternative designs use conical cavities, shaped dielectric filled cavities and lossy cavities. Each has its own advantages and disadvantages. In general an input VSWR of a cavity-backed spiral over an octave bandwidth is of the order 2:1.

Finally either form of spiral can improve the low frequency performance for a finite diameter structure by the inclusion of loads at the ends of the arms.

In general spirals are normally used in relatively low-power applications due principally to the feed arrangement required.

21.5.2 TEM Feeds

Conceptually, we start by considering a standard rectangular waveguide as shown in Figure 21.8(a). The fundamental mode of propagation is the TE_{10} mode, which has a cut-off wavelength of twice the width of the aperture, so $\lambda_c = 2a$ and the waveguide acts as a high pass filter.

If we remove the side walls of the guide, it becomes a parallel plate structure and the modal characteristics change from TE to TEM mode. A TEM (transversal electromagnetic) mode has two key properties for UWB use: first this mode has no low frequency cut-off, and secondly an inherent field distribution which naturally matches that of free space. Furthermore the fundamental TEM mode travels at the speed of light irrespective of wavelength and hence TEM structures are inherently nondispersive.

To use these modes as a broadband feed two parallel plates are arranged in a flare configuration, Figure 21.9. In the throat region a dominant TEM mode is excited via some type of transition, often into

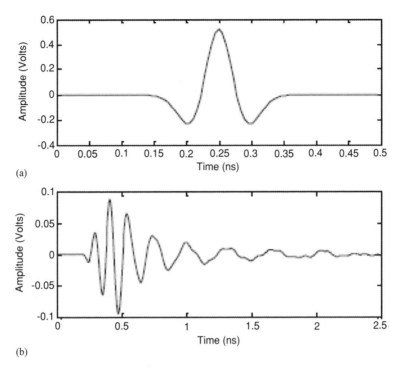

Figure 21.7 Response of an Archimedean spiral antenna under impulse conditions. (a) Excitation voltage; (b) E-field at a distant on-axis point

a coaxial probe. The plates are then flared to produce a radiating aperture consistent with the radiation pattern requirements. In plan view the width of the plates is normally tapered to provide a smooth impedance transition to free space [14]. This type of simple horn has some disadvantages as a reflector feed, principally in the size of the horn required for a particular pattern requirement and in pattern symmetry issues.

In fact, references [15] and [16] show that the requisite TEM mode can be supported using wire, rather than plate structures. Typically either two or four conically tapered wires are used. As this implementation

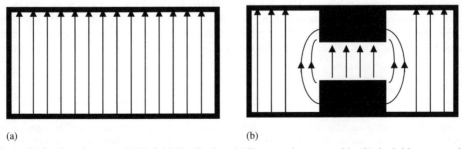

Figure 21.8 Dominant modal E-field distribution: (a) Rectangular waveguide; (b) dual ridge waveguide

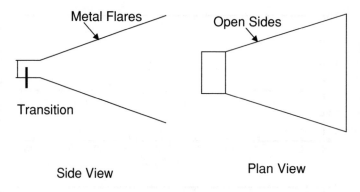

Figure 21.9 Simple TEM horn

is fundamental to the IRA (impulse radiating antenna) we will leave further discussion to the next section.

In order to provide more flexibility over the radiation pattern performance so-called TEM horns (otherwise known as dual ridge waveguide horns) may be used. Figure 21.10 illustrates such a configuration. Strictly these are not pure TEM mode devices due to the presence of the metallic side walls but in some sense are extended bandwidth waveguide devices.

To explain their operation initially we compare the simple waveguide horn with the dual ridge waveguide structure. As previously noted, a rectangular waveguide has a dominant mode which is transverse electric in type (TE_{10}). The maximum useful range of bandwidth is between the cut-off frequency of this mode and the frequency at which the next order higher mode will start to propagate.[5] This gives a useful frequency range of about 2:1. In a conventional waveguide horn as long as the taper starts where the waveguide will only support the dominant mode a linear taper will not cause mode coupling along the length of the horn, hence the field distribution at the radiation aperture will be like the dominant mode but with the addition of a phase 'cap' introduced by the taper [16], Figure 21.11.

For UWB, working techniques have been examined to decrease the lower working limit of the waveguide. This is achieved by the introduction of two ridges as shown in Figure 21.10(b).[6] The ridges concentrate the field around the ridges themselves. In between the ridges a 'parallel plate-like' structure exists which will support a TEM mode between the ridges. The result is a pseudo-TEM structure with extended low frequency cut-off compared to a simple plain waveguide.

Reference [17] shows the cut-off for the TE_{10} mode in ridged waveguide decreases in frequency whereas the next highest mode, the TE_{20}, has increased cut-off. The TE_{30} is largely unaffected by the ridges. The modal cut-off frequencies depend essentially on the ratio of the width of the ridge to the width of the waveguide and the ratio of the gap between the ridges to the height of the guide. If we define the useful possible bandwidth as the difference between the dominant mode cut-off and TE_{30} cut-on

---

[5] This is an oversimplification as the impedance of the waveguide changes rapidly close to cut-off. The actual useful frequency range is therefore somewhat reduced.

[6] A four-ridge waveguide may also be used in the square waveguide, so that two polarisations may be supported and both two and four ridges can be used in the circular waveguide. For simplicity we limit our discussion here to the simple two-ridge type.

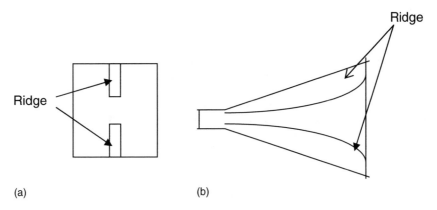

Figure 21.10 Dual ridge TEM horn: (a) Front view showing radiating aperture; (b) cut-through show ridge configuration

frequencies we get a bandwidth of approximately 8:1 as a maximum. Over this bandwidth the major difficulty becomes stability of the input impedance and the transition design at the horn throat so that VSWR better than 2:1 over this frequency range is difficult to achieve.

Having launched the mode at the throat of the horn the horn flare is linear to the aperture following normal horn design guidelines. The ridges, however, are shaped to provide a smooth impedance transition to free space. Typically the ridge follows an exponential curve, Figure 21.10. We note here that the overall match of the horn is limited by the input waveguide but also by the size of the radiating aperture in terms of wavelengths at the lowest frequency of operation.

To first order the resulting radiation patterns can be assumed to be those of a waveguide aperture without ridges. As with a standard horn the resultant pattern depends on the taper angle of the horn and the horn aperture. Accordingly, a short wide angle horn – a so-called scalar horn – can be designed where

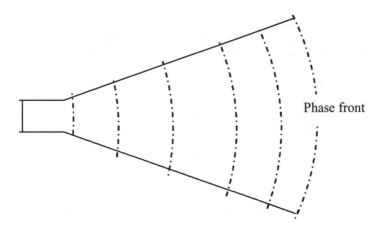

Figure 21.11 Linear horn antenna development of phase error 'cap' due to the presence of a flare

the radiation pattern variation over a wide bandwidth [17] is minimised. A typical variation of beam width of 2:1 over a 4:1 bandwidth has been reported.

As the taper sets the position of the phase centre, the position of the phase centre is well defined over large frequency bands, hence the feed introduces little dispersion to an applied impulse, principally that due to directivity changes across the frequency band rather than phase centre movement.

For detailed design, a full numerical simulation is needed using a full-wave simulation technique (such as FDTD) and has been successfully applied by a number of authors (for example [14]).

Dual ridge horns can be designed successfully to operate over wide bandwidths and matched into dual ridge waveguide feeders for high-power applications. When dealing with high power, a point to note is that the field concentration around the ridges can significantly reduce the power-handling capability of the waveguide.

21.5.3 Impulse Radiating Antenna (IRA)

The impulse radiating antenna (IRA) has been widely reported in the specialist literature ([18] to [20]). Originally suggested by Baum [15], this structure utilises the properties of a TEM structure together with a parabolic reflector to provide extremely wideband performance. Frequency ranges of 200:1 (for example 10 MHz to 2 GHz) have been reported, crucially with low dispersion hence preserving the pulse width radiated of extremely short pulses.[7]

There have been a number of different variants of IRA introduced in the literature. Here we use the TEM-fed reflector type. The reader is referred to references [20] to [21] for other variants.

We start then by considering a TEM horn. As noted above this has inherently broadband properties, principally low dispersion and no low frequency cut-off. At the lowest frequency of operation a significant aperture is needed to provide high directivity and a good impedance match. At the highest frequency of operation the aperture becomes large in electrical terms.

Unfortunately, due to the horn taper the radiating aperture includes for a significant phase 'cap' which increases the radiating beamwidth and decreases the prompt aperture efficiency of the antenna.

The IRA development starts with the observation that the essential TEM modal structure does not need wide metal plates as discussed above. In fact, the mode can be supported by a number of wire-like structures. A TEM horn can be constructed from two conical shaped wires as shown in Figure 21.12. If we are to achieve a good match to free space the characteristic impedance of the balanced transmission line should approximate 400 ohms hence as the structure flares the wire diameter is increased to maintain a constant impedance, resulting in conical shaped wires. As discussed for the dual ridge horn, a spherical phase cap occurs across the aperture due to the taper. This phase cap reduces the aperture efficiency of the antenna.

To correct for the phase cap, one could use a lens extremely close to the antenna, and indeed such structures have been proposed. An alternative approach, as first suggested by Baum, is to introduce a parabolic reflector extremely close to the radiating aperture of the TEM feed. This will, in ray-optic terms, correct for the spherical phase cap caused by the feed taper. The purpose of using a wire implementation of the TEM feed is now clear: the feed horn will present only limited blockage to reflector aperture [22], hence efficient radiation will occur off the reflector.

---

[7] Note the radiated E-field of the IRA on boresight still has the characteristic of differentiating the applied source voltage – however, the pulse closely follows this structure with little additional dispersion.

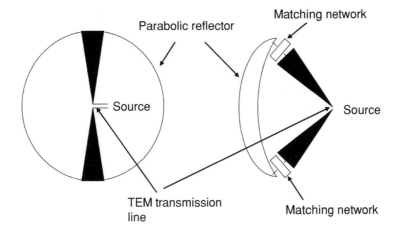

Figure 21.12 Reflector impulse radiating antenna (IRA)

This description is reasonably valid at high frequencies where the dish is at least a few wavelengths across. At low frequencies, however, it is far too small in electrical terms to be acting in a ray-optic sense. In fact, the IRA behaves in a very different manner at low frequencies.

The low frequency analysis of the IRA recognises that the dish is acting more like a small current loop at the end of a balanced wire transmission line. Such a structure can be made to radiate as a balanced transmission-line wave antenna (BTW) [6]. A BTW antenna uses a balanced transmission line with matched loads at either end. Radiation will occur from such a structure at frequencies when the transmission-line wires are separated by a significant proportion of a wavelength. It can then be shown that the radiation will preferentially occur in the direction of the source end of the wire provided the transmission line is properly terminated. The radiated field on axis is then the same as a loop with an area equal to the cross-sectional area of the transmission line as viewed from the side. Farr [6] gives a value for the electric field in the frequency domain on axis of this magnetic loop model at distance R to be:

$$E(\omega) = K_1 (j\omega)^2 \frac{e^{-jkR}}{R} V(\omega),$$

where

$$K_1 \cong \frac{A}{4\pi c^2}$$

where A is the area of the current loop formed by the feed arms and c is the speed of light.

Off-axis behaviour is approximately $(1 + \cos(\theta))$ with backlobe suppression of the order 20 to 30 dB with respect to peak measured in practice.

For the IRA to look like a balanced transmission line at low frequencies, the arms of the TEM feeder must be loaded with a matched impedance. Approximately, the matching circuit should present 200 ohms to each leg. Due to the presence of the dish this is only approximate and the matching impedance should be optimised depending on IRA configuration.

This low frequency approximation to the behaviour of the device is reasonable up to a frequency where the length of the transmission line is less than or equal to $\lambda/(2\pi)$ [19]. At highest frequencies an optical-based analysis (using physical optics) becomes appropriate. Reference [6] gives an approximate expression for the on-axis field distribution in this region to be:

$$E(\omega) = K_2(j\omega)\frac{e^{-jkR}}{R}V(\omega)$$

where $K_2 \cong \frac{0.9D}{4\pi c}$ for a 400 ohm TEM feed line, D being the dish diameter.

At intermediate frequencies where neither model is accurate a moment method formulation was used in [6]. It was seen that the predicted fields on axis followed a similar trend to the formulation given above, with the results producing a somewhat pessimistic approximation at low frequencies.

Overall the IRA possess a near unique ability to radiate extremely broad bandwidths with extremely low distortion.

21.6 Summary

In this chapter we have concentrated on the problems involved with impulsive transmission from large electrical apertures, in particular reflector antennas. Key points noted include:

- In general the on-axis field strength in the time domain is the derivative of the source voltage time response.
- That relectors have a 'fill-time' which can be important when considering radiation performance.
- That symmetrical feeds with roughly constant beamwidth are used to provide low pulse dispersion off reflector antennas.
- For impulse transmission minimising phase centre movement over the required frequency band becomes important.
- Spiral antennas provide good circularly polarised performance but do produce some phase centre movement and resulting pulse dispersion. In general equi-angular geometries are slightly less dispersive than Archemedian spirals.
- TEM horns are a particularly important class of feeds for impulsive transmission. The dual ridge horn provides the possibility of broadband performance and may be used in high-power designs.
- A combination of a TEM mode tapered transmission line with a parabolic reflector is the basis of the impulse radiating antenna (IRA), capable of supporting an extremely broadband of frequencies with low pulse distortion.

References

[1] Astanin, Kostylev, Zinoviev and Pasmurov, *Radar Target Characteristics*, CRC Press, 1994.
[2] Skolnik, *Introduction to Radar Systems*, McGraw-Hill, 2001.
[3] Agee, Scholfield, Prather and Burger, Powerful ultra-wide band RF emitters: status and challenges, *SPIE* 1998.
[4] Buchenauer, Tyo and Schoenberg, Prompt aperture efficiencies of impulse radiating antennas with arrays as an application, *IEEE Transactions on Antennas and Propagation*, **49**(87), 2001.
[5] Rudge, Milne, Olver and Knight, *Handbook of Antenna Design Volume 1*, Peter Pereginus, 1982.
[6] Farr, Analysis of the impulse radiating antenna, *Sensor and Simulation Notes*, Note 329, 24 July, 1991.
[7] Hansen and Kramer, Correction to short pulse excitation of reflector antennas, *IEE Proc. H*, **139**(1), 1992.
[8] Hansen, Short pulse excitation of reflector antennas, *IEE Proc. A*, **134**, 1987.
[9] Skulkin and Turchin, Transient field calculation of aperture antennas, *IEEE Transactions of Antennas and Propagation*, **47**(5), 1999.

[10] de Oliveira and Helier, Time response of a parabolic reflector antenna to a generalized Gaussian pulse, *11th International Symposium on Antenna Technology and Apply Electromagnetics (ANTEM 2005)*, Saint-Malo, 2005.
[11] Rego, Hasselmann and Moreria, Time-domain analysis of a reflector antenna illuminated by a Gaussian pulse, *Journal of Microwaves and Optoelectronics*, **1**(4), 1999.
[12] Zhang and Brown, Complex multipath effects in UWB communication channels, *IEE Proceedings on Communications, Special Issue on Ultra Wideband Systems, Technologies and Applications*, 2006.
[13] Rumsey, *Frequency Independent Antennas*, Academic Press, 1966.
[14] Lui, Fan and Liu, Pulse radiation antenna feeded with a face to face TEM horn, *Proc. 5th International Symposium on Antennas, Propagation & EM Theory*, IEEE, 2000.
[15] Baum, Radiation of impulse-like transient fields, *Sensor and Simulation Notes*, Note 321, 25 November, 1989.
[16] Olver, Clarricoates, Kishk and Shafai, *Microwave Horns and Feeds*, The Institution of Electrical Engineers, 1994.
[17] Helszajn and Caplin, Impedance and propagation in ridge waveguide, *Microwave Engineering Europe*, May 1997.
[18] Baum, Configurations of a TEM feed for an IRA, *Sensor and Simulation Notes*, Note 327, 27 April, 1991.
[19] Tyo, Optimization of the TEM feed structure for four-arm reflector impulse radiating antennas, *IEEE Transactions on Antennas and Propagation*, **49**(4), 2001.
[20] Baum, The conical transmission line as a wave launcher and terminator for cylindrical transmission line, *Sensor and Simulation Notes*, XXXI, 16 January, 1967.
[21] Farr and Baum, Impulse radiating antennas with two refracting or reflecting surfaces, *Sensor and Simulation Notes*, Note 379, May, 1995.
[22] Rahmat-Samii, Analysis of blockage effects on TEM-FED paraboloidal reflector antennas, *Sensor and Simulation Notes*, Note 347, 25 October, 1992.

Bibliography

Barrett, History of ultra-wideband (UWB) radar and communications, Pioneers and Innovators Progress in *Electromagnetics Symposium 2000 (PIERS2000)*, Cambridge, MA, July 2000.

Bennett, Time-domain electromagnetics and its application, *Proceedings of the IEEE*, **66**(3), 1978.

Schantz, *Dispersion and UWB Antennas, IEEE Symposium on Ultra-Wideband*, IEEE, 2004.

Tyo, Farr, Schoenberg, Bowen and Altgilbers, Effect of aperture feed and reflector configuration on the time- and frequency-domain radiation patterns of reflector impulse radiating antennas, *IEEE Transactions on Antennas and Propagation*, **52**(7), 2004.

Sabath and Garbe, Impact of near field dispersion on time domain susceptibility tests, *Advances in Radio Science*, 1, 43–7, 2003.

Tyo, Farr, Bowen and Schoenberg, Off-boresight radiation from impulse radiating antennas, *IEEE International Symposium on Antennas and Propagation*, Colombus, Ohio, USA, June 2003.

Schiavone, Wahid, Palaniappan, Tracy, Van Doorn and Lonske, Outdoor propagation analysis of ultra wide band signals, *IEEE International Symposium on Antennas and Propagation*, Colombus, Ohio, USA, June 2003.

Li, Hagness, Choi and van der Weide, Numerical and experimental investigation of an ultrawideband ridged pyramidal horn antenna with curved launching plane for pulse radiation, *IEEE Antennas and Wireless Propagation Letters*, **2**, 2003.

Index

Accuracy
   Location, 390–1, 405–6
   Ranging, 398ff
Adaptive sidelobe canceller, 67
Alamouti scheme, 100
Angular spread, 376, 379
Antenna
   Antipodal, 155, 157, 158
   Aperture, 42
   Balanced, 43
   Bandwidth, 41
   Bicone, 164, 286, 299–302, 320–1, 426
   Bowtie, 45–6, 167
   Diamond dipole, 176
   Dipole, 42–45, 111, 442
   Directivity, 41
   Disc monopole, 120
   Discone, 149
   Elliptical dipole, 171
   Effective area, 38
   Efficiency, 425–8, 441
   Four-square, 233
   Frequency-independent, 45–6, 111, 288
   Horn, 147–8, 443, 428, 444, 462
   Horn-shaped self-complementary, 333–57
   Hybrid, 442–3
   Impedance, *see* Impedance
   Impulse radiating, 466–8
   Key to, 27
   Log periodic, 387, 461
   Matching, *see* Matching
   Monopole, 44–5, 112
   Omni-directional, 116, 153
   Patch, 70, 444
   Planar inverted cone, 333–57
   Polarisation, 460
   Printed, 3, 45, 150, 333, 336
   Reflector, 148, 453, 454–63
   Resistive loading, 425, 441
   Roll, 148, 153
   Self-complementary, 333
   Slot, 132, 429
   Spiral, 48, 234, 429, 461–2
   Vivaldi, 155, 429
   Wideband, 45, 234
   Wire, 42
Antenna array, 49ff, 69ff, 221, 389ff, 433ff, 448
Array
   Active, 108, 235
   Adaptive, 74
   Beamforming, *see* Beamforming
   Broadside, 53, 221
   Circular, 395
   Efficiency, 454, 460–1, 466
   End-fire, 53, 221, 391–2
   Factor, 56ff, 238
   Gain, 65
   Linear, *see* Linear arrays
   Orientation, 379
   Passive, 108, 235
   Pattern multiplication, 55, 223
   Phased, 230
   Scanning, 224
   Series fed, 236
AWGN channel, 89

Ultra-wideband Antennas and Propagation for Communications, Radar and Imaging   Edited by B. Allen, M. Dohler, E. E. Okon, W. Q. Malik, A. K. Brown and D. J. Edwards
© 2007 John Wiley & Sons, Ltd

BAN, *see* Wireless body area network
Beam steering, 78
Beamforming, 75ff, 241, 394, 398
BEAMLOC, 401
Blass network, 238
Body area network, *see* Wireless body area network
Broadside array, *see* Array
Butler matrix, 238

Capacity, 364–70, 375–6
Central limit theorem, 92
Channel impulse response, 268, 310–13, 319–24, 336
Channel model,
    Delta-K, 317, 325
    Saleh-Valenzuela, 310–19, 325–7
    Single-Poisson, 318, 321
    Split-Poisson, 319, 321
    Two-Cluster Poisson, 319, 321
Channel state information at receiver, 93, 102, 362, 365–74, 377, 381
Channel state information at transmitter, 362, 366–75, 378, 380–1
Chi-square distribution, 92
CIR, *see* Channel impulse response
Circular array, 63, 71
Circular polarisation, *see* Polarisation
CLEAN algorithm, 312
Coaxial transition, 428, 443
Coexistence, 221
Combining, *see* Diversity
Conductivity, 26
Conformal array, 72
Consumer applications, 197
Correlation, 310, 323–4, 364, 367, 375–80
Coverage range, 371, 375, 381
CPW, *see* Waveguide
CSIR, *see* Channel state information at receiver
CSIT, *see* Channel state information at transmitter
Curl, 26

Delay-line, 109, 242
Delay spread, 204, 241
Delta-K model, *see* Channel model
Dielectric
    Constant, 392, 426, 439–40
    Properties of materials, 392, 423, 439, 445
    Loading, 428–9, 462

Digital versatile disk, 208
Dipole, *see* Antennas
Direct-sequence code division multiple access, 2,6, 8–9, 14, 107, 214
Directive gain, 41
Directivity, 41–43, 45, 224, 340, 345, 457, 466
Divergence, 26
Diversity
    Receive, 93
    Combining, 95
    Transmit, 100
    MIMO, 102
Dolph-Tschebyscheff distribution, 59ff
DS-CDMA, *see* Direct-sequence code division multiple access

E-plane, 123, 124
Edge ray, 457
Effective area, 39
Effective isotropic radiated power, 107
Effective length, 38, 46
EIRP, *see* Effective isotropic radiated power
Eigenspectrum, 377–8
Electric field intensity, 26
Electric flux density, 26
Electric scalar potential, 28
Electrodynamics, 25
End-fire array, *see* Array
Equal gain combining, 98
Equalisation, 241
Equivalent circuit, 37–8

Fading, *see* Fast fading *and* Shadowing
Far field, 22, 30, 33–6, 39, 45, 287, 306, 343
Fast fading, 91, 268
FCC, *see* Federal Communications Commission
FDTD, *see* Finite-difference time-domain
Federal Communications Commission, 2–3, 12, 107, 197, 154, 266
Feed, 125, 459
Fidelity, 151
Filters
    Band pass, 453
    Matched, 392, 402–5, 424
    Pre-filtering, 399
First null beamwidth, 40–1
Finite-difference time-domain, 164, 197, 199, 296–7, 322, 346–53, 356
Finite length dipole, *see* Antennas

FNBW, *see* First null beamwidth
Fractional bandwidth, 67
Fraunhofer region, *see* Far field
Free-space pathloss model, *see* Pathloss model
Fresnel region, 33, 395, 410, 419
Frequency hopping, 10
Frequency regulations, 12
Friis' transmission formula, 30, 39, 283–5, 287
Full-wave, 234

Gamma distribution, 92
Gaussian pulse, 27, 285, 288, 322–3, 349, 405, 458
Geolocation systems, 5
GPR, *see* Ground probing radar
Grating lobes, 83, 231, 246
Green's functions, 234
Ground probing radar, 419, 424, 430, 432, 440
Group delay, 160

H-plane, 123, 124
Half power beamwidth, 40–1
Harmonic representation, 29
Harmonics, 116
HDR, *see* High data rate
Helmholtz equations, 29, 421
Hertzian dipole, 31, 35, 42, 69
High data rate, 2, 9, 14–15, 93–4, 309, 324, 364, 367
Horn-shaped self-complementary antenna, *see* Antennas
HPBW, *see* Half power beamwidth
HSCA, *see* Horn-shaped self-complementary antenna
Huygens' principle, 42

IEEE 802.15.3a, *see* IEEE 802.15 models
IEEE 802.15.4a, *see* IEEE 802.15 models
IEEE 802.15 models, 2–3, 14, 303ff, 324ff, 325ff, 353
Imaging
  Breast cancer, 438
  High resolution, 418–9
Impedance, 30, 37–8, 41, 45, 332–4
Impulse radio, 2–3, 6–9, 14, 23, 147, 265, 284, 332, 339
Impulse response, 204
Interference, 392, 400, 405
Intrinsic impedance, 32

IR, *see* Impulse radio
Isotropic, 221

Large-scale fading, *see* Pathloss
Laplace operator, 29
LDR, *see* Low data rate
Linear arrays, 56, 58, 70, 221
Linear polarisation, *see* Polarisation
Line-of-sight, 34, 269, 279, 285–6, 292–3, 298–306, 322–26, 340
Location systems, *see* Geolocation systems
Lorentz gauge, 28
LOS, *see* Line-of-sight
Lossy media, 426
Low data rate, 309, 325

Magnetic field intensity, 26
Magnetic vector potential, 28
MAI, *see* Multiple-access interference
Mainlobe, 40–3
Marcum Q function, *see* Q function
Matching, 38–9, 287, 333–4
Material characteristics, *see* Materials
Materials
  Cinder block, 262, 270–1, 273, 280
  Double glazing, 262, 270–1, 273, 280
  Fibreboard, 262, 270–1, 273, 280
  Glass, 262, 270–1, 273, 280
  Melamine chipboard, 262, 270–1, 273, 280
  Metal, 262, 270–1, 273, 280
  Plasterboard, 262, 270–1, 273, 280
  Plywood, 262, 270–1, 273, 280
  Red brick, 262, 270–1, 273, 280
  Softwood, 262, 270–1, 273, 280
  Venetian blind, 262, 270–1, 273, 280
Maximal-ratio combining, 96
Maxwell's equations, 25–32, 37, 267, 284, 286, 291, 296
MB-OFDM, *see* Multiband OFDM
Medium-scale fading, *see* Shadowing
Measurements
  On-body, 335
  MIMO, 367
Method of moments, 234
Microstrip, 236
MIMO, *see* Multiple-input multiple-output
MISO, *see* Multiple-input single-output
Monopole, *see* Antennas
Multiband OFDM, 14, 107

Multipath, 90, 390–1, 394, 405
Multipath fading, 321–2, 379
Multiple-access interference, 2, 6, 10
Multiple-input multiple-output, 3, 23, 361–81
Multiple-input single-output, 362–374
Mutual coupling, 65, 74, 234

Nakagami-m channel, 92, 313, 327, 380
Near field, *see* Fresnel region
Network analyser, *see* Vector network analyser
NLOS, *see* Non-line-of-sight
Noise temperature, 459–60
Non-centrality parameter, 92
Non-line-of-sight, 269, 298–306, 320–1, 324–6, 345
Null, 225, 248
Null steering, 81

OBS, *see* Obstructed line of sight
Obstructed line of sight, 269
OFDM, *see* Orthogonal frequency division modulation
Orthogonal frequency division modulation, 2, 6, 9–11, 14–17, 22, 214, 381
On-body channel measurements, 322–357
Outage, 89

PAN, *see* Wireless personal area network
Pathloss model, 267, 284ff
 deterministic, 262, 267, 284ff, 332, 346
 breakpoint, 291–6, 299, 342–3
 dual-slope, 293–6, 342–3
 empirical, 297ff
 free-space,
 IEEE standard models, 304–5
 WBAN, 341, 348ff
PCB, *see* Printed circuit board
PDP, *see* Power delay profile
Permittivity, 26
Perfectly matched layer, 200
Personal area network, *see* Wireless personal area network
Phase centre, 50, 387, 424, 428–9, 451, 454, 459–62, 466–8
PICA, *see* Antennas
Planar array, 62, 72, 72
Planar inverted cone antenna, *see* Antennas
Plane wave, 33, 395, 406

PML, *see* Perfectly matched layer
Poisson process, 312, 317–21, 326
Polarisation, 33–35, 270, 392, 421, 429, 444, 454, 458, 461
Power delay profile, 313, 317, 326, 345
Power density, 35–9
Power gain, 41–2
Printed antenna, *see* Antennas
Printed circuit board, 178
Propagation model, *see* Pathloss model
Pulse position modulation (PPM), 7

Q-function, 93

Radar, 2–3, 11, 452, *see also* Ground probing radar
Radiated power, 36
Radiation efficiency, 41–2
Radiation intensity, 39–42
Radiation resistance, 37, 43, 45
Radiation pattern, 39, 169, 397
 Directivity, *see* Antenna
 Half-power beamwidth, 41, 69, 71, 86
 Mainlobe, 40, 71
 Null, 40, 41
 Sidelobe, 40, 71, 86, 398
Radio astronomy, 459
Rake receiver, 3, 5, 7–8, 91, 319–23, 354
Ray-tracing, 216, 262, 296, 346–7, 456
Rayleigh fading, 92, 325
Rayleigh pulse, 156
Rectangular arrays, 62
Reflection coefficient, 66, 404
Resonance, 115
Retarded potentials, 29
Return loss, 115
Rice distribution, 92
RMS delay spread, 311, 314–17, 324–6, 345–6
Rumney's principle, 45

S parameters, *see* Scattering parameters
Saleh-Valenzuela model, *see* Channel Model
Scattering parameters, 179, 182
Selection combining, 99
Shadowing, 268, 301–3, 304–5, 345
Shannon capacity, *see* Capacity
Sidelobe, 40
SIMO, *see* Single-input multiple-output

Single-input multiple-output, 362–74
Single-input single-output, 362–4, 368–74
Single-Poisson channel model, *see* Channel model
SISO, *see* Single-input single-output
Small-scale channel modelling, *see* Channel model
Small-scale fading, *see* Channel model
Smart antennas, *see* Array
Smith's chart, 114
Spatial channel modelling, 361–81
Spatial correlation, 375ff
Spatial filtering, 73
Spatio-temporal, 197, 201
Split-Poisson channel model, *see* Channel model
Standardisation, 2, 13–16, 262, 306
Standing waves, 121
Steering vector, 75
Substrate, 112
System level modelling, 332, 353–4

Tapped delay line, 242
Taylor distribution, 62
TDL, *see* Tapped delay line
TEM, *see* Transverse electromagnetic
TH, *see* Time-hopping
Time-hopping, 2, 6–10
Transfer function, 150, 202, 454
Transmission lines, 30
Transmit-MRC, 100
Transverse electromagnetic, 147

Travelling wave, 121, 425, 428
Two-Cluster Poisson channel model, *see* Channel model

UCA, *see* Circular array
ULA, *see* Uniform array
Ultra-wideband, 1ff
Uniform array, 56, 70, 379
Uniform circular array, *see* Circular array
Uniform linear array, *see* Uniform array
UWB, *see* Ultra-wideband
UWB Forum, 2, 9, 14

Vector network analyzer, 121, 337, 342, 364
VNA, *see* Vector network analyser
Voltage standing wave ratio, 462, 465
VSWR, *see* Voltage standing wave ratio

Wave equation, 29
Waveguide
    Coplanar, 112
    Dual ridged, 452, 463–6
WBAN, *see* Wireless body area network
WLAN, *see* Wireless lcoal area network
Weight vector, 75, 78
Window functions, 85
Wireless body area network 14, 262, 306, 325, 331ff, 356
Wireless local area network, 197
Wireless personal area network, 197, 281, 303, 324

TK 7871.67 .U45 U44 2007

Ultra-wideband